Antibiotic Resistance

Antibiotic Resistance

Mechanisms and New Antimicrobial Approaches

Kateryna Kon, MD, PhD
Department of Microbiology, Virology, and Immunology
Kharkiv National Medical University
Kharkiv, Ukraine

Mahendra Rai, PhD
Biotechnology Department
SGB Amravati University
Maharashtra, India

ELSEVIER

AMSTERDAM • BOSTON • HEIDELBERG • LONDON
NEW YORK • OXFORD • PARIS • SAN DIEGO
SAN FRANCISCO • SINGAPORE • SYDNEY • TOKYO

Academic Press is an imprint of Elsevier

Academic Press is an imprint of Elsevier
125 London Wall, London EC2Y 5AS, UK
525 B Street, Suite 1800, San Diego, CA 92101-4495, USA
50 Hampshire Street, 5th Floor, Cambridge, MA 02139, USA
The Boulevard, Langford Lane, Kidlington, Oxford OX5 1GB, UK

British Library Cataloguing-in-Publication Data
A catalogue record for this book is available from the British Library.

Library of Congress Cataloging-in-Publication Data
A catalog record for this book is available from the Library of Congress.

ISBN: 978-0-12-803642-6

For Information on all Academic Press publications
visit our website at https://www.elsevier.com/

Publisher: Sara Tenney
Acquisition Editor: Linda Versteeg-buschman
Editorial Project Manager: Halima Williams
Production Project Manager: Lucía Pérez
Designer: Victoria Pearson

Typeset by MPS Limited, Chennai, India

Contents

List of Contributors

K. Ács
Institute of Pharmacognosy, Medical School, University of Pécs, Pécs, Hungary

C.N. Aguilar
Food Research Department, School of Chemistry, Universidad Autónoma de Coahuila, Saltillo, Coahuila, CP, Mexico

J. Aguirre-Joya
Food Research Department, School of Chemistry, Universidad Autónoma de Coahuila, Saltillo, Coahuila, Mexico

A. Amin
Hospitalist Program, University of California, Irvine Medical Center, Irvine, CA, United States

Z. Aumeeruddy-Elalfi
Department of Health Sciences, Faculty of Science, University of Mauritius, Réduit, Mauritius

M. Ballal
Enteric Diseases Division, Kasturba Medical College, Manipal, Karnataka, India

I. Barbu
Microbiology Immunology Department, Faculty of Biology, University of Bucharest, Bucharest, Romania

C. Bejenaru
Department of Vegetal & Animal Biology, Faculty of Pharmacy, University of Medicine and Pharmacy of Craiova, Craiova, Romania

L.E. Bejenaru
Department of Pharmacognosy & Phytotherapy, Faculty of Pharmacy, University of Medicine and Pharmacy of Craiova, Craiova, Romania

T. Bencsik
Institute of Pharmacognosy, Medical School, University of Pécs, Pécs, Hungary

V.G. Beoletto
Department of Microbiology and Immunology, Universidad Nacional de Río Cuarto, Río Cuarto, Argentina

V. Bhattaram
Department of Animal Science, University of Connecticut, Storrs, CT, United States

C.E. Boucher
Department of Microbial, Biochemical and Food Biotechnology, University of the Free State, Bloemfontein, Republic of South Africa

R.R. Bragg
Department of Microbial, Biochemical and Food Biotechnology, University of the Free State, Bloemfontein, Republic of South Africa

M.E. Carezzano
Department of Microbiology and Immunology, Universidad Nacional de Río Cuarto, Río Cuarto, Argentina

M.L. Chávez-González
Food Research Department, School of Chemistry, Universidad Autónoma de Coahuila, Saltillo, Coahuila, CP, Mexico

M.C. Chifiriuc
Microbiology Immunology Department, Faculty of Biology, University of Bucharest, Bucharest, Romania; Research Institute of the University of Bucharest, Life, Environmental and Earth Sciences, Bucharest, Romania

N.B. Cobacho
Centro de Análises Proteômicas e Bioquímicas, Programa de Pós-Graduação em Ciências Genômicas e Biotecnologia, Universidade Católica de Brasília, Brasília, DF, Brazil

E. Coetsee
Department of Microbial, Biochemical and Food Biotechnology, University of the Free State, Bloemfontein, Republic of South Africa

S.S. da Silva
Department of Biotechnology, School of Engineering of Lorena, University of São Paulo, Lorena, São Paulo, Brazil

M. de las Mercedes Oliva
Department of Microbiology and Immunology, Universidad Nacional de Río Cuarto, Río Cuarto, Argentina

L.A. de Lima
Centro de Análises Proteômicas e Bioquímicas, Programa de Pós-Graduação em Ciências Genômicas e Biotecnologia, Universidade Católica de Brasília, Brasília, DF, Brazil

K. de Sousa Oliveira
Centro de Análises Proteômicas e Bioquímicas, Programa de Pós-Graduação em Ciências Genômicas e Biotecnologia, Universidade Católica de Brasília, Brasília, DF, Brazil

M.S. Demo
Department of Microbiology and Immunology, Universidad Nacional de Río Cuarto, Río Cuarto, Argentina

S.C. Dias
Centro de Análises Proteômicas e Bioquímicas, Programa de Pós-Graduação em Ciências Genômicas e Biotecnologia, Universidade Católica de Brasília, Brasília, DF, Brazil

P.F. Seke Etet
Department of Basic Health Sciences, College of Applied Medical Sciences, Qassim University, Buraydah, Saudi Arabia

M. Farahna
Department of Basic Health Sciences, College of Applied Medical Sciences, Qassim University, Buraydah, Saudi Arabia

A.C. Flores-Gallegos
Food Research Department, School of Chemistry, Universidad Autónoma de Coahuila, Saltillo, Coahuila, Mexico

O.L. Franco
Centro de Análises Proteômicas e Bioquímicas, Programa de Pós-Graduação em Ciências Genômicas e Biotecnologia, Universidade Católica de Brasília, Brasília, DF, Brazil; S-Inova Biotech, Pós-Graduação em Biotecnologia, Universidade Católica Dom Bosco, Campo Grande, MS, Brazil

A. Gade
Nanobiotechnology Laboratory, Department of Biotechnology, SGB Amravati University, Amravati, Maharashtra, India

A.M. Grumezescu
Department of Science and Engineering of Oxide Materials and Nanomaterials, Faculty of Applied Chemistry and Materials Science, University Politehnica of Bucharest, Bucharest, Romania

V. Grumezescu
Lasers Department, National Institute for Lasers, Plasma and Radiation Physics, Bucharest, Romania; Department of Science and Engineering of Oxide Materials and Nanomaterials, Faculty of Applied Chemistry and Materials Science, University Politehnica of Bucharest, Bucharest, Romania

A. Gurib-Fakim
CIDP Research & Innovation, BioPark, Socota Phoenicia, Mauritius

P.S. Haddad
Exact and Earth Sciences Department, Universidade Federal de São Paulo, Diadema, SP, Brazil

D.H. Halat
Microbiology Laboratory, School of Pharmacy, Saint-Joseph University, Beirut, Lebanon; Rodolphe Mérieux Laboratory, Beirut, Lebanon; Department of Pharmaceutical Sciences, School of Pharmacy, Lebanese International University, Beirut and Bekaa, Lebanon

A.M. Holban
Department of Science and Engineering of Oxide Materials and Nanomaterials, Faculty of Applied Chemistry and Materials Science, University Politehnica of Bucharest, Bucharest, Romania; Microbiology Immunology Department, Faculty of Biology, University of Bucharest, Bucharest, Romania

Gy. Horváth
Institute of Pharmacognosy, Medical School, University of Pécs, Pécs, Hungary

L. Hsieh
Hospitalist Program and Division of Infectious Diseases, University of California, Irvine Medical Center, Irvine, CA, United States

A. Ingle
Nanobiotechnology Laboratory, Department of Biotechnology, SGB Amravati University, Amravati, Maharashtra, India

D.P. Karumathil
Department of Animal Science, University of Connecticut, Storrs, CT, United States

B. Kocsis
Institute of Medical Microbiology and Immunology, Medical School, University of Pécs, Pécs, Hungary

K. Kon
Department of Microbiology, Virology, and Immunology, Kharkiv National Medical University, Kharkiv, Ukraine

V. Lazar
Microbiology Immunology Department, Faculty of Biology, University of Bucharest, Bucharest, Romania

J.-Y. Lee
Department of Microbial, Biochemical and Food Biotechnology, University of the Free State, Bloemfontein, Republic of South Africa

G.A. Lutzu
Key Laboratory of Biofuels, Qingdao Institute of Bioenergy and Bioprocess Technology, Chinese Academy of Sciences, Qingdao, Shandong, People's Republic of China

M.F. Mahomoodally
Department of Health Sciences, Faculty of Science, University of Mauritius, Reduit, Mauritius

J.M. Marioli
Department of Chemistry, Universidad Nacional de Río Cuarto, Río Cuarto, Argentina

L. Meyburgh
Department of Microbial, Biochemical and Food Biotechnology, University of the Free State, Bloemfontein, Republic of South Africa

M.M. Michel
Food Research Department, School of Chemistry, Universidad Autónoma de Coahuila, Saltillo, Coahuila, Mexico

G.D. Mogoşanu
Department of Pharmacognosy & Phytotherapy, Faculty of Pharmacy, University of Medicine and Pharmacy of Craiova, Craiova, Romania

J. Morlett-Chávez
Food Research Department, School of Chemistry, Universidad Autónoma de Coahuila, Saltillo, Coahuila, Mexico

C.A. Moubareck
Microbiology Laboratory, School of Pharmacy, Saint-Joseph University, Beirut, Lebanon; Rodolphe Mérieux Laboratory, Beirut, Lebanon; Department of Natural Science and Public Health, College of Sustainability Sciences and Humanities, Zayed University, Dubai, United Arab Emirates

D. Nagaonkar
Nanobiotechnology Laboratory, Department of Biotechnology, SGB Amravati University, Amravati, Maharashtra, India

A.H. Nwabo Kamdje
Department of Biomedical Sciences, Faculty of Sciences, University of Ngaoundéré, Ngaoundéré, Cameroon

A.E. Oprea
Department of Science and Engineering of Oxide Materials and Nanomaterials, Faculty of Applied Chemistry and Materials Science, University Politehnica of Bucharest, Bucharest, Romania

P. Paralikar
Nanobiotechnology Laboratory, Department of Biotechnology, SGB Amravati University, Amravati, Maharashtra, India

A. Parsaeimehr
Laboratory of Biosystems Engineering, Key Laboratory of Agrobiology of Jiangsu Province, Institute of Biotechnology, Jiangsu Academy of Agricultural Sciences, Nanjing, Jiangsu, People's Republic of China

M.T. Pelegrino
Exact and Earth Sciences Department, Universidade Federal de São Paulo, Diadema, SP, Brazil

R.C. Popescu
Department of Science and Engineering of Oxide Materials and Nanomaterials, Faculty of Applied Chemistry and Materials Science, University Politehnica of Bucharest, Bucharest, Romania; Department of Life and Environmental Physics, Horia Hulubei National Institute of Physics and Nuclear Engineering, Bucharest, Romania

G.M. Preciado
Food Research Department, School of Chemistry, Universidad Autónoma de Coahuila, Saltillo, Coahuila, Mexico

M. Rai
Nanobiotechnology Laboratory, Department of Biotechnology, SGB Amravati University, Amravati, Maharashtra, India

R. Rodríguez-Herrera
Food Research Department, School of Chemistry, Universidad Autónoma de Coahuila, Saltillo, Coahuila, CP, Mexico

D.K. Sarkis
Microbiology Laboratory, School of Pharmacy, Saint-Joseph University, Beirut, Lebanon; Rodolphe Mérieux Laboratory, Beirut, Lebanon

A.B. Seabra
Exact and Earth Sciences Department, Universidade Federal de São Paulo, Diadema, SP, Brazil

M.G. Shurygin
Irkutsk Scientific Centre of Surgery and Traumatology, Irkutsk, Russia; Irkutsk Scientific Centre SB RAS, Irkutsk, Russia

I.A. Shurygina
Irkutsk Scientific Centre of Surgery and Traumatology, Irkutsk, Russia; Irkutsk Scientific Centre SB RAS, Irkutsk, Russia

B.G. Sukhov
A. E. Favorsky Irkutsk Institute of Chemistry SB RAS, Irkutsk, Russia

C. Theron
Department of Microbial, Biochemical and Food Biotechnology, University of the Free State, Bloemfontein, Republic of South Africa

A. Upadhyay
Department of Animal Science, University of Connecticut, Storrs, CT, United States

I. Upadhyaya
Department of Animal Science, University of Connecticut, Storrs, CT, United States

W.A. van der Westhuizen
Department of Microbial, Biochemical and Food Biotechnology, University of the Free State, Bloemfontein, Republic of South Africa

L. Vecchio
Department of Basic Health Sciences, College of Applied Medical Sciences, Qassim University, Buraydah, Saudi Arabia

K. Venkitanarayanan
Department of Animal Science, University of Connecticut, Storrs, CT, United States

S.L. Villarreal-Morales
Food Research Department, School of Chemistry, Universidad Autónoma de Coahuila, Saltillo, Coahuila, Mexico

Preface

Resistance to antimicrobials has been described involving all types of infectious agents and continues to be a growing challenge in modern medicine. A common problem in acute care hospitals is the use of potent, broad-spectrum antibiotics, as well as inappropriate or suboptimal therapy, all of which contribute to the development of antibiotic resistance. One of the solutions to this issue, which may prolong the effectiveness of currently available antibiotics is antimicrobial stewardship strategies that encourage the judicious use of these antibiotics. Moreover, the potential of emerging and enabling technologies like nanotechnology should be harnessed to tackle the global problem of multidrug-resistance (MDR). These issues are discussed in this book.

Understanding the fundamental mechanisms of antibiotic resistance is a key step for the discovery of effective methods to cope with the resistance. This book discusses up-to-date knowledge about the mechanisms of antibiotic resistance in different types of bacteria and application of this knowledge into developing new and optimal treatment strategies. In addition, it explores alternative antimicrobial approaches in fighting microbial resistance. Several chapters are devoted to the potential of nanotechnology in combating MDR bacteria, to the role of magnetite-based nanomaterials in developing alternative therapeutic and preventive approaches against pathogens, to different polymer-containing metal nanocomposites as promising antimicrobial agents, and to the role of nanooxides and nitric oxides in fighting antibiotic-resistant bacteria. Alternative natural sources of antimicrobial agents are also discussed in detail, including plant-derived products, such as essential oils (EOs) and plant extracts, bacteriophages, bacteriocin-producing bacteria, marine-derived bacteria, fungi, sponges, algae, corals, mollusks, and other invertebrates.

The potential audience for the book include researchers in microbiology, biotechnology, pharmacology, nanotechnology, and infection control; students of medical, pharmaceutical, and biological faculties; and clinicians dealing with infections in various locations.

CHAPTER

ANTIMICROBIAL STEWARDSHIP: HOSPITAL STRATEGIES TO CURB ANTIBIOTIC RESISTANCE

1

L. Hsieh[1] and A. Amin[2]

[1]Hospitalist Program and Division of Infectious Diseases, University of California, Irvine Medical Center, Irvine, CA, United States [2]Hospitalist Program, University of California, Irvine Medical Center, Irvine, CA, United States

INTRODUCTION

The discovery of antibiotics may be one of the greatest achievements in medicine. Antibiotic treatments have improved clinical outcomes from infections, leading to the reduction of morbidity and mortality in surgical, transplant, cancer, and critical care patients. With the use of potent broad-spectrum antibiotics, selective pressures have made antibiotic resistance an urgent worldwide concern. Increasing numbers of hospital-acquired infections are now caused by multidrug-resistant pathogens, making treatment progressively difficult and antibiotic choice increasingly limited. The Centers for Disease Control and Prevention (CDC) data showed that in 2010, 55.7% of patients discharged from 325 hospitals received antibiotics during their hospital stay, and 37.2% of those antibiotic prescriptions were unnecessary or could be improved, which is in line with previous studies in the literature on antibiotic utilization in acute care hospitals.[1] Inappropriate antimicrobial use can lead to the selection of resistant pathogens, *Clostridium difficile* infections, antibiotic-induced toxicities, and adverse drug reactions, all of which have a significant negative impact on patient morbidity, mortality, and health-care costs. It has been recognized that reduction of antibiotic use would slow the growth of or alleviate the problem of antibiotic resistance and *C. difficile* infections. In 1997, the Society for Healthcare Epidemiology of America (SHEA) and Infectious Diseases Society of America (IDSA) published guidelines for the prevention of antimicrobial resistance in hospitals, coining the term *antimicrobial stewardship.*[2] In addition to a comprehensive infection control program and surveillance of bacterial resistance, these guidelines advocated for the judicious use of antimicrobials in order to slow or prevent the development of antimicrobial resistance.

Since 2006, the CDC has launched multiple campaigns to target antibiotic resistance and improve health care. In each of these campaigns, the CDC emphasized the need for judicious antibiotic utilization in acute care hospitals. In 2007, the IDSA and SHEA published guidelines for developing antimicrobial stewardship programs (ASPs) at an institutional level.[3] In this document, IDSA and SHEA define the primary goal of ASPs to be the improvement of clinical outcomes while minimizing the unintended consequences of antimicrobial use, including emergence of resistance, toxicity, and selection of pathogenic organisms such as *C. difficile.* SHEA and IDSA, along with the Pediatric Infectious Disease Society (PIDS), issued a policy statement in 2012

Antibiotic Resistance. DOI: http://dx.doi.org/10.1016/B978-0-12-803642-6.00001-0

recommending that ASPs be mandated through regulatory channels.[4] Specifically, this policy statement advocated for the Centers for Medicare and Medicaid Services (CMS) to require participating institutions to implement ASPs. This document further detailed the minimum requirements for an effective ASP and made recommendations on expansion of stewardship efforts to the ambulatory setting, education, and research. In 2014, the CDC recommended that all acute care hospitals implement ASPs to combat the worsening problems of antibiotic resistance and *C. difficile* infections and released a document called "Core Elements of Hospital Antibiotic Stewardship Programs" to aid hospitals in this goal.[5]

In the United States, California remains the first and only state to pass legislation on antimicrobial stewardship. In 2008, Senate Bill 739 (SB 739) mandated California acute care hospitals to put processes in place to monitor the judicious use of antibiotics. It was left to each institution to develop its own procedures to comply with this mandate. The regulatory conditions were vague, and hospitals made various efforts to fulfill these requirement. In a web-based survey of 422 California acute care hospitals in 2010–11, 50% of the 223 respondents had a current ASP, and 30% reported planning to implement an ASP.[6] This survey was subject to reporting bias, however, as those hospitals with active ASPs were more likely to respond. It was reported that SB 739 encouraged some of these hospitals to initiate an ASP. Those that did not adopt ASPs were likely to be smaller or rural hospitals, and lack of resources and administrative support were the most frequently cited barriers to doing so. Interestingly, many hospitals that reported not having an official ASP did have some stewardship processes in place, such as formulary restriction and antimicrobial oversight by pharmacists. In Sep. 2014, California Senate Bill 1311 (SB 1311) further required hospitals to implement an official policy on antimicrobial stewardship and to establish a multidisciplinary stewardship committee by Jul. 2015.[7] This bill was much more prescriptive in its requirements than its predecessor, driving home the minimum requirements for an effective stewardship program in acute care hospitals. In Mar. 2015, the White House released the National Action Plan for Combating Antibiotic-Resistant Bacteria. This 63-page action plan aims to guide programs of the US government, public health, health care, and agriculture in a common effort to address the urgent challenge of mounting antibiotic resistance. The plan sets 1-, 3-, and 5-year milestones for each of five goals, the first of which is to slow the emergence of resistant bacteria and prevent the spread of infections. By 2020, the action plan seeks to establish ASPs in all acute care hospitals and improved antibiotic stewardship across all health-care settings, with a reduction of inappropriate antibiotic use by 20% in inpatient settings.[8] This illustrates the commitment of the United States to meet the critical problem of antimicrobial resistance.

Globally, antimicrobial stewardship is gaining momentum as the challenge of antibiotic resistance became an urgent worldwide concern. In a 2012 international survey of 660 hospitals from 67 countries from 6 continents on inpatient antimicrobial stewardship, 58% of respondents had an ASP and 22% planned to implement one.[9] Academic medical centers accounted for almost 50% of respondents. Reducing antimicrobial resistance was the most frequently cited goal for ASPs across all continents. Formulary restriction was practiced in 81% of hospitals, while 64% had postprescription review as part of the stewardship strategy. Two-thirds of hospitals in North America and Europe had ASPs, with 80% of European countries having antimicrobial stewardship standards at the national level. It was noted that ASPs were more established in Sweden (100%), United States (88%), France (81%), and the United Kingdom (77%) than other countries.[10] However, it should be noted since hospitals with active ASPs were more likely to participate in this survey, the results are subject to selection bias. Progress is being made in the implementation and expansion of stewardship programs across the United States and Europe. The degree of legislative requirements differ significantly among countries, their primary objectives are different as well. In France, the primary goal is to reduce antibiotic resistance. Reduction of health-care-associated infections is the main impetus in the United Kingdom. For the United States, the principal driver is to improve clinical outcomes. The utilization of electronic medical records, physician order entry, and data warehousing as part of antimicrobial stewardship is much more common in the United States than in France or the United Kingdom.

MOVING FROM COST TO QUALITY AND PATIENT SAFETY

The CDC estimates that at least 2 million patients each year acquire serious infections with drug-resistant bacteria in the United States, accounting for at least 23,000 deaths and significant morbidity and health-care costs.[11] In addition, the CDC estimates that 150,000 cases per year go to US Emergency Departments for antimicrobial-related adverse events.[12] Effective ASPs can improve patient care and be cost effective. They have shown reductions in antimicrobial use with annual savings of $200,000–900,000, making the programs self-supporting.[13,14] Most of the data from the literature measured direct pharmacy acquisition costs. When taken together with the impact on length of stay, readmission rates, and avoidance of potential adverse drug reactions, the financial impact of ASPs is even greater. The cost effectiveness of ASPs is often cited to garner administration support in a climate of cost-conscious medical care. However, the main goal of antimicrobial stewardship should be to improve patient care and optimize clinical outcomes. More and more, antimicrobial stewardship has become the focus of quality and patient safety improvement efforts.

Appropriate antimicrobial utilization is tied to quality patient care. In a cohort study of 500 randomly selected hospitalized patients with an antimicrobial course, Filice et al. found that diagnostic accuracy correlated to optimal antimicrobial use.[15] Diagnostic accuracy in turn was closely tied to the quality of clinical evidence at the time of initial diagnosis. Accuracy was generally poor for the diagnoses of pneumonia and urinary tract infection, which are extremely common in the inpatient setting. In this study, the appropriateness of antibiotics was judged by a group of four infectious disease physicians. It should be noted that each reviewer's responses was compared with those of the other three reviewers, and agreement was 69–72%.

It is generally accepted that routine treatment of asymptomatic bacteriuria is inappropriate.[16] It is a substantial contributor to antibiotic overuse in hospitalized and nursing home patients, particularly among patients with urinary catheters. In a study at two Veterans Affairs health-care systems, Trautner et al. introduced a streamlined diagnostic algorithm for catheter-associated urinary tract infection versus asymptomatic bacteriuria.[17] This intervention significantly decreased the inappropriate ordering of urine cultures, thereby decreasing the inappropriate treatment of asymptomatic bacteriuria. At the same time, the study did not find undertreatment of true catheter-associated urinary tract infections during the intervention period.

Regardless of appropriateness, all antibiotic use exerts selective pressures that contribute to the development of resistance and other unintentional consequences. It is apparent that not only is the choice of antibiotics important, the duration of therapy may be equally important. In a retrospective cohort study of 7792 hospitalized adult patients who received 2 or more days of antibiotics, dose-dependent increases in the risk of *C. difficile* infection was associated with the cumulative dose, number of antibiotics, and days of antibiotic exposure.[18] These studies illustrate that appropriate and judicious use of antimicrobials is directly tied to quality patient care and clinical outcomes. Hence, a program such as antimicrobial stewardship, which aids clinicians in this goal, has a major impact on quality and patient safety.

ELEMENTS OF ASPs

ANTIMICROBIAL STEWARDSHIP TEAM

The core members of the ASP should include a physician leader and a clinical pharmacist, both of whom with training in infectious diseases, antimicrobial stewardship, or both, with the patient at the center. Larger medical centers can employ infectious disease specialists in this role. Hospitalists can also be ideal leaders in this position, given their involvement in inpatient care and quality improvement.[5] Administration and medical staff support are paramount to the success of ASPs. Physician buy-in is essential for compliance with stewardship team recommendations. Collaboration among the staff in infection control, hospital epidemiology, quality improvement, microbiology laboratory, and information technology (IT) is important for a comprehensive and efficient program.

It should be noted that the structure of the ASP should correspond to available resources and the needs of the individual institution. Even in resource-limited settings, reducing inappropriate antibiotic use is an effective way to improve quality and patient safety by decreasing antibiotic-resistant pathogens, *C. difficile* infection, and antibiotic-related adverse events. In a hospital in Utah without an infectious disease physician, a pharmacist-led ASP made substantial decreases in the use of the four most commonly used antimicrobial agents, with statistically significant reduction in length of stay for community-acquired-pneumonia patients and substantial cost savings.[19] In a review of ASPs in community hospitals, Ohl et al. described the benefits of ASPs and illustrated several case studies of various ASP structures and strategies, utilizing an institution's available resources and support.[20] The available evidence indicates that antimicrobial stewardship is a worthwhile and cost-effective endeavor regardless of hospital size or location.

ANTIMICROBIAL STEWARDSHIP STRATEGIES FOR ACUTE CARE HOSPITALS (TABLE 1.1)

KEY STEWARDSHIP INTERVENTIONS

Two key interventions were recommended in the IDSA/SHEA guidelines as the cornerstones of an effective ASP: (1) formulary restriction with preauthorization and (2) prospective audits with intervention and feedback.[3] Antimicrobial restriction, either through formulary limitations at the level of the institution's Pharmacy and Therapeutic Committee or requirement for justification of use, has been shown to be effective in reducing the use of targeted antibiotics. Some hospitals require

Table 1.1 Antimicrobial Stewardship Strategies for Acute Care Hospitals

	Patient-Based	Systems-Based
Key interventions	• Prospective auditing with feedback • 48-h antibiotic time-out	• Formulary restriction
Supplemental interventions		
Education	• Physician feedback	• Institutionwide conferences • Clinical pathways • Practice guidelines
Pharmacy	• Dosing optimization • Intravenous to oral conversion • Prolonged infusion	• Monitor for adverse events
Laboratory	• Rapid diagnostics with stewardship intervention	• Cascade reporting • Antibiograms
Information technology	• Order sets • Decision support for antibiotic choice • Alert for antibiotic time-outs	• Electronic order forms • Drug–bug mismatch decision support • Monitor antimicrobial utilization

the physician to fill out the indication at the time of order entry, while others require prior authorization through an approval process. A retrospective study was done at New York-Presbyterian Hospital, a 700-bed academic teaching hospital in New York City, to assess the appropriateness of using Gram-negative antibiotics in a setting of extensive antimicrobial resistance.[21] This hospital had an active ASP in place for 10 years prior to the study, comprising of formulary restriction, hospital-specific guidelines, and institutionwide education, but not postprescription review. Two snapshot dates were reviewed. The majority of the antimicrobials were used as part of an empiric regimen, with one-third of the patients being in the intensive care unit. In this study, 26% of the antibiotics were not optimal; the most common reason for this was that the spectrum of activity was too broad. This study illustrated the importance of deescalation interventions such as prospective auditing.

Prospective auditing with feedback to the prescriber has been demonstrated to improve appropriate antibiotic prescription. The principal function of this intervention is the deescalation of an empiric antimicrobial regimen based on culture results, thereby focusing therapy on the most effective agent for treating the infection. Other recommendations may include dosing optimization, discontinuation of antibiotics, and monitoring for adverse events. The reviews are done by a physician or clinical pharmacist who is not part of the treatment team. In a multicenter study of five tertiary care academic hospitals in the United States, prospective auditing with feedback was shown to reduce antimicrobial use, although the degree varied by institution, with those centers with established ASPs reducing antimicrobial use significantly.[22] It was also noted that the utilization reduction was not sustained in the follow-up period, suggesting that continued prospective auditing and feedback are required to maintain judicious use of antibiotics. It is likely that because this study was conducted in academic centers, the turnover of resident physicians required continued auditing and feedback, as senior residents graduate and new interns entered each year. Various feedback mechanisms have been described in the research literature, ranging from one-on-one education sessions to placing a feedback sheet at the front of the patient's chart. In a single-center study at a tertiary care teaching hospital comparing distinct modes of communication, there was no statistically significant difference in compliance with stewardship recommendations made by direct telephone calls, notes in the medical record, or text pager messages.[23] Higher acceptance of recommendations was noted with the deescalation of antibiotics than discontinuation of them, suggesting that physicians were more comfortable with changing or focusing therapy than stopping it. This finding is consistent with the pervasive practice of continuing antibiotics "just in case" and the erroneous perception that antibiotics cause no harm.

In addition to the two key interventions recommended by IDSA and SHEA, the CDC recommended antibiotic "time-outs" at 48 h after the initiation of antibiotics by the treating physician.[5] While prospective auditing is done by a member of the ASP team, antibiotic time-outs are performed by the treating team. Antibiotics are often started in an empiric fashion, especially in the critically ill. At 48 h, the treating physician should consider whether antibiotics are still indicated, or whether an alternative diagnosis other than infection has been made. Culture results should be available by this time, allowing the physician to deescalate the antimicrobial regimen based on culture results in order to target the pathogen with the most appropriate therapy for the site of infection. The dose, route of administration, and duration of antibiotics should also be considered. This intervention lies in the hands of the individual prescriber. Each clinician should prescribe antimicrobials in a thoughtful, evidence-based manner.

SUPPLEMENTAL INTERVENTIONS

EDUCATION

Education is an important element of ASPs. These include educational conferences, teaching sessions for residents, and written guidelines. The development of evidence-based practice guidelines and clinical pathways that incorporate local resistance patterns can be helpful in influencing prescribing habits. Not only would education influence prescriber behavior, it can lead to buy-in for acceptance of other stewardship strategies, such as prospective auditing with feedback. However, education alone, without the key interventions discussed previously, have been shown to be only minimally effective.[3] In academic centers, where residents order the majority of antibiotics, it is essential to provide education on an ongoing basis, as senior residents graduate and new interns arrive each year. Antimicrobial stewardship should become part of medical education in order to institute a culture of judicious antibiotic use. In addition, as local susceptibility data change, the medical staff should be made aware of these trends in order to select the most appropriate empiric antimicrobial regimen. In a study that analyzed four years of bacterial susceptibility data and prescribing trends from two West London tertiary referral hospitals, their affiliated renal units, and the surrounding community practices served by a single laboratory, the authors found significant differences in antibiotic susceptibility within and between the hospitals, with substantial year-to-year fluctuations among most drug–bug combinations.[24] As one would expect, more resistant organisms were found in the critical care units and renal cohorts than in the medical wards and community cohorts. These trends likely are related to the greater frequency of antibiotic use in critical care units and renal cohorts. Through education, resistance trends can be communicated to the clinicians. In addition to medical management of infections, education efforts should include emphasis on the dire nature of antibiotic resistance and the impact of inappropriate antimicrobial utilization on quality and patient safety. It should be stressed that antimicrobials are not benign, and its use should be thoughtful and evidence based.

PHARMACY STRATEGIES

Certain pharmacy-driven strategies can enhance an ASP. Optimization of antimicrobial dosing is an important part of improving clinical outcomes while minimizing toxicity. Consideration of patient attributes (such as age, weight, and renal function), the particular organism and its minimal inhibitory concentration, site of infection, and drug characteristics (pharmacokinetics and pharmacodynamics) are essential to the optimization of antimicrobial therapy. A simple strategy is the automatic conversion from intravenous to oral antibiotics for those drugs with good oral bioavailability. This would improve patient safety by decreasing the incidence of catheter-associated infections and have a positive effect on health-care costs.

Prolonged or extended infusion of intravenous β-lactam antibiotics optimizes the time-dependent bactericidal activity of this class of antimicrobials. Achieving the target serum levels is necessary for treatment success, especially in critically ill patients with difficult-to-treat infections. Several pharmacokinetic–pharmacodynamic studies using Monte Carlo simulation have been done for piperacillin–tazobactam and cefepime.[25] Prolonging the infusion increased the likelihood of reaching optimal target serum levels, while a lower total daily dose of the drug was used. This dosing scheme is likely not necessary to treat highly susceptible pathogens. Robust clinical outcome data related to this dosing strategy are lacking. Those institutions that have implemented routine prolonged infusion of β-lactam antibiotics have been able to demonstrate financial benefits in terms of lower total daily dose and drug acquisition costs. Logistical issues must be considered, including dedicated time through an intravenous line and the stability of certain β-lactam antibiotics. In a critically ill patient with severe sepsis or septic shock, prolonged or extended infusion of antibiotics may not be practical during the resuscitation phase of clinical management.

Another important pharmacy function is the monitoring of antibiotic-related adverse events and toxicities, which directly affects patient safety. The pharmacist on the antimicrobial stewardship team is in an optimal position for adverse event reporting. The US Food and Drug Administration (FDA) MedWatch program (the FDA Adverse Event Reporting System) was introduced in 1993.[26] It is the largest federal voluntary reporting system used to report observed or suspected adverse events, product quality, and therapeutic failures for medications, biologics, and medical devices. Historically, postmarketing surveillance of antimicrobials has discovered adverse events not previously documented in clinical trials, resulting in the withdrawal of several antibiotics.

INCORPORATING IT

In 2009, the Health Information Technology for Economic and Clinical Health Act was signed to encourage physicians and hospitals to implement electronic health records (EHRs) in the United States.[27] As more and more hospitals adopt EHRs, this element of technology can be used to enhance antimicrobial stewardship efforts. EHRs and add-on clinical decision support systems

(CDSSs) have capabilities to improve patient safety and enhance efficiency. EHR and CDSS platforms differ with customizable features to varying degrees, but most have capabilities that can be utilized in ASPs.[28] EHRs offer the ability to look at patient data from a central remote location and to provide lists of patients on specific antimicrobials, making the reviewing of medical records highly efficient. Electronic order forms can be utilized to require clinical indications for antimicrobial prescriptions, while order sets for certain clinical syndromes can provide decision support to treating physicians and guide antibiotic prescription. At the time of the initial prescription, physicians have the opportunity to start the most effective therapy for the patient.

In a pilot study to incorporate a point-of-prescription tool to improve antibiotic prescribing, four academic medical centers in conjunction with the CDC formed a collaborative to develop a daily rounding flowsheet that incorporated the core principles of antimicrobial stewardship.[29] The study found that they were able to engage unit-based providers in stewardship efforts. Alerts can be incorporated as a reminder for antimicrobial time-outs. Most EHRs and CDSSs provide preprogrammed drug–bug mismatch decision support, utilizing microbiological susceptibility results and inpatient pharmacy data. CDSSs also can track antimicrobial utilization, microbial susceptibility, and generate antibiograms. As more and more hospitals adopt EHRs and CDSSs become more sophisticated, opportunities exist to incorporate technology into antimicrobial stewardship efforts.

LABORATORY STRATEGIES

Laboratory strategies can be used to supplement antimicrobial stewardship efforts. Current conventional culture-based methods to isolate and identify a pathogen, followed by susceptibility testing, can take 72 h or more. The time that passed before effective antibiotics were used can affect clinical outcomes. This has been shown for Gram-negative septicemia as well as septic shock patients, where the delay in the timely initiation of antibiotics directly correlated with mortality.[30,31] Automated alerts for positive blood cultures coupled with stewardship interventions can ensure that the patient is prescribed effective antibiotics sooner. In a study of three hospitals at the Detroit Medical Center with active ASPs, active alerts were sent to the pharmacist, who then made stewardship interventions as indicated. This decreased the time it took to begin appropriate therapy and had a positive impact on length of stay and mortality.[32]

Direct susceptibility testing by disk diffusion provides antibiotic susceptibility for the whole sample rather than individual pure colonies. These results are generally available 24 h sooner than conventional culture methods. There is research that supports its accuracy and provides physicians with early microbiological results to direct antibiotic therapy. In a study of 123 clinical samples comparing conventional antimicrobial susceptibility testing with direct susceptibility testing, Coorevits et al. showed an overall agreement of over 86%.[33] In 89% of discordant cases, direct susceptibility testing showed a more resistant result than conventional methods, mainly due to the presence of a resistant organism in a mixed culture. It should be noted that direct susceptibility testing by disk diffusion has been criticized by several microbiology societies because the bacterial inoculum is not standardized. Hence, the results should be interpreted with a critical eye, preferably by an infectious disease physician.

Technical advances in the area of microorganism detection have led to the commercial availability of several rapid molecular assays.[34] Multiplex polymerase chain reaction (PCR) uses a fluorescent labeled probe with more than one set of primers to amplify pieces of target DNA, detecting multiple organisms and resistance genes. After implementation of a FilmArray Blood Culture Identification (BCID) panel (BioFire Diagnostics, Salt Lake City, UT) at an academic tertiary care hospital in Omaha, NE, Southern and colleagues compared its performance to conventional culture methods using clinical blood culture isolates during the first 30 days of implementation.[35] This PCR panel identifies 19 pathogens as well as genes for *mecA* (methicillin resistance), *vanA/B* (vancomycin resistance), and bla_{KPC} (carbapenem resistance) directly from positive blood cultures. Overall, the FilmArray BCID showed a sensitivity of 75.7% and a specificity of 100%. When considering only on-panel organisms, the sensitivity increased to 94.3%. The BCID also failed to detect some blood pathogens. In addition, the assay had difficulty with organisms that were not on the panel, in polymicrobial cultures, and in positive blood cultures from patients already on antibiotics. The BCID panel was found to be rapid and relatively reliable, showing promise for clinical practice. Nanosphere's Verigene blood culture Gram-positive (BC-GP) and Gram-negative (BC-GN) assays (Northbrook, IL) use a nanoparticle probe technology. Positive blood cultures undergo nucleic acid extraction and PCR amplification, followed by the hybridization of target DNA to capture oligonucleotides on a microarray. The Verigene BC-GP identifies 12 Gram-positive organisms and 3 resistance genes, while the Verigene BC-GN identifies 9 Gram-negative genuses and species, as well as multiple resistance markers. A multicenter, prepost, quasi-experimental study was conducted at five acute care community hospitals to measure the effect of rapid testing using the Verigene BC-GP in combination with antimicrobial stewardship intervention.[36] The sensitivity and specificity of the Verigene BC-GP was 100% sensitive and specific as opposed to conventional culture methods. Rapid identification with ASP intervention significantly improved the mean time to targeted therapy and lower median length of stay, while mortality rates were unchanged. The Verigene BC-GN was evaluated in a single-center study conducted at an 850-bed tertiary care medical center.[37] The investigators confirmed the diagnostic accuracy of the BC-GN, finding a sensitivity of 97.1% and a specificity of 99.5%. Using a theoretical stewardship intervention design, they found a significant difference in potential times to both effective and optimal antibiotic treatment compared to standard care. However, a significant limitation to this study was that the stewardship interventions were only theoretical, with a presumed 100% acceptance rate, and the results were potential improvements.

In a study of 219 patients with Gram-negative bacteremia, Perez et al. investigated the use of matrix-assisted laser desorption ionization time-of-flight mass spectrometry (MALDI-TOF MS) combined with direct susceptibility testing on positive blood cultures in conjunction with real-time notification and stewardship intervention. This resulted in a 46-h reduction in time to optimal antibiotic therapy, with a significant reduction on length of stay and hospital costs.[38] In a related study, Perez et al. investigated the clinical effects of rapid identification and direct susceptibility testing on patients with multidrug-resistant Gram-negative bacteremia.[39] The use of MALDI-TOF MS directly on positive blood cultures significantly reduced the mean time to identification by 26 h. The additional use of direct susceptibility testing that was set up at the time of blood culture positivity reduced the average time to final identification and susceptibility from 47 to 29 h. Combining rapid laboratory identification of the pathogen and direct susceptibility data with real-time antimicrobial stewardship interventions, the average time to optimal antibiotic therapy was reduced from 81 h during the preintervention period to 23 h in the intervention period. These reductions translated to improvements in 30-day mortality and mean hospital length of stay.

It should be noted that this study was done at a 1000-bed quaternary-care academic hospital with a high rate of antibiotic resistance. A total of 69% of study patients were on ineffective therapy at the time of blood culture positivity. In this setting, antibiotic susceptibility is difficult to predict, making the rapid diagnostic techniques more crucial in patient care and likely resulting in higher impact. Interestingly, almost half of the interventions were made based on the identification of the organism alone, before direct susceptibility results were available. It should be emphasized that rapid results are of little value if these results are not acted upon in a timely manner. In these studies, real-time antimicrobial stewardship interventions were used in conjunction with rapid diagnostics to achieve the improved outcomes.

Cascade reporting of antimicrobial susceptibilities can be an effective strategy to influence prescriber habits.[40] In cascade reporting, the antimicrobial agent with the narrowest spectrum within each class is released first. The broader-spectrum antibiotics are released only if the organism is resistant to the narrower-spectrum agents. All the resistant results are released (Fig. 1.1).

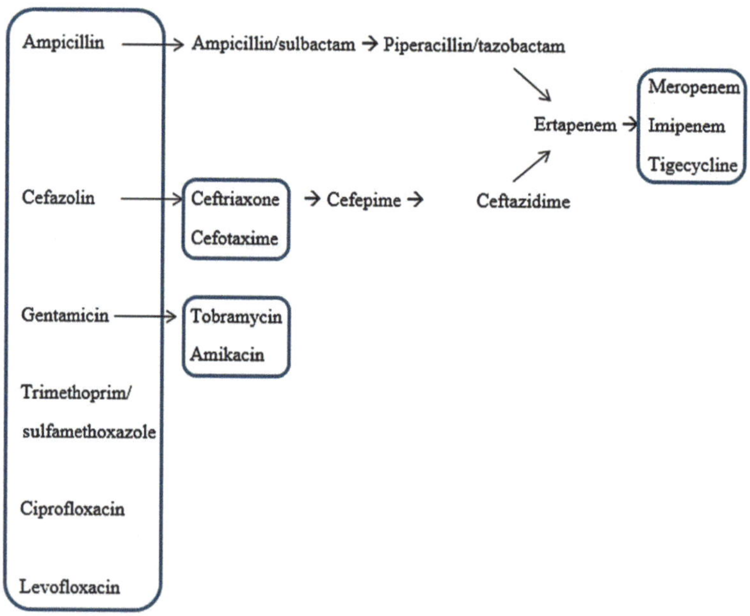

FIGURE 1.1

Sample selective reporting cascade of antibiotics for Enterobacteriaceae. The left column of agents are routinely released on susceptibility reports. If the bacterium is resistant to one or more antimicrobial, agents further on the cascade are released per this algorithm. For example, ceftriaxone and cefotaxime are released if the bacterium is resistant to cefazolin. Nonsusceptible agents are released regardless of their location on the cascade.

Clinical judgment remains the most valuable tool when decisions regarding antibiotic use are made. However, clinical presentations of infection are often nonspecific; it is often difficult to distinguish between bacterial causes of infection (when antibiotics are indicated), viral etiology, or noninfectious causes of the patient's symptoms. In such cases, biomarkers have become an attractive tool, with procalcitonin, a precursor of calcitonin, being the most widely studied candidate. In a meta-analysis including seven clinical trials and 1075 intensive care unit patients with severe sepsis or septic shock, Prkno et al. noted that procalcitonin-guided protocols can be used to safely reduce the duration of antibiotic therapy.[41] In another study, the Procalcitonin to Reduce Antibiotic Treatment Algorithm (PRORATA) trial showed a 23% reduction in antibiotic exposure with 2.7 more antibiotic free days in the procalcitonin algorithm arm.[42] However, mortality was 3.8% higher than the control group.

It should be noted that in the PRORATA trial, protocol compliance was suboptimal, with 53% of patients managed outside of the algorithm. In a prospective, single-blind, randomized, controlled trial conducted in 11 intensive care units in Australia, Shehabi and colleagues investigated whether a procalcitonin-guided algorithm with a low cutoff value of 0.1 ng/mL can reduce antibiotic utilization compared to standard care.[43] In this study, almost 400 patients admitted to the intensive care unit with presumed sepsis were included, with about 200 patients in each of the procalcitonin and standard care arms. Procalcitonin was measured daily, and the proportion of study days where the procalcitonin algorithm was not followed was less than 3%, indicating very high algorithm compliance. The authors found that a strategy based on a procalcitonin algorithm with a cutoff value of 0.1 ng/mL did not significantly reduce antibiotic utilization or time to antibiotic discontinuation when compared to standard care.

Of note, the initial procalcitonin level correlated with higher severity of illness and likelihood of subsequent positive blood culture and a slow decline of procalcitonin level over the first 72 h was associated with hospital and 90-day all-cause mortality. In a Swiss multicenter noninferiority randomized controlled trial involving both academic and nonacademic hospitals, a procalcitonin-guided algorithm was compared with standard care in the treatment of lower respiratory tract infections.[44] This study found that the procalcitonin algorithm was not inferior to treatment based on clinical guidelines in terms of composite overall adverse outcomes occurring within 30 days following presentation to the emergency department, with a statistically significant reduction in antibiotic exposure. This trial included patients with different severities of lower respiratory tract infections, 93% of whom were hospitalized. The community-acquired-pneumonia patients required hospital admission, thereby accounting for the relatively high antibiotic utilization in this group. The procalcitonin algorithm decreased the duration of antibiotics by about 3 days in this group. In acute bronchitis, upper respiratory tract infections, and chronic obstructive pulmonary disease exacerbations, where the indication for antibiotics was less clear cut, the procalcitonin algorithm decreased the initiation of antibiotics up to 75%.

These studies illustrate the many unanswered questions regarding the use of procalcitonin in antibiotic stewardship. Further studies must be undertaken to better delineate the role of this biomarker in antibiotic initiation, deescalation, and duration, in conjunction with sound clinical judgment and a comprehensive ASP.

PENICILLIN SKIN TESTING

Up to 10% of patients report having a penicillin or other β-lactam allergy, but a vast majority of these reports are inaccurate. In those with a true immunoglobulin E–mediated allergy, hypersensitivity does not persist over time. Only about 10% of patients with reported penicillin allergies have a positive penicillin skin test. Patients with reported penicillin allergies often receive broader-spectrum, suboptimal, or more toxic antibiotics than those without reported penicillin allergies.[45] In a small 1999 study of penicillin skin testing in hospitalized patients, 28 patients were enrolled who had a reported penicillin allergy and the need for antibiotic therapy.[46] Of these patients, 89% tested negative on penicillin skin testing, resulting in a significant reduction in vancomycin, fluoroquinolone, and clindamycin use without a negative impact on patient safety or outcome. In a review article on the use of penicillin skin testing in a variety of health-care settings, Unger and colleagues cited evidence that penicillin skin testing in the wards or the intensive care unit was well tolerated, with 89–95% of patients with a reported history of penicillin allergy testing negative.[47] In those studies that were prospective in nature, antibiotic regimens were switched to β-lactam in the setting of a negative penicillin skin test, resulting in a reduction in the use of alternative antibiotics. As an antimicrobial stewardship strategy, a penicillin skin testing protocol can be implemented to identify those patients with a true β-lactam allergy, thereby reserving the broader-spectrum antibiotics for multidrug-resistant pathogens.

ANTIBIOTIC CYCLING

The term *antibiotic cycling* refers to the routine scheduled substitution of a specific antibiotic on an institution's formulary in an attempt to preserve the diversity of antibiotic prescriptions and minimize bacterial selection pressure. In a retrospective study of Gram-negative susceptibility in a surgical intensive care unit 6 years into a monthly antibiotic cycling protocol, there was improved susceptibility in *Pseudomonas aeruginosa* and *Escherichia coli* isolates, while there were no significant changes in *Klebsiella pneumoniae* and *Enterobacter cloacae* susceptibility.[48] The susceptibility profiles of Gram-negative organisms in the medical intensive care unit (MICU) where antibiotic cycling was not done did not change significantly during this period. It is unclear how much of the improvement in Gram-negative susceptibility was due to antibiotic cycling alone. Concurrent infection control practices improvements may have contributed to these results. In a prospective before-and-after cohort study of 1172 MICU patients, four classes of antibiotics with Gram-negative activity for empiric use were cycled every 3–4 months over a 2-year period. The frequency of resistant Gram-negative infections seen in this study was about 30%. It was found that routine antibiotic cycling did not significantly change the risk of receiving inappropriate empiric antibiotic therapy for the treatment of intensive care unit infections.[49] At the present time, there is insufficient evidence for the routine use of antimicrobial cycling to reduce resistance over time.

FACILITATORS AND BARRIERS OF ASPs

Communications is the cornerstone of an effective ASP. In both the preauthorization process and prospective auditing with feedback, a nonconfrontational approach and the development of collegial relationships are important. In a telephone interview of ASP members from 21 large academic medical centers with established ASPs, the face-to-face style of communicating ASP recommendations was preferred.[50] Members also stressed the importance of being collaborators of patient care rather than being seen as the "antibiotics police." As antimicrobial stewardship moves from the cost-cutting arena into the realm of quality and patient safety, this collaboration becomes more important.

Some centers have developed and used antimicrobial stewardship ward rounds to enable face-to-face discussions on the more challenging cases.[45] The treating physicians should be made to feel that the stewardship team is there to help provide patients with optimal care, rather than just trying to cut costs. Inclusion of nonstewardship program clinicians on committees related to the program's activities is an effective way to encourage physician buy-in, while identifying potential barriers. Conflict management is an essential aspect in the successful implementation of stewardship intervention. It is important to garner support for antimicrobial stewardship efforts, especially within clinical specialty areas where antimicrobial use is high, such as critical care, oncology, and transplant services. ASPs often face a lack of resources in the form of personnel and time, since many of the stewardship interventions are time consuming.

MEASUREMENTS AND BENCHMARKS

The true impact of ASPs on clinical outcome has been difficult to measure.[51] Intuitively, it makes sense that the judicious use of antibiotics would lead to a reduction in antibiotic resistance, *C. difficile* rates, and antibiotic-associated adverse events, hence improving patient outcomes. However, these outcomes are challenging to measure and causal relationship is problematic to substantiate. A study of the impact of a comprehensive ASP in the intensive care unit setting was unable to show statistical relationships with *P. aeruginosa* resistance rates despite changes in antibiotic utilization.[52]

In China, where a national campaign and antimicrobial stewardship policy were enacted to promote judicious antibiotic use, a retrospective study was done over a 6-year period to evaluate its effectiveness in an 1800-bed tertiary care teaching hospital in Shanghai.[53] The hospital developed a computer-assisted program, and antibiotics were divided into three levels of restrictions based on clinical guidelines. During the intervention period, antibiotic utilization of some of the broad-spectrum-restricted antibiotics decreased, but the study was unable to demonstrate any significant improvement in antimicrobial susceptibility among bacteria. Most stewardship studies are single center, making it difficult to detect statistically significant differences in clinical outcomes, such as mortality and length of stay. Changes in bacterial resistance and *C. difficile* incidence occur over time, and other interventions such as infection control practices may also affect these rates. Antimicrobial utilization and cost are commonly used metrics in ASPs. These data are easy to capture and important to administrators. In a 2011 survey, the IDSA Emerging Infections Network found that 83% of administrators emphasized the importance of cost savings, while 63–72% of physicians cited patient outcomes as the most essential goal for an ASP.[54] Process measures such as acceptance rate of antimicrobial stewardship recommendations and the number of stewardship interventions assess whether an ASP intervention was carried out, but they do not measure whether the intervention affected clinical outcome or antibiotic resistance.

The most commonly accessible measure of an institution's antimicrobial resistance rate is the hospital antibiogram. An antibiogram lists the proportion of organisms that are susceptible to a hospital's formulary antibiotics over a given period of time. Most hospitals update their antibiograms annually, possibly with the help of EHRs or add-on CDSSs. It usually reflects the susceptibility of clinical isolates collected from the entire institution. In the United States, this includes the first isolate per patient per month, as recommended by the Clinical and Laboratory Standards Institute (CLSI). The antibiogram can provide guidance for empiric antibiotic therapy for certain resistant pathogens and help direct formulary decisions. Because antibiograms usually report the percent susceptibility of each antibiotic-pathogen combination, it is unable to monitor multidrug resistance effectively. In a review of eight studies that evaluated antimicrobial stewardship interventions, Schulz et al. found that changes in antibiotic utilization may or may not be reflected in changes in susceptibility as measured by the antibiogram in individual hospitals.[55] Hence, although the antibiogram has its role in antimicrobial stewardship, it does not serve as an effective measure of stewardship interventions.

The core function of an effective ASP is to promote the judicious and appropriate use of antimicrobials. As alluded to earlier in the chapter, a standardized definition of appropriate antibiotic use is lacking, making measurement of such use difficult. In an observational, retrospective, cohort study conducted at four sites in the United States, DePestel and colleagues compared four definitions of appropriateness of select antibiotics in the treatment of suspected or documented infections caused by methicillin-resistant *Staphylococcus aureus*, vancomycin-resistant enterococci, and *P. aeruginosa*.[56] These definitions were (1) study-site specific, (2) supported by in vitro susceptibility data, (3) antimicrobial used for indication with literature support, and (4) reflecting the opinion of the principal investigator. This study found significant inconsistency when these definitions were applied to determine the appropriateness of antibiotic therapy, particularly between principal investigator opinion (clinical judgment) and the other definitions.

Of note, nosocomial pneumonia and urinary tract infection had the highest disparity across the definitions for appropriate use of antibiotics. The CDC recognizes this dilemma and acknowledges the need for standardized assessment of appropriate antibiotic utilization, and has developed worksheets to help physicians assess the appropriateness of antibiotic use for various clinical syndromes, including community-acquired pneumonia and urinary tract infections.[57] It is often difficult to distinguish colonization from infection in these syndromes, leading to overtreatment. Furthermore, quality studies should be undertaken to determine the optimal duration of treatment of various infectious syndromes.

REFERENCES

1. Fridkin S, Baggs J, Fagan R, Gaill S, Pollack L, Malpiedi P, et al. Vital signs: improving antibiotic use among hospitalized patients. *MMWR Morb Mortal Wkly Rep* 2014;**63**(9):194–200.
2. Shlaes DM, Gerding DN, John Jr JF, Craig WA, Bornstein DL, Duncan RA, et al. Society for Healthcare Epidemiology of America and Infectious Diseases Society of America Joint Committee on the Prevention of Antimicrobial Resistance: guidelines for the prevention of antimicrobial resistance in hospitals. *Infect Control Hosp Epidemiol* 1997;**18**:275–91.
3. Dellit TH, Owens RC, McGowan Jr JE, Gerding DN, Weinstein RA, Burke JP, et al. Infectious Diseases Society of America and the Society for Healthcare Epidemiology of America guidelines for developing an institutional program to enhance antimicrobial stewardship. *Clin Infect Dis* 2007;**44**:159–77.
4. SHEA, IDSA, PIDS. Policy statement on antimicrobial stewardship by the Society for Healthcare Epidemiology of America (SHEA), the Infectious Diseases Society of America (IDSA) and the Pediatric Infectious Diseases Society (PIDS). *Infect Control Hosp Epidemiol* 2012;**33**(4):322–7.
5. CDC. *Core elements of hospital antibiotic stewardship programs*. Atlanta, GA: US Department of Health and Human Services, CDC. <http://www.cdc.gov/getsmart/healthcare/implementation/core-elements.html/>; 2014 [accessed 03.06.15].
6. Trivedi KK, Rosenberg J. The state of antimicrobial stewardship programs in California. *Infect Control Hosp Epidemiol* 2013;**34**(4):379–84.
7. <https://www.cdph.ca.gov/programs/hai/Pages/AntimicrobialStewardshipProgramInitiative.aspx/> [accessed 02.06.15].
8. <https://www.whitehouse.gov/sites/default/files/docs/national_action_plan_for_combating_antibotic-resistant_bacteria.pdf/> [accessed 04.06.15].
9. Howard P, Pulcini C, Hara GL, West RM, Gould IM, Harbarth S, et al. An international cross-sectional survey of antimicrobial stewardship programmes in hospitals. *J Antimicrob Chemother* 2015;**70**:1245–55.
10. Trivedi KK, Dumartin C, Gilchrist M, Wade P, Howard P. Identifying best practices across three countries: hospital antimicrobial stewardship in the United Kingdom, France, and the United States. *Clin Infect Dis* 2014;**59**(S3):S170–8.
11. File Jr TM, Srinivasan A, Bartlett JG. Antimicrobial stewardship: importance for patient and public health. *Clin Infect Dis* 2014;**59**(S3):S93–6.
12. Powers JH. Antimicrobial drug development-the past, the present, the future. *Clin Microbiol Infect* 2004;**10**:23–31.
13. Ruttimann S, Keck B, Hartmeier C, Maetzel A, Bucher HC. Long-term antibiotic cost savings from a comprehensive intervention program in a medical department of a university-affiliated teaching hospital. *Clin Infect Dis* 2004;**38**:348–56.
14. Standiford HC, Chan S, Tripoli M, Weekes E, Forrest GN. Antimicrobial stewardship at a large tertiary care academic medical center: cost analysis before, during, and after a 7-year program. *Infect Control Hosp Epidemiol* 2012;**33**:338–45.
15. Filice GA, Drekonja DM, Thurn JR, Hamann GM, Masoud BT, Johnson JR. Diagnostic errors that lead to inappropriate antimicrobial use. *Infect Control Hosp Epidemiol* 2015;**36**(8):949–56.
16. Nicolle LE, Bradley S, Colgan R, Rice JC, Schaeffer A, Hooton TM. Infectious Diseases Society of America; American Society of Nephrology; American Geriatric Society. Infectious Disease Society of America guidelines for the diagnosis and treatment of asymptomatic bacteriuria in adults. *Clin Infect Dis* 2005;**40**(5):643–54.
17. Trautner BW, Grigoryan L, Petersen NJ, Hysong S, Cadena J, Patterson JE, et al. Effectiveness of an antimicrobial stewardship approach for urinary catheter associated asymptomatic bacteriuria. *JAMA Intern Med* 2015;**175**(7):1120–7. Available from: http://dx.doi.org/10.1001/Jamainternmed.2015.1878.

18. Stevens V, Dumyati G, Fine LS, Fisher SG, Van Wijngaarden E. Cumulative antibiotic exposure over time and the risk of *Clostridium difficile* infection. *Clin Infect Dis* 2011;**53**(1):42–8.
19. Waters CD. Pharmacist-driven antimicrobial stewardship program in an institution without infectious diseases physician support. *Am J Health Syst Pharm* 2015;**72**(6):466–8.
20. Ohl CA, Dodds Ashley ES. Antimicrobial stewardship programs in community hospitals: the evidence base and case studies. *Clin Infect Dis* 2011;**53**(S1):S23–8.
21. Vora NM, Kubin CJ, Furuya EY. Appropriateness of gram-negative agent use at a tertiary care hospital in the setting of significant antimicrobial resistance. *Open Forum Infect Dis* 2015;**2**(1):ofv009. Available from: http://dx.doi.org/10.1093/ofid/ofv009.
22. Cosgrove SE, Seo SK, Bolon MK, Sepkowitz KA, Climo MW, Diekema DJ. Evaluation of postprescription review and feedback as a method of promoting rational antimicrobial use: a multicenter intervention. *Infect Control Hosp Epidemiol* 2012;**33**(4):374–80.
23. Cosgrove SE, Patel A, Song X, Miller RE, Speck K, Banowetz A, et al. Impact of different methods of feedback to clinicians after postprescription antimicrobial review based on the centers for disease control and prevention's 12 steps to prevent antimicrobial resistance among hospitalized adults. *Infect Control Hosp Epidemiol* 2007;**28**:641–6.
24. Moore LSP, Freeman R, Gilchrist MJ, Gharbi M, Brannigan ET, Donaldson H, et al. Homogeneity of antimicrobial policy, yet heterogeneity of antimicrobial resistance: antimicrobial non-susceptibility among 108707 clinical isolates from primary, secondary, and tertiary care patients in London. *J Antimicrob Chemother* 2014;**69**:3409–22.
25. George JM, Towne TG, Rodvold KA. Prolonged infusions of β-lactams antibiotics: implications for antimicrobial stewardship. *Pharmacotherapy* 2012;**32**(8):707–21.
26. Hoffman C, Khadem T, Schweighardt A, Brown J. New thoughts on the "Forgotten" aspect of antimicrobial stewardship: adverse event reporting. *Pharmacotherapy* 2015;**35**(1):59–63.
27. Blumenthal D. Launching HITECH. *N Engl J Med* 2010;**362**:382–5.
28. Forrest GN, Van Schooneveld TC, Kullar R, Schulz LT, Duong P, Postelnick M. Use of electronic health records and clinical decision support systems for antimicrobial stewardship. *Clin Infect Dis* 2014;**59**(S3): S122–33.
29. Hamilton KW, Gerber JS, Moehring R, Anderson DJ, Calderwood MS, Han JH, et al. Point-of prescription intervetions to improve antimicrobial stewardship. *Clin Infect Dis* 2015;**60**(8):1252–8.
30. Micek ST, Welch EC, Khan J, Pervez M, Doherty JA, Reichley RM, et al. Resistance to empiric antimicrobial treatment predicts outcome in severe sepsis associated with gram-negative bacteremia. *J Hosp Med* 2011;**6**(7):405–10.
31. Kumar A, Roberts D, Wood KE, Light B, Parrillo JE, Sharma S, et al. Duration of hypotension before initiation of effective antimicrobial therapy is the critical determinant of survival in human septic shock. *Crit Care Med* 2006;**34**:1589–96.
32. Pogue JM, Mynatt RP, Marchaim D, Zhao JJ, Barr VO, Moshos J. Automated alerts coupled with antimicrobial stewardship intervention lead to decreases in length of stay in patients with gram-negative bacteremia. *Infect Control Hosp Epidemiol* 2014;**35**(2):132–8.
33. Coorevits L, Boelens J, Claeys G. Direct susceptibility testing by disk diffusion on clinical samples: a rapid and accurate tool for antibiotic stewardship. *Eur J Clin Microbiol Infect Dis* 2015;**34**:1207–12.
34. Bauer KA, Perez KK, Forrest GN, Goff DA. Review of rapid diagnostic tests used by antimicrobial stewardship programs. *Clin Infect Dis* 2014;**59**(S3):S134–45.
35. Southern TR, VanSchooneveld TC, Bannister DL, Brown TL, Crismon AS, Buss SN, et al. Implementation and performance of the BioFire FilmArray Blood Culture identification panel with antimicrobial treatment recommendations for bloodstream infections at a midwestern academic tertiary hospital. *Diagn Microbiol Infect Dis* 2015;**81**:96–101.

36. Box MJ, Sullivan EL, Ortwine KN, Parmenter MA, Quigley MM, Aguilar-Higgins LM, et al. Outcomes of rapid identification for gram-positive bacteremia in combination with antibiotic stewardship at a community-based hospital system. *Pharmacotherapy* 2015;**35**(3):269–76.
37. Bork JT, Leekha S, Heil EL, Zhao L, Badamas R, Johnson JK. Rapid testing using the Verigene Gram-negative blood culture nucleic acid test in combination with antimicrobial stewardship intervention against Gram-negative bacteremia. *Antimicrob Agents Chemother* 2015;**59**(3):1588–95.
38. Perez KK, Olsen RJ, Musick WL, Cernoch PL, Davis JR, Land GA, et al. Integrating rapid pathogen identification and antimicrobial stewardship significantly decreases hospital costs. *Arch Pathol Lab Med* 2013;**137**:1247–54.
39. Perez KK, Olsen RJ, Musick WL, Cernoch PL, Davis JR, Peterson LE, et al. Integrating rapid diagnostics and antimicrobial stewardship improves outcomes in patients with antibiotic resistant gram-negative bacteremia. *J Infect* 2014;**69**:216–25.
40. Al-Tawfiq JA, Momattin H, Al-Habboubi F, Dancer SJ. Restrictive reporting of selected antimicrobial susceptibilities influences clinical prescribing. *J Infect Public Health* 2015;**8**(3):234–41 http://dx.doi.org/10.1016/j.jiph.2014.09.004.
41. Prkno A, Wacker C, Brunkhorst FM, Schlattman P. Procalcitonin-guided therapy in intensive care unit patients with severe sepsis and septic shock—a systematic review and meta-analysis. *Crit Care* 2014;**17**:R291.
42. Bouadma L, Luyt C-E, Tubach F, Cracco C, Alvarez A, Schwebel C, et al. Use of procalcitonin to reduce patients' exposure to antibiotics in intensive care units (PRORATA trial): a multicentre randomised controlled trial. *Lancet* 2010;**375**:463–74.
43. Shehabi Y, Sterba M, Garrett PM, Rachkonda KS, Stephens D, Harrigan P. Procalcitonin algorithm in critically ill adults with undifferentiated infection or suspected sepsis. *Am J Respir Crit Care Med* 2014;**190**(10):1102–10.
44. Schuetz P, Christ-Crain M, Thomann R, Falconnier C, Wolbers M, Widmer I, et al. Effect of procalcitonin-based guidelines vs standard guidelines on antibiotic use in lower respiratory tract infections. *JAMA* 2009;**302**(10):1059–66.
45. Hamilton KW, Fishman NO. Antimicrobial stewardship interventions: thinking inside and outside the box. *Infect Dis Clin North Am* 2014;**28**:301–13.
46. Harris AD, Sauberman L, Kabbash L, Greineder DK, Samore MH. Penicillin skin testing: a way to optimize antibiotic utilization. *Am J Med* 1999;**2**:166–8.
47. Unger NR, Gauthier TP, Cheung LW. Penicillin skin testing: potential implications for antimicrobial stewardship. *Pharmacotherapy* 2013;**33**(8):856–67.
48. Sarraf-Yazdi S, Sharpe M, Bennett KM, Dotson TL, Anderson DJ, Vaslef SN. A 9-year retrospective review of antibiotic cycling in a surgical intensive care unit. *J Surg Res* 2012;**176**:E73–8.
49. Merz LR, Warren DK, Kollef MH, Fridkin SK, Fraser VJ. The impact of antibiotic cycling program on empirical therapy for gram-negative infections. *Chest* 2006;**130**:1672–8.
50. Pakyz AM, Moczygemba LR, VanderWielen LM, Edmond MB, Stevens MP, Kuzel AJ. Facilitators and barriers implementing antimicrobial stewardship strategies: results from a qualitative study. *Am J Infect Control* 2014;**42**:S257–63.
51. Ashley ESD, Kaye KS, DePestel DD, Hermsen ED. Antimicrobial stewardship: philosophy versus practice. *Clin Infect Dis* 2014;**59**(S3):S112–21.
52. Slain D, Sarwari AR, Petros KO, McKnight RL, Sager RB, Mullett CJ. Impact of a multimodal antimicrobial stewardship program on *Pseudomonas aeruginosa* susceptibility and antimicrobial use in the intensive care unit setting. *Crit Care Res Pract* 2011. Article ID 416426, http://dx.doi.org/10.1155/2011/41626.
53. Guo W, He Q, Wang Z, Wei M, Yang Z, Du Y. Influence of antimicrobial consumption on gram-negative bacteria in inpatients receiving antimicrobial resistance therapy from 2008–2013 at a tertiary hospital in Shanghai, China. *Am J Infect Control* 2015;**43**:358–64.

54. Johannsson B, Beekmann SE, Srinivasan A, Hersh AL, Laxminarayan R, Polgreen PM. Improving antimicrobial stewardship: the evolution of programmatic strategies and barriers. *Infect Control Hosp Epidemiol* 2011;**32**:367–74.
55. Schulz LT, Fox BC, Polk RE. Can the antibiogram be used to assess microbiologic outcomes after antimicrobial stewardship interventions? A critical review of the literature. *Pharmacotherapy* 2012;**32**(8):668–76.
56. DePestel DD, Eiland EH, Lusardi K, Destache CJ, Mercier RC, McDaneld PM, et al. Assessing appropriateness of antimicrobial therapy: in the eye of the interpreter. *Clin Infect Dis* 2014;**59**(S3):S154–61.
57. Centers for Disease Control and Prevention. *Get smart for healthcare.* <http://www.cdc.gov/getsmart/healthcare/>; 2014 [accessed 18.06.15].

CHAPTER

MECHANISMS OF ANTIBACTERIAL RESISTANCE: SHEDDING SOME LIGHT ON THESE OBSCURE PROCESSES?

2

K. de Sousa Oliveira[1], L.A. de Lima[1], N.B. Cobacho[1], S.C. Dias[1] and O.L. Franco[1,2]

[1]Centro de Análises Proteômicas e Bioquímicas, Programa de Pós-Graduação em Ciências Genômicas e Biotecnologia, Universidade Católica de Brasília, Brasília, DF, Brazil [2]S-Inova Biotech, Pós-Graduação em Biotecnologia, Universidade Católica Dom Bosco, Campo Grande, MS, Brazil

INTRODUCTION

DISCOVERING THE FIRST ANTIBIOTICS

In 1877, the French microbiologists Louis Pasteur and Jules Francois Joubert published the results of a classic experiment using anthrax bacilli showing that they did not grow significantly in animals coinfected with other pathogenic bacteria.[1] This occurred because (1) the anthrax bacillus could form spores that were resistant to high temperature, pressure, and pure oxygen; (2) the blood of a dead animal that has putrefied for more than 16 h can kill a living animal not being exposed to the anthrax; and (3) growth of the anthrax bacillus was delayed by the existence of other microscopic organisms, similar to those found in the blood during putrefaction, in a process described by Pasteur as *la lutte pour la vie* (the battle for life). In a putrefied animal, different bacteria develop in the blood, removing the anthrax bacillus. In another experiment, the culture used to inoculate a guinea pig was exposed to filtration that removed the bacillus. The culture filtrate was totally avirulent, suggesting that the bacillus was essential to cause anthrax.[2]

Different microorganisms such as *Pseudomonas aeruginosa* or *Streptococcus erysipelatis* (*Streptococcus pyogenes*) have long been used to antagonize the infective properties of other bacteria.[3] Later, this antagonism between two different microorganisms was recognized as the ability of some bacteria to generate chemical substances capable of killing other microorganisms.[4]

Antimicrobials are one of the greatest successful forms of chemotherapy in the history of medicine. At the beginning of the new "antibiotic era," between 1904 and 1908, physician Paul Ehrlich and biologist Alexander Fleming were extremely important for antibiotics development. Ehrlich's ideas were related to the creation of a "magic bullet" that would involve disease-causing microbes without affecting the host. In 1904, Ehrlich joined chemist Alfred Bertheim and bacteriologist Sahachiro Hata to produce hundreds of organoarsenic derivatives of a high concentration of the toxic drug atoxyl and tested them in syphilis-infected rabbits. In 1909, they found a 6th compound in the

Antibiotic Resistance. DOI: http://dx.doi.org/10.1016/B978-0-12-803642-6.00002-2

600th series tested, which cured syphilis-infected rabbits and exhibited important potential for the treatment of patients.[5] Due to the severe side effects, a more soluble and less toxic form of this compound, called *salvarsan* (*arsphenamine* or *Ehrlich 606*), was developed. Known as neosalvarsan, this became the most often recommended drug until it was replaced by penicillin in the 1940s.[6]

The discovery of antibiotics really started in 1928. Fleming noted an unusual colony of mold in several petri dishes containing *Staphylococcus* bacteria. Around the mold, no perceptible *Staphylococcus* was to be found. The unexpected mold colony had somehow diffused the bacteria around it. A few small semitransparent groups, defined by his assistant as "ghosts," were identified as penicillin.[7] Fleming wrote his first paper on penicillin in 1929, mentioning the use of additives in bacteriological media to selectively isolate the bacterium *Bacillus influenzae* (*Haemophilus influenzae*), which was assumed to cause influenza. Over time, Fleming demonstrated that penicillin was bacteriostatic, inhibiting the growth of microbes, and even bactericidal. He induced lytic changes in the bacteria, disrupting bacteriolytic and recognizing afterward that penicillin is easily diffusible. Additional tests showed that penicillin completely repressed the growth of *Staphylococcus* in a dilution up to 1 in 1000 when tested on human blood. Fleming then injected it into animals and observed no toxicity.[8,9] However, the drug was not made accessible for medical use at first, being largely restricted to the military.[10]

Through the early days of antibiotics research, other drugs such as sulfa, namely sulfonamidochrysoidine (KI-730, Prontosil), was synthesized by chemists Josef Klarer and Fritz Mietzsch. It was then tested by German bacteriologist Gerhard Domagk for antibacterial activity in different types of diseases.[11] Sulfanilamide was cheap to produce, off patent, and easy to modify. For these reasons, many companies started the mass production of sulfonamide derivatives.[12]

Moreover, combinations of drugs were seen to produce potent synergistic effects. Trimethoprim, for example, a powerful antibacterial agent, was selected for combination with sulfonamides such as sulfadiazine.[13] Around 1940, Donald Devereux Woods and Paul Fildes suggested that many chemotherapeutic agents can produce action by important effects of bacterial metabolites. Woods analyzed the structural similarity between sulfanilamide and para-aminobenzoic acid (PABA) and observed that both compete for a critical enzymatic site. This competitive nature was confirmed when graded concentrations of PABA reduced the bacteriostatic effects of sulfanilamide.[14]

At the same time, new antibiotics were discovered by Waksman and Woodruff, who established techniques to identify and isolate actinomycin by 1940,[15] followed by the isolation of streptothricin.[16] Even though penicillin continued to be widely used as a therapeutic agent in humans for almost 70 years, several Gram-positive pathogens, including *S. pyogenes*, are susceptible to this antibiotic.[17]

The technology used to find new antibiotics has changed over the last 75 years. Several drugs are being developed to cure diseases that have no effective antimicrobial treatment as structure-based drug design or using natural product isolation for the synthesis of new antimicrobial agents.

BACTERIAL RESISTANCE

Antimicrobial resistance (AMR) is defined as the resistance of microorganisms to an antimicrobial agent to which they were at first sensitive.[18] This natural evolutionary phenomenon, enhanced by the misapplication of antimicrobial medicines and the global spread of AMR, mainly affects unhealthy and debilitated patients, giving rise to superbugs. AMR inflicts high costs in the public

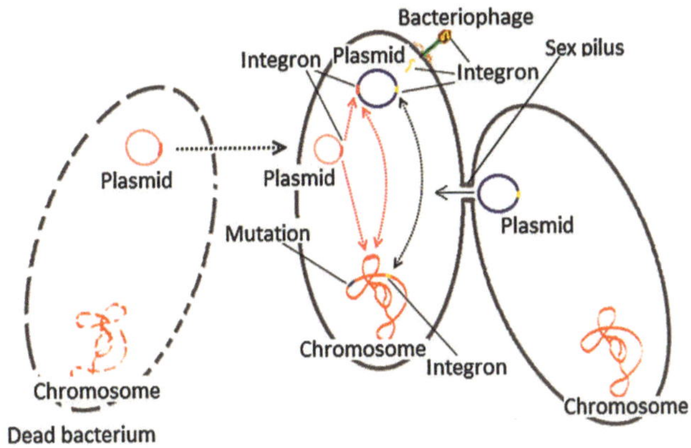

FIGURE 2.1

Uncommon mechanisms of antibiotic resistance. One of the mechanisms of resistance is the integration of DNA from dead bacteria on the genome and plasmids of living cells, which also can happen with integrons coming from bacteriophages. Mutations happen all the time, and a simple, single nucleotide change can alter the gene and, by vertical gene transfer, grant some kind of resistance to the microorganism. The plasmid from one bacterium, with any type of integron, can easily be transferred to another bacterium through horizontal gene transfer by conjugation.

health sectors of all countries, and many researchers are involved in searching for greater understanding of resistance and ways to mitigate it. A wide range of antibiotics have been faced with the threat of resistance in recent decades, and this resistance may be generated and transmitted in many different ways (Fig. 2.1).[18]

Through horizontal gene transfer, for example, mobile integrons carried on transposons permit pathogens to share resistance mechanisms. For organisms resistant to one antibiotic, the gaining of a transposon that transports several antibiotic resistance cassettes offers the organism resistance to numerous other antibiotics.[17] Another case of natural resistance is measured frequently by the incidence of natural mutations within chromosomally located genes that later are spread vertically as the bacteria replicate.[19,20]

In other cases, intrinsic resistance occurs that refers to the presence of genes in bacterial genomes that could produce a resistance phenotype. Any of the additional genetic elements originating in bacteria are capable of obtaining resistance genes and promoting their transmission, and the type of element involved differs with the pathogen's genus. It is already known that there are clear differences between Gram-positive and Gram-negative bacteria in this regard.[21]

Bacterial resistance to antibiotics was documented as early as the beginning of the antibiotic era. Within 20 years, the development of dangerous and resistant strains was already happening with disturbing regularity. The main cause was and continues to be a lack of public knowledge about antibiotics, resulting in their overuse despite recent stricter controls on their prescription and purchase worldwide.[22,23]

Unquestionably, self-medication affects the quality of an effective therapy; the correct diagnosis by a medical specialist often would avoid the use of last-line antimicrobials. Human use (and misuse) of antibiotics has noticeably placed an unnatural selective pressure on bacteria, which has favored their accelerated evolutionary process.[22,24] Furthermore, compounds and conditions that occur in these communities may offer additional selection pressures. Certainly, most antibiotics are made from strains of fungi and bacteria that occur naturally in all environments. Most antibiotic-producing strains transfer genes encoding resistance to the antibiotics that they yield, and these genes typically originated in the same gene cluster as the antibiotic biosynthesis pathway genes.[25] In addition, antibiotics produced in the environment may apply selective pressure on neighboring organisms.[26]

Another problem lies in the use of antibiotics in animal feedstock, which has contributed to the spread of resistance[22] causing infection in domesticated animals that need antibiotic therapy.[27]

In addition, antibiotics are used for prophylactic purposes in farmed animals, as well as for even wider and less targeted treatment in aquaculture and horticulture.[12] Attempts to reverse this trend are taking place: for example, in Scandinavian countries, new programs have sought better defensive routines and the appropriate use of antimicrobials, along with a removal of antibiotic growth promoters from farmed animals. These actions have led to a decrease in the use of antimicrobials and work against further AMR.[28] Since 2006, other European Union countries have been applying similar measures to limit the incidence and distribution of antibiotic resistance from agricultural sources.[12]

INFECTION CONTROL

In spite of greater understanding of antibiotic resistance mechanisms, procedures to control resistance have been unsuccessful in many cases.[29] However, infection control programs can cause a reduction in the occurrence of infections in patients and staff. Several antimicrobial-resistant organisms, including methicillin-resistant *Staphylococcus aureus* (MRSA) and vancomycin-resistant *Enterococci* (VRE), are mainly spread among patients and health-care professions. Correct handling procedures for contaminated material would be effective in reducing the transmission of organisms, as well as of strains that are not antimicrobial resistant. Hand decontamination for health-care workers can be done with basic products like soap and water, or an antiseptic or antimicrobial solution. Likewise, sterilization and disinfection of patient care equipment is similarly effective for antimicrobial-resistant and vulnerable organisms.[30,31] The goal of an infection control program is to limit nosocomial infections in patients and staff. Among the proposals for action, there are tight controls on antibiotic use by humans, no distribution of antibiotics without a medical prescription, and measured therapeutic use in animal husbandry and agriculture. For example, while the Netherlands and Scandinavia have decreased resistance levels, it is clear that it is difficult to impose a limit on antibiotic usage on a global scale. In many developing nations, antibiotic use is still somewhat unrestrained. Normally used antimicrobials are relatively cheap in these nations, frequently costing one-tenth to one-thirtieth of the price charged in developed nations, even though these drugs are not essential.[21]

Despite growing awareness of antibiotic resistance and programs to combat it, during the period between 2009 and 2012, growth in the percentages of resistant *Escherichia coli* isolates was detected throughout Europe. This was especially noted with respect to the cephalosporins, aminoglycosides, and combined resistance to antimicrobial groups, in which a majority of the countries of Europe reported a significant rise. Thereby, high-level aminoglycoside resistance in *Enterococcus faecalis*, for example, seems constant in Europe, but it is usually very high, with most countries reporting percentages more than 25% in recent years. MRSA in the most European countries describe percentages less than 20%. In general, the European Union/European Economic Area population-weighted average of MRSA was 17.8% in 2012, and it fell significantly from 2009 to 2012.[32]

The public in most countries still needs to be made more aware of antibiotic resistance, and standards of hygiene in society (as well as in hospitals) need to be boosted dramatically. Furthermore, there is a need to have actions that deal with the foundations of resistance, and this involves understanding the mechanism of resistance in microorganisms that are associated with a range of antibiotics.[22,29]

MECHANISMS OF ANTIBACTERIAL RESISTANCE

There are many mechanisms of resistance in bacteria. Of these, five are the most frequently observed, showing high prevalence in clinical isolates. They are enzymatic inhibition, penicillin-binding protein (PBP) modifications, porin mutations, efflux pumps, and target changes, and all five are described further later in this chapter, as well as being summarized in Fig. 2.2.[33–38]

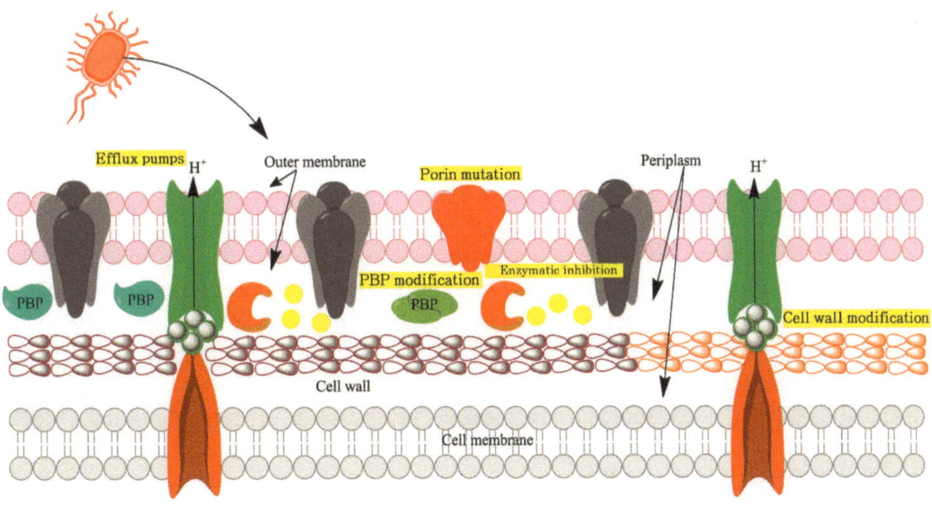

FIGURE 2.2

The most common mechanisms of bacterial resistance in Gram-negative bacteria. Such as enhanced efflux pumps, porin mutation, mutation in the penicillin-binding proteins (PBPs), enzymatic inhibition, and target modification of the drug. (Represented by modification of the bacterial cell wall.)

ENZYMATIC INHIBITION

The most common mechanism of resistance in bacteria is enzymatic inhibition. This mechanism is based on several strategies for modifying the structure of antibacterial compounds: with hydrolysis, a type of reaction that occurs mainly with β-lactam agents; transference of functional groups (acyl, phosphoryl, thiol, nucleotidil, ADP-ribosyl, glycosyl), which occurs with a lot of antibacterials, such as aminoglycoside, chloramphenicol, rifamycin, and lincosamide; and other chemical modifications (redox, lyase), which occur with tetracycline, rifamycin, and streptogramin.[33,39]

Among the vast abundance of enzymes that can modify antibacterial compounds, β-lactamases are a major problem in the treatment of Gram-negative bacteria.[40,41] These include penicillinases, which confer resistance against penicillins, AmpC cephalosporinases (eg, MOXs, MIR, FOX family, CMI family, and others), and which are also able to hydrolyze penicillins and many cephalosporins that can avoid clinical β-lactamase inhibitors (eg, clavulanic acid, sulbactam, and tazobactam). They also include extended-spectrum β-lactamases (eg, SHV-1, TEM-1, TEM-2, CTX-Ms, and others), which are able to hydrolyze penicillins and all cephalosporins. This group of enzymes confers resistance against penicillins and all cephalosporins, including cefotaxime and ceftazidime; moreover, many producers of TEM and SHV display coresistance to tetracyclines, sulfonamides, and aminoglycosides. The majority of the CTX-M producers are also resistant to fluoroquinolones. Finally, the carbapenemases (eg, IMP family, VIM family, KPCs, OXAs, and others), which are enzymes capable of inactivating all β-lactam agents except aztreonam, have been divided into three classes, metallo-β-lactamases (class B) and serine carbapenemases (classes A and D).[41–44]

In Gram-positive bacteria, β-lactamases do not represent a real problem. Only penicillinases were recognized with enzyme inhibition mechanisms, which in turn confer resistance only against penicillins.[45] Mechanisms such as the acquisition of new PBPs (PBP2a, PBP2′ mutation/overexpression of enterococcal PBP4 or PBP5), and cell wall modifications (increased thickness and modified peptidoglycan target (d-Ala-d-Lac or d-Ala-d-Ser depsipeptides)) need more attention because these mechanisms confer resistance against last-line antimicrobials, such as cephalosporins and vancomycin.[36,45,46]

PBP MODIFICATIONS

PBPs are important proteins involved in the construction of peptidoglycan, which is the major constituent of bacterial cell walls.[34] These enzymes catalyze the glycan strand (transglycosylation) and the cross-linking between glycan chains (transpeptidation).[34,47] However, some PBP classes did not have transglycosylation activity, such as B PBPs and low-molecular-mass PBPs.[47,48]

The transpeptidase active site is the target of β-lactam agents.[49] These compounds mimic the D-Ala-D-Ala dipeptide in peptidoglycan and form a very stable acyl-enzyme complex, leading to enzyme inactivation.[49,50]

Among the different modified PBPs, some of them have high prevalence, including PBP4 and PBP5, which confer resistance to penicillins; and PBP2x and PBP1a, which are responsible for conferring variable resistance to penicillins and other β-lactams, both of chromosomal origin.[45] However, the most alarming is PBP2a (also called PBP2′), a modified protein that confers resistance to penicillins and cephalosporins.[51] This protein is the product of the gene *mecA* and the homologous genes *mecB* and *mecC*, all of plasmid origin.[51] These modified PBPs change the active site, causing the β-lactam agents to lose or diminish their affinity with the target protein, promoting resistance.[51,52]

PORIN MODIFICATIONS

Gram-negative bacteria have a membrane outside the cell wall, the outer membrane, which consists of a lipid bilayer. The main constituent of this bilayer is the lipopolysaccharide, and due to its hydrophobicity, the passage of hydrophilic compounds is very difficult; thus, porins or outer membrane porins (Omps), which are proteins that aid in the passage of hydrophilic solutes across lipid bilayer membranes, are required.[35,53] Many factors affect the ability of the drug to pass through porins, such as charge, shape, and size.[53]

There are some typical porins, such as OmpF, OmpC, and OmpE.[42] Each bacterial species produces specific porins, and the loss or impairment of one or more Omps is a common contributing factor in establishing resistance (eg, loss of OprD in *P. aeruginosa* confers resistance to imipenem and meropenem; in other species, loss of OmpF can lead to multidrug-resistant (MDR) organisms).[42,54] This phenomenon results in an increase in minimum inhibitory concentrations to hydrophilic antimicrobials and reduces the choices of antibacterial therapeutics in clinical practice.[42,54,55]

A reduction in porin production is characteristic of some bacteria, such as *P. aeruginosa*, which gives low susceptibility to β-lactam agents.[56] In some strains, it is possible to observe porin exchange, which promotes a reduction or loss of affinity of the antibacterial with these proteins, which then lose their ability to overcome the outer membrane and enter the cell (eg, OmpK35 to OmpK36 in *Klebsiella pneumoniae* isolated from a patient using antibiotic therapy).[54,56]

Many studies have demonstrated that selective pressure exerted by the prolonged use of antibiotics is an important factor in the appearance of MDR bacteria, and the modification of porins is an important factor in this process.[55,56] The most common mechanisms are involved in decreasing porin expression and mutations, which in turn prevent the antibiotic from entering the cell.[56]

EFFLUX PUMPS

A highly efficient mechanism of resistance is the production of an efflux pump, a proton-dependent system that effects an active removal of the antibiotic from inside the cell.[36] There are five families of membrane-spanning efflux proteins (Table 2.1), including major facilitators (MFs), small multidrug resistance (SMR), resistance nodulation cell division (RND), ATP-binding cassette (ABC), and multidrug and toxic compound extrusion (MATE).[57] On the one hand, drug efflux from Gram-positive bacteria is commonly mediated by a single cytoplasmic membrane-located transporter of the MF, SMR, or ABC families. On the other hand, Gram-negative bacteria are more complex due to the presence of an outer membrane.[58]

The MF family consists of membrane transport proteins, with 12–14 transmembrane domains (TMDs),[59] implicated in the antiport, symport, or uniport of many substances.[60] In MF and SMR family transporters, the propulsion force for drug efflux appear to be an electrochemical potential of H^+ over the cell membrane.[59]

All members of this family have three conserved motifs: motif A, which acts as a cytoplasmic gate controlling the passage of the substrate to and from the cytoplasm; motif B, which is involved in energy coupling; and motif C, which determines the orientation of the unoccupied substrate-binding site and thus commands the direction of transport. The best characterized protein in this family is the tetracycline transporter (TetB), from *E. coli*, which has been shown to function as an electroneutral antiport system, catalyzing the exchange of a tetracycline-divalent-metal-cation complex for a proton.[60]

Table 2.1 Drug Resistance Profiles of Efflux Pump Transporters

Family	Gene	Organism	Resistance
MF	*bcr*	*Escherichia coli*	Bicyclomycin Sulfathiazole
	norA	*Staphylococcus aureus*	Chloramphenicol Quinolones
	qacA	*S. aureus*	Ethidium bromide Benzalkonium Cetrimide Chlorhexidine Pentamidine
	bmr	*Bacillus subtilis*	Chloramphenicol Quinolones
	tetB	*E. coli*	Tetracyclines
SMR	*emrE*	*E. coli*	Ethidium bromide QAC Lipophilic cations β-Lactams Cephalosporins Aminoglycosides Erythromycin Sulfadiazine Tetraphenylphosphonium (TPP)
	hsmr	*Halobacterium salinarum*	Ethidium bromide, QAC
	qacG	*Staphylococcus* sp.	Ethidium bromide, QAC
	smr	*S. aureus*	Ethidium bromide, QAC
	qacC	*S. aureus*	Ethidium bromide, QAC
	qacE	*Klebsiella aerogenes*	Ethidium bromide, QAC
RND	*acrAB-TolC*	*E. coli*	Macrolides Fluoroquinolones β-Lactams Tetracyclines Chloramphenicol Rifampicin Novobiocin
	mexAB-OprM	*Pseudomonas aeruginosa*	β-Lactams Chloramphenicol Macrolides Novobiocin Quinolones Sulfonamides Tetracyclines Trimethoprim

Table 2.1 Drug Resistance Profiles of Efflux Pump Transporters *Continued*

Family	Gene	Organism	Resistance
ABC	*lmrA*	*Lactococcus lactis*	Daunomycin
			Ethidium bromide
			Nigericin
			Valinomycin
			Aminoglycosides
			Cephalosporins
			Penicillins
			Chloramphenicol
			Lincosamides
			Macrolides
			Quinolones
			Tetracyclines
MATE	*norM*	*Vibrio parahaemolyticus*	Ciprofloxacin
			Norfloxacin
			Kanamycin
			Streptomycin
			Berberine
	pmpM	*P. aeruginosa*	Ciprofloxacin
			Norfloxacin
			Fradiomycin
			Chlorhexidine
			Acriflavine
			Rhodamine
			TPP

The SMR transporters have 100–140 amino acid residues, and compared to other proton-dependent transporters, SMRs are small. Analysis of the structure of SMRs shows four transmembrane hydrophobic domains that are connected by flexible hydrophilic segments.[61,62] This confers resistance to a variety of quaternary ammonium compounds (QAC), besides other lipophilic cations, presenting drug efflux via an electrochemical proton gradient.[63]

Phylogenetic characterization and genome sequencing studies of the SMR family revealed three subclasses: small multidrug pumps, the suppressor of groEL mutation proteins (SUGs), and paired SMR proteins (PSMRs). The first subclass is characterized by its ability to confer multidrug resistance against Gram-negative, Gram-positive bacteria, and Archaea from the expression of a single gene. SUG demonstrated isogenic transport activity and the potential to import and export a very narrow variety of these substrates in overaccumulation studies. PSMR is a distinct subclass due to the requirement for both copies of each SMR homolog to be simultaneously expressed in order to confer a drug resistance phenotype.[63]

The EmrE transporter from *E. coli*, which is the most studied SMR member, consists of 110 residues and is an asymmetric and antiparallel homodimer where the helices in each protomer are organized in a linear fashion. This assembly allows conformational switching between inward-open and outward-open states, an ability that is necessary for the transport of substrate or protons across the cellular membrane.[61]

SMR protein multimerization is suggested as a requirement for active drug transport by this class, particularly for members of the PSMR subclass. However, monomeric proteins can bind drugs with strong affinity,[64] and the transport mechanism of this family has not yet been elucidated.[63] Yerushalmi and Schuldiner[65] suggested that an EmrE trimer is a functional oligomeric form, and in this model, two of the three EmrE E14 residues deprotonate upon the approach of a single, positively charged drug molecule. In that case, the cationic substrate binds within a hydrophobic pocket formed by the trimer. A subsequent conformational change in the protein complex opens the pocket to face the periplasmic side of the membrane while closing off the cytoplasmic pocket exposure. Then two protons from within the periplasm move into the binding pocket, catalyzing the release of the substrate into the periplasm. Reprotonation within the binding pocket relaxes the trimer complex back, restarting the cycle.[65]

In Gram-negative bacteria, the tripartite RND class is probably the most important one for resistance. This system consists of a cell membrane–spanning pump (AcrB and MexB), an outer membrane pore (TolC and OprM), and a periplasmic adapter protein (AcrA and MexA) that joins both. The most studied members are AcrAB-TolC from *E. coli* and MexAB-OprM from *P. aeruginosa*.[36] The transporter AcrB is a large protein containing more than 1000 residues, and according to the crystallographic study by Murakami and coworkers,[66] it was found to exist as a trimer. On this trimer, each subunit is composed of a TMD with 12 transmembrane helices and a periplasmic domain with the same size, composed of a porter domain deporting the drugs and the TolC docking domain. This transporter captures the drug molecules from the periplasm, showing a wide substrate specificity; for example, AcrAB-TolC can pump out basic dyes, such as acriflavine and ethidium; antibiotics, such as β-lactams, tetracyclines, chloramphenicol, and rifampin, except aminoglycosides; detergents, like sodium dodecyl sulfate (SDS) and Triton X-100; and even simple solvents, such as hexane and heptanes.[67]

Antibiotic efflux is conjugated with a vectorial proton influx into the cell. The pumps are trimers that recognize a broad array of small molecules in at least two cavities ("cave" and "groove").[68] Murakami and coworkers (2006) published an asymmetric trimer model of AcrB in which the periplasmic domains of each subunit assume a unique conformation, called Access, Binding, and Extrusion.[69] Only one subunit (the binding type) binds the drug molecule in this pocket, and this led to the proposal, confirmed by biochemical studies, that the transition of molecules across the outer membrane pore may occur by a trimer rotation mechanism.[36,67] It is also proposed that each subunit goes through a cycle of conformational alterations, facilitated in turn by the complementary alterations in neighboring subunits. In support of this, Takatsuka and Nikaido[68] showed that the inactivation of only one subunit causes a loss in the function of the entire trimer, and if there is a defect in the proton relay network of one subunit, the pumping action by the entire trimer comes to halt.

ABC transporters are ubiquitous adenosine triphosphate (ATP)–dependent transmembrane pumps, receptors, and ion channels that have been found in all three kingdoms of life.[70] The first ABC transporter was identified, in 1996, from *Lactococcus lactis*. Although this transport system has a wide range of substrates, all members of this family share a common four-domain architecture, which consists of two TMDs. These form the ligand-binding sites and provide specificity, and two nucleotide-binding domains and ATP hydrolyze to drive the translocation of the bound ligand.[71] These can be regarded as the molecular motor that transforms chemical energy into mechanical work, containing all diagnostic sequence motifs, the C-loop, and the H-loop, as well as the Walker A and B motifs.[72]

ABC efflux pumps are a unidirectional gate that is remodeled after drug release and ATP hydrolysis.[72] When a drug molecule is released from one of the multiple drug-binding sites, it leaves the drug-binding pocket, having been expelled from the cell, which prevents it from returning. A polytopic protein is the major challenge in attempts to understand the transport cycle of this system, but it is agreed that several conformational changes happen during one turn of the catalytic cycle. The atomic structure of many ABC transporters supports the proposal that the high-affinity drug-binding conformation switches upon ATP binding and hydrolysis to a drug release structure of lower affinity.[73]

MATE is an energy-dependent efflux system that has 12 putative TMDs. This mediates resistance to dyes, hydrophilic fluoroquinolones, and aminoglycosides[74] by an H^+ or Na^+ gradient across the membrane, but the coupling mechanism is not well understood.[75] The overall properties of the MATE family have not been completely determined.[76] NorM, a member of this family, consists of 456 amino acid residues and is very rich in hydrophobic residues. It possesses 12 hydrophobic regions that may be TMDs,[59] arranged as two sheaves of six transmembrane helices forming a large internal cavity open to the extracellular space.[75] In addition to several MATE transporters, NorM is a Na^+/drug antiporter and cannot utilize H^+ as a coupling cation. However, the transporters PmpM, from *P. aeruginosa*, and AbeM, from *Acinetobacter baumannii*, can use H^+ as well.[77]

MOLECULAR MODIFICATION OF ANTIBIOTIC TARGETS

Most antibiotics affect the protein synthesis process targeting the ribosome,[37] and differences between the structure of this account for the selective action of antibiotics in bacterial, archaeal, and eukaryotic cells. Even among species, slight variations in the ribosomal structure may lead to idiosyncratic, species-specific interactions among the drugs and their targets.[78]

Antibiotics that target the translational machinery of the bacterial cell are potent inhibitors of prokaryotic pathogens.[79] Nevertheless, over decades of clinical use, these pathogens have become resistant to antibiotics that inhibit protein synthesis.[78] A notable mechanism of resistance is the modification of the molecular target of antibiotics. Commonly, this can arise through point mutations in selected genes, resulting in relatively rapid and easy resistance with a minimal impact on microbe fitness. Relatively small changes in an amino acid sequence alter the protein structure sufficiently to impede antibiotic binding and action. For example, single mutations in target genes such as gyrA provide high-level resistance, while interactive mutations in the same gene can increase the level of resistance. Furthermore, target modification can arise from catalytic resistance strategies. An example is the ribosome methyltransferase, where Erm enzymes modify the 23S ribosomal RNA (rRNA) of the large subunit of the ribosome at position A2058 (in *E. coli*). This confers resistance to macrolides, lincosamides, and type B streptogramins.[36,78]

Alterations in rRNA, like posttranscriptional modification and mutations of rRNA that confer resistance, and biochemical footprinting experiments that localize the binding sites of an antibiotic have been found to play an ever-more-important role in clinical and other forms of naturally occurring antibiotic resistance.[80] Erythromycin inhibits protein synthesis by its effect on ribosome function. The first reports of ribosomal structural changes in erythromycin-resistant mutants of *E. coli* described ribosomal protein alteration, notably of proteins L4 and L22. It has been shown that the methylation of adenine in 16S rRNA was required for proficiency to bind to kasugamycin.[81] Indeed, Lai and colleagues showed that specific methylation of 23S rRNA in *S. aureus* conferred resistance to the macrolide-lincosamide-streptogramin family of antibiotics.[82]

In addition, daptomycin (DAP), which is a cycle lipopeptide from *Streptomyces roseosporus*,[83] showed activity against many Gram-positive organisms,[84] besides being the only antibiotic with in vitro activity against VRE. However, its use is threatened by the emergence of resistance during therapy.[85] This peptide is anionic, so its antibacterial activity is closely dependent on the presence of physiological levels of free calcium ions, which gives it many properties of the large class of cationic antimicrobial peptides. DAP disrupts Gram-positive cytoplasmic membrane function, causing leakage of potassium ions and ultimately leading to the loss of membrane potential and cell death. In addition, studies using artificial membranes have shown that it can act directly on the lipid bilayer in the absence of any bacterial protein or other surface components.[83]

As yet, the mechanism of DAP resistance is not completely understood. In *S. aureus*, a correlation has been observed with a relatively more positive surface charge. In this regard, a major factor in the cell envelope charge is the phospholipid composition of the cell membrane, such as the negative phospholipid cardiolipin, and the positively charged amino derivatives of phosphatidylglycerol. Modification in phospholipid content leads to increased synthesis and translocation of the positive charge from the inner to the outer leaflet of the cell membrane.[86] That in turn causes an electrostatic repulsion between the positively charged cell surface and the calcium-DAP molecule complex, and this is believed to prevent the initial interaction of DAP with the cell membrane.[85]

Most mutations that alter the sensibility to DAP have demonstrated that they directly affect the membrane lipid composition. Interestingly, depending on the species, the specific membrane modification change: in *Enterococci*, for example, the resistance is associated with changes in cardiolipin synthesis, while in *Bacillus subtilis*, resistance may be linked to the overall phosphatidylglycerol content due to mutations in pgsA. Mutations conferring DAP resistance on *S. aureus* have been reported in the *yycG* gene product, a histidine kinase believed to coordinate peptidoglycan synthesis and cell division.[83] In *E. faecalis*, substitutions in three proteins are involved in resistance. This resistance in *E. faecalis* occurs by mutation on three proteins: LiaF, GdpD, and Cls. LiaF is a member of the regulatory system that controls the cell envelope response to stress in Gram-positive bacteria and in *B. subtilis*; GdpD is a protein involved in phospholipid metabolism; and Cls is involved on cardiolipin synthase.[38,39]

CONCLUDING REMARKS

Many years of increasing bacterial resistance have generated an unquestionable dilemma for the antibiotics research field. Developments in technology have provided the essential tools for finding new antibiotic classes and enhancing already known ones to combat the mainly Gram-negative resistant pathogens. The discovery and use of antibiotics, antibiotic resistance markers, and mobile elements such as plasmids have been the basis of genetic engineering and molecular biology that ultimately resulted in remarkable success in the human genome and other sequencing projects. Likewise, it is necessary to learn about targeting the pathogens and limiting the indiscriminate use of antimicrobials and other practices that increase the development of novel resistance mechanisms. Infections initiated by resistant microorganisms frequently fail to respond to conventional treatment, leading to continued illness, higher risk of death, and inflated costs. For these reasons, screening and further validation of novel antimicrobial compounds with unusual activities are urgently needed to ensure the survival of human populations.

ACKNOWLEDGMENTS

This work was supported by the Coordination for the Improvement of Higher Education Personnel (CAPES), National Counsel of Technological and Scientific Development (CNPq), Foundation to Support the Development of Education, Science, and Technology of the State of Mato Grosso do Sul (FUNDECT), and Foundation for Research Support of the Distrito Federal (FAPDF).

REFERENCES

1. Pasteur L, Joubert J. Charbon et septicemie. *C R Acad Sci* 1877;**85**:101–15.
2. Sams ER, Whiteley M, Turner KH. 'The battle for life': Pasteur, anthrax, and the first probiotics. *J Med Microbiol* 2014;**63**(Pt 11):1573–4.
3. Foster JW, Woodruff HB. Antibiotic substances produced by bacteria. *Ann NY Acad Sci* 1946;**48**(2): 87–98.
4. Bush K. The coming of age of antibiotics: discovery and therapeutic value. *Ann NY Acad Sci* 2010;**1213**: 1–4.
5. Ehrlich P, Hata S. *Die experimentelle chemotherapie der spirilosen*. Berlin: Julius Springer; 1910.
6. Mahoney JF, Arnold RC, Harris A. Penicillin treatment of early syphilis-a preliminary report. *Am J Public Health Nations Health* 1943;**33**(12):1387–91.
7. Bendiner E. Alexander Fleming: player with microbes. *Hosp Pract* 1989;**24**(2):283–5.
8. Ligon BL. Penicillin: its discovery and early development. *Semin Pediatr Infect Dis* 2004;**15**(1):52–7.
9. Bennett JW, Chung KT. Alexander Fleming and the discovery of penicillin. *Adv Appl Microbiol* 2001;**49**:163–84.
10. Alharbi SA, Wainwright M, Alahmadi TA, Salleeh HB, Faden AA, Chinnathambi A. What if Fleming had not discovered penicillin? *Saudi J Biol Sci* 2014;**21**(4):289–93.
11. Domagk G. Ein Beitrag zur Chemotherapie der bakteriellen Infektionen. *Dtsch Med Wochenschr* 1935;**61** (7):250–3.
12. Aminov RI. A brief history of the antibiotic era: lessons learned and challenges for the future. *Front Microbiol* 2010;**1**:134.
13. Miert ASJPAMV. The sulfonamide-diaminopyrimidine story. *J Vet Pharmacol Ther* 1994;**17**(4):309–16.
14. Woods DD. The relation of *p*-aminobenzoic acid to the mechanism of action of sulphanilamide. *Br J Exp Pathol* 1940;**21**(2):74–90.
15. Waksman SA, Woodruff HB. The soil as a source of microorganisms antagonistic to disease-producing bacteria. *J Bacteriol* 1940;**40**(4):581–600.

16. Waksman SA, Woodruff HB. Streptothricin, a new selective bacteriostatic and bactericidal agent, particularly active against gram-negative bacteria. *Exp Biol Med* 1942;**49**(2):207–10.
17. Bradley JS. Which antibiotic for resistant Gram-positives, and why? *J Infect* 2014;**68**(Suppl. 1):S63–75.
18. Jindal AK, Pandya K, Khan ID. Antimicrobial resistance: a public health challenge. *Med J Armed Forces India* 2015;**71**(2):178–81.
19. Martinez JL, Baquero F. Mutation frequencies and antibiotic resistance. *Antimicrob Agents Chemother* 2000;**44**(7):1771–7.
20. Cox G, Wright GD. Intrinsic antibiotic resistance: mechanisms, origins, challenges and solutions. *Int J Med Microbiol* 2013;**303**(6–7):287–92.
21. Davies J, Davies D. Origins and evolution of antibiotic resistance. *Microbiol Mol Biol Rev* 2010;**74**(3): 417–33.
22. Fair RJ, Tor Y. Antibiotics and bacterial resistance in the 21st century. *Perspect Med Chem* 2014;**6**: 25–64.
23. Goossens H, Ferech M, Vander Stichele R, Elseviers M, Group EP. Outpatient antibiotic use in Europe and association with resistance: a cross-national database study. *Lancet* 2005;**365**(9459):579–87.
24. Smith DW. Decreased antimicrobial resistance after changes in antibiotic use. *Pharmacotherapy* 1999;**19** (8 Pt 2):129S–32S discussion 33S–37S.
25. Hopwood DA. How do antibiotic-producing bacteria ensure their self-resistance before antibiotic biosynthesis incapacitates them? *Mol Microbiol* 2007;**63**(4):937–40.
26. Allen HK, Donato J, Wang HH, Cloud-Hansen KA, Davies J, Handelsman J. Call of the wild: antibiotic resistance genes in natural environments. *Nat Rev Microbiol* 2010;**8**(4):251–9.
27. Lyon BR, Skurray R. Antimicrobial resistance of *Staphylococcus aureus*: genetic basis. *Microbiol Rev* 1987;**51**(1):88–134.
28. Bengtsson B, Wierup M. Antimicrobial resistance in Scandinavia after ban of antimicrobial growth promoters. *Anim Biotechnol* 2006;**17**(2):147–56.
29. Cunha BA. Effective antibiotic-resistance control strategies. *Lancet* 2001;**357**(9265):1307–8.
30. Millar MR, Keyworth N, Lincoln C, King B, Congdon P. 'Methicillin-resistant' *Staphylococcus aureus* in a regional neonatology unit. *J Hosp Infect* 1987;**10**(2):187–97.
31. Nicolle LE. Infection control programmes to contain antimicrobial resistance. In: Organization WH, editor. United States; 2001.
32. European Centre for Disease Prevention and Control E. *Antimicrobial resistance surveillance in Europe*. Stockholm: European Centre for Disease Prevention and Control; 2013.
33. Bhullar K, Waglechner N, Pawlowski A, Koteva K, Banks ED, Johnston MD, et al. Antibiotic resistance is prevalent in an isolated cave microbiome. *PLoS One* 2012;**7**(4):e34953.
34. Sun S, Selmer M, Andersson DI. Resistance to beta-lactam antibiotics conferred by point mutations in penicillin-binding proteins PBP3, PBP4 and PBP6 in *Salmonella enterica*. *PloS One* 2014;**9**(5):e97202.
35. Vila J, Marti S, Sanchez-Cespedes J. Porins, efflux pumps and multidrug resistance in *Acinetobacter baumannii*. *J Antimicrob Chemother* 2007;**59**(6):1210–15.
36. Wright GD. Molecular mechanisms of antibiotic resistance. *Chem Commun* 2011;**47**(14):4055–61.
37. Sohmen D, Harms JM, Schlunzen F, Wilson DN. Enhanced SnapShot: antibiotic inhibition of protein synthesis II. *Cell* 2009;**139**(1). 212-e1.
38. Diaz L, Tran TT, Munita JM, Miller WR, Rincon S, Carvajal LP, et al. Whole-genome analyses of *Enterococcus faecium* isolates with diverse daptomycin MICs. *Antimicrob Agents Chemother* 2014;**58** (8):4527–34.
39. Wright GD. Bacterial resistance to antibiotics: enzymatic degradation and modification. *Adv Drug Deliv Rev* 2005;**57**(10):1451–70.

40. Walsh TR, Weeks J, Livermore DM, Toleman MA. Dissemination of NDM-1 positive bacteria in the New Delhi environment and its implications for human health: an environmental point prevalence study. *Lancet Infect Dis* 2011;**11**(5):355–62.
41. Paterson DL. Resistance in Gram-negative bacteria: Enterobacteriaceae. *Am J Infect Control* 2006;**34**(5 Suppl. 1):S20–8 discussion S64–73.
42. Gootz TD. The global problem of antibiotic resistance. *Crit Rev Immunol* 2010;**30**(1):79–93.
43. Canton R, Coque TM. The CTX-M beta-lactamase pandemic. *Curr Opin Microbiol* 2006;**9**(5):466–75.
44. Cornaglia G, Akova M, Amicosante G, Canton R, Cauda R, Docquier JD, et al. Metallo-beta-lactamases as emerging resistance determinants in Gram-negative pathogens: open issues. *Int J Antimicrob Agents* 2007;**29**(4):380–8.
45. Rossolini GM, Mantengoli E, Montagnani F, Pollini S. Epidemiology and clinical relevance of microbial resistance determinants versus anti-Gram-positive agents. *Curr Opin Microbiol* 2010;**13**(5):582–8.
46. Cornaglia G. Fighting infections due to multidrug-resistant Gram-positive pathogens. *Clin Microbiol Infect* 2009;**15**(3):209–11.
47. Sauvage E, Kerff F, Terrak M, Ayala JA, Charlier P. The penicillin-binding proteins: structure and role in peptidoglycan biosynthesis. *FEMS Microbiol Rev* 2008;**32**(2):234–58.
48. Pinho MG, Kjos M, Veening JW. How to get (a)round: mechanisms controlling growth and division of coccoid bacteria. *Nat Rev Microbiol* 2013;**11**(9):601–14.
49. Yoneyama H, Katsumata R. Antibiotic resistance in bacteria and its future for novel antibiotic development. *Biosci Biotechnol Biochem* 2006;**70**(5):1060–75.
50. Pimenta AC, Fernandes R, Moreira IS. Evolution of drug resistance: insight on TEM beta-lactamases structure and activity and beta-lactam antibiotics. *Mini Rev Med Chem* 2014;**14**(2):111–22.
51. Becker K, Ballhausen B, Kock R, Kriegeskorte A. Methicillin resistance in *Staphylococcus* isolates: the "mec alphabet" with specific consideration of mecC, a mec homolog associated with zoonotic *S. aureus* lineages. *Int J Med Microbiol* 2014;**304**(7):794–804.
52. Pantosti A, Sanchini A, Monaco M. Mechanisms of antibiotic resistance in *Staphylococcus aureus*. *Future Microbiol* 2007;**2**(3):323–34.
53. Livermore DM, Woodford N. The beta-lactamase threat in Enterobacteriaceae, *Pseudomonas* and *Acinetobacter*. *Trends Microbiol* 2006;**14**(9):413–20.
54. Kaczmarek FM, Dib-Hajj F, Shang W, Gootz TD. High-level carbapenem resistance in a *Klebsiella pneumoniae* clinical isolate is due to the combination of bla(ACT-1) beta-lactamase production, porin OmpK35/36 insertional inactivation, and down-regulation of the phosphate transport porin phoe. *Antimicrob Agents Chemother* 2006;**50**(10):3396–406.
55. Chopra I. The increasing use of silver-based products as antimicrobial agents: a useful development or a cause for concern? *J Antimicrob Chemother* 2007;**59**(4):587–90.
56. Pages JM, James CE, Winterhalter M. The porin and the permeating antibiotic: a selective diffusion barrier in Gram-negative bacteria. *Nat Rev Microbiol* 2008;**6**(12):893–903.
57. Nishino K, Yamaguchi A. Analysis of a complete library of putative drug transporter genes in *Escherichia coli*. *J Bacteriol* 2001;**183**(20):5803–12.
58. Stavri M, Piddock LJ, Gibbons S. Bacterial efflux pump inhibitors from natural sources. *J Antimicrob Chemother* 2007;**59**(6):1247–60.
59. Morita Y, Kodama K, Shiota S, Mine T, Kataoka A, Mizushima T, et al. NorM, a putative multidrug efflux protein, of *Vibrio parahaemolyticus* and its homolog in *Escherichia coli*. *Antimicrob Agents Chemother* 1998;**42**(7):1778–82.
60. Paulsen IT, Brown MH, Skurray RA. Proton-dependent multidrug efflux systems. *Microbiol Rev* 1996;**60**(4):575–608.

61. Banigan JR, Gayen A, Cho MK, Traaseth NJ. A structured loop modulates coupling between the substrate-binding and dimerization domains in the multidrug resistance transporter EmrE. *J Biol Chem* 2015;**290**(2):805–14.
62. Grinius L, Dreguniene G, Goldberg EB, Liao CH, Projan SJ. A staphylococcal multidrug resistance gene product is a member of a new protein family. *Plasmid* 1992;**27**(2):119–29.
63. Bay DC, Rommens KL, Turner RJ. Small multidrug resistance proteins: a multidrug transporter family that continues to grow. *Biochim Biophys Acta* 2008;**1778**(9):1814–38.
64. Winstone TL, Jidenko M, le Maire M, Ebel C, Duncalf KA, Turner RJ. Organic solvent extracted EmrE solubilized in dodecyl maltoside is monomeric and binds drug ligand. *Biochem Biophys Res Commun* 2005;**327**(2):437–45.
65. Yerushalmi H, Schuldiner S. A model for coupling of H(+) and substrate fluxes based on "time-sharing" of a common binding site. *Biochemistry* 2000;**39**(48):14711–19.
66. Murakami S, Nakashima R, Yamashita E, Yamaguchi A. Crystal structure of bacterial multidrug efflux transporter AcrB. *Nature* 2002;**419**(6907):587–93.
67. Takatsuka Y, Chen C, Nikaido H. Mechanism of recognition of compounds of diverse structures by the multidrug efflux pump AcrB of *Escherichia coli. Proc Natl Acad Sci USA* 2010;**107**(15):6559–65.
68. Takatsuka Y, Nikaido H. Covalently linked trimer of the AcrB multidrug efflux pump provides support for the functional rotating mechanism. *J Bacteriol* 2009;**191**(6):1729–37.
69. Murakami S, Nakashima R, Yamashita E, Matsumoto T, Yamaguchi A. Crystal structures of a multidrug transporter reveal a functionally rotating mechanism. *Nature* 2006;**443**(7108):173–9.
70. Davidson AL, Dassa E, Orelle C, Chen J. Structure, function, and evolution of bacterial ATP-binding cassette systems. *Microbiol Mol Biol Rev* 2008;**72**(2):317–64 table of contents.
71. Schmitt L, Tampe R. Structure and mechanism of ABC transporters. *Curr Opin Struct Biol* 2002;**12**(6): 754–60.
72. Gupta RP, Kueppers P, Schmitt L, Ernst R. The multidrug transporter Pdr5: a molecular diode? *Biol Chem* 2011;**392**(1–2):53–60.
73. Mehla J, Ernst R, Moore R, Wakschlag A, Marquis MK, Ambudkar SV, et al. Evidence for a molecular diode-based mechanism in a multispecific ATP-binding cassette (ABC) exporter: SER-1368 as a gatekeeping residue in the yeast multidrug transporter Pdr5. *J Biol Chem* 2014;**289**(38):26597–606.
74. Putman M, van Veen HW, Konings WN. Molecular properties of bacterial multidrug transporters. *Microbiol Mol Biol Rev* 2000;**64**(4):672–93.
75. He X, Szewczyk P, Karyakin A, Evin M, Hong WX, Zhang Q, et al. Structure of a cation-bound multidrug and toxic compound extrusion transporter. *Nature* 2010;**467**(7318):991–4.
76. Su CC, Long F, McDermott G, Shafer WM, Yu EW. Crystallization and preliminary X-ray diffraction analysis of the multidrug efflux transporter NorM from *Neisseria gonorrhoeae. Acta Crystallogr Sect F Struct Biol Cryst Commun* 2008;**64**(Pt 4):289–92.
77. Kuroda T, Tsuchiya T. Multidrug efflux transporters in the MATE family. *Biochim Biophys Acta* 2009;**1794**(5):763–8.
78. Dunkle JA, Xiong L, Mankin AS, Cate JH. Structures of the *Escherichia coli* ribosome with antibiotics bound near the peptidyl transferase center explain spectra of drug action. *Proc Natl Acad Sci USA* 2010;**107**(40):17152–7.
79. Bulkley D, Innis CA, Blaha G, Steitz TA. Revisiting the structures of several antibiotics bound to the bacterial ribosome. *Proc Natl Acad Sci USA* 2010;**107**(40):17158–63.
80. Weisblum B. Erythromycin resistance by ribosome modification. *Antimicrob Agents Chemother* 1995;**39**(3):577–85.
81. Helser TL, Davies JE, Dahlberg JE. Change in methylation of 16S ribosomal RNA associated with mutation to kasugamycin resistance in *Escherichia coli. Nat New Biol* 1971;**233**(35):12–14.

82. Lai CJ, Dahlberg JE, Weisblum B. Structure of an inducibly methylatable nucleotide sequence in 23S ribosomal ribonucleic acid from erythromycin-resistant *Staphylococcus aureus*. *Biochemistry* 1973;**12**(3):457–60.
83. Pogliano J, Pogliano N, Silverman JA. Daptomycin-mediated reorganization of membrane architecture causes mislocalization of essential cell division proteins. *J Bacteriol* 2012;**194**(17):4494–504.
84. Munita JM, Tran TT, Diaz L, Panesso D, Reyes J, Murray BE, et al. A liaF codon deletion abolishes daptomycin bactericidal activity against vancomycin-resistant *Enterococcus faecalis*. *Antimicrob Agents Chemother* 2013;**57**(6):2831–3.
85. Tran TT, Panesso D, Mishra NN, Mileykovskaya E, Guan Z, Munita JM, et al. Daptomycin-resistant *Enterococcus faecalis* diverts the antibiotic molecule from the division septum and remodels cell membrane phospholipids. *MBio* 2013;**4**:4.
86. Arias CA, Panesso D, McGrath DM, Qin X, Mojica MF, Miller C, et al. Genetic basis for in vivo daptomycin resistance in enterococci. *N Engl J Med* 2011;**365**(10):892–900.

CHAPTER

3

SIGNALING PATHWAYS SUSTAINING ANTIBIOTIC RESISTANCE: LESSONS FROM METHICILLIN-RESISTANT *STAPHYLOCOCCUS AUREUS*

P.F. Seke Etet[1], A.H. Nwabo Kamdje[2], L. Vecchio[1], M. Farahna[1] and M.F. Mahomoodally[3]

[1]Department of Basic Health Sciences, College of Applied Medical Sciences, Qassim University, Buraydah, Saudi Arabia [2]Department of Biomedical Sciences, Faculty of Sciences, University of Ngaoundéré, Ngaoundéré, Cameroon [3]Department of Health Sciences, Faculty of Science, University of Mauritius, Reduit, Mauritius

INTRODUCTION

The incidence of multidrug-resistant bacteria or superbugs has been increasing worldwide. Multidrug-resistant bacteria constitute a major threat to public health as these lethal pathogens of fish, birds, and mammals cause difficult-to-treat health care–associated or community-acquired infections in humans.[1,2] Common clinically relevant multidrug-resistant bacteria include multidrug-resistant (MDR) *Mycobacterium tuberculosis*, extended spectrum β-lactamase-producing bacteria (ESBL), vancomycin-resistant enterococci (VRE), multidrug-resistant *Acinetobacter baumannii* (MRAB), vancomycin-resistant *Staphylococcus aureus* (VRSA), and methicillin-resistant *S. aureus* (MRSA).[3–8]

S. aureus is a leading cause of deadly bacterial infections worldwide, with a death rate of about 20% of cases in industrialized countries,[9,10] and even more in developing countries.[11–14] A recent systematic review addressing the economic burden of MRSA revealed that MRSA infection management carries important direct and indirect costs for hospitals, such as longer lengths of stay, with wide ranges in increasing costs emerging from patients' variability, health-care systems, and the designs of the available studies.[15] Alarmingly, the study also revealed that most authoritative reports for health management and clinical decisions did not consider important variables such as infection type, antimicrobial susceptibility, and resistance emergence. Moreover, the economic implications of growing MRSA incidence in community patients and farm animal lineages[16,17] suggest that the burden of MRSA infections could even be more considerable than reported, particularly in developing countries.[13,14,18] The emergence of MRSA infection in dairy animals is also of great concern for livestock.[19,20]

S. aureus has been constituting a serious threat to public health since the 1960s, when the first MRSA strains emerged.[21] MRSA are also resistant to a number of other antibiotic classes and

Antibiotic Resistance. DOI: http://dx.doi.org/10.1016/B978-0-12-803642-6.00003-4

antibacterial agents, including antiseptics.[22–24] Notably, the β-lactam antibiotic class, which includes methicillin, constitutes the broadest and probably most widely used class of antibiotics.[25,26] Methicillin-susceptible *S. aureus* (MSSA) and MRSA strains cause both skin and systemic infections with severe complications such as bacteremia, endocarditis, osteomyelitis, and pneumonia.[10,27]

A number of frontline antibiotics no longer inhibit some strains of bacteria due to extreme resistance. Such bacteria are termed as multidrug-resistant (MDR) bacteria (or superbugs), and a significant fraction of these microorganisms are pathogenic.[7,28,29] In particular, MRSA, a major notorious pathogenic causing widespread hospital- and community-associated infections around the world, has been acquiring high resistance to a number of antibiotics. Consequently, MRSA has been raising serious concerns in the medical and scientific communities about the risk of an epidemic outbreak. Recent studies attempting to understand the biology of the pathogen and uncover new pharmacological targets in order to develop better therapeutics have addressed specific aspects of biochemical changes in cell wall processes and molecules,[25,26,30,31] metabolic pathways,[32–34] and oxidative stress response.[35–38]

The current understanding of biological processes and signaling molecules accounting for the development of resistance in bacteria are herein summarized and discussed, with a particular emphasis on evidence from resistant and persistent variants of *S. aureus*.

EMERGENCE OF ANTIBIOTIC-RESISTANT BACTERIA

RESISTANT BACTERIA ARE COMMON IN URBAN ENVIRONMENTS

S. aureus has asymptomatically colonized about 30% of healthy individuals (body surface, oropharynx, and gastrointestinal tract[39,40]) in many industrialized countries in its commensal forms, which have the potential to start infections.[19,41,42] A huge body of environmental studies, translational research reports, and community-based clinical data warrants against community-acquired MRSA nasal colonization. *S. aureus* strains resistant to many classes of antibiotics, which are known opportunistic pathogens causing life-threatening diseases in humans,[43,44] are commonly reported high rates of nasal colonization in healthy pediatric populations worldwide.[6–8] Clinically relevant MRSA strains can be acquired from many sources, rising concern and warranting surveillance. For instance, a phylogenetic analysis assessing antibiotic-resistant bacteria in airborne bacterial communities from hog farm and spray field sites revealed the presence of *Staphylococcus* spp. in both types of bioaerosol community,[45] evidencing potential health risks to farmers and their relatives exposed to swine bioaerosol. In addition, a number of recent reports provided strong evidence that dogs and probably other companion animals can be MRSA carriers,[46,47] pointing out the need for continued surveillance of these pathogens in companion animals to prevent increased incidence in humans (Fig. 3.1).

S. aureus commensal forms that asymptomatically colonize body surfaces have an heterogeneous expression of resistance, with a fraction of naturally resistant phenotypes termed as *persisters*, as well as a fraction termed as *small colony variants (SCVs)*, which are cells with a high ability to mutate into resistant forms in the presence of host antibacterial strategies and various antibiotics.[48–50] Even more alarming, studies of staphylococcal and nonstaphylococcal bacterial isolates revealed that bacteria-harboring genes associated with antibiotic resistance (eg, *mecA* and *mecC*)

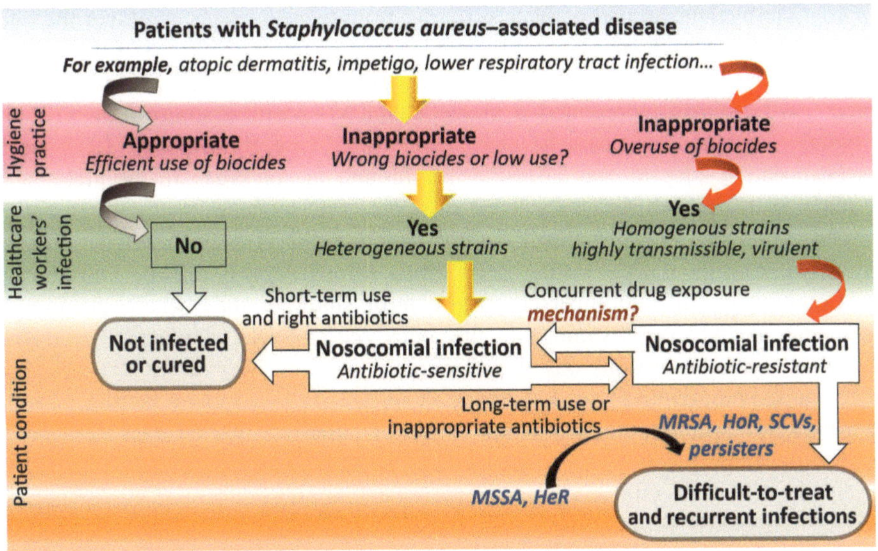

FIGURE 3.1

Hygiene and risk of *Staphylococcus aureus* nosocomial infection.

are common in urban environments, including public buses,[42] households and schools,[51,52] and hospitals.[53,54] The brown-banded cockroach *Supella longipalpa*, one of the numerous insects known to carry pathogenic bacteria in urban environments, was reported as a reservoir of drug-resistant bacteria.[55] Bacteria carrying resistance genes are common in surface waters[56,57] and dairy farms.[19,58] In addition, the techniques currently used to remove antibiotic-resistant bacteria in advanced wastewater treatments are not very effective,[59,60] increasing the risk of infection for the community. All together, these data show that strains of antibiotic-resistant bacteria, particularly MRSA, are endemic in communities worldwide.

HIGH MRSA PREVALENCE REPORTED IN HOSPITALS OF DEVELOPING COUNTRIES: IS LOW HYGIENE THE MAJOR CULPRIT?

Lower prevalence of MRSA was reported in hospitals of industrialized countries like Canada[61] or the Netherlands[62] than in developing countries lacking adequate surveillance systems such as Rwanda,[11] Gabon[12,13], Nigeria,[14] Algeria, Egypt, Morocco, Senegal, Tunisia,[18] Nepal,[63] Iran,[7,64] and China.[65] It was proposed that the importance of appropriate staff knowledge of hygiene practices may explain lower MRSA prevalence in the former countries, whereas the lack of such knowledge could explain the relatively higher prevalence of antimicrobial resistance among common bacterial isolates in health-care facilities of the latter countries.[14,61,62,64] Many diseases are commonly associated with *S. aureus* infection. For instance, staphylococci are commonly found in the sputum of 1- to 3-month-old infants with lower respiratory tract infection,[66] in the bloodstream of pediatric hematology/oncology patients,[67] and in atopic dermatitis and impetigo patients.[68]

Thus, inappropriate hygiene around patients suffering from these diseases can favor the infection of health-care workers. An outbreak of *cfr*-mediated linezolid-resistant *Staphylococcus epidermidis* was recently reported in a tertiary referral center in the Republic of Ireland, for the first time in that country.[69] All isolates represented a single clonal *cfr*-positive strain with ST2 multilocus sequence typing, staphylococcal cassette chromosome *mec* type III (*SCCmec* III). The infection prevention intervention adopted was effective and limited the outbreak to the affected intensive care unit (ICU), partly due to staff knowledge of appropriate hygiene practices and antimicrobial susceptibilities at the time of the outbreak.[69] This underlies the pivotal importance of educational intervention as a strategy for preventing nosocomial infection with multiple antibiotic-resistant pathogens. Comparably, in a multicenter case–control study in Italy where the prevalence of linezolid-resistant staphylococcal strains was the focus, the key role of infection control measures in the prevention of these infections was also reported.[70]

However, the hypothesis that poor hygiene in hospitals of developing countries is the cause of higher frequency of resistant antibiotics is undermined by the fact that the real incidence of resistant bacteria in developing countries is controversial. Notably, a recent cross-sectional study using an automated identification system in Kenya revealed a low prevalence of MRSA, suggesting that the subjectivity associated with the interpretation of manual phenotypic tests commonly used in most African countries for the identification of staphylococci may result in misidentification of coagulase negative staphylococci as *S. aureus*.[71] This misidentification may account for the particularly high incidence of potentially resistant bacteria reported in the hospitals of these countries, rather than inappropriate staff knowledge and practice of hygiene. Also sustaining the hypothesis that, contrary to numerous available reports, resistant bacteria incidence is actually lower in developing countries, community-acquired bacteremia rates are low in a number of these countries.[18,72,73] If confirm, this would indicate that the allegedly high prevalence of *S. aureus* reported in a number of studies in hospitals of developing countries are probably due to contamination in the health-care facilities. In addition, the expansion of the existing testing for MRSA to screening of incoming patients from endemic countries with less hygienic conditions and an overuse of antibiotics revealed that these patients faced a high risk of being contaminated with antibiotic-resistant bacteria when transferred to hospitals of industrialized countries.[74] Returning home, such infected patients may transmit MRSA to other patients and health-care workers. Further studies in health-care facilities of both industrialized and developing countries are necessary to understand the impact of differences in hygiene practices on the emergence of resistant bacteria.

ORIGIN OF RESISTANCE: LONG-TERM USE OF ANTIBIOTICS

Although nonantibiotic biocides may represent promising alternatives to antibiotics in monotherapy[75,76] or combination therapy,[77] pharmacological and microbiological data are insufficient and large-scale clinical trials are lacking. Consequently, antibiotics still represent the most common and best treatment option for most bacterial infections. Conditions and diseases necessitating long-term prophylactic use of antibiotics are numerous. For instance, in cystic fibrosis, a common life-shortening hereditary disease among Caucasian-descent young children,[78] prophylactic antibiotics are prescribed in an attempt to prevent chronic *S. aureus* infection and subsequent severe pulmonary damage.[79]

Unfortunately, the rise of antibiotic resistance has been posing a serious challenge for therapy in many bacterial infections, as observed in nosocomial urinary tract infections, life-threatening infections accounting for up to 40% of all health care–associated infections,[80] where the choice of antibiotics has been limited by the rising rate of antibiotic resistance.[81] *S. aureus* resistance correlating with long-term use of antibiotics was also reported in cystic fibrosis, where among other serious adverse effects, it also is associated with infection with *Pseudomonas aeruginosa*, another pathogenic bacterium causing lung damage.[79] Furthermore, antibiotic usage during the last 3 months was reported as a risk factor for nasal carriage of multidrug-resistant *S. aureus*, a substantial source of human infection, among healthy children in Iran.[7] Due to these observations and many similar ones, it is now widely accepted that bacterial resistance to antibiotics may result from an increase in the selective pressure on the population of bacteria targeted, which emerges from antibiotic long-term use and in turn favors the thriving of resistant bacteria.[28,29]

It was proposed that "for each antimicrobial–pathogen combination, an antimicrobial concentration range exists in which selective amplification of single-step, drug-resistant mutants occurs."[82] This hypothesis is known as the *mutant selection window hypothesis*. It implies that exposure of bacteria to upper-boundary concentrations of antibiotics (so-called mutant prevention concentrations) may prevent resistance-favoring mutations. A number of reports sustaining this hypothesis has emerged in the last decade.[33,83–85] However, the mutant selection window hypothesis is still controversial. For instance, using in vitro and in vivo techniques, Mei et al.[86] recently tested the mutant selection window hypothesis by exposing *S. aureus* to fosfomycin, a broad-spectrum antibiotic frequently associated with bacterial resistance.

Surprisingly, regardless of the fosfomycin dosage, the selection of resistant mutants was not observed either during antibacterial treatment and 48 h after fosfomycin treatment termination. In addition, no difference was found between in vitro isolated mutant and sensitive parental strain, indicating the absence of fitness cost of fosfomycin resistance in *S. aureus* (ATCC 29213). Moreover, agar plate determinations did not fit the mutant selection window for fosfomycin treatment of rabbits infected with *S. aureus* ATCC 29213.[86] These observations indicate that mechanisms triggering and driving *S. aureus* mutant selection are complex and may depend on the antibiotic and on the strain considered, among other possible factors. Further research is needed to unravel the exact mechanism of mutant selection, considering the implications for bacterial infection treatment. On the same hand, a recent report revealed reversible antibiotic tolerance induction in *S. aureus* following concurrent drug exposure.[26] More specifically, in this study, MRSA behaved like vancomycin-intermediate *S. aureus* (VISA) upon exposure to the polymyxin antibiotic colistin, which also acts on bacterial membranes and cell walls as do methicillin and vancomycin, but with a different mechanism of action (displacement of bacterial counter ions in the lipopolysaccharide).[87,88] It can be concluded from this interesting finding that it may be possible to resensitize resistant strains by exposure to bactericidal agents acting on the same target as the drug for which resistance was developed, but in another fashion. The mechanisms of reversible antibiotic tolerance induction by concurrent drug exposure should be further investigated, considering the implications for bacterial-resistant infection therapy.

CLINICAL IMPORTANCE OF METHICILLIN AND VANCOMYCIN RESISTANCE

Renamed *meticillin* in 2005 to match the International Pharmacopeia Guidelines, the β-lactam antibiotic family member methicillin is a penicillin-class, narrow-spectrum antibiotic historically used to treat *S. aureus* and other Gram-positive bacteria. Like most β-lactam antibiotics, methicillin mediates its bactericidal effects by targeting penicillin-binding proteins (PBPs) of bacterial cell walls like peptidoglycans, which are pivotal for bacterial cell wall synthesis.[30,31] Its wide use led to the discovery (and eventually the development[25,26]) of resistant strains, reported for the first time in 1961.[21]

S. aureus methicillin resistance would mainly emerge from the expression of the resistance gene *mecA*, which encodes PBP2a, a PBP that is resistant to β-lactam action (low affinity), possibly acquired via horizontal gene transfer, as suggested in early studies[89–92] and elegantly demonstrated in a recent study where horizontal infection control strategies decreased nosocomial MRSA infection and eliminated bacteremia in a surgical ICU without the need of active surveillance.[93] In addition to the presence of *mecA* gene, mutations in genes encoding chromosomal factors known as *fem* factors (eg, *femA*, *femB*, *femX*) are necessary for the expression of methicillin resistance in MRSA.[94–96] The clinical importance of methicillin resistance emerge from the fact that infections with methicillin-resistant bacteria are difficult to treat, as these pathogens are also resistant to a number of non-β-lactam antibiotics and various antibacterial agents,[22,23] including oxacillin, the penicillinase-resistant β-lactam that replaced methicillin in clinical use.[97–99] As discussed in "MRSA Metabolism Changes and Antibiotic Resistance" section, findings from studies addressing the metabolism of MRSA, VRSA, and other resistant bacteria have provided new insights in the mechanisms that account for the resistance of bacterial cells and the maintenance of bacterial colonies in the presence of antibiotics.

Resistance to the glycopeptide antibiotic vancomycin is also of the utmost clinical importance, as this drug was the last resort to treat resistance to other antibiotics until the last two decades, when irrefutable proof of resistance to vancomycin was provided.[100,101] Vancomycin resistance was shown in enterococci referred to as VRE, and in *S. aureus* strains termed as VISA or VRSA.[3–6] Vancomycin acts by inhibiting the second stage of bacterial cell wall synthesis.[102]

Understanding the mechanisms of resistance to such a last-resort antibiotic developed by bacteria may allow for anticipating the resistance strategies to future antibiotic drugs. Recently, a study assessing the mechanism of action of oritavancin, a semisynthetic antibiotic derivative of the glycopeptide antibiotic chloroeremomycin, underlined a mechanism of resistance to vancomycin in resistant staphylococci and enterococci: the loss of hydrogen bonds crucial for vancomycin binding in the cell wall, resulting in a decrease in binding affinity of the antibiotic and related agents.[103] Similar resistance to other naturally occurring antibiotics acting on the *S. aureus* cell wall was reported. For instance, resistance to the lipopeptide antibiotic lysostaphin is due to modifications in biosynthetic pathways for the cell wall assembly, reducing the ability of this bacteriolytic enzyme to cleave the pentaglycine cross-bridges of peptidoglycan.[104] Moreover, resistance to the lipopeptide daptomycin (DAP), an antibiotic whose usage increased drastically (as vancomycin) due to satisfactory bactericidal activity in MRSA strains,[105,106] constitute a notorious example. DAP penetrates and accumulates in the bacterial cell membrane, redirecting the localization of cell division and cell wall synthesis proteins and leading to marked cell wall defects that cause cell death.[105] DAP resistance results from adaptive changes in metabolic pathways induced in the resistant *S. aureus* cell membrane by the presence of the antibiotic.[106]

The mechanisms of resistance of vancomycin, methicillin, DAP, and lysostaphin show that *S. aureus* can modify biosynthesis mechanisms, as well as molecular interactions and structure in its cell wall, to resist the action of antibiotics. Aberrant changes were reported in numerous molecules involved in cell wall processes, including, for instance, *femA* mutations favoring *mecA*-promoted methicillin resistance that is associated with the presence of aggregated triglycyl bridge peptidoglycan units in the cell wall.[107] Surprisingly, MRSA can also use cell wall thickening as an adaptive resistance mechanism to prevent protein synthesis inhibitors, such as aminoglycoside antibiotic amikacin, from entering the cell.[108] All together, these findings show that membrane and cell wall are pivotal players in the bacterial strategy to overcome environmental pressure.

Notably, although vancomycin-resistant strains are often methicillin-resistant,[8,109] only a small fraction of clinically relevant vancomycin-resistant strains (namely, methicillin-resistant strains) also acquire a high level of methicillin resistance (namely, vancomycin resistance),[110] due to the antagonistic effects of *vanA* and *mecA* resistance determinants in *S. aureus*.[103]

MRSA METABOLISM CHANGES AND ANTIBIOTIC RESISTANCE

METABOLIC PATHWAY CHANGES FAVOR *S. AUREUS* RESISTANCE

Despite the development of rapid antimicrobial susceptibility tests[111,112] and promising reports of new antibiotics killing methicillin-resistant pathogens without detectable resistance,[113,114] understanding the biological bases of bacterial resistance to antibiotics still presents greater potential for preventing this phenomenon. Numerous reports of metabolic changes associating with MRSA resistance are available, including a recent in silico study where the metabolic pathways of MRSA strain 252 were compared with human cells, together with a subtractive genomics analysis for the retrieval of proteins only from unique metabolic pathways.[115] This approach allowed to shortlist a few unique essential proteins that can be targeted to eliminate MRSA from the numerous metabolic pathways unique to MRSA identified. Among these molecules, 12 essential enzymes with no corresponding deposition in the Protein Data Bank (ie, with crystal structures still to be solved) were discovered. Then, out of those 12 enzymes, the authors specifically concentrated on glutamate synthase, an enzyme pivotal in bacterial glutamate, alanine, aspartate, and nitrogen metabolism,[116–118] and generated a 3D structure by homology modeling validated by multiple analysis tools.[115] As shown by previous studies assessing the effect of changes in glutamate synthesis, it can be deduced that changes in glutamate synthase play a major role in bacterial resistance. Notably, since glutamate is a major osmoprotector in *S. aureus* and various bacteria,[119,120] increases in its synthesis resulting from increases in cellular glutamate synthase levels (caused by glutamate synthase gene overexpression caused by activating mutations) may contribute to processes allowing bacteria to become resistant to antibiotics targeting metabolic processes. The major role that glutamate plays in bacterial wall synthesis[121,122] also suggests that increased synthesis of glutamate may increase bacterial resistance. Notably, the inhibition of amidation of glutamate residues of the peptidoglycan of *S. aureus*, via inhibition of peptidoglycan amidation and cell wall biosynthesis pivotal enzymes MurT and GatD, resulted in decreases in growth rate and β-lactam resistance, with an increased sensitivity to lysozyme that caused peptidoglycan degradation.[123]

In addition, among other metabolic genes, mutations in the initiation-condensing enzyme (FabH) and elongation-condensing enzyme (FabF/B), two essential components of the fatty acid biosynthetic pathway, can make *S. aureus* resistant to multiple natural products acting by inhibiting this enzyme.[32,33] A study of 20 independently isolated platensimycin-, platencin-, and thiolactomycin-resistant clones revealed no mutation in these genes.[34] However, missense mutations in the *fabH* of resistant strains were observed—notably, altered amino acids clustered in and around the FabH active-site tunnel, which resulted in a decrease in enzyme catalytic activity, which in turn sufficed to make *S. aureus* resistant to a number of antibiotics that act by binding to FabF. Thus, antibiotics targeting FabH and FahH/FabF downstream targets may abrogate such resistance.

Moreover, metabolic changes favor the development of resistance and the presence of bacterial persisters and SCVs, which account for the chronicity, redundancy, and antibiotic recalcitrance of *S. aureus* infections.

BACTERIAL PERSISTERS

Challenges associated with the elimination of bacterial persisters

As first described in 1944, thanks to their ability to survive the antibiotic action of penicillin,[124] persisters are major players in persistent chronic and resistant *S. aureus* infections in humans and animals. Early in vitro studies revealed that persisters of *P. aeruginosa*, *Escherichia coli*, and *S. aureus* to β-lactam antibiotics at bactericidal concentrations survive by modifying metabolic processes in their cell wall, allowing them to increase their adhesion to the surface of the culture vessel, and subsequently to multiply despite the presence of β-lactam antibiotics in the medium.[124–126] Experimental evidence strongly suggests that persisters are present in most bacterial species. The existence of persisters was confirmed in clinical studies assessing the effects of various antibiotics (pefloxacin, imipenem, fluoroquinolones, ampicillin, ofloxacin, etc.) in the treatment of nosocomial lung infections caused by multiresistant *Acinetobacter anitratus*, *P. aeruginosa*, *S. aureus*, and numerous Enterobacteriaceae.[127–130] Persister cells are wild-type phenotypic variants with metabolic particularities that allow them to achieve a dormant state associated with multidrug resistance and tolerance without mutations. Basically, resistance mechanisms of these cells would result in preventing antibiotics from hitting their targets (for instance, preventing β-lactam from binding in the cell wall), while on the other hand, tolerance mechanisms of persisters would consist of shutting down the cellular targets whose inhibition is mandatory for antibiotics to mediate their bactericidal effects.[50] In addition, bacterial biofilm produced in response to adverse conditions would limit the access of the immune system components to persisters, allowing them to escape the immune response,[49,131,132] further sustaining the recalcitrance and fueling the relapse of bacterial infections.

Our understanding of the mechanism of persister formation is incomplete due to the redundancy of pathways of persister formation. In addition, these pathways may differ between bacterial species, as indicated by the existence of differences in the metabolic and cellular behavior of persisters from different species in cultures. For instance, there are significantly fewer multidrug-tolerant persisters of cultures of the opportunistic pathogen *P. aeruginosa* than those of *E. coli* and *S. aureus* cultures, and their number increased in response to quorum-sensing-related signaling molecules (phenazine pyocyanin, paraquat, and the acyl-homoserine lactone 3-OC12-HSL), unlike the number of persisters in *E. coli* or *S. aureus* cultures.[48]

Metabolic pathways sustaining persister formation

The increasing number of reports on metabolic pathways sustaining persister formation have provided insights into the biology of these organisms. Isolation of *E. coli* persisters by sorting green fluorescent protein–expressing cells with diminished translation and population revealed a gene expression profile of these cells. The hallmark of this profile was the downregulation of biosynthetic pathways, which was to be expected considering the dormant nature of these cells, and the overexpression of toxin/antitoxin modules that appeared to inhibit essential functions, such as translation, that contribute to persister formation.[49,50] Sustaining this hypothesis, in these studies, ectopic expression of toxins like the global inhibitor of translation RelE, the messenger RNA (mRNA) interferase MazF, and the serine/threonine-protein kinase HipA produced multidrug-tolerant cells. On the other hand, overexpression cloning of genomic libraries and antibiotic tolerance selection demonstrated the existence of persister genes such as *glpD* (encoding aerobic sn-glycerol-3-phosphate dehydrogenase), *plsB* (encoding aerobic glycerol-3-phosphate acyltransferase), and *GlpF* (encoding a glycerol uptake facilitator), among other genes that provide persisters with the ability to undergo metabolic changes allowing persistence.[133–135]

Bacterial toxin/antitoxin systems would be activated in response to the low supply of nutrients in biofilm environments. In response to starvation, Clp or Lon protease activation results in the degradation of antitoxin proteins. Then, liberated toxins (eg, MazF, HipA, and RelE) may induce growth arrest directly or indirectly by inducing mRNA cleavage or via the stringent response-signaling pathway alarmone ppGpp (guanosine pentaphosphate or tetraphosphate)–dependent signal transduction.[134,136–139] Notably, the synthesis of ppGpp via RelA is triggered by the occurrence of uncharged transfer RNA due to amino acid starvation (nutrient limitation).[140]

Nutrient limitation may also induce CspD-mediated inhibition of DNA replication by favoring the synthesis of the second messenger cyclic adenosine monophosphate (cAMP) by adenylate cyclase, which in turn may activate the expression of both *cspD* and *relA* through the cAMP/cAMP receptor protein (Crp) complex.[141] The DksA/ppGpp complex may also modulate RNA polymerase activity, causing metabolic flux alterations resulting in a decreased tricarboxylic acid cycle (TCA cycle) activity (thus, aerobic respiration) and increased persistence.[142] TCA cycle activity modulation is a hallmark of persister metabolism[139]; thus, it represents a possible target for persister eradication. Other metabolic characteristics of persisters include (1) the deprivation of pyruvate and acetyl-CoA from the TCA cycle by activating acetoin and triglyceride synthesis pathways[139,143]; (2) the synthesis of the toxic metabolite methylglyoxal, which activates the methylglyoxal pathway, an offshoot of glycolysis that inhibit bacterial growth[144]; and (3) the upregulation of enzymes counteracting the antibacterial activity of reactive oxygen species (ROS).[145]

Other major mechanisms favoring the formation of persisters include reactive cysteine residue phosphorylation in the global transcriptional regulator A (MgrA), a member of the family of SarA (staphylococcal accessory regulator A)/MarR (multiple antibiotic resistance regulator) virulence proteins, mediated in *S. aureus* by the kinase–phosphatase pair Stk1–Stp1[146]; and the activation of amino acid synthesis and motility pathways.[135] Identifying pivotal persister formation pathways will open new avenues for developing high-potency and high-efficacy antibiofilm drugs, which used in combination with conventional antibiotics will allow effective treatment of chronic and resistant bacterial infections. Strategies for eradicating bacterial persisters based on the exclusive features of their metabolism were proposed, such as the use of aminoglycosides in combination with specific metabolites to treat *E. coli* and *S. aureus* biofilms,[147,148] highlighting the importance of the metabolic environment for effective antibiotic treatment.

SMALL COLONY VARIANTS

Biology of SCVs

Contrary to persisters that escape antibiotics without undergoing genetic changes, SCVs are slow-growing (dormant) mutants that can escape the bactericidal nature of antibiotics thanks to their multidrug tolerance and their ability to hide inside host cells[149] and eventually modulate host defenses.[114,150] A number of key concepts of SCV biology have emerged in the last few decades. Experimental evidence sustains that, despite reduced virulence,[151] SCVs are major players in *S. aureus* pathogenesis. The dormant state of SCVs (metabolic inactivity) allows them to be protected from the antibiotic actions of agents that are only active against growing cells such as oxacillin.[152]

SCVs from natural and experimental animal infections possess similar phenotypic and transcriptomic properties to those of human SCVs, as shown, for instance, in studies of dairy cows with a history of chronic mastitis.[149,153] Like human SCVs, SCVs isolated from these animals formed non-hemolytic and colorless small colonies after 48 h in culture, with typical metabolic hallmarks also shared by human SCVs, such as failure to ferment mannitol, low prevalence of coagulase-positive cells, and ability to revert to the virulent parental phenotype.[153] These observations suggest that dairy animals and pets probably play a role in disseminating SCVs to humans.

A study addressing the role of antibiotics in the generation/selection for persisters using in vitro experiments with *S. aureus* and mathematical models revealed considerable variation in levels of persistence among the antibiotics tested (ie, oxacillin, ciprofloxacin, gentamicin, and vancomycin) according to drug concentrations, and independent of cultures and mixtures of independent cultures.[154] The study also revealed data consistent with the hypothesis that the rate of production of persisters is low and that persisters grow slowly in the presence of antibiotics.[148,155] Key findings on changes associated with the SCV phenotype in clinically relevant *S. aureus* strains include the identification of virulence genes conferring dormancy to these cells and the ability to return to the wild-type phenotype. Such genes include various mutant metabolic genes and dominant regulators of SCVs, such as the sigma B (*SigB*) gene.[156,157] Increased SigB activity associates with the production of virulence-associated surface capsular polysaccharide and decreased antibiotic susceptibility in *S. aureus* strains with intermediate vancomycin susceptibility (VISA).[158,159] Virulence genes also include the MarR family members, including (1) teicoplanin-associated locus regulator (TcaR), which regulates gene expression of proteins on the intercellular adhesion locus involved in staphylococci poly-*N*-acetylglucosamine biosynthesis, and whose deletion or silencing decreases poly-*N*-acetylglucosamine biosynthesis and promotes biofilm formation[160]; and (2) the *marR* repressor gene of *E. coli*, which encodes a repressor of the marRAB operon and whose silencing or deletion in the presence of antibiotics or other adverse change in the medium results in marRAB operon transcription and multiple antibiotic resistance.[161,162]

Changes in metabolic pathways typically observed in SCVs

Menadione- and hemin-auxotrophic strains are the two most common SCV phenotypes.[163,164] Point mutations in the menadione biosynthesis genes *menB*, *menC*, *menE*, and *menF* of SCVs that were auxotrophic for menadione were reported to favor SCV formation.[165,166] Mutations included bp deletion from nucleotides 55 to 63, a frameshift mutation inducing premature stop codon at position 230, and point mutations (eg, Gly to Val amino acid substitution at codon 233 in *menB*).[165,167]

Complementation of the mutant alleles of these menadione biosynthesis genes restored wild-type colony morphology and growth characteristics in cultures, but also strain virulence in mice.[166]

Frequent auxotrophies typically observed in SCVs are based on two major mechanisms: (1) deficiencies in electron transport (menadione- and hemin-auxotrophs); and (2) defective thymidylate biosynthesis (thymidine-dependent SCVs).[38,168] Alterations in hemin biosynthesis are due to mutations in biosynthesis genes, such as *hemB*, which are associated with reduced phagocytosis, slower extracellular growth and, consequently, abrogation of vancomycin bactericidal effects.[169] Similarly, thymidine dependency in clinical *S. aureus* SCVs may be due to nonsynonymous point mutations or deletions of *thyA*, the gene encoding for thymidylate synthase.[170] Of particular interest for auxotrophic SCV pharmacological targeting, a recent study revealed that metabolic features of central carbon metabolism are common in SCVs, regardless of their auxotrophism. Notably, the study was aimed at assessing all possible metabolic differences between sets of clinically relevant thymidine-, menadione-, and hemin-auxotrophic SCVs, nonauxotrophic mutants exhibiting the SCV phenotype, and the corresponding parental strains (normal phenotype).[37] Notably, decreases in fractions of (13)C2-aspartate and glutamate and absence of (13)C3-glutamate were observed in all SCVs, indicating a reduced carbon flux via the TCA cycle. Similarly, *acnA*, the gene encoding for the TCA cycle pivotal enzyme aconitase, was downregulated in all the SCV phenotypes studied.[168] These findings strongly suggest that TCA cycle targeting constitute a good target for SCV eradication in spite of SCV genetic backgrounds.

Other metabolic changes typically observed in the SCV phenotype include (1) downregulation of a number of *S. aureus* active metabolism genes such as the α-hemolysin gene (*hlY*) and coagulase gene (*coa*), and of the effector molecule of the staphylococcal Agr quorum-sensing system RNAIII (controls bacterial virulence); and (2) upregulation of genes involved in glycolytic and arginine-deiminase pathways and capsular biosynthesis.[165–167]

MRSA AND PERSISTER RESISTANCE TO OXIDATIVE STRESS

Methicillin-resistant S. aureus

A report by Posada et al.[35] explored the role of bacillithiol, the *S. aureus* counterpart to the glutathione system of mammalian cells, in the survival of MRSA under oxidative stress. Mutants with the deletion of bacillithiol biosynthesis genes such as *bshA* and *fosB*, which encode molecules pivotal for this process (a glycosyltransferase and a bacillithiol-*S*-transferase, respectively) displayed a 16- to 60-fold reduction in fosfomycin resistance,[35] suggesting that these genes are involved in such resistance. Moreover, mutations of *fosB* and *bshA* in some clinical MRSA isolates restored H_2O_2 sensitivity and were associated with a marked decrease in nicotinamide adenine dinucleotide phosphate (NADPH) levels (marker of the thioredoxin antioxidant system[36]) compared with wild types. Deletion of *fosB* led to a decrease in bacillithiol levels and increased expressions of genes involved in staphyloxanthin synthesis were observed in the isogenic *bshA* mutant under thiol stress conditions; and SH1000, a naturally bacillithiol-deficient strain, survived less well than its bacillithiol-producing isogenic counterpart.

All together, these findings suggest that bacillithiol signaling plays a major role in the survival of *S. aureus* under oxidative stress, at least partly through a FosB-dependent mechanism, independent of staphyloxanthin,[35] a virulence factor protecting MRSA from H_2O_2, superoxide radical, hypochloride, hydroxyl radical, and neutrophil killing.[171] Bacillithiol and staphyloxanthin pathways and related antioxidant systems may also explain how, unlike their methicillin-sensitive counterparts, some MRSA strains are able to resist photodynamic inactivation[172]; that is, protoporphyrin IX-dependent cell inactivation and death signaling triggered by the production of ROS following exposure to a wavelength light exciting photosensitizer molecules.[173] In addition, the thioredoxin system, which includes thioredoxin, and NADPH are pivotal in the defense against oxidative stress and in DNA synthesis.[174] Notably, bacterial thioredoxin reductase is structurally and functionally different from mammalian enzymes, and pharmacological targeting showed promising bactericidal effects in MRSA that lacks bacillithiol.[36] This effect was even more marked in bacteria lacking glutathione and glutaredoxin, such as *Helicobacter pylori*, in which a thioredoxin system is essential for DNA synthesis.[36,175]

Small colony variants

A recent report by Painter et al.[176] revealed that *S. aureus* SCVs are generated via the SOS response. Addressing why the exposure of *S. aureus* to sublethal concentrations of H_2O_2 leads to a dose-dependent specific increase in gentamicin-resistant SCVs, these authors found that H_2O_2 exposure causes a bacteriostasis in wild-type cells, during which *S. aureus* SCVs appeared spontaneously within the population. Interestingly, this transformation occurred via a mutagenic DNA repair pathway, including polymerase V, recombinase A, and the DNA double-strand break repair protein RexAB.[176] The study also revealed a more marked resistance of gentamicin-resistant SCVs to the toxic effects of H_2O_2, compared to wild-type, clinically relevant *S. aureus*. These findings were partly explained by a marked increase in catalase activity in SCVs. Such a change in catalase may also account for the ability of SCV to survive in the lethal acidic environment of the phagolysosome.[153,177]

MECHANISM OF ACTIONS OF NONMETABOLIC GENES PIVOTAL FOR MRSA RESISTANCE

Nonmetabolic genes conferring bacterial resistance are either genes encoding for antibiotic target proteins with low affinity or efflux proteins eliminating antibiotic molecules from the bacterial cell. These genes are either spontaneous, induced (adverse environmental changes including the presence of host antibacterial agents and antibiotics), or acquired (horizontal gene transfer via conjugation, transformation, or transduction) genetic mutations. Notably, antibiotic-resistance genes are frequently found in plasmids, facilitating horizontal gene transfer. Notably, several antibiotic resistance genes reside on transmissible plasmids, facilitating their transfer.

EXAMPLES OF MAJOR ANTIBIOTIC RESISTANCE GENES: *MECA* AND *CFR*

SCCmec type XI gene mecA, mecC

S. aureus methicillin resistance associates with the expression of the SCCmec gene type XI *mecA*, which encodes the β-lactam-insensitive PBP called PBP2a. Typically, most *mecA*-expressing MRSA isolates from health-care and community infections have a heterogeneous expression of resistance to β-lactam (HeR), with a very small portion of the population expressing resistance to β-lactams; however, exposure of these isolates to subinhibitory concentrations of β-lactams can result in high-level homotypic resistance (HoR).[97,178,179] This phenomenon is termed as *Her–Hor selection*. Reports by the Rosato group[97,180,181] revealed that HeR–HoR selection by β-lactams is associated with increased activating mutations of the *mecA* gene and *lexA/recA* regulators of the SOS response, a conserved regulatory bacterial network induced in response to DNA damage (for review, see Refs 182,183). These authors showed that mutational inactivation of *lexA* repressor results in a decrease in both mutation rate and oxacillin resistance in the HoR cells, with oxacillin resistance restauration at high levels observed in the wild-type strain following complementation of the *lexA* mutant strain.[180,181] This finding supports the hypothesis that an increase in the expression of *mecA* concomitantly with mutations triggering β-lactam-mediated SOS response genes is a pivotal mechanistic determinant of the selection of a highly oxacillin-resistant population. Furthermore, using a combinatorial approach of global biochemical profiling using gas chromatography–mass spectrometry and liquid chromatography–mass spectrometry, as well as microarray analyses, these authors also investigated changes in metabolic pathways (and the metabolome) associated with β-lactam-mediated HeR–HoR selection in clinically relevant heterogeneous MRSA.[97] The study revealed that when exposed to β-lactam pressure, HeR-MRSA clinical strains pay a heavy metabolic price to survive, redirecting the energy production to supply the cell wall synthesis and other metabolic processes that contribute to the survival response in the presence of these antibiotics.

Notably, unique features present in the HoR derivative selected by oxacillin (compared to a HeR parental strain) included a significant decrease in fermentative pathways and concomitant increases in TCA cycle intermediates, whose silencing via inactivation of the TCA cycle enzyme *cis*-aconitase gene in the HeR strain abolished β-lactam-mediated HeR/HoR selection.[97] Therefore, demonstrating altered TCA cycle activity would be pivotal during the HeR/HoR selection. Mechanisms accounting for TCA cycle–mediated survival of HeR-MRSA under β-lactam antibiotic exposure would mainly include the production of ROS, which in turn temporally affects DNA integrity, resulting in the activation of the SOS response, which enhances mutagenesis.[179]

In a recent report by Ballhausen et al.,[99] the antibiotic resistance SCCmec type XI gene *mecC*, which shares 70% similarity to *mecA*,[184] was suggested as the major genetic determinant that confers methicillin resistance to MRSA. In this study, a *mecC* knockout revealed a considerable reduction in the minimum inhibitory concentrations (MICs) of the β-lactam antibiotics oxacillin and cefoxitin compared to wild-type MRSA strains. Wild-type *mecC* restoring by complementation of the mutants in trans restored resistance to oxacillin and cefoxitin; *mecC* appeared to mediate oxacillin and cefoxitin resistance regardless of the strain's genetic background, with MIC values comparable to *mecA* MRSA strains. In addition, further supporting the hypothesis that a functional β-lactam-dependent regulatory system is active in MRSA strains possessing SCCmec type XI, the authors also showed that *mecC* expression is inducible by oxacillin.[99]

Chloramphenicol–florfenicol resistance (cfr) *gene*

The *cfr* gene encodes a 23S rRNA methyltransferase reported to confer resistance to protein synthesis inhibitors such as antibiotics of the phenicol class (eg, chloramphenicol, florfenicol), pleuromutilin-class, macrolide–lincosamide–streptogramin B antibiotics, and recently the synthetic antibacterial agent of the oxazolidinone class linezolid.[185,186] These agents kill bacteria by inhibiting bacterial protein synthesis, mainly by binding to bacterial ribosomes.[185,187–189] The Cfr protein confers resistance by methylation of the 23S rRNA position A2503.[190] In addition, linezolid-resistance mechanism studies also revealed G2576T mutation in domain V of 23S rRNA and aminoacid changes in L3 and L4 ribosomal protein mutation as major mechanisms of resistance of linezolid-resistant staphylococcal strains.[69,70,191,192]

EFFLUX GENES

An early study performed in three distinct geographical locations identified *mefE* as a mobile macrolide (eg, erythromycin, clarithromycin, roxithromycin) efflux gene that can confer antibiotic resistance to a variety of Gram-positive bacteria, including *Streptococcus pneumoniae*, *Corynebacterium jeikeium*, *Micrococcus luteus*, *Enterococcus faecalis*, and the viridans group streptococci.[193] This finding raised concerns about the possibility of transferring macrolide (and eventually other antibiotic) resistance between oropharyngeal Gram-positive organisms. Other macrolide resistance genes were also found in the oropharyngeal Gram-positive organisms, including other *mef* genes (eg, *mefA*) and erythromycin (Erm) resistance genes (eg, *ermB*).[193,194] These macrolide resistance genes were also reported in *S. aureus* and other staphylococci, where they were associated with resistance to protein synthesis inhibitor agents (ie, antibiotics acting by binding prokaryotic 70S ribosomes to stop protein synthesis in bacterial cells), such as macrolides, lincosamides, type B streptogramins, aminoglycoside, and tetracycline.[187–189] These results indicate that the efflux gene and mutations inhibiting protein synthesis inhibitor action may collaborate in the same bacterial cell to confer resistance. Also, in support of this hypothesis, a study addressing the mechanism of resistance to a novel class of potent antibiotics called the *bis-indole class* revealed the collaboration between a mutant resistance gene and an efflux gene in *S. aureus* NCTC 8325.[195] More specifically, resistant mutants had a truncated open reading frame of *mepR*, a gene encoding a MarR-like repressor with reduced susceptibility to bis-indoles. The aberrant MepR regulated the expression of *mepA*, a gene encoding an efflux pump of the multidrug and toxic compound extrusion family.[196] Reverse transcription polymerase chain reaction experiments showed that *mepA* overexpression is required for resistance, while complementation in trans of mutant *mepR* (to the wild type), abrogated resistance.[195,197] Similarly, the resistance genes *mgrA* and *norG* were reported to bind specifically to the promoter of genes encoding a number of efflux pumps, causing overexpressions of multidrug efflux pumps NorA, NorB, NorC, and AbcA, which are associated with significant increases in quinolone resistance levels in *S. aureus*.[198] Efflux genes reported on also include *marA*, which reduces antibiotic accumulation by acting at several target genes in the bacterial cell.[161,162]

CONCLUSIONS

There is increasing concern about the menace posed by antibiotic resistance (and, more broadly, multidrug-resistant bacteria) to health, food production, and economic prosperity. It is generally agreed that a sound understanding of the molecular mechanism and chemistry of resistance is primordial in the fight against this invisible foe. Faced with such alarming threats, the

World Health Organization presented a draft global action plan on antimicrobial resistance at the 68th World Health Assembly in Geneva in May 2015, in an endeavor to ensure that we can continue to safely and effectively treat and prevent infectious diseases. Indeed, antimicrobial resistance is a multifactorial problem that must be addressed promptly, and multiple sectors and disciplines will need to work together on any sustainable solutions against antibiotic resistance. However, one promising avenue that has attracted much attention is antimicrobial discovery from natural products, as the vast majority of antibiotics in clinical use originate from natural products or semisynthetic derivatives. Nonetheless, it is becoming apparent that antibiotics act as selection pressure to induce resistance, but from a Darwinian perspective, bacterial resistance is an inevitable consequence of the evolutionary process of life since to date, no antimicrobial has escaped from bacterial resistance.

REFERENCES

1. Custovic A, Smajlovic J, Tihic N, Hadzic S, Ahmetagic S, Hadzagic H. Epidemiological monitoring of nosocomial infections caused by *Acinetobacter baumannii*. *Med Arch* 2014;**68**:402–6.
2. Bassetti M, Righi E. Development of novel antibacterial drugs to combat multiple resistant organisms. *Langenbecks Arch Surg* 2015;**400**:153–65.
3. Bhattacharya PK. Emergence of antibiotic-resistant bacterial strains, methicillin-resistant *Staphylococcus aureus*, extended spectrum beta lactamases, and multi-drug resistance is a problem similar to global warming. *Rev Soc Bras Med Trop* 2014;**47**:815–16.
4. Porto JP, Batistao DW, Ribas RM. Authors' reply: emergence of antibiotic-resistant bacterial strains, methicillin-resistant *Staphylococcus aureus* and extended spectrum beta-lactamases, and multi-drug resistance are problems similar to global warming. *Rev Soc Bras Med Trop* 2014;**47**:817–18.
5. Dozois A, Thomsen I, Jimenez-Truque N, Soper N, Pearson A, Mohamed-Rambaran P, et al. Prevalence and molecular characteristics of methicillin-resistant *Staphylococcus aureus* among skin and soft tissue infections in an emergency department in Guyana. *Emerg Med J* 2014;**32**(10):800–3.
6. Shetty V, Trumbull K, Hegde A, Shenoy V, Prabhu R, Sumathi K, et al. Prevalence of community-acquired methicillin-resistant *Staphylococcus aureus* nasal colonization among children. *J Clin Diagn Res* 2014;**8**:DC12–15.
7. Erami M, Soltani B, Taghavi AA, Moravveji A, Haji RM, Soltani S, et al. Nasal carriage and resistance pattern of multidrug resistant *Staphylococcus aureus* among healthy children in Kashan, Iran. *Iran Red Crescent Med J* 2014;**16**:e21346.
8. Alzoubi HM, Aqel AA, Al-Sarayreh SA, Al-Zayadneh E. Methicillin-resistant *Staphylococcus aureus* nasal carriage among primary school-aged children from Jordan: prevalence, antibiotic resistance and molecular characteristics. *J Egypt Public Health Assoc* 2014;**89**:114–18.
9. Klevens RM, Morrison MA, Nadle J, Petit S, Gershman K, Ray S, et al. Invasive methicillin-resistant *Staphylococcus aureus* infections in the United States. *JAMA* 2007;**298**:1763–71.
10. Holland TL, Arnold C, Fowler Jr. VG. Clinical management of *Staphylococcus aureus* bacteremia: a review. *JAMA* 2014;**312**:1330–41.
11. Ntirenganya C, Manzi O, Muvunyi CM, Ogbuagu O. High prevalence of antimicrobial resistance among common bacterial isolates in a tertiary healthcare facility in Rwanda. *Am J Trop Med Hyg* 2015; **92**:865–70.

12. Kouegnigan Rerambiah L, Ndong JC, Mbakob Mengue MP, Medzegue S, Elisee-Ndam M, Mintsa-Ndong A, et al. Antimicrobial profiles of bacterial clinical isolates from the Gabonese National Laboratory of Public Health: data from routine activity. *Int J Infect Dis* 2014;**29**:48–53.
13. Schaumburg F, Alabi AS, Frielinghaus L, Grobusch MP, Kock R, Becker K, et al. The risk to import ESBL-producing Enterobacteriaceae and *Staphylococcus aureus* through chicken meat trade in Gabon. *BMC Microbiol* 2014;**14**:286.
14. O'Malley SM, Emele FE, Nwaokorie FO, Idika N, Umeizudike AK, Emeka-Nwabunnia I, et al. Molecular typing of antibiotic-resistant *Staphylococcus aureus* in Nigeria. *J Infect Public Health* 2015;**8**:187–93.
15. Antonanzas F, Lozano C, Torres C. Economic features of antibiotic resistance: the case of methicillin-resistant *Staphylococcus aureus*. *Pharmacoeconomics* 2015;**33**:285–325.
16. Papich MG. Antibiotic treatment of resistant infections in small animals. *Vet Clin North Am Small Anim Pract* 2013;**43**:1091–107.
17. Soe MM, Edwards JR, Sievert DM, Ricks PM, Magill SS, Fridkin SK. Evaluating state-specific antibiotic resistance measures derived from central line-associated bloodstream infections, national healthcare safety network, 2011. *Infect Control Hosp Epidemiol* 2015;**36**:54–64.
18. Zerouali K, Ramdani-Bouguessa N, Boye C, Hammami A. Multicentric study in five African countries of antibiotic susceptibility for three main pathogens: *Streptococcus pneumoniae*, *Staphylococcus aureus*, and *Pseudomonas aeruginosa*. *J Chemother* 2014 (in press). Available from: http://dx.doi.org/10.1179/1973947814Y.0000000220
19. Perencevich EN, Skov R, Kluytmans J. Identifying livestock-associated methicillin-resistant *Staphylococcus aureus* in the United States. *JAMA Intern Med* 2014;**174**:824–5.
20. Wang D, Wang Z, Yan Z, Wu J, Ali T, Li J, et al. Bovine mastitis *Staphylococcus aureus*: antibiotic susceptibility profile, resistance genes and molecular typing of methicillin-resistant and methicillin-sensitive strains in China. *Infect Genet Evol* 2015;**31C**:9–16.
21. Jevons M. "Celbenin"-resistant staphylococci. *Br Med J* 1961;**1**:124–5.
22. Das B, Sarkar C, Schachter J. Oritavancin—a new semisynthetic lipoglycopeptide agent to tackle the challenge of resistant gram positive pathogens. *Pak J Pharm Sci* 2013;**26**:1045–55.
23. Pasberg-Gauhl C. A need for new generation antibiotics against MRSA resistant bacteria. *Drug Discov Today Technol* 2014;**11**:109–16.
24. Lee L, Teh L, Zainuddin Z, Salleh M. The genome sequence of a type ST239 methicillin-resistant *Staphylococcus aureus* isolate from a Malaysian Hospital. *Stand Genomic Sci* 2014;**9**:933–9.
25. Cordova-Guerrero J, Hernandez-Guevara E, Ramirez-Zatarain S, Nunez-Bautista M, Ochoa-Teran A, Muniz-Salazar R, et al. Antibacterial activity of new oxazolidin-2-one analogues in methicillin-resistant *Staphylococcus aureus* strains. *Int J Mol Sci* 2014;**15**:5277–91.
26. Haaber J, Friberg C, McCreary M, Lin R, Cohen SN, Ingmer H. Reversible antibiotic tolerance induced in *Staphylococcus aureus* by concurrent drug exposure. *MBio* 2015;**6** e02268–14.
27. Diekema DJ, Pfaller MA, Schmitz FJ, Smayevsky J, Bell J, Jones RN, et al. Survey of infections due to *Staphylococcus* species: frequency of occurrence and antimicrobial susceptibility of isolates collected in the United States, Canada, Latin America, Europe, and the Western Pacific region for the SENTRY Antimicrobial Surveillance Program, 1997–1999. *Clin Infect Dis* 2001;**32**:S114–32.
28. Ternent L, Dyson RJ, Krachler AM, Jabbari S. Bacterial fitness shapes the population dynamics of antibiotic-resistant and -susceptible bacteria in a model of combined antibiotic and anti-virulence treatment. *J Theor Biol* 2015;**372**:1–11.
29. Reeve SM, Gainza P, Frey KM, Georgiev I, Donald BR, Anderson AC. Protein design algorithms predict viable resistance to an experimental antifolate. *Proc Natl Acad Sci USA* 2015;**112**:749–54.
30. Schwartz MA, Pflug GR. Model catalysts which simulate penicillinase. II. Mechanism of hydrolysis of penicillins catalyzed by catechol. *J Pharm Sci* 1967;**56**:1459–64.

31. Stapleton PD, Taylor PW. Methicillin resistance in *Staphylococcus aureus*: mechanisms and modulation. *Sci Prog* 2002;**85**:57–72.
32. Wang J, Kodali S, Lee SH, Galgoci A, Painter R, Dorso K, et al. Discovery of platencin, a dual FabF and FabH inhibitor with in vivo antibiotic properties. *Proc Natl Acad Sci USA* 2007;**104**:7612–16.
33. Zheng Z, Parsons JB, Tangallapally R, Zhang W, Rock CO, Lee RE. Discovery of novel bacterial elongation condensing enzyme inhibitors by virtual screening. *Bioorg Med Chem Lett* 2014;**24**:2585–8.
34. Parsons JB, Yao J, Frank MW, Rock CO. FabH mutations confer resistance to FabF-directed antibiotics in *Staphylococcus aureus*. *Antimicrob Agents Chemother* 2015;**59**:849–58.
35. Posada AC, Kolar SL, Dusi RG, Francois P, Roberts AA, Hamilton CJ, et al. Importance of bacillithiol in the oxidative stress response of *Staphylococcus aureus*. *Infect Immun* 2014;**82**:316–32.
36. Lu J, Vlamis-Gardikas A, Kandasamy K, Zhao R, Gustafsson TN, Engstrand L, et al. Inhibition of bacterial thioredoxin reductase: an antibiotic mechanism targeting bacteria lacking glutathione. *FASEB J* 2013;**27**:1394–403.
37. Kriegeskorte A, Grubmuller S, Huber C, Kahl BC, von EC, Proctor RA, et al. *Staphylococcus aureus* small colony variants show common metabolic features in central metabolism irrespective of the underlying auxotrophism. *Front Cell Infect Microbiol* 2014;**4**:141.
38. Morelli P, De AA, Manno G, Marchese A, Bassi M, Lobello R, et al. Characterization of *Staphylococcus aureus* small colony variant strains isolated from Italian patients attending a regional cystic fibrosis care centre. *New Microbiol* 2015;**38**:235–43.
39. Levy RM, Huang EY, Roling D, Leyden JJ, Margolis DJ. Effect of antibiotics on the oropharyngeal flora in patients with acne. *Arch Dermatol* 2003;**139**:467–71.
40. Sasan M, Donyadide N, Askari E, Naderi-Nasab M. Invasive community-acquired *Staphylococcus aureus* among pediatric population of Eastern Iran. *Iran J Microbiol* 2014;**6**:84–6.
41. Gorwitz RJ, Kruszon-Moran D, McAllister SK, McQuillan G, McDougal LK, Fosheim GE, et al. Changes in the prevalence of nasal colonization with *Staphylococcus aureus* in the United States, 2001–2004. *J Infect Dis* 2008;**197**:1226–34.
42. Conceicao T, Diamantino F, Coelho C, de LH, Aires-de-Sousa M. Contamination of public buses with MRSA in Lisbon, Portugal: a possible transmission route of major MRSA clones within the community. *PLoS One* 2013;**8**:e77812.
43. von Eiff C, Becker K, Machka K, Stammer H, Peters G. Nasal carriage as a source of *Staphylococcus aureus* bacteremia. Study Group. *N Engl J Med* 2001;**344**:11–16.
44. Zacharioudakis IM, Zervou FN, Ziakas PD, Mylonakis E. Meta-analysis of methicillin-resistant *Staphylococcus aureus* colonization and risk of infection in dialysis patients. *J Am Soc Nephrol* 2014; **25**:2131–41.
45. Arfken AM, Song B, Sung JS. Comparison of airborne bacterial communities from hog farm and spray field. *J Microbiol Biotechnol* 2015;**25**:709–17.
46. Jang Y, Bae D, Cho JK, Bahk GJ, Lim SK, Lee YJ. Characterization of methicillin-resistant *Staphylococcus* spp. isolated from dogs in Korea. *Jpn J Vet Res* 2014;**62**:163–70.
47. Rheinwald M, Hartmann K, Hahner M, Wolf G, Straubinger RK, Schulz B. Antibiotic susceptibility of bacterial isolates from 502 dogs with respiratory signs. *Vet Rec* 2015;**176**:357.
48. Moker N, Dean CR, Tao J. *Pseudomonas aeruginosa* increases formation of multidrug-tolerant persister cells in response to quorum-sensing signaling molecules. *J Bacteriol* 2010;**192**:1946–55.
49. Lewis K. Persister cells and the riddle of biofilm survival. *Biochemistry (Mosc)* 2005;**70**:267–74.
50. Lewis K. Multidrug tolerance of biofilms and persister cells. *Curr Top Microbiol Immunol* 2008;**322**: 107–31.
51. Uhlemann AC, Knox J, Miller M, Hafer C, Vasquez G, Ryan M, et al. The environment as an unrecognized reservoir for community-associated methicillin resistant *Staphylococcus aureus* USA300: a case–control study. *PLoS One* 2011;**6**:e22407.

52. Fritz SA, Hogan PG, Singh LN, Thompson RM, Wallace MA, Whitney K, et al. Contamination of environmental surfaces with *Staphylococcus aureus* in households with children infected with methicillin-resistant *S. aureus*. *JAMA Pediatr* 2014;**168**:1030–8.
53. Dibah S, Arzanlou M, Jannati E, Shapouri R. Prevalence and antimicrobial resistance pattern of methicillin resistant *Staphylococcus aureus* (MRSA) strains isolated from clinical specimens in Ardabil, Iran. *Iran J Microbiol* 2014;**6**:163–8.
54. Diawara I, Bekhti K, Elhabchi D, Saile R, Elmdaghri N, Timinouni M, et al. *Staphylococcus aureus* nasal carriage in hemodialysis centers of Fez, Morocco. *Iran J Microbiol* 2014;**6**:175–83.
55. Vazirianzadeh B, Dehghani R, Mehdinejad M, Sharififard M, Nasirabadi N. The first report of drug resistant bacteria isolated from the brown-banded cockroach, *Supella longipalpa*, in Ahvaz, South-western Iran. *J Arthropod Borne Dis* 2014;**8**:53–9.
56. Concepcion Porrero M, Harrison EM, Fernandez-Garayzabal JF, Paterson GK, Diez-Guerrier A, Holmes MA, et al. Detection of mecC-methicillin-resistant *Staphylococcus aureus* isolates in river water: a potential role for water in the environmental dissemination. *Environ Microbiol Rep* 2014;**6**:705–8.
57. Seyedmonir E, Yilmaz F, Icgen B. mecA gene dissemination among staphylococcal and non-staphylococcal isolates shed in surface waters. *Bull Environ Contam Toxicol* 2015;**95**:131–8.
58. Gomez P, Lozano C, Gonzalez-Barrio D, Zarazaga M, Ruiz-Fons F, Torres C. High prevalence of methicillin-resistant *Staphylococcus aureus* (MRSA) carrying the mecC gene in a semi-extensive red deer (*Cervus elaphus hispanicus*) farm in Southern Spain. *Vet Microbiol* 2015;**177**:326–31.
59. Luddeke F, Hess S, Gallert C, Winter J, Gude H, Loffler H. Removal of total and antibiotic resistant bacteria in advanced wastewater treatment by ozonation in combination with different filtering techniques. *Water Res* 2015;**69**:243–51.
60. Scheurer M, Hess S, Luddeke F, Sacher F, Gude H, Loffler H, et al. Removal of micropollutants, facultative pathogenic and antibiotic resistant bacteria in a full-scale retention soil filter receiving combined sewer overflow. *Environ Sci Process Impacts* 2015;**17**:186–96.
61. Williams V, Simor AE, Kiss A, McGeer A, Hirji Z, Larios OE, et al. Is the prevalence of antibiotic resistant organisms changing in Canadian hospitals? A comparison of point-prevalence survey results in 2010 and 2012. *Clin Microbiol Infect* 2015;**21**:553–9.
62. Diederen BM, Wardle CL, Krijnen P, Tuinebreijer WE, Breederveld RS. Epidemiology of clinically relevant bacterial pathogens in a burn center in the Netherlands between 2005 and 2011. *J Burn Care Res* 2015;**36**:446–53.
63. Ansari S, Nepal HP, Gautam R, Shrestha S, Neopane P, Rimal B, et al. Childhood septicemia in Nepal: documenting the bacterial etiology and its susceptibility to antibiotics. *Int J Microbiol* 2014; **2014**:452648.
64. Shahkarami F, Rashki A, Rashki GZ. Microbial susceptibility and plasmid profiles of methicillin-resistant *Staphylococcus aureus* and methicillin-susceptible *S. aureus*. *Jundishapur J Microbiol* 2014;**7**:e16984.
65. Sun C, Chai W, Pan Y, Zhou Y. Analyses of microbiology and antibiotic susceptibility in prosthetic knee infections. *Zhonghua Yi Xue Za Zhi* 2014;**94**:3575–8.
66. Zuo MF, Liu HL, Zhu ML, Shu QZ, Jiang L. Pathologic bacterial distribution and antibiotic resistance in induced sputum of infants aged from 1 to 3 months with lower respiratory tract infection. *Zhongguo Dang Dai Er Ke Za Zhi* 2014;**16**:1226–30.
67. Al-Mulla NA, Taj-Aldeen SJ, El SS, Janahi M, Al-Nasser AA, Chandra P. Bacterial bloodstream infections and antimicrobial susceptibility pattern in pediatric hematology/oncology patients after anticancer chemotherapy. *Infect Drug Resist* 2014;**7**:289–99.
68. Salah LA, Faergemann J. A retrospective analysis of skin bacterial colonisation, susceptibility and resistance in atopic dermatitis and impetigo patients. *Acta Derm Venereol* 2015;**95**:532–5.

69. O'Connor C, Powell J, Finnegan C, O'Gorman A, Barrett S, Hopkins KL, et al. Incidence, management and outcomes of the first cfr-mediated linezolid-resistant *Staphylococcus epidermidis* outbreak in a tertiary referral centre in the Republic of Ireland. *J Hosp Infect* 2015;**90**:316–21.
70. Russo A, Campanile F, Falcone M, Tascini C, Bassetti M, Goldoni P, et al. Linezolid-resistant staphylococcal bacteraemia: a multicentre case–case–control study in Italy. *Int J Antimicrob Agents* 2015;**45**: 255–61.
71. Omuse G, Kabera B, Revathi G. Low prevalence of methicillin resistant *Staphylococcus aureus* as determined by an automated identification system in two private hospitals in Nairobi, Kenya: a cross sectional study. *BMC Infect Dis* 2014;**14**:669.
72. Isendahl J, Manjuba C, Rodrigues A, Xu W, Henriques-Normark B, Giske CG, et al. Prevalence of community-acquired bacteraemia in Guinea-Bissau: an observational study. *BMC Infect Dis* 2014; **14**:715.
73. Rosenthal ME, Mediavilla J, Chen L, Sonnenfeld J, Pierce L, Shannon A, et al. Molecular epidemiology of *Staphylococcus aureus* in post-earthquake northern Haiti. *Int J Infect Dis* 2014;**29**:146–51.
74. Seifert J, Frank M, Koln T, Beniers K, Kramer A, Ekkernkamp A, et al. Epidemiology of multidrug-resistant organisms in travellers: results of a 2-year screening in a German level 1 trauma center. *Unfallchirurg* 2015;**45**:255–61.
75. Connell S, Li J, Shi R. Synergistic bactericidal activity between hyperosmotic stress and membrane-disrupting nanoemulsions. *J Med Microbiol* 2013;**62**:69–77.
76. Obad J, Suskovic J, Kos B. Antimicrobial activity of Ibuprofen: new perspectives on an "Old" non-antibiotic drug. *Eur J Pharm Sci* 2015;**71**:93–8.
77. Ojeda-Sana AM, Repetto V, Moreno S. Carnosic acid is an efflux pumps modulator by dissipation of the membrane potential in *Enterococcus faecalis* and *Staphylococcus aureus*. *World J Microbiol Biotechnol* 2013;**29**:137–44.
78. Klinger-Strobel M, Lautenschlager C, Fischer D, Mainz JG, Bruns T, Tuchscherr L, et al. Aspects of pulmonary drug delivery strategies for infections in cystic fibrosis—where do we stand? *Expert Opin Drug Deliv* 2015;1–24.
79. Smyth AR, Walters S. Prophylactic anti-staphylococcal antibiotics for cystic fibrosis. *Cochrane Database Syst Rev* 2014;11 CD001912
80. Kalsi J, Arya M, Wilson P, Mundy A. Hospital-acquired urinary tract infection. *Int J Clin Pract* 2003;**57**:388–91.
81. Sultan A, Rizvi M, Khan F, Sami H, Shukla I, Khan HM. Increasing antimicrobial resistance among uropathogens: Is fosfomycin the answer? *Urol Ann* 2015;**7**:26–30.
82. Drlica K, Zhao X. Mutant selection window hypothesis updated. *Clin Infect Dis* 2007;**44**:681–8.
83. Firsov AA, Vostrov SN, Lubenko IY, Drlica K, Portnoy YA, Zinner SH. In vitro pharmacodynamic evaluation of the mutant selection window hypothesis using four fluoroquinolones against *Staphylococcus aureus*. *Antimicrob Agents Chemother* 2003;**47**:1604–13.
84. Zinner SH, Lubenko IY, Gilbert D, Simmons K, Zhao X, Drlica K, et al. Emergence of resistant *Streptococcus pneumoniae* in an in vitro dynamic model that simulates moxifloxacin concentrations inside and outside the mutant selection window: related changes in susceptibility, resistance frequency and bacterial killing. *J Antimicrob Chemother* 2003;**52**:616–22.
85. Ni W, Song X, Cui J. Testing the mutant selection window hypothesis with *Escherichia coli* exposed to levofloxacin in a rabbit tissue cage infection model. *Eur J Clin Microbiol Infect Dis* 2014;**33**:385–9.
86. Mei Q, Ye Y, Zhu YL, Cheng J, Chang X, Liu YY, et al. Testing the mutant selection window hypothesis in vitro and in vivo with *Staphylococcus aureus* exposed to fosfomycin. *Eur J Clin Microbiol Infect Dis* 2015;**34**:737–44.

87. Ouberai M, El GF, Bussiere A, Riou M, Alsteens D, Lins L, et al. The *Pseudomonas aeruginosa* membranes: a target for a new amphiphilic aminoglycoside derivative? *Biochim Biophys Acta* 2011;**1808**:1716–27.
88. Cai Y, Cao X, Aballay A. Whole-animal chemical screen identifies colistin as a new immunomodulator that targets conserved pathways. *MBio* 2014;**5**:e01235–14.
89. Musser JM, Kapur V. Clonal analysis of methicillin-resistant *Staphylococcus aureus* strains from intercontinental sources: association of the mec gene with divergent phylogenetic lineages implies dissemination by horizontal transfer and recombination. *J Clin Microbiol* 1992;**30**:2058–63.
90. Udou T. Dissemination of nosocomial multiple-aminoglycoside-resistant *Staphylococcus aureus* caused by horizontal transfer of the resistance determinant (aacA/aphD) and clonal spread of resistant strains. *Am J Infect Control* 2004;**32**:215–19.
91. Hurdle JG, O'Neill AJ, Mody L, Chopra I, Bradley SF. In vivo transfer of high-level mupirocin resistance from *Staphylococcus epidermidis* to methicillin-resistant *Staphylococcus aureus* associated with failure of mupirocin prophylaxis. *J Antimicrob Chemother* 2005;**56**:1166–8.
92. Kriegeskorte A, Peters G. Horizontal gene transfer boosts MRSA spreading. *Nat Med* 2012;**18**:662–3.
93. Traa MX, Barboza L, Doron S, Snydman DR, Noubary F, Nasraway Jr. SA. Horizontal infection control strategy decreases methicillin-resistant *Staphylococcus aureus* infection and eliminates bacteremia in a surgical ICU without active surveillance. *Crit Care Med* 2014;**42**:2151–7.
94. Berger-Bachi B, Strassle A, Gustafson JE, Kayser FH. Mapping and characterization of multiple chromosomal factors involved in methicillin resistance in *Staphylococcus aureus*. *Antimicrob Agents Chemother* 1992;**36**:1367–73.
95. Berger-Bachi B, Tschierske M. Role of fem factors in methicillin resistance. *Drug Resist Updat* 1998;**1**:325–35.
96. Kim SJ, Chang J, Singh M. Peptidoglycan architecture of Gram-positive bacteria by solid-state NMR. *Biochim Biophys Acta* 2015;**1848**:350–62.
97. Keaton MA, Rosato RR, Plata KB, Singh CR, Rosato AE. Exposure of clinical MRSA heterogeneous strains to beta-lactams redirects metabolism to optimize energy production through the TCA cycle. *PLoS One* 2013;**8**:e71025.
98. Yen TY, Sung YJ, Lin HC, Peng CT, Tien N, Hwang KP, et al. Emergence of oxacillin-resistant *Staphylococcus lugdunensis* carrying staphylococcal cassette chromosome mec type V in central Taiwan. *J Microbiol Immunol Infect* 2014 (in press). Available from: http://dx.doi.org/10.1016/j.jmii.2014.11.018
99. Ballhausen B, Kriegeskorte A, Schleimer N, Peters G, Becker K. The mecA homolog mecC confers resistance against beta-lactams in *Staphylococcus aureus* irrespective of the genetic strain background. *Antimicrob Agents Chemother* 2014;**58**:3791–8.
100. Bayer AS, Lam K. Efficacy of vancomycin plus rifampin in experimental aortic-valve endocarditis due to methicillin-resistant *Staphylococcus aureus*: in vitro–in vivo correlations. *J Infect Dis* 1985;**151**:157–65.
101. van Hal SJ, Fowler Jr. VG. Is it time to replace vancomycin in the treatment of methicillin-resistant *Staphylococcus aureus* infections? *Clin Infect Dis* 2013;**56**:1779–88.
102. Watanakunakorn C. Mode of action and in-vitro activity of vancomycin. *J Antimicrob Chemother* 1984;**14D**:7–18.
103. Munch D, Engels I, Muller A, Reder-Christ K, Falkenstein-Paul H, Bierbaum G, et al. Structural variations of the cell wall precursor lipid II and their influence on binding and activity of the lipoglycopeptide antibiotic oritavancin. *Antimicrob Agents Chemother* 2015;**59**:772–81.
104. Grundling A, Missiakas DM, Schneewind O. *Staphylococcus aureus* mutants with increased lysostaphin resistance. *J Bacteriol* 2006;**188**:6286–97.
105. Pogliano J, Pogliano N, Silverman JA. Daptomycin-mediated reorganization of membrane architecture causes mislocalization of essential cell division proteins. *J Bacteriol* 2012;**194**:4494–504.

106. Gaupp R, Lei S, Reed JM, Peisker H, Boyle-Vavra S, Bayer AS, et al. *Staphylococcus aureus* metabolic adaptations during the transition from a daptomycin susceptible phenotype to a daptomycin nonsusceptible phenotype. *Antimicrob Agents Chemother* 2015;**59**:4226–38.
107. Sharif S, Kim SJ, Labischinski H, Chen J, Schaefer J. Uniformity of glycyl bridge lengths in the mature cell walls of fem mutants of methicillin-resistant *Staphylococcus aureus*. *J Bacteriol* 2013;**195**:1421–7.
108. Yuan W, Hu Q, Cheng H, Shang W, Liu N, Hua Z, et al. Cell wall thickening is associated with adaptive resistance to amikacin in methicillin-resistant *Staphylococcus aureus* clinical isolates. *J Antimicrob Chemother* 2013;**68**:1089–96.
109. Mirza HC, Sancak B, Gur D. The prevalence of vancomycin-intermediate *Staphylococcus aureus* and heterogeneous VISA among methicillin-resistant strains isolated from pediatric population in a Turkish university hospital. *Microb Drug Resist* 2015;**21**(5):537–44.
110. Howden BP, Johnson PD, Ward PB, Stinear TP, Davies JK. Isolates with low-level vancomycin resistance associated with persistent methicillin-resistant *Staphylococcus aureus* bacteremia. *Antimicrob Agents Chemother* 2006;**50**:3039–47.
111. Tawil N, Mouawad F, Levesque S, Sacher E, Mandeville R, Meunier M. The differential detection of methicillin-resistant, methicillin-susceptible and borderline oxacillin-resistant *Staphylococcus aureus* by surface plasmon resonance. *Biosens Bioelectron* 2013;**49**:334–40.
112. Choi J, Yoo J, Lee M, Kim EG, Lee JS, Lee S, et al. A rapid antimicrobial susceptibility test based on single-cell morphological analysis. *Sci Transl Med* 2014;**6** 267ra174.
113. Ling LL, Schneider T, Peoples AJ, Spoering AL, Engels I, Conlon BP, et al. A new antibiotic kills pathogens without detectable resistance. *Nature* 2015;**517**:455–9.
114. Kwan BW, Chowdhury N, Wood TK. Combatting bacterial infections by killing persister cells with mitomycin C. *Environ Microbiol* 2015;**17**(11):4406–14.
115. Uddin R, Saeed K, Khan W, Azam SS, Wadood A. Metabolic pathway analysis approach: identification of novel therapeutic target against methicillin resistant *Staphylococcus aureus*. *Gene* 2015;**556**:213–26.
116. Balderrama-Subieta A, Quillaguaman J. Genomic studies on nitrogen metabolism in *Halomonas boliviensis*: metabolic pathway, biochemistry and evolution. *Comput Biol Chem* 2013;**47**:96–104.
117. Toya Y, Hirasawa T, Morimoto T, Masuda K, Kageyama Y, Ozaki K, et al. 13 C-metabolic flux analysis in heterologous cellulase production by *Bacillus subtilis* genome-reduced strain. *J Biotechnol* 2014;**179**:42–9.
118. Ortiz-Marquez JC, Do NM, Curatti L. Metabolic engineering of ammonium release for nitrogen-fixing multispecies microbial cell-factories. *Metab Eng* 2014;**23**:154–64.
119. Ruzal SM, Sanchez-Rivas C. Effect of glutamate synthase (GOGAT) activity on *Bacillus subtilis* spore properties. *Curr Microbiol* 2003;**47**:208–13.
120. Lunse CE, Schmidt MS, Wittmann V, Mayer G. Carba-sugars activate the glmS-riboswitch of *Staphylococcus aureus*. *ACS Chem Biol* 2011;**6**:675–8.
121. Johnston C, Bootsma HJ, Aldridge C, Manuse S, Gisch N, Schwudke D, et al. Co-inactivation of GlnR and CodY regulators impacts pneumococcal cell wall physiology. *PLoS One* 2015;**10**:e0123702.
122. Morayya S, Awasthy D, Yadav R, Ambady A, Sharma U. Revisiting the essentiality of glutamate racemase in *Mycobacterium tuberculosis*. *Gene* 2015;**555**:269–76.
123. Figueiredo TA, Sobral RG, Ludovice AM, Almeida JM, Bui NK, Vollmer W, et al. Identification of genetic determinants and enzymes involved with the amidation of glutamic acid residues in the peptidoglycan of *Staphylococcus aureus*. *PLoS Pathog* 2012;**8**:e1002508.
124. Bigger JW. The bactericidal action of penicillin on *Staphylococcus pyogenes*. *Ir J Med Sci* 1944;**19**:587–94.
125. Gwynn MN, Webb TL, Rolinson GN. Regrowth of *Pseudomonas aeruginosa* and other bacteria after the bactericidal action of carbenicillin and other beta-lactam antibiotics. *J Infect Dis* 1981;**144**:263–9.

126. Woolfrey BF, Lally RT, Ederer MN. Influence of technical factor variations during inoculum preparation on the agar dilution plate-count method for quantitation of *Staphylococcus aureus* oxacillin persisters. *Antimicrob Agents Chemother* 1986;**30**:792–3.
127. Giamarellou H, Mandragos K, Bechrakis P, Pigas K, Bilalis D, Sfikakis P. Pefloxacin versus imipenem in the therapy of nosocomial lung infections of intensive care unit patients. *J Antimicrob Chemother* 1990;**26B**:117–27.
128. Tomiyama M. The effect of cefaclor and cefixime on nasopharyngeal pathogens in children. *Nihon Jibiinkoka Gakkai Kaiho* 1995;**98**:659–68.
129. Keren I, Kaldalu N, Spoering A, Wang Y, Lewis K. Persister cells and tolerance to antimicrobials. *FEMS Microbiol Lett* 2004;**230**:13–18.
130. Nguyen HA, Grellet J, Paillard D, Dubois V, Quentin C, Saux MC. Factors influencing the intracellular activity of fluoroquinolones: a study using levofloxacin in a *Staphylococcus aureus* THP-1 monocyte model. *J Antimicrob Chemother* 2006;**57**:883–90.
131. Singh R, Ray P, Das A, Sharma M. Role of persisters and small-colony variants in antibiotic resistance of planktonic and biofilm-associated *Staphylococcus aureus*: an in vitro study. *J Med Microbiol* 2009;**58**:1067–73.
132. Dawson CC, Intapa C, Jabra-Rizk MA. "Persisters": survival at the cellular level. *PLoS Pathog* 2011;**7**: e1002121.
133. Han J, He L, Shi W, Xu X, Wang S, Zhang S, et al. Glycerol uptake is important for L-form formation and persistence in *Staphylococcus aureus*. *PLoS One* 2014;**9**:e108325.
134. Germain E, Roghanian M, Gerdes K, Maisonneuve E. Stochastic induction of persister cells by HipA through (p)ppGpp-mediated activation of mRNA endonucleases. *Proc Natl Acad Sci USA* 2015; **112**:5171–6.
135. Shan Y, Lazinski D, Rowe S, Camilli A, Lewis K. Genetic basis of persister tolerance to aminoglycosides in *Escherichia coli*. *MBio* 2015;**6**:e00078–15.
136. Traxler MF, Summers SM, Nguyen HT, Zacharia VM, Hightower GA, Smith JT, et al. The global, ppGpp-mediated stringent response to amino acid starvation in *Escherichia coli*. *Mol Microbiol* 2008;**68**:1128–48.
137. Bokinsky G, Baidoo EE, Akella S, Burd H, Weaver D, Alonso-Gutierrez J, et al. HipA-triggered growth arrest and beta-lactam tolerance in *Escherichia coli* are mediated by RelA-dependent ppGpp synthesis. *J Bacteriol* 2013;**195**:3173–82.
138. Maisonneuve E, Castro-Camargo M, Gerdes K. (p)ppGpp controls bacterial persistence by stochastic induction of toxin–antitoxin activity. *Cell* 2013;**154**:1140–50.
139. Prax M, Bertram R. Metabolic aspects of bacterial persisters. *Front Cell Infect Microbiol* 2014;**4**:148.
140. Nguyen D, Joshi-Datar A, Lepine F, Bauerle E, Olakanmi O, Beer K, et al. Active starvation responses mediate antibiotic tolerance in biofilms and nutrient-limited bacteria. *Science* 2011;**334**:982–6.
141. Uppal S, Shetty DM, Jawali N. Cyclic AMP receptor protein regulates cspD, a bacterial toxin gene, in *Escherichia coli*. *J Bacteriol* 2014;**196**:1569–77.
142. Li Z, Nimtz M, Rinas U. The metabolic potential of *Escherichia coli* BL21 in defined and rich medium. *Microb Cell Fact* 2014;**13**:45.
143. Kobylarz MJ, Grigg JC, Sheldon JR, Heinrichs DE, Murphy ME. SbnG, a citrate synthase in *Staphylococcus aureus*: a new fold on an old enzyme. *J Biol Chem* 2014;**289**:33797–807.
144. Ferguson GP, Battista JR, Lee AT, Booth IR. Protection of the DNA during the exposure of *Escherichia coli* cells to a toxic metabolite: the role of the KefB and KefC potassium channels. *Mol Microbiol* 2000;**35**:113–22.
145. Brynildsen MP, Winkler JA, Spina CS, MacDonald IC, Collins JJ. Potentiating antibacterial activity by predictably enhancing endogenous microbial ROS production. *Nat Biotechnol* 2013;**31**:160–5.

146. Sun F, Ding Y, Ji Q, Liang Z, Deng X, Wong CC, et al. Protein cysteine phosphorylation of SarA/MgrA family transcriptional regulators mediates bacterial virulence and antibiotic resistance. *Proc Natl Acad Sci USA* 2012;**109**:15461–6.
147. Allison KR, Brynildsen MP, Collins JJ. Metabolite-enabled eradication of bacterial persisters by aminoglycosides. *Nature* 2011;**473**:216–20.
148. Lebeaux D, Chauhan A, Letoffe S, Fischer F, de RH, Beloin C, et al. pH-mediated potentiation of aminoglycosides kills bacterial persisters and eradicates in vivo biofilms. *J Infect Dis* 2014;**210**:1357–66.
149. Tuchscherr L, Medina E, Hussain M, Volker W, Heitmann V, Niemann S, et al. *Staphylococcus aureus* phenotype switching: an effective bacterial strategy to escape host immune response and establish a chronic infection. *EMBO Mol Med* 2011;**3**:129–41.
150. Lewis K. Persister cells. *Annu Rev Microbiol* 2010;**64**:357–72.
151. Sandberg A, Lemaire S, Van BF, Tulkens PM, Hughes D, von EC, et al. Intra- and extracellular activities of dicloxacillin and linezolid against a clinical *Staphylococcus aureus* strain with a small-colony-variant phenotype in an in vitro model of THP-1 macrophages and an in vivo mouse peritonitis model. *Antimicrob Agents Chemother* 2011;**55**:1443–52.
152. Mirani ZA, Aziz M, Khan SI. Small colony variants have a major role in stability and persistence of *Staphylococcus aureus* biofilms. *J Antibiot (Tokyo)* 2015;**68**:98–105.
153. Atalla H, Gyles C, Mallard B. *Staphylococcus aureus* small colony variants (SCVs) and their role in disease. *Anim Health Res Rev* 2011;**12**:33–45.
154. Johnson PJ, Levin BR. Pharmacodynamics, population dynamics, and the evolution of persistence in *Staphylococcus aureus*. *PLoS Genet* 2013;**9**:e1003123.
155. Lechner S, Patra P, Klumpp S, Bertram R. Interplay between population dynamics and drug tolerance of *Staphylococcus aureus* persister cells. *J Mol Microbiol Biotechnol* 2012;**22**:381–91.
156. Senn MM, Bischoff M, von EC, Berger-Bachi B. sigmaB activity in a *Staphylococcus aureus* hemB mutant. *J Bacteriol* 2005;**187**:7397–406.
157. Mitchell G, Fugere A, Pepin GK, Brouillette E, Frost EH, Cantin AM, et al. SigB is a dominant regulator of virulence in *Staphylococcus aureus* small-colony variants. *PLoS One* 2013;**8**:e65018.
158. Jansen A, Szekat C, Schroder W, Wolz C, Goerke C, Lee JC, et al. Production of capsular polysaccharide does not influence *Staphylococcus aureus* vancomycin susceptibility. *BMC Microbiol* 2013;**13**:65.
159. Boyle-Vavra S, Li X, Alam MT, Read TD, Sieth J, Cywes-Bentley C, et al. USA300 and USA500 clonal lineages of *Staphylococcus aureus* do not produce a capsular polysaccharide due to conserved mutations in the cap5 locus. *MBio* 2015;**6** e02585–14.
160. Chang YM, Ho CH, Chen CK, Maestre-Reyna M, Chang-Chien MW, Wang AH. TcaR–ssDNA complex crystal structure reveals new DNA binding mechanism of the MarR family proteins. *Nucleic Acids Res* 2014;**42**:5314–21.
161. Sulavik MC, Gambino LF, Miller PF. The MarR repressor of the multiple antibiotic resistance (mar) operon in *Escherichia coli*: prototypic member of a family of bacterial regulatory proteins involved in sensing phenolic compounds. *Mol Med* 1995;**1**:436–46.
162. Khan SA, Sung K, Nawaz MS. Detection of aacA-aphD, qacEdelta1, marA, floR, and tetA genes from multidrug-resistant bacteria: comparative analysis of real-time multiplex PCR assays using EvaGreen ((R)) and SYBR((R)) Green I dyes. *Mol Cell Probes* 2011;**25**:78–86.
163. Kahl B, Herrmann M, Everding AS, Koch HG, Becker K, Harms E, et al. Persistent infection with small colony variant strains of *Staphylococcus aureus* in patients with cystic fibrosis. *J Infect Dis* 1998;**177**:1023–9.
164. Yagci S, Hascelik G, Dogru D, Ozcelik U, Sener B. Prevalence and genetic diversity of *Staphylococcus aureus* small-colony variants in cystic fibrosis patients. *Clin Microbiol Infect* 2013;**19**:77–84.

165. Lannergard J, von EC, Sander G, Cordes T, Seggewiss J, Peters G, et al. Identification of the genetic basis for clinical menadione-auxotrophic small-colony variant isolates of *Staphylococcus aureus*. *Antimicrob Agents Chemother* 2008;**52**:4017–22.
166. Dean MA, Olsen RJ, Long SW, Rosato AE, Musser JM. Identification of point mutations in clinical *Staphylococcus aureus* strains that produce small-colony variants auxotrophic for menadione. *Infect Immun* 2014;**82**:1600–5.
167. Lannergard J, Cao S, Norstrom T, Delgado A, Gustafson JE, Hughes D. Genetic complexity of fusidic acid-resistant small colony variants (SCV) in *Staphylococcus aureus*. *PLoS One* 2011;**6**:e28366.
168. Kriegeskorte A, Block D, Drescher M, Windmuller N, Mellmann A, Baum C, et al. Inactivation of thyA in *Staphylococcus aureus* attenuates virulence and has a strong impact on metabolism and virulence gene expression. *MBio* 2014;**5**:e01447–14.
169. Garcia LG, Lemaire S, Kahl BC, Becker K, Proctor RA, Denis O, et al. Pharmacodynamic evaluation of the activity of antibiotics against hemin- and menadione-dependent small-colony variants of *Staphylococcus aureus* in models of extracellular (broth) and intracellular (THP-1 monocytes) infections. *Antimicrob Agents Chemother* 2012;**56**:3700–11.
170. Chatterjee I, Kriegeskorte A, Fischer A, Deiwick S, Theimann N, Proctor RA, et al. In vivo mutations of thymidylate synthase (encoded by thyA) are responsible for thymidine dependency in clinical small-colony variants of *Staphylococcus aureus*. *J Bacteriol* 2008;**190**:834–42.
171. Clauditz A, Resch A, Wieland KP, Peschel A, Gotz F. Staphyloxanthin plays a role in the fitness of *Staphylococcus aureus* and its ability to cope with oxidative stress. *Infect Immun* 2006;**74**:4950–3.
172. Grinholc M, Rapacka-Zdonczyk A, Rybak B, Szabados F, Bielawski KP. Multiresistant strains are as susceptible to photodynamic inactivation as their naive counterparts: protoporphyrin IX-mediated photoinactivation reveals differences between methicillin-resistant and methicillin-sensitive *Staphylococcus aureus* strains. *Photomed Laser Surg* 2014;**32**:121–9.
173. Nakonieczna J, Michta E, Rybicka M, Grinholc M, Gwizdek-Wisniewska A, Bielawski KP. Superoxide dismutase is upregulated in *Staphylococcus aureus* following protoporphyrin-mediated photodynamic inactivation and does not directly influence the response to photodynamic treatment. *BMC Microbiol* 2010;**10**:323.
174. Lu J, Holmgren A. The thioredoxin antioxidant system. *Free Radic Biol Med* 2014;**66**:75–87.
175. Hong Y, Lai YT, Chan GC, Sun H. Glutathione and multidrug resistance protein transporter mediate a self-propelled disposal of bismuth in human cells. *Proc Natl Acad Sci USA* 2015;**112**:3211–16.
176. Painter KL, Strange E, Parkhill J, Bamford KB, Armstrong-James D, Edwards AM. *Staphylococcus aureus* adapts to oxidative stress by producing H_2O_2-resistant small colony variants via the SOS response. *Infect Immun* 2015;**83**:1830–44.
177. Nguyen HA, Denis O, Vergison A, Theunis A, Tulkens PM, Struelens MJ, et al. Intracellular activity of antibiotics in a model of human THP-1 macrophages infected by a *Staphylococcus aureus* small-colony variant strain isolated from a cystic fibrosis patient: pharmacodynamic evaluation and comparison with isogenic normal-phenotype and revertant strains. *Antimicrob Agents Chemother* 2009;**53**:1434–42.
178. Fernandez R, Paz LI, Rosato RR, Rosato AE. Ceftaroline is active against heteroresistant methicillin-resistant *Staphylococcus aureus* clinical strains despite associated mutational mechanisms and intermediate levels of resistance. *Antimicrob Agents Chemother* 2014;**58**:5736–46.
179. Rosato RR, Fernandez R, Paz LI, Singh CR, Rosato AE. TCA cycle-mediated generation of ROS is a key mediator for HeR-MRSA survival under beta-lactam antibiotic exposure. *PLoS One* 2014;**9**:e99605.
180. Cuirolo A, Plata K, Rosato AE. Development of homogeneous expression of resistance in methicillin-resistant *Staphylococcus aureus* clinical strains is functionally associated with a beta-lactam-mediated SOS response. *J Antimicrob Chemother* 2009;**64**:37–45.

181. Plata KB, Riosa S, Singh CR, Rosato RR, Rosato AE. Targeting of PBP1 by beta-lactams determines recA/SOS response activation in heterogeneous MRSA clinical strains. *PLoS One* 2013;**8**:e61083.
182. Baharoglu Z, Mazel D. SOS, the formidable strategy of bacteria against aggressions. *FEMS Microbiol Rev* 2014;**38**:1126–45.
183. Zgur-Bertok D. DNA damage repair and bacterial pathogens. *PLoS Pathog* 2013;**9**:e1003711.
184. Harrison EM, Paterson GK, Holden MT, Morgan FJ, Larsen AR, Petersen A, et al. *Staphylococcus xylosus* isolate with a new mecC allotype. *Antimicrob Agents Chemother* 2013;**57**:1524–8.
185. Long KS, Poehlsgaard J, Kehrenberg C, Schwarz S, Vester B. The Cfr rRNA methyltransferase confers resistance to phenicols, lincosamides, oxazolidinones, pleuromutilins, and streptogramin A antibiotics. *Antimicrob Agents Chemother* 2006;**50**:2500–5.
186. Mendes RE, Deshpande LM, Castanheira M, DiPersio J, Saubolle MA, Jones RN. First report of cfr-mediated resistance to linezolid in human staphylococcal clinical isolates recovered in the United States. *Antimicrob Agents Chemother* 2008;**52**:2244–6.
187. Moosavian M, Shoja S, Rostami S, Torabipour M, Farshadzadeh Z. Inducible clindamycin resistance in clinical isolates of *Staphylococcus aureus* due to erm genes, Iran. *Iran J Microbiol* 2014;**6**:421–7.
188. Wipf JR, Schwendener S, Nielsen JB, Westh H, Perreten V. The new macrolide–lincosamide–streptogramin B resistance gene erm(45) is located within a Genomic Island in *Staphylococcus fleurettii*. *Antimicrob Agents Chemother* 2015;**59**:3578–81.
189. Schwendener S, Perreten V. New shuttle vector-based expression system to generate polyhistidine-tagged fusion proteins in *Staphylococcus aureus* and *Escherichia coli*. *Appl Environ Microbiol* 2015; **81**:3243–54.
190. Kehrenberg C, Schwarz S. Distribution of florfenicol resistance genes fexA and cfr among chloramphenicol-resistant *Staphylococcus* isolates. *Antimicrob Agents Chemother* 2006;**50**:1156–63.
191. Lozano C, Aspiroz C, Gomez-Sanz E, Tirado G, Fortuno B, Zarazaga M, et al. Characterization of methicillin- and linezolid-resistant *Staphylococcus epidermidis* and *S. haemolyticus* strains in a Spanish hospital. *Enferm Infecc Microbiol Clin* 2013;**31**:136–41.
192. Alonso M, Marin M, Iglesias C, Cercenado E, Bouza E, de Viedma DG. Rapid identification of linezolid resistance in *Enterococcus* spp. based on high-resolution melting analysis. *J Microbiol Methods* 2014;**98**:41–3.
193. Luna VA, Coates P, Eady EA, Cove JH, Nguyen TT, Roberts MC. A variety of gram-positive bacteria carry mobile mef genes. *J Antimicrob Chemother* 1999;**44**:19–25.
194. Rodriguez-Avial I, Rodriguez-Avial C, Culebras E, Benitez A, Picazo JJ. Distribution of mef(A) and erm(B) genes in macrolide-resistant blood isolates of viridans group streptococci. *J Antimicrob Chemother* 2001;**47**:727–8.
195. Opperman TJ, Williams JD, Houseweart C, Panchal RG, Bavari S, Peet NP, et al. Efflux-mediated bis-indole resistance in *Staphylococcus aureus* reveals differential substrate specificities for MepA and MepR. *Bioorg Med Chem* 2010;**18**:2123–30.
196. Kaatz GW, DeMarco CE, Seo SM. MepR, a repressor of the *Staphylococcus aureus* MATE family multidrug efflux pump MepA, is a substrate-responsive regulatory protein. *Antimicrob Agents Chemother* 2006;**50**:1276–81.
197. Birukou I, Seo SM, Schindler BD, Kaatz GW, Brennan RG. Structural mechanism of transcription regulation of the *Staphylococcus aureus* multidrug efflux operon mepRA by the MarR family repressor MepR. *Nucleic Acids Res* 2014;**42**:2774–88.
198. Truong-Bolduc QC, Hooper DC. The transcriptional regulators NorG and MgrA modulate resistance to both quinolones and beta-lactams in *Staphylococcus aureus*. *J Bacteriol* 2007;**189**:2996–3005.

CHAPTER

TRENDS IN ANTIMICROBIAL RESISTANCE AMONG ENTERIC PATHOGENS: A GLOBAL CONCERN

4

M. Ballal

Enteric Diseases Division, Kasturba Medical College, Manipal, Karnataka, India

INTRODUCTION

The harmful effects of infectious diseases in many developing countries is considerable, and within these countries, economically disadvantaged persons are most likely to contract communicable diseases and least likely to have access to appropriate treatment. Many bacterial and parasitic diseases could be treated with inexpensive antimicrobial agents until the emergence and spread of resistant organisms.[1] Treatment has recently been made more expensive and less successful.[1] Antimicrobial resistance (AMR) is a growing public health threat that is of broad concern to all countries and multiple sectors of society.[2] Governments around the world are also increasingly paying attention to this problem, which has become a serious obstruction to the achievements of modern-day medicine.[2] World health leaders have projected antibiotic-resistant organisms as "nightmare bacteria" that "pose a catastrophic threat" to people in every country in today's world.[3]

Each year in the United States, at least 2 million people acquire serious infections with bacteria that are resistant to one or more of the antibiotics designed to treat those infections. At least 23,000 people die each year as a direct result of these antibiotic-resistant infections.[3] Many more die from other conditions that were complicated by an antibiotic-resistant infection.[3] Drug resistance is a large and growing problem in infections including malaria, tuberculosis (TB), HIV infections, and respiratory and diarrheal diseases.[1] Very high rates of resistance have been observed in bacteria that cause common health-care-associated and community-acquired infections, such as urinary tract infections and pneumonia, in all World Health Organization (WHO) regions.[2]

Multidrug-resistant (MDR), Gram-negative bacilli are becoming a major threat in hospital settings.[4] The ESKAPE pathogens (ie, *Enterococcus faecium*, *Staphylococcus aureus*, *Klebsiella pneumoniae*, *Acinetobacter* spp., *Pseudomonas aeruginosa*, and *Enterobacter* spp.) are considered to be the most emerging threats in AMR. Of these, the majority are Gram-negative bacilli.[2] *P. aeruginosa* and *Acinetobacter baumannii* are the two most important clinical pathogens in worldwide health-care settings associated with significant mortality, especially in immunocompromised hosts.[4]

The high proportions of resistance to third-generation cephalosporins have been reported for *Escherichia coli* and *K. pneumoniae*. Carbapenems are the antibacterials that are more expensive

Antibiotic Resistance. DOI: http://dx.doi.org/10.1016/B978-0-12-803642-6.00004-6

and not available in resource-restricted settings, causing the bacteria to further accelerate the development of resistance.[4] *K. pneumoniae* resistant to carbapenems reported from various countries is found to be 54%. Resistance to fluoroquinolones among *E. coli* has been reported. In addition, the resistance to fluoroquinolones has been observed worldwide among two of the major causes for bacterial diarrhea being nontyphoidal *Salmonella* (NTS) and *Shigella.* Their resistance pattern was comparatively lower than in *E. coli.* Higher resistance in the case of NTS is a great concern, as it can lead to the worst patient outcomes.[2]

Resistance equally compromises the management of acute respiratory infections, sexually transmitted diseases, and diseases spread via the fecal–oral route like typhoid, cholera, dysentery, and other diarrheal diseases.[5]

KEY PROBLEMS OF RESISTANCE IN HOSPITALS AND COMMUNITIES

Antibiotic-resistant infections add considerable and avoidable costs to the already overburdened health-care system. In most cases, antibiotic-resistant infections require prolonged and costlier treatments, extend hospital stays, necessitate additional doctor visits and health-care use, and result in greater disability and death compared with infections that are easily treatable with antibiotics. The total cost of antibiotic resistance to the world economy has been difficult to calculate. Estimates vary but have ranged as high as $20 billion in excess direct health-care costs, with additional costs to society (in lost productivity and other negative impacts) as high as $35 billion a year.[6,7]

The use of antibiotics is the single most important factor leading to antibiotic resistance around the globe.[6] Antibiotics are among the most commonly prescribed drugs used in human medicine. However, up to 50% of all the antibiotics prescribed for people are not needed or are not optimally effective as prescribed.[6,7] Antibiotics are also commonly used in food animals to prevent, control, and treat disease and to promote the growth of food-producing animals.[8] The use of antibiotics for promoting growth is not necessary, and this practice should be ended.[8,9] It is difficult to directly compare the amount of drugs used in food animals with the amount used in humans, but there is evidence that more antibiotics are used in food production.[8,9]

The other major factor in the growth of antibiotic resistance is the spreading of resistant strains of bacteria from person to person or from nonhuman sources in the environment like food to persons.[3] There are four core actions that will help fight these deadly infections:

- Preventing infections and preventing the spread of resistance
- Tracking resistant bacteria
- Improving the use of today's antibiotics
- Promoting the development of new antibiotics and developing new diagnostic tests for resistant bacteria[3]

Bacteria will inevitably find ways of resisting the antibiotics we develop, which is why aggressive action is needed now to keep new resistance from developing and to prevent the resistance that already exists from spreading.[3] Antibiotic resistance in health-care settings is a significant threat to public health. In addition, many times, patients in medical settings such as hospitals and long-term care facilities are already vulnerable due to weak immune systems and underlying illnesses.[10]

For these patients, contracting an antibiotic-resistant infection is a dangerous threat. By preventing antibiotic resistance in health-care settings, patients' lives are better protected and their health can be safeguarded. In addition, health-care facilities, systems, insurers, and patients can save money that otherwise would have been spent on more complex issues and medications needed to manage antibiotic-resistant infections.[10]

The Centers for Disease Control and Prevention (CDC) work to prevent antibiotic resistance in health-care settings by providing a system to track resistance and prescribing patterns at the national, regional, and local levels; providing guidance to health-care facilities interested in better antibiotic use; and working to prevent all patient infections through infection control guidelines, assistance implementing these guidelines, and laboratory expertise. Next is an example of how the CDC is working to prevent antibiotic resistance in health-care settings.[3]

TRACKING

The CDC's National Healthcare Safety Network (NHSN) is used by health-care facilities to electronically report infections, antibiotic use, and antibiotic resistance. Data currently submitted by hospitals to NHSN allow facilities, states, and regions the ability to track and benchmark antibiotic resistance in bacteria responsible for many health-care-associated infections. As more hospitals submit data to the new NHSN Antibiotic Use and Resistance Module, they will be able to track and benchmark antibiotic resistance in all bacteria, as well as track antibiotic usage. This information will allow facilities to target areas of concern, to make needed improvements, and to track the success of their efforts. In addition, NHSN allows the CDC to perform and report national assessments of antibiotic resistance.[3,6]

Antibiotic-resistant infections outside the hospital setting were rare until recently. The CDC works to prevent antibiotic resistance in the community by providing systems to track infections and changes in resistance; improving prescribing at national, regional, and local levels; and limiting or interrupting the spread of infections.[3] Here are some examples of the strategies the CDC uses to prevent antibiotic resistance in communities.

WHAT CAUSES DRUG RESISTANCE?

The resistance problem can be seen as an equation with two main components: the antibiotic or antimicrobial drug, which inhibits susceptible organisms and selects the resistant ones; and the genetic resistance determinant in microorganisms selected by the antimicrobial drug. Drug resistance emerges only when the two components come together in an environment or host, which can lead to a clinical problem.[4] Selected resistance genes and their hosts spread and propagate under continued antimicrobial selection to amplify and extend the problem to other hosts and other geographic locations. There are more than 15 classes of antibiotics whose targets are involved in essential physiological or metabolic functions of the bacterial cell (Table 4.1). None has escaped a resistance mechanism. Millions of kilograms of antimicrobials are used each year in the prophylaxis and treatment of people, animals, and agriculture globally, driving the resistance problem by killing susceptible strains and selecting those that are resistant.[4]

But how do bacteria acquire resistance? Drug resistance is mobile—the genes for resistance traits can be transferred among bacteria of different taxonomic and ecological groups by means of mobile

Table 4.1 Major Antibiotic Families and Their Mechanisms of Action[3]

Mechanism of Action	Antibiotic Families
Inhibition of cell wall synthesis	Penicillins, cephalosporins, carbapenems, daptomycin, monobactams, glycopeptides
Inhibition of protein synthesis	Tetracyclines, aminoglycosides, oxazolidonones, streptogramins, ketolides, macrolides, lincosamides
Inhibition of DNA synthesis	Fluoroquinolones
Competitive inhibition of folic acid synthesis	Sulfonamides, trimethoprim
Inhibition of RNA synthesis	Rifampin Other metronidazoles

Data from CDC. Antibiotic Resistance Threats in the United States, 2013. *US Department of Health & Human Services, Centers for Disease Control and Prevention; 2013.*

genetic elements such as bacteriophages, plasmids, naked DNA, or transposons. These genes are generally directed against a single family or type of antibiotic, although multiple genes (each bearing a single drug-resistant trait) can accumulate in the same organism, and like the antibiotics themselves, resistance mechanisms vary. In the absence of plasmids and transposons (which generally mediate high-level resistance), a stepwise progression from low-level to high-level resistance occurs in bacteria through sequential mutations in chromosomes. Strains of *E. coli* and other Enterobacteriaceae have evolved increasing resistance to fluoroquinolones, the result of mutations in the target enzymes and an increase in the expression of membrane proteins that pump the drugs out of the cell.[10,11]

COSTS OF RESISTANCE

Diarrheal illness rarely requires antibiotic treatment, and this illness can be stopped by hygienic living environments. However, antimicrobials, which are widely available over the counter and through other unregulated outlets in a number of developing countries, remains the backbone of empirical antibiotic therapy. This extensively automatic use of antibiotics damages the potency of the cost-effective and obtainable antimicrobials. This can cause a major problem when antibiotic treatment is needed.[4] Enteric pathogens generally acquire resistance when exposed to resistant normal flora that colonize intestinal tracts of humans and animals or there is possibility that resistance is acquired from other environmental sources before infecting the host. Several studies from developing countries show an alarming swing in multiple resistance among the prime enteric pathogens, such as *E. coli*, *Klebsiella* spp., *Salmonella* spp., *Vibrio cholerae*, and *Shigella* spp., to nearly all generally available antibiotics. Hence, it is essential that this trend should be altered.[7] Antibiotic-resistant infections double the duration of hospital stays, double the mortality rate, make treatments less effective, make resistant infections more severe, probably double morbidity, and presumably increase the costs compared to drug-susceptible infections.

Salmonella Typhi causes approximately 21.7 million illnesses worldwide. In the United States, it causes approximately 5700 illnesses and 620 hospitalizations each year. Most illnesses occur in people who travel to some parts of the developing world where the disease is common. Travel-associated infections are more likely to be antibiotic-resistant. The CDC has observed some level of resistance to ciprofloxacin in two-thirds of *Salmonella* Typhi samples tested. NTS causes approximately 1.2 million illnesses, 23,000 hospitalizations, and 450 deaths each year in the United States. The CDC has identified resistance to ceftriaxone in about 3% of NTS tested and some level of resistance to ciprofloxacin in about 3%. About 5% of NTS tested by the CDC are resistant to five or more types of drugs. *Shigella* causes approximately 500,000 diarrheal illnesses, 5500 hospitalizations, and 40 deaths each year in the United States. The CDC has observed resistance to ciprofloxacin in 1.6% of the *Shigella* cases tested and resistance to azithromycin in approximately 3%. *Campylobacter* is estimated to cause approximately 1.3 million infections, 13,000 hospitalizations and 120 deaths each year in the United States. The CDC has noted that resistance to ciprofloxacin in almost 25% of *Campylobacter* tested and resistance to azithromycin in about 2%. Each year, *V. cholerae* causes an estimated 3–5 million cases of illness and over 100,000 deaths around the world. Resistance to tetracycline and other antimicrobial agents in *V. cholerae* have been demonstrated in both endemic and epidemic cholera settings. A likely risk factor for AMR is the widespread use of antibiotics, including mass distribution for prophylaxis in asymptomatic individuals.[3,7]

RESISTANT BACTERIA ACCUMULATE MULTIPLE RESISTANCE DETERMINANTS

Long-term use of a single antibiotic (ie, for more than 10 days) will select for bacteria that are resistant not only to that antibiotic, but to several others. Under continued antimicrobial selection, the susceptible intestinal flora, skin flora, or both may become colonized by organisms that are resistant not only to the ingested drug, but also to other structurally unrelated drugs. In animals, MDR emerged after the application of subtherapeutic levels of tetracyclines in feed. Within days, chickens began excreting tetracycline-resistant *E. coli*; by 2 weeks, the excreted *E. coli* were resistant to several antibiotics. This phenomenon reflects the linkage of different resistance genes on the same transposon or plasmid. Bacteria that are already resistant to one growth inhibitory agent seem to be favored in acquiring additional resistance traits from other bacteria sharing the environment: it was from the doubly resistant (penicillin and tetracycline) strains of *Neisseria gonorrhoeae* that the new fluoroquinolone-resistant strains emerged.[8,12]

Loss of resistance is slow. Resistant bacteria may rapidly appear in the host or environment after antibiotic use, but they are slow to be lost, even in the absence of the selecting antibiotic. This phenomenon reflects the minimal survival cost to the emerging resistant strains.[10,11] In addition, resistance genes are often linked with genes specifying resistance to other antimicrobials or toxic substances on the same plasmids. The presence of MDR plasmids ensures maintenance of the plasmids, so long as any of the resistances provides a survival advantage to the host bacterium.[1,12]

GENETICS AND SPREAD OF DRUG RESISTANCE

Drug-resistance genes can be spread from one bacterium to another through various mechanisms, such as plasmids, bacteriophages, naked DNA, or transposons. Some transposons contain integrons—more complex transposons that contain a site for integrating different antibiotic-resistance genes and other gene cassettes in tandem for expression from a single promoter.[9,10] Originally discovered among Gram-negative bacteria, integrons have since been located in Gram-positive commensal flora—a newly found reservoir of these unique genetic elements. Chromosomal genes can be also transferred: they are acquired by one bacterium through the uptake of naked DNA released from another microorganism. This transfer process is called *transformation*.[9,10] Bacteria themselves are mobile and can easily travel from person to person and between countries. Thus, countries and citizens worldwide have become part of a global microbial ecology, sharing and spreading the consequences of AMR.[1,4,5]

ECOLOGY OF ANTIBIOTIC RESISTANCE

The impact of the drug selection process can be largely confined to the individual taking the antibiotic if widespread antibiotic usage is absent.[10] After therapy, the selected resistant commensal strains will eventually be diluted, and their growth will be suppressed by the return of natural drug-susceptible competitors. If, however, whole populations are being treated with the same class of antibiotic, susceptible strains will have little opportunity to recolonize their niche and resistant strains will acquire an important advantage. The resulting ecological imbalance produces a potentially serious environmental pool of resistance genes. Ecologically, it is the density of antibiotic usage that enhances resistance selection and its effects.[11] The selection density involves the total amount of antibiotic being applied to a geographically defined number of individuals in a setting, whether it is the home, daycare center, hospital, or farm. Each individual becomes a "factory" of resistant bacteria that enter the environment. The disparity between resistance rates in the local community and those in city hospitals reflects differential ecological effects of antibiotic use.[12] The result of the selective pressure will reflect the number of individuals who are contributing resistant bacteria to that environment and the residual number of surviving, susceptible bacteria. The ecological effects of antibiotics make them unique therapeutic agents. High numbers of MDR bacteria were found in the intestinal flora of ambulatory individuals in the Boston area, even though none of the study participants had recently taken an antibiotic.[12,13] In Nepal, resistance rates in individuals were found to correlate more with the total community use of antibiotics than with the individual's own use. In addition, the selection of resistance continues because antimicrobials persist, largely intact, in natural environments. Antimicrobials in wastewater are being reported with increasing frequency and are potentially important contributors to the environmental selection of antibiotic-resistant organisms.[13] The findings suggest that one approach to the antibiotic resistance problem could be to design drugs that self-destruct after treatment, thereby removing a contributing factor in the propagation of resistance.[1,11]

BIOLOGICAL MECHANISMS OF RESISTANCE

Resistance mechanisms vary. Some are directed at the antibiotic itself: for example, enzymes such as β-lactamases destroy penicillins and cephalosporins. The modifying enzymes inactivate chloramphenicol and aminoglycosides such as streptomycin and gentamicin. Others target how the drug is transported; for example, an active efflux of drug mediates resistance to tetracyclines, chloramphenicol, and fluoroquinolones. A third type of mechanism alters the intracellular target of the drug (eg, the ribosome, metabolic enzymes, or proteins involved in DNA replication or cell wall synthesis), making the drug unable to inhibit a vital function in the microbial cell. The same kind of drug-resistance mechanism can be specified by many different genes. For example, the β-lactamases now number in the hundreds, and more than 20 different resistance determinants mediate an efflux of tetracyclines. In addition, more than one type of mechanism may provide resistance to the same antibiotic; for example, tetracycline resistance can be effected by either efflux or ribosome protection. Although most fluoroquinolone resistance stems from chromosomal mutations in the gyrase target or from drug efflux, a plasmid-mediated resistance to fluoroquinolones was recently described. MDR can be specified by chromosomal genes for regulatory proteins such as MarA and SoxS. These proteins promote drug resistance by controlling the expression of other chromosomal genes such as those involved in drug efflux.[3–5,7,11]

USE OF ANTIBIOTICS IN FOOD ANIMALS AND AGRICULTURE

There has been considerable debate concerning the relationship between antimicrobial use in animals and the problem of resistance in people. The chronic use of subtherapeutic amounts of antibiotics for growth promotion in food animals has been banned in the European Union, but it continues in the United States. Despite their low-level application, the antibiotics select determinants mediating high-level, clinically relevant resistance. Enteric organisms such as *Salmonella*, *Campylobacter*, *Listeria*, *Enterococci*, and some strains of *E. coli* are propagated primarily among animals and subsequently infect people. The transfer may occur through the food chain or through animal handlers. If the organisms are MDR, the emergence of their resistance results principally from the use and overuse of antibiotics in the animals. Overall, animal contributions to the resistance problem in human infections are small but not insignificant; they have a major role if enteric organisms are involved.[8,9]

HOW CAN WE MANAGE AND PREVENT DRUG RESISTANCE?

Every year, millions of people become sick from foodborne and other enteric infections.[2] These infections are mild and actually do not require treatment; however, antibiotics can be a savior in severe infections. The prevention of AMR in case of enteric infections requires a multifaceted approach and cooperation because bacteria that cause infections such as salmonellosis and campylobacteriosis have animal reservoirs, while other bacteria, such as those that cause shigellosis and typhoid fever, have human reservoirs. To prevent antibiotic-resistant foodborne infections, the CDC works closely with state and local health departments; with the US Food and Drug Administration (FDA), which regulates antibiotics, many foods, animal feed, and other products; and with the US Department of Agriculture (USDA), which regulates meat, poultry and egg products. In 1996, the National Antimicrobial Resistance Monitoring System (NARMS) was established as a collaboration among the CDC, FDA, USDA, and state and local public health departments. This national public health surveillance system tracks antibiotic resistance among *Salmonella*, *Campylobacter*, and other bacteria transmitted commonly through food. NARMS tests bacteria from humans (CDC) retail meats (FDA), and food-producing animals (USDA) in the United States. The primary objectives of the NARMS program are to do all of the following[1]:

- Monitor trends in antibiotic resistance among enteric bacteria from humans, retail meats, and food-producing animals
- Disseminate information on antibiotic resistance to promote interventions that reduce antibiotic resistance among foodborne bacteria
- Conduct research to better understand the emergence, persistence, and spread of antibiotic resistance
- Provide data that assist the FDA in making decisions about approving safe and effective antibiotic drugs for animals

The CDC reference laboratory conducts antibiotic susceptibility testing on isolates from sporadic cases and outbreaks of illness. The laboratory also confirms and studies bacteria that have new antibiotic resistance patterns. NARMS provides information about patterns of emerging resistance among enteric pathogens to stakeholders, including federal regulatory agencies, policymakers, consumer advocacy groups, industry, and the public to guide public health prevention and policy efforts that protect people from resistant infections.[1–3]

IMPROVING ANTIBIOTIC USE

Antibiotics are widely used in food-producing animals, and according to data published by the FDA, more kilograms of antibiotics were sold in the United States for food-producing animals than for people. This use contributes to the emergence of antibiotic-resistant bacteria in food-producing animals. Resistant bacteria in food-producing animals are of particular concern because these animals serve as carriers.[8]

Resistant bacteria can contaminate the foods that come from those animals, and people who consume these foods can develop antibiotic-resistant infections. Antibiotics must be used judiciously in humans and animals because both uses contribute to not only the emergence, but also the persistence and spread of antibiotic-resistant bacteria.[8,9]

Scientists around the world have provided strong evidence that antibiotic use in food-producing animals can harm public health through the following sequence of events[1]:

- The use of antibiotics in food-producing animals allows antibiotic-resistant bacteria to thrive, while susceptible bacteria are suppressed or die.
- Resistant bacteria can be transmitted from food-producing animals to humans through the food supply.
- Resistant bacteria can cause infections in humans.
- Infections caused by resistant bacteria can result in adverse health consequences for humans.

Because of the link between antibiotic use in food-producing animals and the occurrence of antibiotic-resistant infections in humans, antibiotics should be used in food-producing animals only under veterinary oversight and only to manage and treat infectious diseases, not to promote growth. The CDC encourages and supports efforts to minimize the inappropriate use of antibiotics in humans and animals, including the FDA's strategy to promote the judicious use of antibiotics that are important in treating humans. The CDC supports the FDA's plan to implement draft guidance in 2013 that will operationalize this strategy. The CDC has also contributed to a training curriculum for veterinarians on prudent antibiotic use in animals.[1]

PREVENTING INFECTIONS

Efforts to prevent foodborne and other enteric infections help to reduce both antibiotic-resistant infections and antibiotic-susceptible infections. CDC activities that help prevent these infections include:

- Estimating how much foodborne illness occurs
- Monitoring trends in foodborne infections
- Investigating outbreaks and sporadic cases of foodborne illness to stop outbreaks and improve prevention
- Attributing illnesses to specific foods and settings
- Tracking and responding to changes in resistance
- Determining the sources of antibiotic-resistant enteric infections
- Educating consumers and food workers about safe food-handling practices
- Identifying and educating groups at high risk for infection
- Promoting proper handwashing
- Strengthening the capacity of state and local health departments to detect, respond to, and report foodborne infections
- Developing better diagnostic tools to rapidly and accurately find sources of contamination
- Providing recommendations for travelers on safe food and clean water[1,13,14]

Tracking the resistance frequency: Local, national, and global surveillance systems of drug susceptibility would help to communicate the current status of resistance in a location, facilitating more appropriate choices of treatment. Such surveillance would alert public health officials to new pathogens and would spur the implementation of control policies.[7]

In this regard, the Alliance for the Prudent Use of Antibiotics has established its Global Advisory on Antibiotic Resistance Data project to synthesize, evaluate, and report on the surveillance data from five large global surveillance systems. Commensal organisms are common reservoirs of antibiotic resistance plasmids, transposons, and genes. *E. coli* and the *Enterococci* of the gut serve as reservoirs from which several antibiotic resistance genes can spread. This concept has been recently sponsored by an Alliance for the Prudent Use of Antibiotics-based Reservoirs of Antibiotic Resistance project that supports studies examining the link between resistance in commensal flora and resistance in clinical isolates.[3,4]

Introduction to new therapeutic approaches: Confronted with a shortage of new antimicrobials, we must use our current drugs more prudently. Reducing and improving use can diminish resistance and permit a drug to resurface eventually as an effective therapy. The appropriate use of the antibiotics not only can help to reverse high resistance frequencies, but also can curb the appearance of resistance to newer agents. Decreasing antibiotic usage in intensive care and other hospital units has shown that susceptible indigenous strains will repopulate the ecological niche in the absence of drug-selective pressure.[2,3]

DRUG RESISTANCE IN ENTERIC PATHOGENS

Microbes collectively include bacteria, viruses, fungi, and parasites. For the past 70 years, antimicrobial drugs such as antibiotics have been successfully used to treat patients with bacterial and infectious diseases. Over time, many infectious microorganisms have adapted to the drugs designed to kill them, making the products less effective.[2] Antimicrobial drug resistance is a large and growing problem among organisms that cause diarrheal disease. Although most diarrheal diseases are self-resolving and should not be treated with antimicrobial agents, invasive or protracted infections require chemotherapy and are typically managed empirically. Although oral rehydration therapy has drastically reduced deaths from the disease, prolonged infectious bouts of diarrhea have long-term consequences for physical and cognitive development. Very few reports have examined the epidemiology of diarrheal pathogens, and even fewer have looked at drug resistance. Infections from resistant bacteria are too common nowadays, and pathogens becoming resistant to multiple types or clones of antibiotics is reported from different parts of the world. Control of fecally-orally transmitted pathogens is inadequate in many developing countries, in particular, in sub-Saharan Africa. Acquired resistance to antimicrobial drugs is becoming more prevalent among *V. cholerae*, *Salmonella* Enteritidis, diarrheagenic *E. coli*, and other pathogens in this region. The poor, who experience most of the infections caused by these organisms, bear the brunt of extended illness and exacerbated proportion of deaths brought about by resistance. Improved antimicrobial drug stewardship is an often cited, but inadequately implemented, intervention for resistance control. Resistance containment also requires improvements in infectious disease control, access to and quality assurance of antimicrobial agents, and diagnostic facilities. Structural improvements along these lines will also enhance disease prevention and control, as well as rational antimicrobial drug use. In addition, more research is needed to identify low-cost, high-impact interventions for resistance control.[15,16]

The adverse effects of infectious diseases in many developing countries is considerable, and within those countries, economically disadvantaged persons are most likely to contract communicable diseases and least likely to access appropriate treatment. Until recently, many bacterial and parasitic diseases could be treated with inexpensive antimicrobial agents, but treatment has recently been made more expensive and less successful by the emergence and spread of resistant organisms. Drug resistance is a large and growing problem in infections that account for most of Africa's disease burden, including malaria, TB, HIV infection, and respiratory and diarrheal diseases. Much of the current discourse on infectious disease and drug resistance as it affects sub-Saharan Africa is limited to the pressing problems associated with HIV, TB, and malaria. Moreover, young children are especially likely to acquire resistant enteric infections, from which they can experience less obvious but long-term adverse effects.[17–19]

AMR IN ENTERIC PATHOGENS IN DEVELOPING COUNTRIES

Bacterial enteric infections exact a heavy toll on human populations, particularly among children and immunosuppressed individuals in developing countries where malnutrition, HIV/AIDS, and poor sanitation abound. Despite the explosion of knowledge on the pathogenesis of enteric diseases during the 2000-2015 years, the number of diarrheal episodes and human deaths reported (especially among poor populations in developing countries) remain of apocalyptic dimensions. With several studies from developing countries showing worrying trends in multiple resistance among key enteric pathogens such as *E. coli*, *Klebsiella*, *Salmonella* spp., *V. cholerae*, and *Shigella* spp. to nearly all commonly available antibiotics, it is imperative that this trend should be reversed.[16] Many countries in the developing world lack a formal surveillance system for antibiotic resistance and often treatment is given empirically based on clinical diagnosis alone. Availability of antibiotics over the counter without prescription has not helped either. To minimize the negative effect on public health, a concerted effort is required in surveillance, public health education and awareness of dangers of resistance and a clear policy on procurement and prudent use of antibiotics in these resource-poor settings.[10,11,17]

COMMON ENTERIC BACTERIA FOR WHICH MDR IS DOCUMENTED

Diarrheal illness rarely requires antimicrobial treatment and can be prevented by improving living conditions; yet antimicrobials, which are widely available over the counter and through other unregulated outlets in a number of developing countries, remain a mainstay of empirical therapy.[14] This widespread injudicious use diminishes the efficacy of affordable and available drugs, which poses a serious problem when antimicrobial treatment is needed. Overuse of antimicrobials also wastes limited financial resources on ineffective medications that expose patients unnecessarily to potential toxicities may prolong illness and can increase risk of death.[17] Enteric pathogens frequently acquire resistance when they are exposed to resistant normal flora that colonize intestinal tracts of humans and animals or may have acquired resistance in other environments before infecting host species. Understanding the prevalent antimicrobial susceptibility patterns could inform guidelines for empirical therapy and also minimize injudicious antimicrobial use. Although there is a problem of antibiotic resistance in enteric pathogens, it is not uniformly bad. Some enteric pathogens remain predictably sensitive to a particular agent and for others problems are developing but at a low prevalence; for a notorious few, however there is certainly the specter of untreatable infections. We concentrate on enteric commensals, enteric pathogens and those pathogens with an enteric reservoir and on problems that may be peculiar to these developing countries.[4,5,17]

KEY PROBLEMS OF RESISTANCE IN HOSPITALS AND COMMUNITIES

Multiply resistant organisms render therapy more precarious and costly and sometimes unsuccessful. Individuals may succumb to MDR infections because all available drugs have failed especially in the developing world. Notable global examples include hospital and community MDR strains of *Mycobacterium tuberculosis*, *E. faecium*, *Enterobacter cloacae*, *K. pneumoniae*, *S. aureus*, *A. baumannii*, and *P. aeruginosa*.[4] In developing countries, MDR enteric disease agents such as *Salmonella* Enteritidis, *Shigella flexneri*, and *V. cholerae* threaten and circumvent public health measures. The commensal reservoir of resistance genes spreads through the same channels undetected, but it provides a ready source of resistance genes and elements that can be transmitted to pathogens.[19] Providing safe water and sanitation to those who cannot afford these capital-intensive options and to public institutions such as schools, health centers, and markets is the single most important intervention for preventing outbreaks and sporadic cases of diarrheal disease, including those caused by resistant organisms.[17] Malnourished or otherwise immunocompromised patients are more likely to have inadequate economic resources, and they become the target of resistant pathogens when these organisms are prevalent. Patients infected with resistant strains pay more for cures, lose more from extended illness in terms of time away from work and other activities (ie, productivity and costs of supportive therapy), and are more likely to be disabled. In wealthy countries, hospitals are often the sites where resistance emerges and then it slowly, but eventually, seeps into the community. In poor countries, often no barrier exists between the hospital and the community. A patient's relatives, who must be on hand to assist overstretched health systems with care, sleep under beds and in hospital corridors. In hospitals, costly infection control measures are often compromised. The potential for organisms to be transmitted into, within, and beyond the hospital is very high.[2,20,21]

DIAGNOSTIC DEVELOPMENT

Diagnostic development represents a potentially powerful strategy to simultaneously improve healthcare delivery and contain resistance. The cause of bloodstream and enteric infections has diversified considerably in recent years, in part due to the definition of previously unrecognized etiologic agents, but also due to the spread of HIV and the emergence of new pathogens. This increased diversity makes syndromic diagnosis of many conditions less accurate, particularly in areas where surveillance does not occur. Better systems are needed to provide laboratory support for serious cases, outbreaks, and routine surveillance. As disease control efforts begin to yield fruit, syndromic diagnosis will become increasingly inaccurate and laboratory diagnosis even more essential.[3,22]

ENTERIC PATHOGENS AND THEIR RESISTANCE

Vibrio cholerae

Cholera toxin (CT)–producing *V. cholerae* cause the characteristic life-threatening gastroenteritis of cholera. At least seven pandemics of the disease, originally designated "Asiatic cholera," have occurred in recent history. The ongoing pandemic has seen the emergence of O139 *V. cholerae* as a non-O1 pandemic strain, and importantly, the emergence and spread of drug-resistant O1 strains. The current focus of the cholera pandemic is Africa, which has seen two-thirds of all cholera outbreaks in the 2005-2015 years. The primary treatment for cholera is rehydration. Most patients will overcome the infection if they are rehydrated promptly and properly, even if they do not receive antimicrobial drugs.[23] Antimicrobial drugs, however, shorten the course of infection and prevent person-to-person transmission, which may be crucial for slowing outbreaks because organisms from infected persons may be more virulent than those acquired in the wild. Antimicrobial agents may also be life-saving for malnourished and other immunocompromised patients who have cholera.[13]

Tetracycline was the empiric drug of choice for cholera in Africa and elsewhere for many years. At the end of the 1970s, however, incompatibility group C tetracycline-resistant plasmids were isolated from *V. cholerae* isolates in Tanzania, Kenya, and other parts of Africa. In each case, resistance emerged during an ongoing epidemic where tetracycline was being used intensively for prophylaxis as well as treatment. Tetracycline has sequentially been replaced by trimethoprim-sulfamethoxazole (TMP-SMX), and more recently quinolones, because of the emergence and spread of resistant strains. Molecular evaluation of more recent resistant *V. cholerae* isolates typically found regionally conserved plasmids, some of which carried class 1 integrons bearing multiple resistance cassettes. A chromosomally integrated transferable resistance element, SXT, has also spread worldwide and has been recently reported from Africa.[13,23]

Emergence of resistance in *V. cholerae* has been linked to increased mortality rates in recent African outbreaks. Similar experiences have also been reported with *Shigella dysenteriae* type 1, another enteric pathogen that causes life-threatening disease and has epidemic potential. The impact of resistance in both pathogens is illustrated by an overwhelming outbreak in Jul. 1994 at the Goma camp, which resulted in the deaths of about 12,000 Rwandan refugees. More recently, Dalsgaard et al. observed a marked increase in the case fatality rate during the 1997–98 phase of a Guinea-Bissau cholera outbreak, compared to an overlapping 1996–97 outbreak.[23] A major feature of the latter wave of cholera was the presence of strains simultaneously resistant to ampicillin, erythromycin, tetracycline, furazolidone, aminoglycosides, trimethoprim, and sulfamethoxazole. These MDR strains were not present during the first wave and probably arose following the acquisition of a 150-kb-pair resistance plasmid bearing a class 1 integron and genes encoding resistance to all antimicrobial agents commonly used in the empiric management of cholera.[23,24]

Cholera is a widespread acute bacterial infection of the intestine caused by ingestion of food or water containing *V. cholerae*, toxigenic serogroups O1 or O139. Symptoms include acute watery diarrhea and vomiting, which can result in severe dehydration or water loss. When left untreated, death can occur rapidly—sometimes within hours. Cholera is transmitted through contaminated food or drinking water, as well as from person to person through the fecal–oral route. Sanitary conditions in the environment play an important role since the *V. cholerae* bacterium survives and multiplies outside the human body and can spread rapidly where living conditions are crowded and water sources unprotected and where there is no safe disposal of feces.[24] These conditions are common in poor countries and in crowded settings such as refugee camps. For example, a major epidemic took place in 1994 in a refugee camp in Goma, Democratic Republic of the Congo. An estimated 58,000–80,000 cases and 23,800 deaths occurred within 1 month (Goma Epidemiology Group 1995).[23]

Cholera is now considered a reemerging disease because infections are appearing in novel communities or communities where the disease had been absent for many years, and the range of areas of endemicity is expanding. To date, nearly 200 serogroups of *V. cholerae* have been recorded, of which only the O1 and O139 strains have been associated with major epidemics due to their ability to express the CT, which is rarely found in non-O1 and non-O139 serogroups. A total of seven cholera pandemics occurred: in 1817–1823, 1826–1837, 1846–1862, 1864–1875, 1883–1896, 1899–1923, and 1961 to today. The seventh pandemic that started in the year Sulawesi, Indonesia still persists to date (in the year 1992 there was an outbreak of *V. cholerae* O139, Bengal strain, but whether it was pandemic or not is still an debatable issue). All seven pandemics started in Asia, with the first six starting in the Indian subcontinent. The seventh pandemic is believed to have originated in Indonesia in the 1960s, and arrived in Africa and South America in the 1970s and in 1991, respectively. The first six pandemics were caused by the classical biotype, which had been replaced by the O1 El Tor biotype (Ogawa and Inaba serovars) by 1961. The O139 serogroup first emerged as a pandemic strain in 1992, possibly through genetic exchange with *V. cholerae* O1 El Tor or with non-O1 strains.[23]

Geographical areas once known to have experienced cholera epidemics can be grouped into three categories: cholera-free zones, endemic areas, and epidemic areas. Cholera-free communities are defined as having no locally acquired infections. In epidemic areas, the disease diminishes after an outbreak, while in endemic areas, the disease does not disappear after an epidemic peak and returns in successive waves. Where cholera is endemic, cases tend to demonstrate distinct seasonal trends. After the sixth pandemic in the 1950s, areas with true cholera endemicity had been reduced to southeastern India and Bangladesh. However, during the current seventh pandemic, the range of endemicity has expanded and includes vast areas on the African continent and South and Central America. Over the 2000-2015 years, several outbreaks of cholera have been documented from Africa and this is thought to be part of the ongoing seventh cholera pandemic.

In a study to characterize 80 strains of *V. cholerae* O1 isolated during the Kenyan epidemic in 1998 and 1999, it was demonstrated that 61 strains from 25 outbreaks belonged to ribotype B27. All the 61 strains showed identical and stable multiple resistances to chloramphenicol, streptomycin, co-trimoxazole and tetracycline and also produced the same cluster type of six randomly amplified DNA patterns. This uniformity of properties among outbreak strains from districts scattered over the entire area of the country provided strong genetic and epidemiological evidence that the predominant strains causing the epidemic had a clonal origin.[24] Identification of strains with traits typical of *V. cholerae* O1 strains active in Somalia from four outbreaks in the North Eastern Province indicated that that province was an epidemic zone where the Kenyan clone and Somali strains were overlapping and presumably competing. This epidemic has since spread into southern Sudan, Uganda, and Tanzania. Ribotype B27 was first identified in Kolkata, India, in 1993 and introduced into the western African country of Guinea-Bissau in 1994. In 1995 and 1996, ribotype B27 was identified among *V. cholerae* O1 strains, causing cholera outbreaks in Senegal. These reports suggest that this emerging ribotype has had a rapid spread into eastern Africa, with reports of sporadic outbreaks occurring in southern Sudan, parts of Somalia, western Kenya, and along the coast as recently as 2006.[25]

AMR in *V. cholerae* was first observed in Tanzania and later in Bangladesh. Resistance reflects the use or misuse of antibiotics in areas where frequent outbreaks of cholera occur or in cholera-endemic zones.[24] Resistance to one or more of the following antibiotics has been observed: tetracycline, ampicillin, kanamycin, streptomycin, sulfonamides, trimethoprim, and gentamicin.[25] Another recent surveillance study done in Angola investigated *V. cholerae* O1 and *Vibrio parahaemolyticus* clinical isolates, as well as *V. cholerae* O1 and *V. cholerae* non-O1 environmental isolates. All clinical isolates of *V. cholerae* O1 were resistant to ampicillin, chloramphenicol, trimethoprim, sulfamethoxazole, and tetracycline. These MDR isolates also contained a large conjugative plasmid (p3iANG) with a set of three class 1 integrons harboring dfrA15, blaP1, and qacH-aadA8 cassettes, which code for resistance to trimethoprim, β-lactams, quaternary ammonium compounds, and aminoglycosides, clustered in a 19-kb region. Chloramphenicol (cat1), kanamycin (aph), sulfonamide (sul2), and tetracycline (tetG) resistance genes were also carried on the plasmid within the same 19-kb region.

A chromosomal integron containing the dfrA15 cassette was also revealed in *V. parahaemolyticus* strains. This study indicates that plasmids and integrons contributed mainly to the circulation of multiple-drug resistance determinants in *Vibrio* strains from Angola.[26,27] Campos-Zahner in 2004 isolated MDR environmental *V. cholerae* from different parts of Brazil between 1991 and 1999. The study reported that clinical and nonclinical O1 strains were more resistant to commonly used antibiotics than environmental non-O1 *V. cholerae* strains.[13,23–27]

Salmonella spp.

The emergence and spread of MDR *Salmonella enterica* subsp. Typhi worldwide has had important consequences for mortality rates from typhoid fever. There are very few reports from Africa; nonetheless, available data suggest that although the problem may not be as intense as in other parts of the world, resistance has emerged and alternatives to current treatment protocols are often not available or sometimes unaffordable. MDR nontyphoidal *Salmonella* spp. have emerged as a global public health threat. In industrialized countries, they are most commonly associated with foodborne gastroenteritis.[3] In parts of sub-Saharan Africa, however, NTS are important causes of life-threatening bacteremia. Studies from Kenya have found that community-acquired NTS are among the top three causes of death among children <5 years of age. Moreover, pulsed-field gel electrophoresis data suggest that most life-threatening diseases are caused by isolates that are clonal in origin. In a recent study, children from poor slums of Kenya were significantly more likely to be infected with MDR NTS than were children from middle-income families. The patterns of resistance among these strains suggest that third-generation cephalosporins should be the drug of choice for empiric management of these infections, but in most cases, these drugs are very expensive.[28,29]

MDR NTS infections

In the research literature, there are an increasing number of reports of cases of MDR foodborne *Salmonella enterica* infections in both developed and developing countries, with few options left for antimicrobial treatment. Examples of increase in resistance to NTS in developing countries, particularly in sub-Saharan Africa, the Indian subcontinent, and Southeast Asia, are exemplified by numerous outbreaks caused by MDR NTS. These are frequently resistant to the newer quinolones, both in hospitals and the community, over wide geographical areas. Although NTS infections are a common cause of self-limiting diarrhea in healthy individuals, bacteremia with or without focal infections can occur as a complication.[16]

In Africa, NTS is among the most common cause of bloodstream infections in children younger than 5 years. In Kenya, for instance, NTS are second only to invasive pneumococcal disease in importance as a cause of bacteremia in children less than 5 years of age with high mortality, especially in malnourished children. Also, in Malawi, NTS was the most common blood culture isolate (40%) in hospitalized children, with a case fatality rate of 24%. In the adult population, bacteremia, relapses, and severe disease are a characteristic of NTS infection in the HIV-infected population.[3,23] A major characteristic feature of NTS in AIDS patients in developing countries is the relapses that occur despite appropriate antibiotic therapy. In Kenya between 1994 and 2004, the two main serotypes of NTS-causing bacteremia in children were *Salmonella* Typhimurium and *Salmonella* Enteritidis, which accounted for 70.8% of all NTS, and these remained stably and almost equally distributed over this period. Genomic characterization of these two main serotypes by pulsed-field gel electrophoresis showed that within each serotype, the strains exhibited a minimal genetic diversity and they appeared to be clonally related.[28] During this period, the NTS were multiply resistant to several commonly available antibiotics but remained fully susceptible to cefotaxime and ciprofloxacin, although the minimal inhibitory concentrations (MICs) of these antibiotics had risen twofold to sixfold compared to fully susceptible strains.

The prevalence of NTS multiply resistant to all commonly available drugs, including ampicillin, streptomycin, co-trimoxazole, chloramphenicol, and tetracycline, rose from 31% in 1993 to 42% at present, with concomitantly higher MICs of each drug, although the difference was not statistically significant ($P > 0.05$). Molecular analysis of these isolates showed that resistance has been encoded on large self-transferable 100–110-kb plasmids among all the commonly isolated NTS serotypes. Higher resistance levels in the urban population may be due to higher usage in health clinics for the treatment of diarrheal and respiratory infections, as well as the misuse of commonly available antimicrobials bought over the counter.

In contrast, studies at a rural district hospital along the coast of Kenya have revealed a remarkably decreasing rate of prevalence of resistance, especially to the most commonly available antimicrobials—amoxicillin, co-trimoxazole, gentamicin, and chloramphenicol, respectively, during the 1994–2005 period. However the trend in decreasing resistance levels in NTS has continued steadily after 2001, suggesting that changes in sampling criteria are not primarily responsible for the effect observed. Similarly, a study that sought to establish the magnitude of bacteremia in severely malnourished children in Uganda evaluated a total of 445 blood specimens that grew bacterial isolates; 58% were Gram-negative, consisting of *Salmonella* Typhimurium (27.6%) and *Salmonella* Enteritidis (11.8%). The isolates were susceptible (≥80%) to ciprofloxacin, ceftriaxone, and gentamicin, with low susceptibility to chloramphenicol, ampicillin (<50%), and co-trimoxazole (<25%).[28,29]

During the 1980s, *Salmonella* Typhimurium definitive phage type (DT) 104 with resistance to ampicillin, chloramphenicol, streptomycin, sulfonamides, and tetracycline (ACSSuT-resistance type) emerged in the United Kingdom and subsequently was disseminated throughout Europe and North America. By the 1990s, both DT 104 and non-DT 104 strains had acquired additional resistance to —TMP-SMX, ciprofloxacin, and extended-spectrum cephalosporins. This phenotype has now been widely disseminated to other regions of the world, including South America, where MDR blaCMY-2 *Salmonella* Typhimurium has become the predominant phenotype of serovar Typhimurium in Mexico and has caused severe enteric and systemic infections in children. Among food animals, the blaCMY-2 phenotype was found only in swine intestines, suggesting that the main selection pressure for MDR *Salmonella* Typhimurium originated and persists in swine production systems in Mexico.[3,28,29]

Diarrheagenic E. coli

Recent data from Gabon, Kenya, Nigeria, Senegal, and Tanzania suggest that resistance among causative organisms of these infections, such as enterotoxigenic, enteropathogenic, and enteroaggregative *E. coli*, is high and appears to be rising. Notable drug-resistant enteropathogenic *E. coli* outbreaks and sporadic cases have been reported from several African countries, including Kenya and Tanzania. The more recently defined enteroaggregative *E. coli* are typically MDR and are one of the most common causes of childhood diarrhea, particularly persistent infections. Antimicrobial drug–resistant diarrheagenic *E. coli* pathotypes, including enteroaggregative *E. coli*, are also emerging as important diarrheal pathogens in AIDS patients.[1,2]

Surveillance in healthy populations has demonstrated that commensals constitute a rich reservoir of genetic material from which pathogens can readily acquire resistance on mobile elements. A long-term study in Nigeria showed that resistance of commensal *E. coli* to almost all agents studied increased quickly over time. In addition, urban residents in Nigeria, Ghana, and Zimbabwe were more likely to carry MDR *E. coli* than were rural or provincial residents.[30] This finding has important consequences in light of the rapid rate of urbanization in these countries and other parts of the continent. Travel networks have become more efficient and are more extensively used. Two overlapping problems are worsening the situation regarding diarrheal disease: the failure to control the spread of diarrheal pathogens due to unclean water, poor sanitation, and malnutrition; and the failure to contain resistant organisms and resistant genes so that when infections occur, they produce more adverse consequences.[12,30]

MDR E. coli from clinical and environmental sources

In a multicenter study to determine the prevalence of antibiotic-resistant fecal *E. coli* from adult volunteers from urban areas in Kenya, Mexico, Peru, and the Philippines and nonurban locations in Mexico, Venezuela, Ghana, Zimbabwe, and the Philippines, the mean age of the volunteers was 35 years; most of them were females.[12] Ciprofloxacin resistance was in the range of 1–63%; the highest percentage was found in the urban populations of Asia and South America. In Peru and the Philippines, the prevalence of gentamicin resistance was >20%. Cefazolin resistance was the highest in the urban Philippines (25%). Higher prevalence of resistance to ampicillin, oxytetracycline, and trimethoprim was found for urban areas compared with nonurban ones of Asia, Africa, and South America, respectively ($P < 0.05$).[3] Antibiotic resistance in fecal *E. coli* from these adult volunteers was emerging for cefazolin, gentamicin, and ciprofloxacin and was high for the older drugs ampicillin, oxytetracycline, trimethoprim, and chloramphenicol.[12]

Ciprofloxacin resistance prevalence showed large variations between the populations and was higher in the urban areas of Asia and South America than in nonurban areas, but was similar in urban and nonurban regions in Africa. The high prevalence of resistance in Asia and South America might arise from the food supply. In both continents, poultry in particular are intensively raised and fluoroquinolones are commonly used during production; they are mixed through the water supply of the whole flock.[16] Similarly, a study showed significant differences in the prevalence of resistance between nonurban and urban areas for ampicillin, oxytetracycline, and trimethoprim. The differences in resistance between urban and nonurban areas were thought to be due to the availability of antibiotics, as in cities, a large variety of often inexpensive antibiotics are available in pharmacies, over the counter, and at market stalls.[16] Furthermore, crowding together with poor hygiene and poor sanitary facilities for sewage disposal in cities might encourage the exchange of antibiotic-resistant bacteria in a population.[3,11]

In a study that characterized antibiotic resistance among diarrheagenic *E. coli* from children less than 5 years of age from Kenya isolates exhibited high-level MDR to WHO recommended antibiotics. Resistance rates to tetracycline, ampicillin, and co-trimoxazole were 70.7%, 65.9%, and 68.3%, respectively, figures that were very similar to resistance prevalence among *E. coli* from healthy children.[16] In another study conducted in Kenya, 74% of persons with bloody diarrhea received antibiotics to which their isolate was not susceptible again, indicating that antibiotic resistance among enteric pathogens within populations in Kenya was high and likely to jeopardize the effective treatment of infections using commonly available antibiotics.[11]

In several studies worldwide, it has been shown that *E. coli* from normal gut flora constitute an important reservoir of resistance genes. These bacteria may act as a reservoir for genes that encode MDR phenotypes, which can be transferred to potential pathogens. The rising prevalence of resistance against antibiotics such as co-trimoxazole and ampicillin could reflect the flow of MDR genes among gut-associated bacteria, which would make a formidable reservoir for antibiotic resistance genes.[16] For instance, in another study, conjugation experiments were used to demonstrate that resistance toward ampicillin, tetracycline, trimethoprim, sulfamethoxazole and chloramphenicol could be transferred among commensal microflora, suggesting that mobile genetic elements may be involved in the dissemination of MDR phenotypes.[11,12] In this respect, *E. coli* is capable of surviving in extraintestinal environments and may acquire other MDR traits from soil and water bacteria. To determine the interaction between resistance determinants in *E. coli* from different ecosystems, two comparative studies were conducted in two districts of central Kenya between 1999 and 2003. Small-scale farmers in these areas grow food crops for subsistence, and a majority keep fewer than three milking cows and goats and 10–300 chickens for commercial purposes.

In each study, the susceptibility of *E. coli* isolated from the feces of healthy children attending mother and child well-clinics in selected hospitals and from the associated environments at their homes to commonly available antibiotics was evaluated. A total of 188 nonduplicate *E. coli* strains were obtained from 256 fecal specimens of children below 5 years of age, while 286 strains were isolated from environmental specimens from the homes of index cases. The environmental strains were from chicken droppings (214; 74.8%), rectal swabs of cattle (47; 16.4%), and water sources (25; 8.7%), among which 18 strains were from boreholes and 4 were from rivers. The first study found that *E. coli* isolates from children were less sensitive to the test drugs than the environmental isolates ($P < 0.05$). Of the *E. coli* isolates from children, 164 (87.2%) were MDR, the commonest resistance being to ampicillin, chloramphenicol, co-trimoxazole, and tetracycline. In contrast, only 26% ($P < 0.05$) of *E. coli* isolates from chickens, and none from cattle or water, were MDR; resistance was usually to streptomycin or tetracycline. Resistance among *E. coli* isolates from chickens was mainly to tetracycline (72%), while the isolates were fully sensitive to most other antibiotics commonly used for treatment of patients, including ampicillin, co-trimoxazole, chloramphenicol, and gentamicin.[12,16] However, all isolates were fully susceptible to third-generation cephalosporins and quinolones. This study strongly suggests that normal intestinal flora of children was more exposed to antibiotics than in livestock in the same setting.

In a second follow-up study conducted in 2003, a total of 344 *E. coli* isolates from healthy children were analyzed for susceptibility to the panel of commonly available antimicrobials. High prevalence of MDR (88.6%) was still observed toward —TMP-SMX, ampicillin, and tetracycline, and this did not differ from the prevalence observed in the 1999–2000 study. Resistance to other antimicrobials, such as nalidixic acid and ciprofloxacin, was rare. In contrast, only 17.4% (24/138) of environmental isolates (18 of these being *E. coli* from chickens) were MDR. It is plausible that the low prevalence of antibiotic-resistant *E. coli* strains from environmental sources may reflect the narrow range of antibiotics applied in poultry rearing and other farming activities in Kenya. In this survey, 70% of the farmers interviewed indicated that they obtained antibiotics such as tetracycline, penicillins, and sulfonamides from pharmacies without prescription for treatment of mastitis in cattle and diarrhea in chickens. However, tetracycline was the most commonly used antibiotic in rearing chickens. High resistance to tetracycline among environmental isolates, therefore, can be attributed to improper use of this drug for prophylaxis in animal husbandry. Compared to previous studies in the same region in 1993, the prevalence of resistance to commonly available antimicrobials in *E. coli* from children has risen from 85.5% in 1993, to 88.6% in 2003, but the difference was not statistically significant ($P < 0.12$). In contrast, the prevalence of the MDR phenotype in *E. coli* from chickens reduced from 26% in 1993 to 17.4% in 2003 ($P < 0.05$), probably a reflection of fewer farmers using antimicrobials for growth promotion. In both studies, resistance in *E. coli* isolates was associated with the presence of 100–120-kb plasmids.[1,12,30,31]

Shigella spp.

Shigella dysenteriae type 1 was first isolated by Kiyoshi Shiga during a severe dysentery epidemic in Japan in 1896, when more than 90,000 cases were described with a mortality rate approaching 30%. Over the subsequent 50 years, the microbiology and epidemiology of *Shigella* spp. were clarified, and the mechanisms by which the microorganism causes disease have been intensively investigated. Humans are the only natural hosts for *Shigella* bacteria, and the transmission predominantly occurs by fecal–oral contact.[14] In Africa, *Shigella* infections still predominate as a cause of sporadic bloody diarrhea, particularly *S. flexneri*. In a study in a rural population in Kenya, it was observed that *Campylobacter* and *Shigella* were isolated with equal frequency from children less than 5 years old with bloody diarrhea.[3,14]

In comparable studies of semiurban Bolivian children less than 5 years old and Thai children 1–10 years of age with bloody diarrhea, *Campylobacter* was isolated at least half as frequently as *Shigella* spp. Shigellosis can occur in sporadic, epidemic, and pandemic forms. Epidemics have been reported from Central American countries, Bangladesh, Sri Lanka, Maldives, Nepal, Bhutan, Myanmar, and from the Indian subcontinent, Vellore, eastern India, and Andaman and Nicobar islands. A plasmid profile of shigellae in Kolkata has shown a correlation between the presence of smaller plasmids and shigellae serotypes, indicating epidemiological changes of the species. For a long time, antibiotics such as ampicillin, co-trimoxazole, chloramphenicol, and tetracycline have been used to treat shigellosis, but their use is increasingly compromised by the emergence of resistance. For example, high prevalence of resistance to ampicillin (82%), chloramphenicol (73%), co-trimoxazole (88%), and tetracycline (97%) was detected in *Shigella* spp. isolated from children in Tanzania. A high level of resistance to the antibiotics most commonly prescribed in Kenyan hospitals was also found.[22]

Further investigation revealed that 74% of persons with bloody diarrhea received antibiotics to which their isolate was not susceptible. Although these data were inadequate to assess the clinical impact of these findings (eg, duration of bloody diarrhea, mortality, or bacterial shedding), nonetheless, strategies to improve prescription practices that use surveillance data to rationally guide more judicious antibiotic use warrant consideration if we are to preserve the effectiveness of commonly available antibiotics.[16,31] Another surveillance study of antibiotic susceptibility in *Shigella* spp. found all isolates multiply resistant to nearly all commonly available antibiotics, including ampicillin, co-trimoxazole, co-amoxyclav, cefuroxime, chloramphenicol, and tetracycline; however, no resistance to ciprofloxacin and minimal resistance to nalidixic acid were reported. Other studies also found little resistance to nalidixic acid and no resistance to ciprofloxacin among *Shigella* spp. in other parts of East Africa. However, in areas where nalidixic acid has been introduced as the drug of choice to treat presumptive shigellosis, a marked increase in corresponding resistance has been observed.

The ease with which antibiotics can be obtained without prescription may add further to selective pressure. Thus, although nalidixic acid is an attractive choice for treating bloody diarrhea where antibiotic resistance limits other options, it should be used ideally only for illnesses most likely caused by *Shigella* or where *Shigella* infection could result in greater morbidity and increased risk of death, such as in immunosuppressed individuals. If nalidixic acid is introduced for routine treatment of bloody diarrhea, surveillance for antimicrobial susceptibility of local bacterial pathogens should be maintained.[1,13,22,31]

Campylobacter spp.

The bacteria *Campylobacter* spp. are part of the normal intestinal flora of poultry, cattle, and a number of other food-producing and domestic animals and are predominantly spread to humans in contaminated food. *Campylobacter jejuni* (and, to a lesser extent, *Campylobacter coli*) are responsible for most cases of campylobacteriosis, which is increasingly one of the most commonly detected bacterial enteric infections in poor resource settings in most developing countries.[13] The incidence of human *Campylobacter* infections has increased markedly in both developed and developing countries worldwide; and more significantly, so has the rapid emergence of antibiotic-resistant *Campylobacter* strains, with evidence suggesting that the use of antibiotics (in particular, the fluoroquinolones) as growth promoters in food animals and the veterinary industry is accelerating this trend.[14]

Although most infections are self-limiting and do not require antimicrobial chemotherapy, in a small proportion of individuals with severe or invasive disease, treatment with erythromycin or a fluoroquinolone may be required. However, the prevalence of quinolone- or macrolide-resistant *C. jejuni* and *C. coli* is increasing worldwide, and several such isolates have been reported from Africa. Other developing world countries have also reported increasing incidences of MDR *Campylobacter* infections, including resistance to fluoroquinolones.[13,32] For instance, MDR was detected in 27.5% of all *C. jejuni* isolated from a large farm in Estonia, all of which were resistant to enrofloxacin. MDR was significantly associated with enrofloxacin resistance ($P < 0.01$) and the use of enrofloxacin on the farms may have led to selection of MDR strains.[32] In a separate study in South Korea that investigated 232 retail stores for raw chicken meat, a total of 317 *Campylobacter* isolates examined showed that resistance to doxycycline was the most common (97.5%), followed by ciprofloxacin (95.9%), nalidixic acid (94.6%), tetracycline (94.6%), enrofloxacin (84.2%), and erythromycin (13.6%). A total of 93.4% showed MDR (ie, resistance to four or more antibiotics).[1,32]

Since bacterial resistance is closely linked to antibiotic use, it may be considered an inevitable antibiotic side effect. The potential of bacteria to spread or share the genetic resistance determinants in hospital wards, community and even globally makes the problem of resistance important not only for biomedical research, but also for societal studies, health administrators, and even governments.[3,13] Human use of antibiotics has undoubtedly contributed to the resistance epidemic all over the world. Resistance in some bacterial species has driven the problem to the margin of there being no available clinically effective treatment for some infections. Although reports from several studies have alerted the medical and lay society for years about the problem of resistance, it has only recently been recognized as a dangerous epidemic.

Several campaigns and projects to limit the antibiotic resistance are underway in the developed world at the moment. At the same time, of course, appropriate antibiotic use is a nonexisting issue in most developing countries where people still buy a few doses of an antibiotic on a street market or over the counter without a prescription. Although there is still a debate about the relationship between animal use of antibiotics and resistance in human pathogens, it is difficult to believe that low concentrations of antibiotics in contact with bacteria in food and in the environment are harmless in the long term.[13] From several studies, especially in Europe, it is estimated that about 20–50% of human and 40–80% agricultural antibiotic use is unnecessary or highly questionable. Studies showing an epidemiological relationship between antibiotic consumption and bacterial resistance to antibiotics for several bacterial-drug combinations provide a scientific basis for the control measures in human medicine.[32]

Clostridium difficile

Although most clinical isolates of toxigenic *Clostridium difficile* are susceptible to metronidazole and vancomycin, there have been few reports from Spain, France, and other European countries mentioning reduced susceptibility to metronidazole. According to a study conducted by Bouza et al., resistance to metronidazole is probably heterogenous.[33] Surveillance for metronidazole resistance in *C. difficile* in other populations needs to be performed to determine whether metronidazole resistance in *C. difficile* is an important or emerging clinical problem. For now, metronidazole remains an inexpensive, highly effective treatment for *C. difficile*–associated diarrhea (CDAD).[34]

CDAD is the most common etiologically defined cause of hospital-acquired diarrhea. It is a confounding complication encountered by patients hospitalized for other illnesses in extended care facilities, acute care areas, and intensive care units. It has become increasingly prevalent in the United States since 2003, when 178,000 cases were diagnosed. The emergence of a new, more virulent strain of *C. difficile* partly explains the recent increase in CDAD. This new strain, the North American pulsed-field gel electrophoresis type I (NAP I), is more virulent and can produce greater quantities of toxin A (16 times more) and toxin B (23 times more) than other strains. Binary toxin, a third toxin, is also produced by NAP I, although its significance is unknown.[33–35]

Researchers in Quebec identified the use of fluoroquinolones as the most important risk factor in the development of this new strain of *C. difficile*. Although stopping the antibiotic may be effective, the reason that the antibiotic was prescribed in the first place should be considered, and monitoring for recurrence or worsening of the underlying infection is essential.[36] Oral metronidazole (250–500 mg, four times a day for 7–14 days) is generally recommended as an initial treatment. Metronidazole also can be given intravenously.[37] Administered by either route, the same dose provides bactericidal levels of the drug in the bowel lumen. Although equally effective as metronidazole when given orally, vancomycin, standard dosage 1 g/day for 10 days, is more expensive and is therefore reserved for patients who cannot tolerate metronidazole are pregnant or breastfeeding or have severe cases of CDAD.

Resistance among some anaerobes has increased significantly over the 1990-2015 years.[38] The patterns of anaerobes' susceptibility and resistance remain virtually inaccessible to laboratorians and clinicians except for those published surveys conducted by a small cadre of research centers scattered worldwide or contained in individual case reports. All anaerobes are resistant to aminoglycosides and sulfamethoxazole trimethoprim.[38] Chloramphenicol resistance is very rare, although there is a clustering of MICs around the breakpoint for some strains. When resistance is found, it is due to inactivation of the drug by nitroreduction or acetyltransferase. In the United States, chloramphenicol is rarely used clinically due to its potential hematopoietic toxicity. *C. difficile* has variable susceptibility with potential resistance to clindamycin, fluoroquinolones, and β-lactams, while metronidazole remains active. *C. difficile* remains universally susceptible to metronidazole and vancomycin (there are no fecal breakpoints, so MIC relevance is speculative). Reports from China, France, and Spain have found clinical *C. difficile* isolates with reduced susceptibility to metronidazole.[39] Wong et al. recovered a clinical *C. difficile* isolate with an MIC of 164 mg/mL. Although the other 99 *C. difficile* isolates tested had an MIC of 2 mg/mL, this report is the first well-documented case of a metronidazole-resistant strain of *C. difficile* in a patient with CDAD.[40] Barbut et al. identified six *C. difficile* isolates with MIC values for metronidazole that ranged from 8 to 32 mg/mL, among 198 isolates recovered from a clinical laboratory in France in 1991 and 1997.[37] Five of these isolates were nontoxigenic strains and were therefore clinically insignificant.

In addition, there was no trend of resistance between the two time periods.[37] A preliminary report from Spain noted an increase of metronidazole resistance in clinical isolates obtained in 1998, compared with those obtained in 1993 (14% vs 6%).[41] In a study conducted in 2008, it has been documented that the metronidazole resistance of toxigenic clinical *C. difficile* isolates is heterogeneous and inducible and is not due to the presence of *nim* genes. According to Edwards, resistance due to *nim* genes in *C. difficile* has not been described to date.[42] In other anaerobic species, such as *Bacteroides* spp., resistance to metronidazole is generally due to the presence of *nim* genes; however, other alternative metronidazole resistance mechanisms have been suggested, such as reduced uptake of metronidazole, reduced nitroreductase activity, and decreased pyruvate-ferredoxin oxidoreductase activity.[42,43] Among the resistant isolates, the authors observed two subpopulations (a majority subpopulation that was susceptible to metronidazole, and a minority subpopulation that was resistant and inducible), and this characteristic was observed in different unrelated *C. difficile* isolates, as determined by ribotyping. Further, it was also demonstrated that resistance to metronidazole in *C. difficile* is unstable. First, the initial MICs determined for fresh isolates decreased after the isolates were thawed; second, the observed increases in MICs against the isolates obtained after induction in the presence of subinhibitory concentrations of metronidazole were followed by reversion to susceptibility levels in the absence of metronidazole. This phenomenon might be similar to the heterogeneous resistance that has been described for *Bacteroides* spp. and *Helicobacter pylori*.[43]

OPPORTUNISTIC ENTERIC INFECTIONS IN HIV-INFECTED ADULTS AND ADOLESCENTS

EPIDEMIOLOGY

Rates of Gram-negative bacterial enteric infections are at least 10-fold higher among HIV-infected adults than in the general population but decline when patients are on antiretroviral therapy (ART). The risk of bacterial diarrhea varies according to CD4 T-lymphocyte (CD4) count and is greatest in individuals with clinical AIDS, <200 CD4 cells/mm, or both. The most common routinely cultured enteric bacteria among HIV-infected adults in the United States are *Salmonella* (particularly the *Salmonella enterica* serotypes Typhimurium and Enteritidis), *Shigella*, and *Campylobacter*. Diarrheagenic *E. coli* (particularly enteroaggregative *E. coli*) may contribute to the burden of diarrheal disease, but their role is poorly understood because diagnosis requires specialized laboratory capacity. *C. difficile*–associated infection (CDI) is common in HIV-infected patients; recent data suggest that low CD4 count (<50 cells/mm^3) is an independent disease risk factor, in addition to the traditional risk factors such as exposure to a health-care facility or to antibiotics. Increased recognition of community-associated CDI in HIV-uninfected individuals suggests that the health-care provider should consider CDI in the evaluation of outpatient diarrheal illnesses.[44–46]

As with bacterial enteric infections in HIV-uninfected persons, the probable source for most enteric infections in HIV-infected patients is the ingestion of contaminated food or water. Sexual activity with the potential for direct or indirect fecal–oral exposure also increases the risk of infection, especially with *Shigella* and *Campylobacter*. HIV-associated alterations in mucosal immunity or intestinal integrity and treatment with acid-suppressive agents may facilitate acquisition of enteric bacterial infections.[44,46]

PATHOGEN-SPECIFIC THERAPY

Salmonella spp.

Immunocompetent hosts who are not HIV-infected often do not require treatment for *Salmonella* gastroenteritis, as the condition is usually self-limited and treatment may prolong the carrier state. In contrast, all HIV-infected patients with salmonellosis should be treated, although no clinical trials have compared antimicrobial therapy with placebos. Notably, HIV infection increases the risk of *Salmonella* bacteremia 20- to 100-fold and mortality as much as 7-fold compared with that in patients who are not HIV-infected.[47] The initial treatment of choice for *Salmonella* infection is a fluoroquinolone. Ciprofloxacin is the preferred agent, and other fluoroquinolones, such as levofloxacin and moxifloxacin, likely would be effective in treating salmonellosis in HIV-infected patients, but they have not been well evaluated in clinical studies. Depending on antibiotic susceptibility, alternatives to fluoroquinolones might include TMP-SMX or expanded-spectrum cephalosporins such as ceftriaxone or cefotaxime.[44]

The optimal duration of therapy for HIV-related *Salmonella* infection has not been defined. For patients with CD4 counts $\geq$200 cells/mm^3 who have mild gastroenteritis without bacteremia, 7–14 days of treatment is reasonable. For the same patients with bacteremia, 14 days is appropriate, provided that clearance of bacteremia is documented. Longer treatment is suggested if bacteremia persists or if the infection is complicated—that is, if metastatic foci are present. For patients with advanced HIV disease (CD4 count <200 cells/mm^3), 2–6 weeks of antibiotics often is recommended. Some patients with *Salmonella* bacteremia may remain febrile for 5–7 days despite effective therapy.[44,48]

HIV-infected patients with *Salmonella* bacteremia, which typically occurs in those with advanced HIV disease, should be monitored clinically for recurrence after treatment. Recurrence may present as bacteremia or as an anatomically localized infection, including intraabdominal, endothelial, urinary tract, soft tissue, bone and joint, lung, or meningeal foci. Secondary prophylaxis should be considered for patients with recurrent *Salmonella* bacteremia, and it might also be considered for patients with recurrent gastroenteritis and in those with CD4 counts <200 cells/mm^3 with severe diarrhea. The value of this secondary prophylaxis has not been established and must be weighed against the risks of long-term antibiotic exposure. Recurrent *Salmonella* bacteremia constitutes an AIDS-defining illness and suppression of HIV replication with ART is expected to decrease the risk of recurrent illnesses. In patients whose *Salmonella* infection is resolved and who have responded to ART with sustained viral suppression and CD4 counts >200 cells/mm^3, secondary prophylaxis for salmonellosis can probably be stopped. Clinicians also should be aware that recurrence may represent development of AMR during therapy.[45]

Shigella spp.

Therapy for *Shigella* infections is recommended both to shorten the duration of illness and to possibly prevent spread of the infection to others. The recommended treatment for shigellosis is with a fluoroquinolone (preferably ciprofloxacin) for 7–10 days. However, ciprofloxacin-resistant *Shigella sonnei* has been reported in the United States and is associated with international travel, homelessness, and men who have sex with men (MSM); ciprofloxacin-resistant shigellosis among MSM appears to be acquired predominantly within the United States rather than during travel.

Depending on antibiotic susceptibilities, alternative agents might include TMP-SMX (7–10 days) or azithromycin (5 days). Azithromycin has not been evaluated in HIV-infected patients with shigellosis, and the therapy suggested is extrapolated from limited data in immunocompetent hosts. Recently, *Shigella* spp. with reduced susceptibility to azithromycin in HIV-infected MSM have been reported. Treatment for patients with *Shigella* bacteremia is less well defined, but extending treatment to at least 14 days is reasonable. Azithromycin is not recommended for treatment of *Shigella* spp. bacteremia. Recurrent infections can occur, particularly in individuals with CD4 counts <200 cells/mm^3, in which case extending antimicrobial therapy for up to 6 weeks is reasonable. As with *Salmonella* infections, suppression of HIV replication with ART is expected to decrease the risk of recurrent shigellosis.[44,48,49]

Campylobacter spp.

The optimal treatment of campylobacteriosis in HIV-infected patients is poorly defined. The culture and susceptibility of *Campylobacter* isolates are recommended. In 2011, 24% of *Campylobacter* isolates in the United States were fluoroquinolone-resistant. For patients with mild disease and CD4 counts >200 cells/mm^3, some clinicians opt to withhold therapy unless symptoms persist for more than several days. For mild-to-moderate campylobacteriosis, initiating therapy with a fluoroquinolone such as ciprofloxacin for 7–10 days (if the organism is sensitive) or azithromycin for 5 days is a reasonable approach. Azithromycin has not been evaluated in HIV-infected patients with campylobacteriosis, and the therapy suggested is extrapolated from limited data in immunocompetent hosts. Patients with *Campylobacter* bacteremia should be treated for at least 14 days using a fluoroquinolone if the isolate is sensitive. Azithromycin is not recommended for treatment of *Campylobacter* bacteremia. Adding a second active agent, such as an aminoglycoside, may be prudent in these patients to limit the emergence of antibiotic resistance.

Antibiotic choice should be guided by antibiotic susceptibility tests. Chronic suppressive or maintenance therapy is not recommended for first-time *Campylobacter* infections in HIV-infected patients. However, recurrent infections can occur, particularly in patients with CD4 counts <200 cells/mm^3. In recurrent disease, extending the length of antimicrobial therapy for 2–6 weeks is reasonable. As with *Salmonella* infections, suppression of HIV replication with ART is expected to decrease the risk of recurrent *Campylobacter* infections.[44,50]

Clostridium difficile

The available data suggest that HIV-infected patients respond to treatment of CDI similarly to HIV-uninfected patients. Guidelines and subsequent updates to guide the treatment of CDI have been published[51–54] and can be consulted for further information. Multivariate analysis of two recent identical, multicenter (91 sites in United States and Canada; 109 sites in Europe), randomized, and double-blind studies involving 537 non-HIV-infected patients with CDI (278 and 259 treated with metronidazole and vancomycin, respectively) found vancomycin to be superior to metronidazole for clinical success (OR 1.575 (1.035,2.396), $P = 0.034$). Stratification by CDI disease severity found 4.0% (mild), 8.3% (moderate), and 12.2% (severe) improved clinical success rates with vancomycin therapy.[55] Given this trial and earlier data,[56] vancomycin is recommended for the treatment of HIV-infected persons with CDI with the possible exception of mild CDI, where treatment with metronidazole may yield clinical success. Treatment of recurrent CDI in HIV-infected patients is the same as in patients who are not HIV-infected. Limited case reports suggest that fecal microbiota therapy may be successful and safe to treat recurrent CDI in HIV-infected patients.[57] The impact of ART on the recurrence of CDI is unknown.[44]

Clostridium difficile infection in pregnancy

Largely due to their young age and overall good health, pregnant women have historically been at low risk for developing CDAD.[58] Infectious disease specialists in Toronto had cited a report of *C. difficile* disease in healthy young women (some of them pregnant) as a sign of an alarming shift of the infection to the community. The report at the Infectious Diseases Society of America (IDSA) meeting came from Judith O'Donnell, MD, of Drexel University in Philadelphia, who described six cases of CDAD in healthy women, one of whom died. Three of the women were pregnant, another had recently given birth, and two others had recently undergone elective hysterectomy.[59]

It is evident from these reports that clinicians must consider CDI in patients with severe diarrhea, even if they do not have the traditional risk factors for CDI, such as antibiotic use or concurrent hospitalizations.

MONITORING OF RESPONSES TO THERAPY AND ADVERSE EVENTS

Patients should be monitored closely for their responses to treatment, defined clinically by improvement in systemic signs and symptoms, resolution of diarrhea, and sterilization of infected tissues or body fluids such as blood. A follow-up stool culture to demonstrate clearance of the organism is not required if clinical symptoms and diarrhea resolve. A follow-up stool culture may be required when public health considerations and state law dictate the need to ensure microbiologic cure, such as in health-care or food service workers. Immune reconstitution inflammatory syndrome has not been described in association with treatment for bacterial enteric pathogens.[44]

SPECIAL CONSIDERATIONS WITH REGARD TO STARTING ART

ART initiation should follow standard guidelines. The presence of a diarrheal illness is relevant only in terms of a patient's ability to ingest and absorb ART. Prompt initiation of ART should be considered regardless of CD4 count; that is the presence of an enteric infection should not delay ART initiation.[44]

MANAGING TREATMENT FAILURE

A follow-up stool culture should be considered for patients who fail to respond clinically to appropriate antimicrobial therapy. In patients with persistent or recurrent diarrhea despite therapy, clinicians should consider other enteric infections in the context of the patient's immune status, and in all cases, the possibility of *C. difficile* or the development of AMR.[44] Observational studies suggest that concentrations of plasma drugs (eg, ciprofloxacin) in HIV-infected patients may be decreased as a result of diarrhea or malabsorption. Coadministration of quinolones with magnesium- or aluminum-containing antacids or with calcium, zinc, or iron should be avoided because these interfere with drug absorption. Although larger prospective studies are needed to determine the impact of severe diarrhea on antibiotic absorption, it is prudent to use intravenous antibiotics in clinically unstable patients.[44]

PREVENTING RECURRENCE

The pharmacologic approach to recurrent enteric infections is covered in the sections on directed therapy for each bacterial species earlier in this chapter. As noted previously, secondary prophylaxis should be considered for patients with recurrent *Salmonella* bacteremia and, in some circumstances, for those with recurrent shigellosis or campylobacteriosis.[44]

SPECIAL CONSIDERATIONS DURING PREGNANCY IN PATIENTS WITH ENTERIC INFECTIONS

The diagnosis of bacterial enteric infection in pregnant women is the same as in women who are not pregnant. Bacterial enteric infections in pregnant women should be managed the same as in women who are not pregnant, with several considerations. Based on the safety profile, expanded-spectrum cephalosporins or azithromycin should be the first-line therapy for bacterial enteric infections during pregnancy if antimicrobials are required, depending on the organism and the results of susceptibility testing. Arthropathy has been noted in the offspring of animals treated with quinolones during pregnancy. However, studies evaluating quinolone use in pregnant women did not find an increased risk of birth defects or musculoskeletal abnormalities. Thus, quinolones can be used in pregnancy for bacterial enteric infections in HIV-infected pregnant women if indicated by susceptibility testing or failure of first-line therapy, as discussed previously. TMP-SMX use in the first trimester should be avoided, if possible, because of an association with an increased risk of birth defects (specifically neural tube, cardiovascular, and urinary tract defects). Neonatal care providers should be informed if maternal sulfa therapy was used near delivery because of the theoretical increased risk to the newborn of hyperbilirubinemia and kernicterus.[44,60]

CONCLUSION

Resistance is encountered with virtually every infectious disease. Few proven mechanisms exist for resistance control, and almost none have been validated in the developing-country setting. The dearth of intervention study data is particularly acute in the context where infectious disease prevalence is high and access to antimicrobial agents is low, which best describes the situation faced by low-income persons in Africa. Most intervention studies in developing countries have focused on relatively inexpensive and easily piloted educational interventions. Educational interventions push against the strong influence of unregulated distribution, sometimes accompanied by unscrupulous counter advertising, and their value has not been evaluated in the long run. Importantly, although educational interventions typically yield positive results, these results are modest. Other methods in fact may be more cost-effective or might boost the value of education. Strategies that have been evaluated and found to deal with the problem of resistance need to be further diversified in poor countries. Safe water and sanitation addressing the imbalance between antimicrobial drug supply and demand and building realistic infrastructure for rational antimicrobial use are priority areas for resistance control that could address the short- and long-term effects of disease on the poor.

REFERENCES

1. CDC. *National Antimicrobial Resistance Monitoring System for Enteric Bacteria (NARMS): human isolates final report, 2011*. Atlanta, GA: U.S. Department of Health & Human Services, CDC; 2013.
2. WHO Library Cataloguing-in-Publication Data. *Antimicrobial resistance: global report on surveillance*. <www.who.int/iris/bitstream/10665/112642/1/9789241564748_eng.pdf>; 2014.
3. CDC. *Antibiotic resistance threats in the United States*. US Department of Health & Human Services, Centers for Disease control and Prevention. <http://www.cdc.gov/drugresistance/threat-report-2013>; 2013.
4. Sosa A, Byarugaba DK, Amabile C, Hsueh PR, Kariuki S, Okeke IN. *Antimicrobial resistance in developing countries*. New York, NY: Springer Science + Business Media, LLC; 2010.
5. Pop-Vicas A, Opal SM. The clinical impact of multidrug-resistant gram-negative bacilli in the management of septic shock. *Virulence* 2014;**5**(1):206–12.
6. World Health Organization (WHO). *WHO global strategy for containment of antibiotic resistance*. Geneva: Organization; 2001. p. 1–99.
7. Levy SB. Microbial resistance to antibiotics. An evolving and persistent problem. *Lancet* 1982;**2**:83–8.
8. European Food Safety Authority. *Estimation of the relative contribution of different food and animal sources to human Salmonella infections in the European Union*. Parma: EFSA. Available at: <http://www.efsa.europa.eu/en/supporting/pub/184e.htm>; July 2011 [accessed 14.10.11].
9. Sandt CH, Fedorka-Cray PJ, Tewari D, Ostroff S, Joyce K, M'ikanatha NM. A comparison of non-typhoidal *Salmonella* from humans and food animals using pulsed-field gel electrophoresis and antimicrobial susceptibility patterns. *PLoS One* 2013;**8**(10):1–10.
10. Okeke IN, Klugman KP, Bhutta ZA, Duse AG, Jenkins P, O'Brien TF, et al. Antimicrobial resistance in developing countries. Part II: strategies for containment. *Lancet Infect Dis* 2005;**5**:568–80.
11. Okeke IN, Edelman R. Dissemination of antibiotic-resistant bacteria across geographic borders. *Clin Infect Dis* 2001;**33**:364–9.
12. Lamikanra A, Okeke IN. A study of the effect of the urban/rural divide on the incidence of antibiotic resistance in *Escherichia coli*. *Biomed Lett* 1997;**55**:91–7.
13. Traveler's Health. <http://wwwnc.cdc.gov/travel/yellowbook/2012/chapter-2-the-pre-travel-consultation/travelers-diarrhea.htm>.
14. Vital Signs. *Making food safer to eat*. <http://www.cdc.gov/VitalSigns/FoodSafety>; June 2011.
15. Scallan E, Hoekstra RM, Angulo FJ, Tauxe RV, Widdowson MA, Roy SL, et al. Foodborne illness acquired in the United States—major pathogens. *Emerg Infect Dis* 2011;**17**:7–15.
16. Hoban DJ, Bouchillon SK, Hawser SP, Badal RE. Trends in the frequency of multiple drug-resistant Enterobacteriaceae and their susceptibility to ertapenem, imipenem, and other antimicrobial agents: data from the study for monitoring antimicrobial resistance trends 2002 to 2007. *Diagn Microbiol Infect Dis* 2010;**66**(1):78–86.
17. Laxminarayan R, Bhutta Z, Duse A, Jenkins P, O'Brien T, Okeke IN, et al. Drug resistance. In: Jamison D, Breman JG, Measham AR, Alleyane G, Claeson M, Evans DB, et al., editors. *Disease control priorities in developing countries*. New York, NY: Oxford University Press; 2006. p. 1031–51.
18. Livermore D. Can better prescribing turn the tide of resistance? *Nat Rev Microbiol* 2004;**2**:73–8.

19. Byarugaba DK. A view on antimicrobial resistance in developing countries and responsible risk factors. *Int J Antimicrob Agents* 2004;**24**:105–10.
20. Weinstein RA. Controlling antimicrobial resistance in hospitals: infection control and use of antibiotics. *Emerg Infect Dis* 2001;**7**:188–92.
21. Barbosa TM, Levy SB. The impact of antibiotic use on resistance development and persistence. *Drug Resist Updat* 2000;**3**:303–11.
22. Okeke IN, Aboderin OA, Byarugaba DK, Ojo KK, Opintan JA. Growing problem of multidrug-resistant enteric pathogens in Africa. *Emerg Infect Dis* 2007;**13**(11):1640–6.
23. *Cholera—Vibrio cholerae infection, general information.* <http://www.cdc.gov/cholera/general>.
24. Towner KJ, Pearson NJ, Mhalu FS, O'Grady F. Resistance to antimicrobial agents of *Vibrio cholerae* El Tor strains isolated during the fourth cholera epidemic in the United Republic of Tanzania. *Bull World Health Organ* 1980;**58**:747–51.
25. Mwansa JC, Mwaba J, Lukwesa C, Bhuiyan NA, Ansaruzamman M, Ramamurthy T, et al. Multiply antibiotic-resistant *Vibrio cholerae* O1 biotype El Tor strains emerge during cholera outbreaks in Zambia. *Epidemiol Infect* 2007;**135**:847–53.
26. Ceccarelli D, Salvia AM, Sami J, Cappuccinelli P, Colombo MM. New cluster of plasmid-located class 1 integrons in *Vibrio cholerae* O1 and a dfrA15 cassette-containing integron in *Vibrio parahaemolyticus* isolated in Angola. *Antimicrob Agents Chemother* 2006;**50**:2493–9.
27. Dalsgaard A, Forslund A, Petersen A, Brown DJ, Dias F, Monteiro S, et al. Class 1 integron-borne, multiple-antibiotic resistance encoded by a 150-kilobase conjugative plasmid in epidemic *Vibrio cholerae* O1 strains isolated in Guinea-Bissau. *J Clin Microbiol* 2000;**38**:3774–9.
28. Cooke FJ, Wain J. The emergence of antibiotic resistance in typhoid fever. *Travel Med Infect Dis* 2004;**2**:67–74.
29. Kariuki S, Revathi G, Kariuki N, Kiiru J, Mwituria J, Hart CA. Characterisation of community acquired non-typhoidal *Salmonella* from bacteraemia and diarrhoeal infections in children admitted to hospital in Nairobi, Kenya. *BMC Microbiol* 2006;**6**:101.
30. Okeke IN, Fayinka ST, Lamikanra A. Antibiotic resistance in *Escherichia coli* from Nigerian students, 1986–1998. *Emerg Infect Dis* 2000;**6**:393–6.
31. Jafari F, Hamidian M, Rezadehbashi M, Doyle M. Prevalence and antimicrobial resistance of diarrheagenic *Escherichia coli* and *Shigella* species associated with acute diarrhea in Tehran, Iran. *Can J Infect Dis Med Microbiol* 2009;**20**(3):56–62.
32. Gupta A, Nelson JM, Barrett TJ, Tauxe RV, Rossiter SP, Friedman CR, et al. Antimicrobial resistance among *Campylobacter* strains, United States, 1997–2001. *Emerg Infect Dis* 2004;**10**:1102–9.
33. Bouza E, Burillo A, Muñoz P. Antimicrobial therapy of *Clostridium difficile*-associated diarrhea. *Med Clin N Am* 2006;**90**:1141–63.
34. Goldstein EJC, Citron DM. Resistance trends in antimicrobial susceptibility of anaerobic bacteria, Part I. *Clin Microb Newsletter* 2011;**33**(1):1–8.
35. Pelleschi ME. *Clostridium difficile*-associated disease: diagnosis, prevention, treatment, and nursing care. *Crit Care Nurse* 2008;**28**:27–35.
36. Huang H, Weintraub A, Fang H, Nord CE. Antimicrobial resistance in *Clostridium difficile*. *Int J Antimicrob Agents* 2009;**34**:516–22.
37. Barbut F, Decre D, Burghoffer B, Lesage D, Delisle F, Lalande V, et al. Antimicrobial susceptibilities and serogroups of clinical strains of *Clostridium difficile* isolated in France in 1991 and 1997. *Antimicrob Agents Chemother* 1999;**43**:2607–11.
38. Teng LJ, Hsueh PR, Tsai JC, Liaw SJ, Ho SW, Luh KT. High incidence of cefoxitin and clindamycin resistance among anaerobes in Taiwan. *Antimicrob Agents Chemother* 2002;**46**(9):2908–13.

39. Johnson S, Sanchez JL, Gerding DN. Metronidazole resistance in *Clostridium difficile. Clin Infect Dis* 2000;**31**(2):625–6.
40. Wong SS, Woo PC, Luk WK, Yuen KY. Susceptibility testing of *Clostridium difficile* against metronidazole and vancomycin by disk diffusion and Etest. *Diagn Microbiol Infect Dis* 1991;**34**:1–6.
41. Peláez T, Cercenado E, Alcalá L, Marín M, Martín-López A, Martínez-Alarcón J, et al. Metronidazole resistance in *Clostridium difficile* is heterogeneous. *J Clin Microbiol* 2008;**46**:3028–32.
42. Edwards DI. Nitroimidazole drugs—action and resistance mechanisms. I. Mechanisms of action. *J Antimicrob Chemother* 1993;**31**:9–20.
43. Henriksen TH, Lerang F, Lia A, Schøyen R, Thoresen T, Berge T. Laboratory handling of *Helicobacter pylori* critically influences the results of in-vitro metronidazole resistance determination. *Clin Microbiol Infect* 2004;**10**:315–21.
44. AIDS info offering information on HIV/AIDS Treatment, Prevention and Research. *Guidelines for the prevention and treatment of opportunistic infections in HIV-infected adults and adolescents.* http://www.aidsinfo.nih.gov; 2015 [accessed 01.11.15].
45. Kuschner RA, Trofa AF, Thomas RJ, et al. Use of azithromycin for the treatment of *Campylobacter enteritis* in travelers to Thailand, an area where ciprofloxacin resistance is prevalent. *Clin Infect Dis* 1995;**21**(3):536–41. Available at: http://www.ncbi.nlm.nih.gov/pubmed/8527539
46. Sanchez TH, Brooks JT, Sullivan PS, et al. Bacterial diarrhea in persons with HIV infection, United States, 1992–2002. *Clin Infect Dis* 2005;**41**(11):1621–7. Available at: http://www.ncbi.nlm.nih.gov/pubmed/16267735
47. Cummings PL, Sorvillo F, Kuo T. Salmonellosis-related mortality in the United States, 1990–2006. *Foodborne Pathog Dis* 2010;**7**(11):1393–9. Available at: http://www.ncbi.nlm.nih.gov/pubmed/20617938
48. Gordon MA, Banda HT, Gondwe M, et al. Non-typhoidal *Salmonella* bacteraemia among HIV-infected Malawian adults: high mortality and frequent recrudescence. *AIDS* 2002;**16**(12):1633–41. Available at: http://www.ncbi.nlm.nih.gov/pubmed/12172085
49. Heiman KE, Karlsson M, Grass J, et al. Notes from the field: *Shigella* with decreased susceptibility to azithromycin among men who have sex with men—United States, 2002–2013. *MMWR Morb Mortal Wkly Rep* 2014;**63**(6):132–3. Available at: http://www.ncbi.nlm.nih.gov/pubmed/24522098
50. Hassing RJ, Melles DC, Goessens WH, Rijnders BJ. Case of *Shigella flexneri* infection with treatment failure due to azithromycin resistance in an HIV-positive patient. *Infection* 2014. Available at: http://www.ncbi.nlm.nih.gov/pubmed/24488332
51. Cohen SH, Gerding DN, Johnson S, et al. Clinical practice guidelines for *Clostridium difficile* infection in adults: 2010 update by the Society for Healthcare Epidemiology of America (SHEA) and the Infectious Diseases Society of America (IDSA). *Infect Control Hosp Epidemiol* 2010;**31**(5):431–55. Available at: http://www.ncbi.nlm.nih.gov/pubmed/20307191
52. Surawicz CM, Brandt LJ, Binion DG, et al. Guidelines for diagnosis, treatment, and prevention of *Clostridium difficile* infections. *Am J Gastroenterol* 2013;**108**(4):478–98. quiz 499. Available at: http://www.ncbi.nlm.nih.gov/pubmed/23439232
53. Bagdasarian N, Rao K, Malani PN. Diagnosis and treatment of *Clostridium difficile* in adults: a systematic review. *JAMA* 2015;**313**(4):398–408. Available at: http://www.ncbi.nlm.nih.gov/pubmed/25626036
54. Leffler DA, Lamont JT. *Clostridium difficile* infection. *N Engl J Med* 2015;**372**(16):1539–48. Available at: http://www.ncbi.nlm.nih.gov/pubmed/25875259
55. Johnson S, Louie TJ, Gerding DN, et al. Vancomycin, metronidazole, or tolevamer for *Clostridium difficile* infection: results from two multinational, randomized, controlled trials. *Clin Infect Dis* 2014;**59**(3):345–54. Available at: http://www.ncbi.nlm.nih.gov/pubmed/24799326

56. Zar FA, Bakkanagari SR, Moorthi KM, Davis MB. A comparison of vancomycin and metronidazole for the treatment of *Clostridium difficile*-associated diarrhea, stratified by disease severity. *Clin Infect Dis* 2007;**45**(3):302–7. Available at: http://www.ncbi.nlm.nih.gov/pubmed/17599306
57. Di Bella S, Gouliouris T, Petrosillo N. Fecal microbiota transplantation (FMT) for *Clostridium difficile* infection: focus on immunocompromised patients. *J Infect Chemother* 2015;**21**(4):230–7. Available at: http://www.ncbi.nlm.nih.gov/pubmed/25703532
58. *C. difficile-associated diarrhea in pregnancy—a complex clinical challenge*. Available at: http://nbharwani.com/2012/medicine-hat-news/c-difficile-associated-diarrhea-in-pregnancy-a-complex-clinical-challenge; May 8, 2012.
59. Rheumatology Network. *IDSA: C. difficile in pregnant women underscores new dangers*. Available at: <http://www.rheumatologynetwork.com/articles/idsa-c-difficile-pregnant-women-underscores-new-dangers#sthash.Nc44U7U5.dpuf>; October 13, 2006.
60. Czeizel AE, Rockenbauer M, Sorensen HT, Olsen J. The teratogenic risk of trimethoprim-sulfonamides: a population based case-control study. *Reprod Toxicol* 2001;**15**(6):637–46. Available at: http://www.ncbi.nlm.nih.gov/pubmed/11738517

CHAPTER

CARBAPENEM-RESISTANT, GRAM-NEGATIVE BACILLI: THE STATE OF THE ART

5

D.H. Halat[1,2,3], D.K. Sarkis[1,2] and C.A. Moubareck[1,2,4]

[1]*Microbiology Laboratory, School of Pharmacy, Saint-Joseph University, Beirut, Lebanon*
[2]*Rodolphe Mérieux Laboratory, Beirut, Lebanon* [3]*Department of Pharmaceutical Sciences, School of Pharmacy, Lebanese International University, Beirut and Bekaa, Lebanon* [4]*Department of Natural Science and Public Health, College of Sustainability Sciences and Humanities, Zayed University, Dubai, United Arab Emirates*

INTRODUCTION

The global spread of antimicrobial-resistant bacteria has been progressive and unremitting since the introduction of antimicrobial agents into clinical medicine more than 70 years ago. Not only the dissemination of antibiotic-resistant, Gram-positive cocci is challenging, but also the growing incidence of antibiotic-resistant, Gram-negative bacilli represents an increasingly pressing issue.[1] Specifically, the development of resistance to β-lactams among Gram-negative bacilli is an urgent contemporary threat to medical disease treatments. Indeed, β-lactam antibiotics are the cornerstone of antimicrobial armamentarium, and it is hard for any other class of antibiotics to exhibit such a large impact on the history of infectious disease. It is estimated that these agents have added at least 10 years to the lifespan of humans. Nowadays they are the most widely utilized antibiotics owing to their comparatively high effectiveness, low cost, ease of delivery, and minimal side effects.[2] Unfortunately, bacteria have evolved sophisticated resistance mechanisms to combat the lethal effects of β-lactam antibiotics, and each new compound has been addressed with a diverse and robust array of resistance determinants. Because carbapenems have the broadest spectra among all β-lactams and are primarily used for infections caused by multidrug-resistant (MDR), Gram-negative bacteria, the emergence and spread of resistance to these agents has become a complicated public health alarm.[3] In clinical terms, carbapenem-resistant, Gram-negative bacilli, including carbapenem-resistant Enterobacteriaceae (CRE), MDR *Pseudomonas aeruginosa*, and MDR *Acinetobacter baumannii* have limited (and sometimes nonexistent) treatment options. This resulted in high rates of mortality, reaching 48% for infections caused by CRE.[4] In a recent meta-analysis, the mortality rate was greater in patients with CRE bacteremia than those with bacteremia due to carbapenem-susceptible Enterobacteriaceae.[5] In epidemiological terms, carbapenem-resistant, Gram-negative bacilli are spreading worldwide,[6] a development largely facilitated by international travel and medical tourism. Genes encoding resistance to carbapenems are often carried on mobile genetic elements that are easily exchanged among Gram-negative bacteria, either within a host or in the environment. Finally, from an economic standpoint, infections caused by carbapenem-resistant pathogens are associated with increased financial costs.[1] For example, a French hospital

Antibiotic Resistance. DOI: http://dx.doi.org/10.1016/B978-0-12-803642-6.00005-8

estimated that it incurred a total additional cost of more than 600,000 euros and more than 1500 additional working hours due to infections caused by carbapenemase-producing Enterobacteriaceae in 2 years.[7] In Colombia, carbapenem-resistant *A. baumannii* was significantly associated with higher total costs of hospitalization than for carbapenem-susceptible *A. baumannii*.[8] Given these considerations, a thorough understanding of the properties and mechanisms of carbapenem resistance is imperative for microbiologists and clinicians. It is urgent to minimize the spread of such resistance through the implementation of rigorous control programs. In this chapter, contemporary trends of carbapenem resistance among Gram-negative bacilli of medical connotation are reviewed, including both enzymatic and nonenzymatic pathways. Also, because such resistance is easily transmissible and its containment is a challenging issue, current knowledge regarding the detection, treatment, and control of such pathogens is presented.

OVERVIEW OF CARBAPENEMS

The four FDA-approved carbapenems (imipenem, meropenem, ertapenem, and doripenem), several of which are shown in Fig. 5.1, belong to the β-lactam group of antibiotics and contain in their chemical structure a fused β-lactam ring and a five-member ring system. They mainly differ from

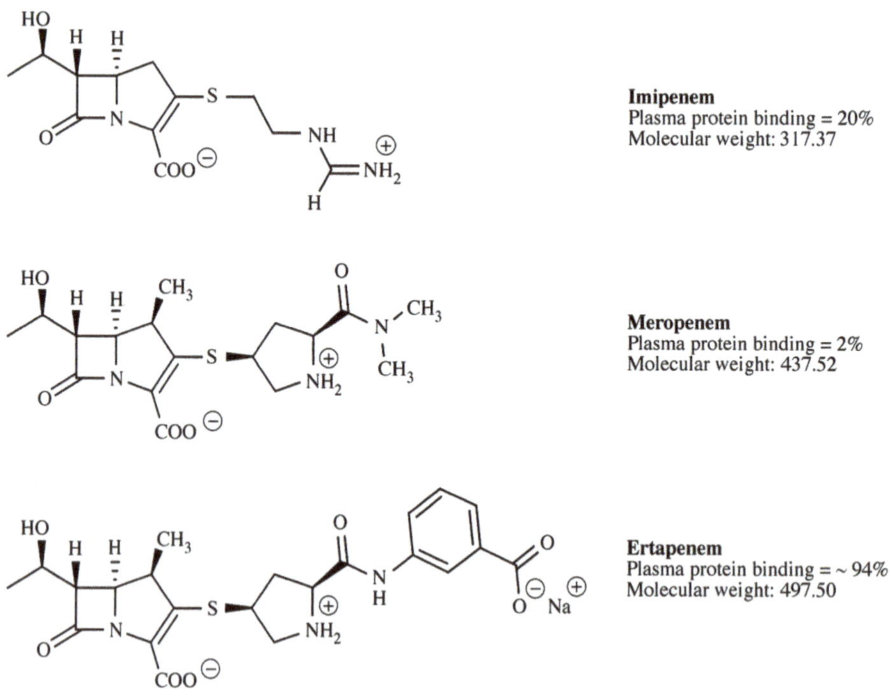

FIGURE 5.1

Chemical structures and some properties of imipenem, meropenem, and ertapenem.

penicillin by being unsaturated and containing carbon instead of sulfur atoms in the β-lactam ring.[9] These compounds not only possess the broader spectrum of activity than most other β-lactams, but also are among the most broad-spectrum antimicrobials available, and are often viewed as last-resort therapy for MDR Gram-negative bacteria.[10] Analogous to other members of the β-lactam class of antibiotics, carbapenems inhibit the terminal step in the formation of peptidoglycan in the bacterial cell wall through inhibiting the action of transpeptidase enzymes. Consequently, they prevent cross-linking of peptidoglycans, resulting in a killing action on susceptible bacteria.[11] Overdependence on carbapenems in clinical practice was especially apparent with the emergence of Enterobacteriaceae producing novel types of β-lactamases like extended-spectrum β-lactamases (ESBLs) and plasmid-mediated AmpC (also termed *Ambler class C*) enzymes.[12] By definition, ESBLs are β-lactamases capable of conferring bacterial resistance to penicillins, first-, second-, and third-generation cephalosporins, and aztreonam by hydrolysis of these antibiotics. However, they have no effect on cephamycins (cefotetan, cefoxitin, and latamoxef) or carbapenems. ESBLs are inhibited by classical β-lactamase inhibitors such as clavulanic acid, sulbactam, and tazobactam.[13] On the other hand, AmpC cephalosporinases provide resistance to both oxyimino and 7-α-methoxy-cephalosporins and monobactams.[14] With such growing spectra of compounds hydrolyzed by ESBLs and AmpC enzymes, carbapenems were the only widely marketed β-lactams that were stable during such hydrolysis, and they retained almost universal activity for around 20 years after the initial introduction of imipenem in 1985; unfortunately, resistance to carbapenems is now accumulating.

CARBAPENEM-RESISTANCE MECHANISMS

Like all β-lactam antibiotics, carbapenems are rendered inactive against bacteria by four mechanisms of resistance: production of carbapenem-hydrolyzing enzymes, porin protein impermeability lesions, active expulsion of drug molecules from bacterial cells by efflux pumps, and utilization of carbapenem-insensitive transpeptidases.[2] The erosion of carbapenem activity becomes even more pronounced when pathogens accumulate more than one of these mechanisms.[12] In addition, carbapenem-resistant, Gram-negative pathogens almost invariably possess other resistance mechanisms that compromise other antibiotic classes, including aminoglycosides, fluoroquinolones, and co-trimoxazole, which drastically limits treatment options.[10] The mechanisms of carbapenem resistance are elaborated next.

PRODUCTION OF CARBAPENEMASES

The most significant development among carbapenem-resistance mechanisms is the gradual proliferation of carbapenemases, a new category of biochemically diverse hydrolyzing enzymes.[12] Although the discovery of β-lactamase genes in bacteria dates back to about 30,000 years preceding the clinical use of antibiotics, and even though the evolution of such enzymes has given rise to about 1300 varieties, newer types of β-lactamases have attracted interest since the beginning of the 1980s. These include ESBLs, AmpC cephalosporinases, and carbapenemases, with the latter group evoking the most apprehension. Among β-lactamases, carbapenemases possess the most extensive substrate profiles of all β-lactamases and are therefore clinically important. They are associated with resistance to other classes of antibiotics (namely, aminoglycosides, fluoroquinolones, and co-trimoxazole). Chromosomal carbapenemases have been recognized as not causing important effects for a long time, but the more malicious forms of these enzymes (whether acquired or imported) are a major threat and form a crucial issue for infection control.[15]

History of carbapenemases

The first carbapenemases were described among Gram-positive bacteria, and unlike other β-lactamases, they were inhibited by ethylene diamine tetraacetic acid (EDTA), establishing them as metalloenzymes, and were later shown to possess at least one zinc atom at their active site to facilitate hydrolysis of the β-lactam ring.[16] In the mid-to-late-1980s, another set of carbapenem-hydrolyzing enzymes emerged among Enterobacteriaceae, which were later found to utilize serine at their active site and to be unaffected by EDTA, but could be treated by clavulanic acid and tazobactam.[17] In 1993, carbapenem resistance in Enterobacteriaceae was described in *Enterobacter cloacae* as a result of the chromosomal carbapenemase NmcA.[18] All such carbapenemases were species-specific and chromosomally derived. However, identification of plasmid-encoded IMP-1 metallo-β-lactamase (MBL) in *P. aeruginosa* from Japan,[19] ARI-1 (OXA-23) in *A. baumannii* from the United Kingdom,[20] and KPC-1 in *Klebsiella pneumoniae* from the United States[21] changed the patterns of carbapenemase dissemination. Later, newly recognized plasmid-encoded carbapenemases proliferated, and the problem of clonal spread of carbapenem resistance became a global issue of interspecies dispersion.[22]

Classification, properties, genetic background, and epidemiology of carbapenemases

The most useful classification of carbapenemases is the molecular scheme of Ambler, in which carbapenemases belong to classes A, B, C, and D. A comparison of the features of carbapenemases and their detection methods is presented in Table 5.1[22–29] and discussed next.

Class A carbapenemases

Class A carbapenemases were first described early in the 1980s.[17] They include *K. pneumoniae* carbapenemase (KPC), *Serratia marcescens* enzyme (SME), nonmetallo enzyme carbapenemase class A (NmcA), imipenem-hydrolyzing (IMI) and Guiana extended-spectrum (GES) families.[23] These enzymes are inhibited by clavulanic acid and tazobactam, but not sulbactam, and their hydrolytic activity requires the presence of a serine active site. Some class A carbapenemases are chromosomally encoded (SME, NmcA, IMI-1, and SFC-1), while others are plasmid-encoded (KPC, GES, IMI-2). The first chromosomal NmcA was identified in *E. cloacae* in 1993.[18] Later, chromosomal class A carbapenemases were isolated in different geographic areas (eg, the United Kingdom, United States, France, and Argentina), but remained relatively rare compared to their plasmid-encoded counterparts. The chromosomal class A carbapenemases possess distinctive resistance profiles with carbapenem resistance coupled to susceptibility to extended-spectrum cephalosporins.[22]

The plasmid-mediated class A carbapenemases hydrolyze all β-lactams, including monobactams, have a high affinity for meropenem and lower affinity for cephamycins and ceftazidime.[23] KPC is the most clinically important of these enzymes, and the first variant, KPC-1, was first described in 1996 in North Carolina,[21] followed by the emergence of a single-amino-acid variant,

Table 5.1 Classification, Comparison of Properties, and Detection Methods of Carbapenemases

Ambler Class	Ambler Class A	Ambler Class B	Plasmid-Mediated Ambler Class C	Ambler Class D	Reference
Active enzyme site	Serine	Zinc	Serine	Serine	23
Geographic epicenter	France, England, United States	Japan, India	South Korea, United States	Turkey, Middle East	24
Important enzymes	Chromosomal: NmcA, SME, IMI-1, SFC-1 Plasmid-borne: KPC, GES, IMI-2	VIM-2, IMP-1, IMP-2, NDM, SPM-1	CMY, FOX, MOX, ACC, DHA	OXA-23, OXA-24, OXA-58, OXA-48, OXA-181	23,25
Most commonly affected species	Enterobacteriaceae, *Acinetobacter baumannii*, *Pseudomonas*	Enterobacteriaceae, *A. baumannii*, *Pseudomonas*	*Klebsiella pneumoniae*, *Salmonella*, *Escherichia coli*	Enterobacteriaceae, *A. baumannii*	26
Affected β-lactam substrates	All β-lactams including monobactams, except cephamycins (cefoxitin, cefotetan) GES does not hydrolyze aztreonam and shows low hydrolysis of carbapenems	All β-lactams except aztreonam Strong hydrolysis of carbapenems	Penicillins, oxyimino cephalosporins, and aztreonam Almost no effect on cefepime and carbapenems except if coupled to porin loss or ESBLs	Oxacillin and cloxacillin No hydrolysis of aztreonam and extended-spectrum cephalosporins; weak hydrolysis of carbapenems	23,27
Modified Hodge test	Positive	Variable	Variable	Positive	28
Common inhibitors	Boronic acid, low inhibition by clavulanate	EDTA, dipicolinic acid	Boronic acid, cloxacillin	None	28
Detection by selective media (CHROMagar, SUPERCARBA)	Positive	Positive	Negative	Positive	29
Detection by Carba NP test	Positive, but not for GES-type enzymes	Positive	Negative	Positive	30

EDTA, ethylene diamine tetraacetic acid; ESBLs, extended-spectrum β-lactamases.

KPC-2, in Maryland.[30] Reports of KPC-2 producing *K. pneumoniae* soon emerged in New York City, creating a noteworthy problem.[31] Concurrent with this, a single-amino-acid variant, KPC-3, with slightly increased hydrolysis of ceftazidime was described in *K. pneumoniae*[32] and *Enterobacter* spp.[33] After rapid expansion within the United States, worldwide reports began to appear from Colombia, China, Greece, Italy, and many other European countries, with vast majority of isolates expressing KPC-2 or KPC-3.[12] KPC hydrolyzes β-lactams of all classes, most effectively nitrocefin, cephalothin, cephaloridine, benzylpenicillin, ampicillin, and piperacillin. Imipenem, meropenem, cefotaxime, ceftazidime, cefoxitin, and aztreonam are hydrolyzed less efficiently.[22] KPC is most frequently encountered in *K. pneumoniae* and has a great potential to spread due to its plasmid location.[34] The bla_{KPC} gene has been mapped to a highly conserved transposon, Tn*4401*, and plasmids carrying the gene bla_{KPC} are of various sizes and may carry additional genes conferring resistance to fluoroquinolones and aminoglycosides.[35] Mortality due to KPC-producing strains is usually high (50% or more) due to limited therapeutic options.[12]

Another class A carbapenemase is GES/IBC family, first described in 2000 with reports of IBC-1 (integron-borne cephalosporinase) from *E. cloacae* in Greece,[36] and GES-1 from *K. pneunomiae*.[37] GES enzymes were originally classified as ESBLs due to the hydrolysis of penicillins and extended-spectrum cephalosporins, but their hydrolytic spectrum was expanded in 2001 to include imipenem, with a report of GES-2 in *P. aeruginosa*.[38] Another carbapenem-hydrolyzing variant, GES-11, was described in 2009 from *A. baumannii*.[39] These two enzymes, in addition to GES-4, -5, -6, and -14, have particular amino acid substitutions at their active sites, making them effective against carbapenems. While GES-4, -5, and -6 have been reported in Enterobacteriaceae, GES-2, -11, and -14 have been restricted to *Pseudomonas* or *Acinetobacter*.[12] Today, GES enzymes have disseminated in South Africa, Japan, Greece, France, and Korea.[24] The gene bla_{GES} has often been identified within class 1 integrons residing on transferrable plasmids.[40]

Class B carbapenemases

Class B carbapenemases, which are the most clinically threatening, are MBLs and are characterized by their ability to hydrolyze all β-lactams and carbapenems except aztreonam (monobactam). This exception is mainly because these carbapenemases bind to aztreonam with low affinity, and positioning of the drug within the active site of the enzyme does not favor hydrolysis. MBLs are resistant to commercially available β-lactamase inhibitors, but they are susceptible to metal ion chelators because MBLs depend upon zinc ions in the active site of the enzyme, so chelation of zinc by metal chelators such as EDTA leads to enzyme inhibition.[22]

The first MBL genes were chromosomally located and isolated from environmental and opportunistic bacteria like *Bacillus cereus*,[41] *Aeromonas*,[42] *Stenotrophomonas maltophilia*,[43] and other nonclinically significant bacteria. However, growing concern about mobile MBLs started to rise since the mid-1990s, most frequently in carbapenem-resistant *Pseudomonas* and *Acinetobacter*, but also in Enterobacteriaceae.[24] The most prevalent types of acquired MBLs are imipenemase (IMP), Verona integrin-encoded metallo-β-lactamase (VIM), German imipenemase (GIM), Sao Paulo metallo-β-lactamase (SPM), and the emerging New Delhi metallo-β-lactamase (NDM) group. The level of carbapenem resistance observed for MBL-producing strains may vary, and the mortality ranges from 18% to 67%.[12]

IMP-type MBLs were the first acquired class B carbapenemases in *Pseudomonas* spp., *Acinetobacter* spp., and Enterobacteriaceae[24] to be identified. Today, IMP-type enzymes are endemic in Japan, Taiwan, the eastern part of China, Greece, and many other countries.[12] Analysis of the bla_{IMP} genetic environment often reveals the presence of class 1 integrons, DNA-based structures harboring antibiotic resistance genes coexpressed from a single promotor. The genes identified among the bla_{IMP}-containing integrons often encode resistance to aminoglycosides and chloramphenicol.[44] Class 1 integrons are not mobile but are usually identified inside transposon structures, thus allowing their spread.[12]

Another MBL family includes VIM enzymes. VIM-1 was first identified in Italy in 1997,[45] and shortly afterward, VIM-2 was reported in France.[46] Up to now, the VIM group has included 37 variants, mainly identified from *P. aeruginosa* and to a lesser extent from Enterobacteriaceae, with genes also corresponding to a class 1 integrons, and with VIM-2 being the MBL most often reported worldwide. Countries with endemic spread of VIM-2 are in southern Europe (Greece, Spain, and Italy) and southeast Asia (South Korea and Taiwan).[47] More geographically restricted MBLs include SPM, in Brazil, GIM in Germany, Seoul imipenemase (SIM-1) in South Korea, Dutch imipenemase (DIM-1), and *S. marcescens* metallo-β-lactamase (SMP), which remain confined to their countries of origin.[24]

Discovered in 2008 in Sweden from an Indian patient previously hospitalized in New Delhi, New Delhi metallo-β-lactamase (NDM-1) is the newest of the MBLs and has attracted worldwide attention.[48] As of 2009, NDM-1 had spread from India and Pakistan via medical tourists to the United Kingdom, the United States, France, Germany, Australia, and the Middle East.[23,49] Most patients infected or colonized with NDM-1 producers had traveled to the Indian subcontinent or have some relationship with India, Pakistan, or Bangladesh, suggesting that the Indian subcontinent is a reservoir of the bla_{NDM-1} gene. Another area of endemicity may be the Balkans, and sporadic recovery from the Middle East suggests an additional reservoir. Like KPC, conveniences of travel quickly propelled this relatively novel MBL into becoming a global public health threat, and with the exception of Central and South America, it has been identified worldwide.[24] Surveillance of tap water and seepage samples from areas around New Delhi suggests that NDM-1 may be environmental as well as being both hospital- and community-acquired.[50]

The rapid spread of NDM-1 highlights the fluidity of gene transfer among bacterial species. This carbapenemase was initially mapped to plasmids isolated from *K. pneumoniae* and *E. coli*. However, plasmid and chromosomal expression was noted in other species of Enterobacteriaceae, as well as in *Acinetobacter* and *P. aeruginosa*.[51] To date, NDM-1 remains the most common of the seven variants (NDM-1 to NDM-7). Bacteria producing NDM-1 hydrolyze all β-lactams and are usually resistant to other classes of antibiotics, including quinolones and aminoglycosides, due to resistance genes harbored on the same plasmid. They remain susceptible to colistin and tigecycline, and to a lesser extent to fosfomycin, although such susceptibility may be short-lived.[52]

Plasmid-acquired class C cephalosporinases

Plasmid-mediated *AmpC* β-lactamases have been found worldwide, with CMY-2 having the broadest geographic distribution; other important enzymes include DHA, ACT, ACC, and CFE.[53] Originally, *AmpC* genes are thought to have transferred from the chromosomal *AmpC* genes of Enterobacteriaceae to mobile genetic elements, facilitating their spread. Consequently, they now can appear in bacteria lacking or poorly expressing a chromosomal *AmpC* gene, such as *K. pneumoniae* and *Proteus mirabilis*. Although AmpC enzymes by themselves cannot hydrolyze carbapenems, they promote carbapenem resistance in isolates with ESBL or defects in permeability.[25] However, CMY-2, the AmpC enzyme with the broadest geographic distribution, appears to possess carbapenemase activity by itself.[54]

In *A. baumannii*, there is an AmpC-type cephalosporinase encoded by *bla*$_{ADC}$ termed *Acinetobacter-derived cephalosporinase (ADC)*, which is intrinsic and confers low-level resistance to carbapenems. By bringing a strong promoter, the presence of an insertion sequence, such as IS*Aba1* or IS*Aba9*, upstream from *bla*$_{ADC}$ can be responsible for an increase in expression of this gene and thus reduction in susceptibility to β-lactams.[55] The presence of such a mechanism may increase the minimum inhibitory concentration (MIC) of carbapenems in isolates of *A. baumannii*–producing carbapenemases.

The most common inhibitors of AmpC enzymes are cloxacillin and boronic acid, although the latter may be less sensitive. Because phenotypic tests cannot differentiate between chromosomal and plasmid-mediated AmpC enzymes, molecular tests are recommended. However, detection of AmpC in *K. pneumoniae*, *Salmonella* spp., or *P. mirabilis* is often confirmatory for plasmid-mediated AmpC because these organisms lack a chromosomal AmpC β-lactamase.[27] In *P. aeruginosa*, the common resistance mechanism against various β-lactams is the hyperproduction of chromosomal AmpC cephalosporinases, which contribute to carbapenem resistance in isolates with upregulated efflux pumps, porin loss, or both.[25] In one report from India, plasmid-mediated *bla*$_{CMY}$ was detected in clinical isolates of *P. aeruginosa* with or without derepressed chromosomal AmpC expression, although this did not elevate the MIC of carbapenems to the resistant range.[56]

Class D carbapenemases

Among the earliest β-lactamases detected, class D or OXA β-lactamases were relatively rare and plasmid-mediated. Their substrate profile was limited to penicillins, but some conferred cephalosporin resistance. However, unlike class A, OXA β-lactamases were able to hydrolyze and confer resistance to oxacillin as well as penicillin. As of 1980, isolates of *A. baumannii*–resistant to carbapenems emerged and manifested plasmid-encoded OXA-23, OXA-40, and OXA-58. Soon afterward, it was found that *A. baumannii* possessed a chromosomally encoded OXA-51, which may confer carbapenem resistance if the genetic environment around the gene promoted its expression.[57]

Oxacillinases are a heterogeneous group of enzymes comprising more than 200 variants, with few possessing some carbapenemase activity, which are referred to as *carbapenem-hydrolyzing class D β-lactamases (CHDLs)*.[12] With the exception of one variant (OXA-163), which has very weak carbapenemase activity and is active toward extended-spectrum cephalosporins, all other CHDLs efficiently inactivate penicillins, first-generation cephalosporins, and β-lactam/β-lactamase combinations, but spare the extended-spectrum cephalosporins. Carbapenem hydrolysis is less than for other carbapenemase classes, including MBLs, and other resistance mechanisms usually are expressed in CHDL-producing organisms to mediate high-level carbapenem resistance. These include the expression of other carbapenemases, alteration in outer membrane proteins (OMPs), increased gene copy number, and amplified drug efflux.[24] CHDLs are not inhibited by clavulanic acid, sulbactam, tazobactam, or EDTA, but they are inhibited to variable levels in vitro by sodium chloride. Different subgroups of CHDLs have been described, the most important of which are those existing in *A. baumannii* and Enterobacteriaceae, and are shown in Table 5.2.[57]

CHDLs can be either intrinsic or acquired. Intrinsic CHDLs mostly exist in *A. baumannii* and include chromosomal OXA-51, OXA-66, and OXA-69 inducing low-level carbapenem resistance.

Table 5.2 Carbapenem-Hydrolyzing Class D Carbapenemases in Gram-Negative Bacilli

Enzyme Group	No. of Enzymes in Group	Representative Examples	Locations
OXA-23-like	19	OXA-23, OXA-27, OXA-49, OXA-73, OXA-134, OXA-146, OXA-225	C, P
OXA-40/24-like	7	OXA-40, OXA-25, OXA-26, OXA-72	C, P
OXA-51-like	95	OXA-51, OXA-64-71, OXA-75-80, OXA-82-84, OXA-86-95, OXA-98-100, OXA-120-128, OXA-172-180, OXA-217, OXA-219, OXA-223, OXA-241, OXA-254	C, P
OXA-58-like	4	OXA-58, OXA-96, OXA-97, OXA-164	C, P
OXA-134a-like	7	OXA-134a, OXA-186-191	C
OXA-143-like	5	OXA-143, OXA-182, OXA-231, OXA-253	P
OXA-213	17	OXA-213	C
OXA-214-like	5	OXA-214, OXA-215	C
OXA-211-like	6	OXA-211, OXA-212, OXA-309	C
OXA-229-like	8	OXA-228-230, OXA-257	C
OXA-235-like	7	OXA-235-237, OXA-278	C
OXA-48-like	11	OXA-48, OXA-48B, OXA-162, OXA-163, OXA-181, OXA-232, OXA-247	C, P

C, chromosome; P, plasmid.

The first acquired oxacillinase with carbapenemase activity was purified from a MDR *A. baumannii* in 1985 from a Scottish patient and first described in 1993.[20] It was designated *Acinetobacter*-resistant to imipenem (ARI-1) and was plasmid-mediated. Sequencing of the enzyme revealed that it belonged to Ambler class D and was later renamed OXA-23.[58] Today, OXA-23 is reported worldwide in the *Acinetobacter* spp. and hydrolyzes oximinocephalosporins, aminopenicillins, piperacillin, oxacillin, and aztreonam in addition to carbapenems. Among the latter, OXA-23 has a higher turnover rate for imipenem than for meropenem, ertapenem, or doripenem. Production of OXA-23 by *A. baumannii* is enough to induce carbapenem resistance, and other synergistic pathways are not necessary to obtain the resistant phenotype. However, high levels of resistance are achieved only when other mechanisms of resistance are present.[57]

Spread of CHDLs among *A. baumannii* is not limited to OXA-23; OXA-58 has been identified initially in France[59] and then worldwide. Many isolates with $bla_{\text{OXA-58}}$ gene concurrently carry insertion sequences IS*Aba1*, IS*Aba2*, or IS*Aba3*, associated with increased carbapenemase production and higher levels of carbapenem resistance.[24]

OXA-48 was originally identified in a carbapenem-resistant *K. pneumoniae* in Turkey,[60] and later, OXA-48 was detected worldwide.[12] This enzyme hydrolyzes penicillins but has a weak activity against carbapenems, third-generation cephalosporins, and aztreonam.[22] The gene $bla_{\text{OXA-48}}$ was mapped to a self-conjugative, 62-kb plasmid, contributing to its spread, and integrated through acquisition of Tn*1999* composite transposon that harbors the insertion sequence IS*1999* participating in mobilization and expression of the gene.[61] More recently, a chromosomal location of $bla_{\text{OXA-48}}$ also has been suggested.[62] In *P. aeruginosa*, and although early class D enzymes like

OXA-11, OXA-13, and OXA-14 have been described and showed ESBL activity,[57] no carbapenem-hydrolyzing OXA enzymes have been described thus far in this organism.

ALTERATIONS IN OMPs

In general, a decrease in the production of OMPs, which allow the transfer of β-lactams through the bacterial outer membrane, lowers the effective concentration of antibiotics inside the bacterial cell, and increases the MIC of antibiotics. Resistant phenotypes may be observed if this mechanism is combined with another one, such as expression of a β-lactamase.[63] Normally, OMPs have evolved in Gram-negative outer membranes to facilitate the uptake of nutrients, and they consist of water-filled channels extending across the membrane; proteins that form such channels are called *porins* and were first characterized in 1976 in *E. coli*.[64] Since then, porins have been identified in Gram-negative bacteria and mycobacteria. It is located on the outer membrane to provide a sieving function monitored by the diameter of channels present in each individual organism. OMPs tend to have exclusion limits approaching the size of many antibiotics, decreasing their diffusion and contributing to intrinsic resistance.

When it comes to carbapenems, resistance attributed to porin loss can reduce drug entry in strains already containing high levels of AmpC or ESBLs.[63] This is a common source of low-level resistance to ertapenem, especially in *Enterobacter* and *Klebsiella*, with resistant variants sometimes selected during carbapenem therapy.[65] In such circumstances, carbapenem resistance is coupled with resistance to aztreonam and third-generation cephalosporins. In *K. pneumoniae*, imipenem resistance can be multifactorial. It can result if high-level AmpC production or extended-spectrum SHV derivative is present along with the loss of a major OMP such as OprK-35 or OprK-36.[25]

More important, alteration of the OMP known as OprD in *P. aeruginosa* affects the ability of carbapenems to reach their target, as OprD is the preferred pathway through the outer membrane of *P. aeruginosa* for this drug class. Loss of OprD in itself increases imipenem MICs into the intermediate or resistant category, while resistance to meropenem and doripenem requires additional mechanisms.[66] Evidence shows that OprD in *P. aeruginosa* could be utilized by imipenem and meropenem but could not be significantly utilized by other β-lactams, quinolones, or aminoglycosides, and its reduced expression is frequently noted in carbapenem-resistant isolates. Mutations, deletions, or insertions in the gene-encoding OprD porin can occur; an example is the insertional inactivation of genes by insertion sequence elements of various sizes like IS*Pa46*. While not providing high-level resistance, loss of OprD function is the major determinant of nonmetallo-β-lactamase-mediated resistance to carbapenem in *P. aeruginosa*, and it is often seen in conjunction with other mechanisms, such as derepressed AmpC cephalosporinases.[67]

Recent reports have demonstrated that *A. baumannii* possesses OMPs that play a role in carbapenem resistance. This organism ordinarily has a smaller size and number of porins than other Gram-negative organisms, contributing to intrinsic outer membrane impermeability, which appears to be responsible for natural resistance to ertapenem. Three porins of reduced expression have been associated with carbapenem resistance in *A. baumannii*: CarO, Omp33–36, and OprD homologue, similar to that of *P. aeruginosa*.[68] However, recent evidence indicates that this homologue is not involved in carbapenem resistance, but it is a specific binding site for magnesium and iron ions, allowing *A. baumannii* to adapt to stress.[69] The most characterized porin in *A. baumannii* is the carbapenem-associated OMP called CarO, whose loss reduces imipenem penetration into the cell. The absence of a meropenem-binding site on CarO suggests that meropenem may penetrate via another type of OMP, which may be Omp33–36, but also could be through OmpF in Enterobacteriaceae, or through OmpD2, OprF, or OprC in *P. aeruginosa.*

Mutational resistance related to a particular porin change has only a minor-to-moderate effect on the overall resistance of the microorganism, leading only to low-level resistance. However, the accumulation of a sequence of independent mutational events affecting various resistance mechanisms can confer a stepwise alteration until high-level resistance is acquired. For example, meropenem-resistant *S. marcescens* strains overproduce β-lactamases and lack the porin OmpF.[70] Similarly, carbapenem-resistant *K. pneumoniae* and *E. coli* that produce carbapenemases also can lack certain OMPs.[71]

EFFLUX PUMP OVERACTIVITY

Efflux pumps constitute between 6% and 18% of all transporters in a given bacterial species and are highly accommodating and broadly unspecific for a wide spectrum of substrates. Their best-known role is the ability to export antibiotics, exogenous noxious compounds, and byproducts of metabolism. Efflux pumps can be divided into two main classes: ATP-binding cassette (ABC) transporters and secondary multidrug transporters, and the major difference between them is the source of energy required for transport (ATP hydrolysis vs proton motive force). Transporters of the ABC family can carry numerous substrates including sugars, amino acids, polysaccharides, and proteins, but are not involved in antibiotic resistance. By contrast, secondary drug transporters include the majority of clinically relevant efflux systems and are classified into four superfamilies based on chemical structure. These include the major facilitator superfamily, small multidrug resistance family, multidrug and toxic compound extrusion family, and resistance nodulation cell division (RND) superfamily. This last group is the most clinically relevant in the context of resistance and has been well characterized in Gram-negative bacteria, and its substrates comprise multiple antibiotics and other compounds.[72]

Efflux pump mutations represent a larger threat in microorganisms that couple them with low permeability of the cell envelope, like Gram-negative pathogens, due to the existence of synergy between these two resistance strategies. In *A. baumannii*, the AdeABC efflux pump, a member of the RND superfamily, has been well characterized.[73] It pumps aminoglycosides, cefotaxime, tetracyclines, erythromycin, chloramphenicol, trimethoprim, and fluoroquinolones. Overexpression of this efflux pump also may confer high-level resistance to carbapenems.[74] In *P. aeruginosa*, 12 RND-type efflux pumps exist, of which MexAB-OprM is by far the most prominent exporter of carbapenems and contributes to reduced susceptibility to meropenem in clinical isolates.[67] This pump drives carbapenem resistance often in the presence of OprD loss, derepressed AmpC enzymes, or both. The MexCD-OprJ and MexXY-tripartite efflux pump systems are additional RND-type pumps in *P. aeruginosa* and they also target and extrude carbapenems except imipenem.[75] Among Enterobacteriaceae, the presence of the RND family pumps AcrAB-TolC, AcrD, ACrE, and AcrF has been well documented, and their overexpression may be relevant to MDR.[76] Efflux pumps have been implicated in carbapenem resistance in *E. coli* and *K. pneumoniae*.[62,77] Moreover, active efflux pumps, together with enzyme production and decreased OMP expression, caused ertapenem resistance in *E. cloacae* isolates from Taiwan.[78]

PRODUCTION OF ALTERED CELL WALL TRANSPEPTIDASES

Although rarely described in Gram-negative bacteria, the alteration of cell wall penicillin-binding proteins (PBPs) may be involved in carbapenem resistance. Conversely, in Gram-positive bacteria (except for *Staphylococci*), which produce a narrow-spectrum penicillinase, clinically important β-lactam resistance arises through the expression of PBPs that bind β-lactams with low affinity.[79]

In carbapenem-resistant, Gram-negative bacteria, the effect of altered PBPs is mostly evident for *A. baumannii*, where reduced expression of PBP2 was among the frequently observed resistance mechanisms to carbapenems.[80] More recently, in Spain, a disruption in the insertion sequence of the gene-encoding PBP6b was identified in an endemic carbapenem-resistant *A. baumannii* strains.[81] Of note, these strains suffered loss of OMPs and production of β-lactamases, illustrating the interplay of several different mechanisms of carbapenem resistance. In Enterobacteriaceae, an early report of PBP2 mutation in *P. mirabilis* provided evidence that such mutation was implicated in imipenem resistance.[82] In *E. coli*, PBP alteration was responsible for resistance to meropenem and doripenem but not to imipenem,[83] whereas conflicting data exist concerning the role of PBP in carbapenem resistance in *P. aeruginosa*.[84,85] The establishing of relative contribution of altered PBPs, control of their expression, and interaction with other mechanisms of carbapenem resistance in Gram-negative bacteria presents formidable challenges.

LABORATORY DETECTION OF CARBAPENEM-NONSUSCEPTIBLE, GRAM-NEGATIVE BACILLI

Carbapenem resistance is an issue expected to preoccupy the scientific community for many years in the future, since few antibiotic alternatives equivalent to carbapenems are readily available. Therefore, prompt and reliable detection of carbapenem-nonsusceptible, Gram-negative organisms is indispensible for therapeutic, epidemiological, and infection control purposes. Such detection can embrace a variety of methods, as summarized in Fig. 5.2. These include antimicrobial susceptibility testing, testing for carbapenemase production like modified Hodge testing, inhibitor-based testing, and molecular characterization of resistance genes by polymerase chain reaction (PCR) sequencing. Besides microbiological and genetic methods, innovative biochemical, analytical, and spectrometric techniques may be exploited. The details of detection methodology have been reviewed extensively.[86,87] Certain shortcomings are associated with each of the available methods, and their accurate application requires adequate standardization.

ANTIMICROBIAL SUSCEPTIBILITY TESTING

An increase in carbapenem MIC or a decrease in inhibition zone diameter are the first causes of suspicion of carbapenem nonsusceptibility, and they render a bacterial isolate eligible for further analysis of carbapenem resistance mechanisms. The carbapenem susceptibility ranges for Enterobacteriaceae, *Pseudomonas*, and *Acinetobacter* are given by the European Committee on Antimicrobial Susceptibility Testing (EUCAST)[88] and Clinical and Laboratory Standards Institute (CLSI)[89] and are shown in Table 5.3. However, the detection of carbapenem nonsusceptibility based only on MIC may not be accurate because certain carbapenemase-producing Enterobacteriaceae may show MIC values within the susceptibility range. Intermediate susceptibility

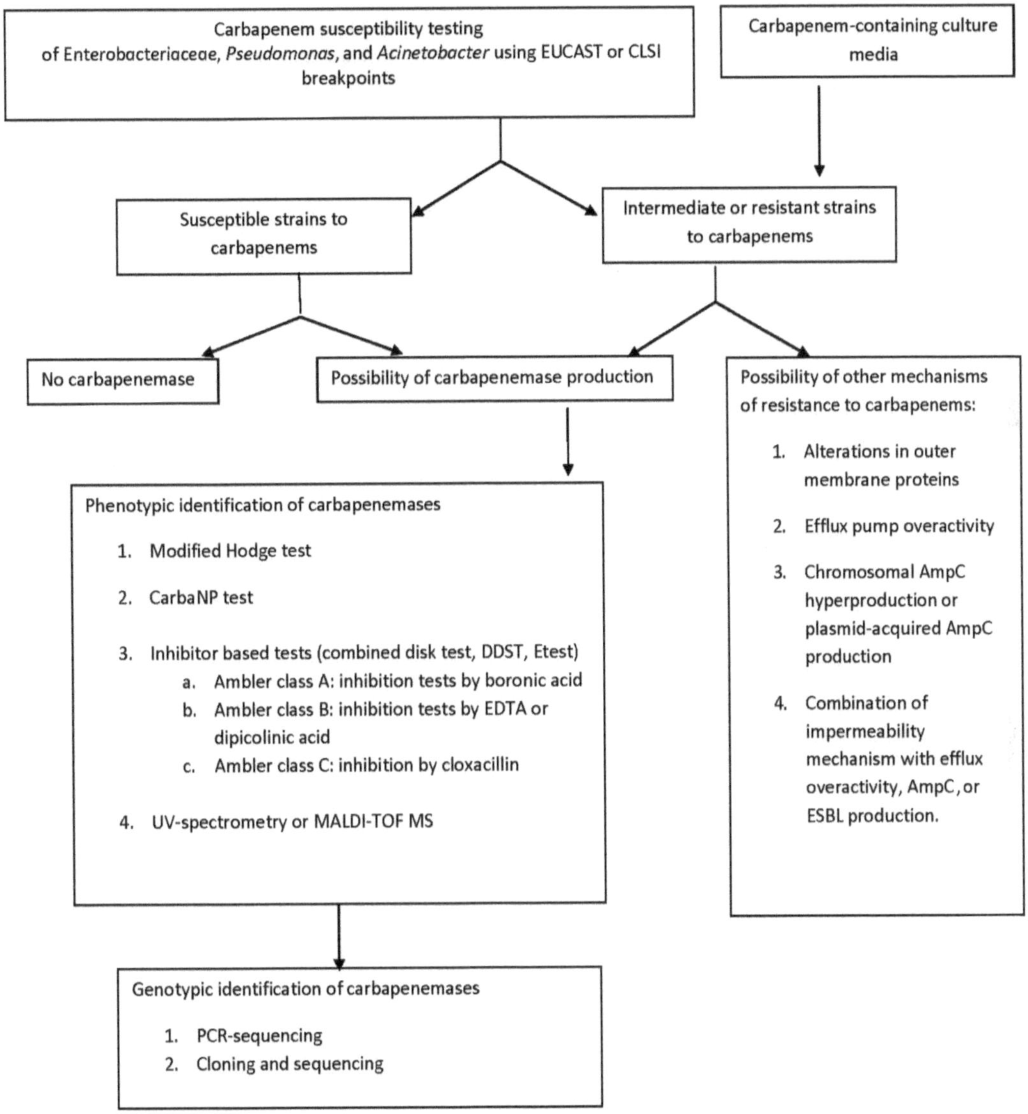

FIGURE 5.2

Simplified scheme for detection of carbapenem-nonsusceptible, Gram-negative bacilli. *DDST*, double disk synergy test; *EDTA*, ethylene diamine tetraacetic acid; *ESBL*, extended-spectrum β-lactamase; *MALDI-TOF MS*, matrix-assisted laser desorption ionization time-of-flight mass spectrometry; *PCR*, polymerase chain reaction.

or even sensitivity to carbapenems has been observed for producers of all types of carbapenemases, especially the OXA-48/OXA-181-producing Enterobacteriaceae that do not coharbor an ESBL (Table 5.4). In such cases, carbapenem MICs are expected to substantially rise only in the presence of an additional resistance mechanism, like permeability lesions due to OMP mutation,

Table 5.3 Breakpoints of Carbapenem Susceptibility for Enterobacteriaceae, *Pseudomonas*, and *Acinetobacter* According to European (EUCAST, 2014) and US (CSLI, 2014) Guidelines

		EUCAST				CLSI			
		MIC Breakpoint (mg/L)		Zone Diameter Breakpoint (mm) for 10 μg Disk		MIC Breakpoint (mg/L)		Zone Diameter Breakpoint (mm) for 10 μg Disk	
Microorganism	**Carbapenem**	**S≤**	**R>**	**S≥**	**R<**	**S≤**	**R≥**	**S≥**	**R≤**
Enterobacteriaceae	Doripenem	1	2	24	21	1	4	23	19
	Ertapenem	0.5	1	25	22	0.5	2	22	18
	Imipenem	2	8	22	16	1	4	23	19
	Meropenem	2	8	22	16	1	4	23	19
Pseudomonas	Doripenem	1	2	25	22	2	8	19	15
	Ertapenem	–	–	–	–	–	–	–	–
	Imipenem	4	8	20	17	2	8	19	15
	Meropenem	2	8	24	18	2	8	19	15
Acinetobacter	Doripenem	1	2	23	20	2	8	18	14
	Ertapenem	–	–	–	–	–	–	–	–
	Imipenem	2	8	23	17	2	8	22	18
	Meropenem	2	8	21	15	2	8	18	14

MIC, minimum inhibitory concentration; S, sensitive; R, resistant.

or simultaneous production of AmpC cephalosporinases or ESBLs. To maximize detection sensitivity and diminish false negatives, it has been proposed to screen enterobacterial isolates for carbapenem nonsusceptibility if they exhibit MICs of ertapenem greater than or equal to 0.5 mg/L or MICs of imipenem or meropenem greater than or equal to 1 mg/L, or to screen any enterobacterial isolate displaying a slight decrease in susceptibility to carbapenems compared with the wild-type organism.[90]

DETECTION USING CARBAPENEM-INCLUDING CULTURE MEDIA

Commercial media incorporating carbapenems are available for the detection of carbapenem-resistant isolates; they inhibit carbapenem-sensitive pathogens and include chromogenic molecules to permit recognition of the Gram-negative carbapenem-resistant genera. For example, the ChromID Carba agar was demonstrated as an efficient screening agar to differentiate CRE.[91] Other media include the Brilliance CRE agar, CHROMagar KPC, and CHROMagar *Acinetobacter*, which are all commercially available and provide reasonable sensitivity and specificity in detecting carbapenem-resistant, Gram-negative pathogens, but they cannot differentiate between carbapenem resistance due to the secretion of carbapenemases or other resistance mechanisms. A more recently

Table 5.4 Ranges of MICs of Carbapenems for Clinical Gram-Negative Bacteria Expressing the Main Carbapenemases

	Carbapenem MIC (mg/L)		
Gram-Negative Bacteria	**Imipenem**	**Meropenem**	**Ertapenem**
Enterobacteriaceae			
KPC (Ambler class A)	0.5 to >32	0.5 to >32	0.5 to >32
IMP/VIM/NDM (Ambler class B)	0.5 to >32	0.5 to >64	0.38 to >32
OXA-48/OXA-181 (Ambler class D)	0.25 to 64	0.5 to 64	0.38 to >32
Pseudomonas			
KPC (Ambler class A)	2 to >64	2 to >64	–
IMP/VIM/NDM (Ambler class B)	>64	>64	–
Acinetobacter			
KPC (Ambler class A)	2 to >64	2 to >64	–
IMP/VIM/NDM (Ambler class B)	1 to >64	1 to >64	–
OXA-23/OXA-40/OXA-58/OXA-143 (Ambler class D)	>32	>32	–

MIC, minimum inhibitory concentration.
Data from:
Nordmann P, Naas T, Poirel L. Global spread of Carbapenemase-producing Enterobacteriaceae. Emerg Infect Dis *October 2011;**17**(10):1791–8.*
Nordmann P, Gniadkowski M, Giske CG, Poirel L, Woodford N, Miriagou V, et al. Identification and screening of carbapenemase-producing Enterobacteriaceae. Clin Microbiol Infect *May 2012;**18**(5):432–8.*

developed medium for the detection of CRE is SUPERCARBA, which consists of Drigalski agar supplemented with ertapenem, zinc sulfate to enhance expression of MBL producers, and cloxacillin to inhibit natural AmpC natural producers like *E. cloacae*, *Enterobacter aerogenes*, *Morganella morganii*, and *S. marcescens*.[28] Theoretically, this medium offers the slight advantage of inhibiting carbapenem-resistant but noncarbapenemase-producing bacteria, giving an insight into probable resistance mechanisms. This medium also detects CRE with low-level resistance, including OXA-48 producers.

TESTS FOR THE DETECTION OF CARBAPENEMASES

Modified Hodge test

Also known as the *cloverleaf test*, this phenotypic method is based upon the inactivation of a carbapenem by a carbapenemase-producing test strain. An indicator organism (usually *E. coli* ATCC 25922 at a turbidity of 0.5 McFarland standards) is used to inoculate on the Mueller–Hinton agar plate surface, and a carbapenem disk is placed at the center. The test strain is heavily streaked from the disk to the plate periphery. After overnight incubation, the cloverleaf-shaped indentation of growth of the tested strain versus the susceptible indicator strain is interpreted as a positive result for carbapenemase production by the tested strain.[92] The modified Hodge test assumes excellent sensitivity for the detection of Ambler classes A and D carbapenemases, but it lacks sensitivity in detecting MBLs; false-positive results occur with the strains that produce the ESBL of type CTX-M with reduced outer membrane permeability, and to a lesser extent for those hyperproducing AmpC cephalosporinases.[93] Regardless of these limitations, this easy and inexpensive test may be incorporated as part of a strategy that facilitates checking carbapenemase activity in outbreak situations.

Biochemical carbapenemase detection by CarbaNP test

A recent biochemical detection method of carbapenemases in Enterobacteriaceae and *Pseudomonas* is the CarbaNP test. The principle is based upon the hydrolysis of the β-lactam ring of imipenem by the tested strain, followed by color change of a pH indicator.[29] Not only does the test detect all known carbapenemases of Ambler classes A, B, and D, but it also identifies virtually any newly emerging carbapenemase, in contrast to molecular methods (described next), is cheap, and needs no special equipment. However, the test cannot differentiate among carbapenemase classes and does not detect GES-type carbapenemases, which have a rather weak carbapenemase activity; it also carries the risk of false negative results in isolates producing OXA-48.[94]

Inhibitor-based tests for carbapenemase detection

Inhibitors of carbapenemase production can be utilized to allow in vitro observation of an increase in inhibition zone diameter (or reduction of the MIC) of the tested isolate in the presence of a carbapenem combined with a carbapenemase inhibitor compared to the same carbapenem alone.[93] Commonly used carbapenemase inhibitors for various classes of carbapenemases are shown in Table 5.1. Verification of synergy between the carbapenem and the inhibitor can be done by combined or double disk tests. Epsilometer test (Etest) strips employing carbapenem and carbapenem combined with EDTA are commercially available as well. The inhibitor also may be incorporated in the culture medium to demonstrate a reduction in carbapenem MIC or an increase in carbapenem inhibition zone diameter.

Genotypic identification of carbapenemases

Molecular techniques remain the standard for precise identification of the genetic background of carbapenem resistance, especially carbapenemases. Most are based on PCR technology, possibly followed by sequencing to know the exact variety of carbapenemase gene rather than just its group (eg, VIM-type, KPC-type, NDM-type, or OXA-type). PCR results are achievable within 4–6 h with high sensitivity and specificity and can be single, multiplex, or quantitative, with the latter even reducing the detection time. Furthermore, DNA hybridization techniques in a microarray format allow the simultaneous detection of many carbapenemases.

The adoption of molecular techniques offers various advantages and tends to replace the widely used phenotypic tests; however, their high cost remains a prohibitive burden for many laboratories, especially in some countries where carbapenem resistance is endemic. Besides the disadvantages of high cost and requirement for trained microbiologists, an additional issue with PCR is that the range of resistance genes to be detected is predefined, and novel carbapenemase genes may stay undetectable unless identification is performed using gene cloning and sequencing.

In cloning experiments, genomic DNA from bacterial strains is cut by restriction enzymes and then allowed to hybridize with a plasmid that has been cut via the same enzyme to create so-called sticky ends. When the fragment of DNA is joined with the cloning vector, the resulting recombinant DNA molecule harboring the carbapenemase gene can be used to transform competent bacterial cells, in which the new gene can be reproduced along with the recipient cell's DNA and purified before sequencing. Several studies have utilized cloning experiments to identify new carbapenemase genes.[39]

Analytical and biochemical detection methods of carbapenemases

Detection of carbapenemase production may be performed using analytical and biochemical methods.[87] For example, spectrophotometric assays can detect the hydrolysis of amide bonds in carbapenems and quantify hydrolysis at wavelengths of 296 nm.[95] In 2011, the first methods for the detection of carbapenemase activity via matrix-assisted laser desorption ionization time-of-flight mass spectrometry (MALDI-TOF MS) were published.[96] In those assays, a fresh bacterial culture is mixed with carbapenem solution, and the mass spectra of the intact molecule and its degradation products are analyzed.

Electrophoresis of carbapenemases through isoelectric focusing, as well as immunodiagnostic methods including immunochromatography and immunofluorescence, also have been employed, and are described elsewhere.[86]

DETECTION OF ALTERATIONS IN OMPs

Genetic mutations leading to loss of production of OMPs or modification of porin size or conductance can have a direct impact on permeability to antibiotics. The most prominent example is the lack of the basic amino-acid-selective OMP OprD in *P. aeruginosa*.[97] Detection of these mutations can be realized by PCR sequencing at the gene level.

Moreover, genetic techniques like real-time PCR may be useful to investigate levels of expression of specific OMPs, like OprD in carbapenem-resistant *P. aeruginosa*. Indeed, low porin expression levels may be responsible for an increase in carbapenem MICs. Imipenem resistance in *P. aeruginosa* has been correlated with a low expression of the OprD gene,[98] and a reduction of OmpF porin expression in *E. cloacae* has been associated with carbapenem resistance in this species.[63]

STUDY OF EFFLUX PUMP OVERACTIVITY

Tests using inhibitors of efflux pumps

Carbapenem-resistant, Gram-negative pathogens may display overexpressed efflux pumps, usually combined with other mechanisms of resistance, and chemical inhibitors of efflux pumps may be used in culture to allow detection of such mechanism. The efflux pump inhibitors phenylalanine-arginin-β-naphthylamide dihydrochloride (PAβN) and carbonyl cyanide *m*-chlorophenylhydrazone (CCCP) may be used at concentrations of 25 μg/mL or 12.5 μM, respectively, in Mueller–Hinton agar to inhibit efflux pumps in Gram-negative bacteria.[78] Carbapenem disks can be tested in parallel on agar plates with and without the inhibitors. A decrease of at least twofold in the MICs of carbapenems in the presence of the inhibitors suggests that active efflux contributes to resistance. This methodology has been used in studies investigating efflux pump overactivity in CRE, *A. baumannii*, and *P. aeruginosa*.[78,99,100]

Real-time PCR for detection of efflux pump overexpression

Overexpression of multidrug efflux pumps like MexAB-OprM in *P. aeruginosa* is considered to play a role in meropenem resistance in this species. Studies based on real-time PCR showed a relationship between efflux pump messenger RNA (mRNA) levels and meropenem resistance.[101] To be able to elucidate such a mechanism in clinical strains, an analysis of the level of transcription of resistance pump genes and a quantification of their mRNA by real-time PCR may be helpful.

OPTIMAL TREATMENT OF INFECTIONS CAUSED BY CARBAPENEM-RESISTANT, GRAM-NEGATIVE BACILLI

Because of the scarcity of convincing clinical data, the preferable choices of antibiotics for the treatment of infections caused by carbapenemase-producing Gram-negative bacilli remain controversial.[102] Current clinical evidence is based on retrospective, observational studies and case reports of severe infections with KPC- or VIM-producing strains. However, the results from these studies are inconsistent, probably due to variations in settings, patient populations, illness severity, type of infections, and causative bacteria, among other factors.[103] Therefore, definitive therapy should always be guided by susceptibility testing, and it is recommended to consult infectious disease specialists.[1]

MONOTHERAPY

Colistin (polymyxin B), tigecycline, and fosfomycin are typically recommended as single-agent therapies for infections caused by carbapenemase-producing, Gram-negative pathogens, with colistin being the backbone agent and is used in substantially high doses (a loading dose and a total standard dose of 9–10 million international units daily divided into two or three doses).[104] However, colistin use is limited by nephrotoxicity and neurotoxicity and is inactive against isolates of *Proteus*, *Providencia*, *Morganella*, and *Serratia*.[1] Tigecycline also has been questioned as a single agent in the treatment of infections caused by carbapenem-resistant, Gram-negative bacilli, but with mortality rates higher than broad-spectrum β-lactams.[105] This agent has no activity against *Pseudomonas*, *Proteus*, *Providencia*, and *Morganella*. In addition, it is only bacteriostatic and achieves poor serum and urine levels and inadequate concentrations at some infection sites, contributing to poor clinical efficacy. Dosing regimens of 75–100 mg at 12-h intervals have been proposed to improve the clinical response to tigecycline.[106] Oral monotherapy with fosfomycin has not been thoroughly evaluated in clinical studies and is probably associated with a high risk of emerging resistance during therapy.[107] Its use, much like nitrofurantoin and some selected aminoglycosides such as gentamycin, is limited to lower urinary tract infections.[1]

COMBINATION THERAPY

Antibiotic combinations have demonstrated a favorable effect and a mortality benefit in infections caused by carbapenem-resistant, Gram-negative bacteria even when the isolates were resistant to the single agents.[103] Double and triple combinations including aminoglycoside, aztreonam, carbapenem, colistin, rifampicin, tigecycline, and fosfomycin have demonstrated synergistic or bactericidal effects.[106] By acting as a detergent that facilitates penetration of other antibiotics across the Gram-negative outer membrane, colistin appears to be a vital component of efficacious combinations.[108] Frequently prescribed combinations include colistin/tigecycline, colistin/carbapenem, colistin/aminoglycoside, and carbapenem/aminoglycoside.

Other double or triple combinations with these antibiotics have been used with variable success. Combinations including fosfomycin might be an option for strains resistant to colistin and other treatments.[109] Data regarding such combinations, however, should be interpreted with caution due to the small sample sizes and methodological limitations of the available studies, and combination therapy for carbapenem-resistant pathogens will require data from larger, randomized clinical trials. Of note, local resistance epidemiology and individual risk factors for resistance, including recent, broad-spectrum antibiotic use, medical procedures, older age, hospitalization, and previous colonization or infection with carbapenem-resistant strains, should guide the choice of empirical antibiotic therapy. Combination empirical therapy including colistin, carbapenem or aminoglycoside might be justified for severely ill patients with suspected carbapenem-resistant strains.[103]

CONTROL AND SURVEILLANCE OF CARBAPENEM-RESISTANT, GRAM-NEGATIVE PATHOGENS

The rise of carbapenem resistance will continue to represent a major infection control threat, awaiting new strategies (development of new antimicrobials, phage therapy, use of antimicrobial peptides, etc.) to combat it. However, unless a breakthrough change in therapy options occurs, a proactive approach aimed at preventing transmission and infections by these organisms may be the only effective way to curtail their dissemination.

PREVENTION AND CONTROL STRATEGIES

Prevention strategies for these organisms are similar to those for other resistant pathogens and include hand hygiene, use of barrier precautions, isolation, and aggressive environmental cleaning. Control practices are usually bundled and should be tailored to the situation and available resources. A guideline that highlights recommended strategies has been published by the Centers of Disease Control and Prevention (CDC).[110] The most important issues are discussed next.

Laboratory detection

Accurately identifying carbapenem-resistant, Gram-negative bacilli in the clinical laboratory is an important first step in prevention. To avoid missing some resistant isolates that may have MIC values within susceptible ranges, lower carbapenem breakpoints were established (Table 5.3).

Laboratory personnel should ensure that clinical and infection control staff are notified about any carbapenem-resistant isolate.

Patient isolation

Single-patient rooms, or if not available, cohorted rooms should be used in health-care facilities to accommodate patients colonized or infected with carbapenem-resistant pathogens. This applies to both acute and long-term care settings. Patients at the highest risk for transmission, such as those with severe disease, comorbidities, incontinence, medical devices, or wounds with uncontrolled drainage, should be given preference for placement in single rooms.

Health-care personnel education

Health-care personnel should be educated about preventing transmission of carbapenem-resistant organisms. This includes adequate hand hygiene with soap and alcohol, contact precautions with patients, and minimizing the use of devices that put patients at risk of carbapenem-resistant infections. This latter point is achievable via the regular review of devices to ensure that they are still required and discontinued promptly when no longer needed.

Antimicrobial stewardship

Multidisciplinary teams, including physicians, pharmacists, and microbiologists, should be encouraged to practice antibiotic stewardship, and it should be extended from inpatients to outpatients. Stewardship programs aim to accomplish a synergy among improving patient outcomes, limiting resistance and superinfections, and decreasing health-care expenditures. This is achievable via administering antibiotics only for appropriate indications and duration and using the narrowest-spectrum antimicrobial whenever possible. Carbapenem restriction has been associated with lower rates of carbapenem-resistant *P. aeruginosa*; however, more research is needed to clarify the effects of such restriction on CRE.[111]

SURVEILLANCE AND MOLECULAR TYPING OF STRAINS

Inpatient facilities should be aware whether carbapenem-resistant pathogens have ever been cultured from patients, and in such cases, whether there is evidence of intrafacility transmission and which units were mostly affected. In addition, facilities should consider collecting information from patients colonized or infected with these organisms like demographics, dates of admission, outcomes, medications, and medical procedures. Screening of epidemiologically linked patients for carriage of carbapenem-resistant bacteria is a primary prevention strategy, and can involve stool, urine, or wound cultures. Molecular typing methods to determine the clonality of carbapenem-resistant isolates are vital for surveillance programs and help in identification of high-risk clones that have potential of dissemination.

Various methods may be used to assess clonal relatedness, including enterobacterial repetitive intergenic consensus–based PCR, pulsed field gel electrophoresis, and multilocus sequence typing (MLST). In areas with long-term persistence of carbapenem resistance, MLST offers the opportunity to track clones and exchange allelic information among different geographic areas. In areas with low prevalence, MLST identifies the emergence of highly epidemic clones.[112] Information on MLST is available via public databases like the Pasteur Institute MLST database and PubMLST, developed by the University of Oxford. These databases tend to provide a common language on microbial strain typing, epidemiology, and evolution.

CONCLUSIONS

Research on carbapenem resistance in Gram-negative bacteria continues. While the spread of these organisms is expected to escalate, further mechanistic, microbiological, and epidemiological studies would be more illustrative of the global status of such a formidable menace with implications for public health and society. The future holds several potential strategies and novel therapeutics to combat carbapenem-resistant pathogens. Examples of these are lytic bacteriophages that attack carbapenem-resistant bacteria, and antimicrobial peptides that are essential components of the innate immune system. One antimicrobial peptide, piscidin, isolated from an aquatic organism, recently demonstrated high activity against carbapenem-resistant *A. baumannii* and NDM-1-producing *K. pneumoniae* in mice experiments.[113] Trials on new, next-generation antimicrobials are also on the horizon. Until such options become attainable, and in light of easy transmission and spread of resistance determinants among these distressing pathogens, advanced diagnostics and effective surveillance apparently remain crucial to mitigate their impact.

REFERENCES

1. Vasoo S, Barreto JN, Tosh PK. Emerging issues in gram-negative bacterial resistance: an update for the practicing clinician. *Mayo Clin Proc* 2015;**90**(3):395–403.
2. Wilke MS, Lovering AL, Strynadka NCJ. Beta-lactam antibiotic resistance: a current structural perspective. *Curr Opin Microbiol* 2005;**8**(5):525–33.
3. Jeon JH, Lee JH, Lee JJ, Park KS, Karim AM, Lee C-R, et al. Structural basis for carbapenem-hydrolyzing mechanisms of carbapenemases conferring antibiotic resistance. *Int J Mol Sci* 2015;**16**(5):9654–92.
4. Akova M, Daikos GL, Tzouvelekis L, Carmeli Y. Interventional strategies and current clinical experience with carbapenemase-producing gram-negative bacteria. *Clin Microbiol Infect* 2012;**18**(5):439–48.
5. Falagas ME, Tansarli GS, Karageorgopoulos DE, Vardakas KZ. Deaths attributable to carbapenem-resistant Enterobacteriaceae infections. *Emerg Infect Dis* 2014;**20**(7):1170–5.
6. Nordmann P, Poirel L. The difficult-to-control spread of carbapenemase producers among Enterobacteriaceae worldwide. *Clin Microbiol Infect* 2014;**20**(9):821–30.
7. Daroukh A, Delaunay C, Bigot S, Ceci JM, Siddhoun N, Bukreyeva I, et al. Characteristics and costs of carbapenemase-producing enterobacteria carriers (2012/2013). *Med Mal Infect* 2014;**44**(7):321–36.
8. Lemos EV, de la Hoz FP, Alvis N, Einarson TR, Quevedo E, Castañeda C, et al. Impact of carbapenem resistance on clinical and economic outcomes among patients with *Acinetobacter baumannii* infection in Colombia. *Clin Microbiol Infect* 2014;**20**(2):174–80.
9. Shahid M, Sobia F, Singh A, Malik A, Khan HM, Jonas D, et al. Beta-lactams and beta-lactamase-inhibitors in current- or potential-clinical practice: a comprehensive update. *Crit Rev Microbiol* 2009;**35**(2):81–108.
10. Harris P, Paterson D, Rogers B. Facing the challenge of multidrug-resistant Gram-negative bacilli in Australia. *Med J Aust* 2015;**202**(5):243–6.

11. Lovering AL, de Castro LH, Lim D, Strynadka NCJ. Structural insight into the transglycosylation step of bacterial cell-wall biosynthesis. *Science* 2007;**315**(5817):1402–5.
12. Nordmann P, Dortet L, Poirel L. Carbapenem resistance in Enterobacteriaceae: here is the storm! *Trends Mol Med* 2012;**18**(5):263–72.
13. Paterson DL, Bonomo RA. Extended-spectrum beta-lactamases: a clinical update. *Clin Microbiol Rev* 2005;**18**(4):657–86.
14. Philippon A, Arlet G, Jacoby GA. Plasmid-determined AmpC-type beta-lactamases. *Antimicrob Agents Chemother* 2002;**46**(1):1–11.
15. Paterson DL, Doi Y. A step closer to extreme drug resistance (XDR) in Gram-negative bacilli. *Clin Infect Dis* 2007;**45**(9):1179–81.
16. Frère J-M, Galleni M, Bush K, Dideberg O. Is it necessary to change the classification of {beta}-lactamases? *J Antimicrob Chemother* 2005;**55**(6):1051–3.
17. Yang YJ, Wu PJ, Livermore DM. Biochemical characterization of a beta-lactamase that hydrolyzes penems and carbapenems from two *Serratia marcescens* isolates. *Antimicrob Agents Chemother* 1990;**34**(5):755–8.
18. Naas T, Nordmann P. Analysis of a carbapenem-hydrolyzing class A beta-lactamase from *Enterobacter cloacae* and of its LysR-type regulatory protein. *Proc Natl Acad Sci USA* 1994;**91**(16):7693–7.
19. Watanabe M, Iyobe S, Inoue M, Mitsuhashi S. Transferable imipenem resistance in *Pseudomonas aeruginosa. Antimicrob Agents Chemother* 1991;**35**(1):147–51.
20. Paton R, Miles RS, Hood J, Amyes SG, Miles RS, Amyes SG. ARI 1: beta-lactamase-mediated imipenem resistance in *Acinetobacter baumannii. Int J Antimicrob Agents* 1993;**2**(2):81–7.
21. Yigit H, Queenan AM, Anderson GJ, Domenech-Sanchez A, Biddle JW, Steward CD, et al. Novel carbapenem-hydrolyzing beta-lactamase, KPC-1, from a carbapenem-resistant strain of *Klebsiella pneumoniae. Antimicrob Agents Chemother* 2001;**45**(4):1151–61.
22. Queenan AM, Bush K. Carbapenemases: the versatile beta-lactamases. *Clin Microbiol Rev* 2007;**20**(3):440–58.
23. El-Herte RI, Kanj SS, Matar GM, Araj GF. The threat of carbapenem-resistant Enterobacteriaceae in Lebanon: an update on the regional and local epidemiology. *J Infect Public Health* 2012;**5**(3):233–43.
24. Patel G, Bonomo RA. "Stormy waters ahead": global emergence of carbapenemases. *Front Microbiol* 2013;**4**:48.
25. Livermore DM, Woodford N. The beta-lactamase threat in Enterobacteriaceae, *Pseudomonas* and *Acinetobacter. Trends Microbiol* 2006;**14**(9):413–20.
26. Kanj SS, Kanafani ZA. Current concepts in antimicrobial therapy against resistant Gram-negative organisms: extended-spectrum beta-lactamase-producing Enterobacteriaceae, carbapenem-resistant Enterobacteriaceae, and multidrug-resistant *Pseudomonas aeruginosa. Mayo Clin Proc* 2011;**86**(3):250–9.
27. Thomson KS. Extended-spectrum-beta-lactamase, AmpC, and Carbapenemase issues. *J Clin Microbiol* 2010;**48**(4):1019–25.
28. Nordmann P, Girlich D, Poirel L. Detection of carbapenemase producers in Enterobacteriaceae by use of a novel screening medium. *J Clin Microbiol* 2012;**50**(8):2761–6.
29. Dortet L, Poirel L, Nordmann P. Rapid detection of carbapenemase-producing *Pseudomonas* spp. *J Clin Microbiol* 2012;**50**(11):3773–6.
30. Smith Moland E, Hanson ND, Herrera VL, Black JA, Lockhart TJ, Hossain A, et al. Plasmid-mediated, carbapenem-hydrolysing beta-lactamase, KPC-2, in *Klebsiella pneumoniae* isolates. *J Antimicrob Chemother* 2003;**51**(3):711–14.
31. Bradford PA, Bratu S, Urban C, Visalli M, Mariano N, Landman D, et al. Emergence of carbapenem-resistant *Klebsiella* species possessing the class A carbapenem-hydrolyzing KPC-2 and inhibitor-resistant TEM-30 beta-lactamases in New York City. *Clin Infect Dis* 2004;**39**(1):55–60.

32. Woodford N, Tierno PM, Young K, Tysall L, Palepou M-FI, Ward E, et al. Outbreak of *Klebsiella pneumoniae* producing a new carbapenem-hydrolyzing class A beta-lactamase, KPC-3, in a New York Medical Center. *Antimicrob Agents Chemother* 2004;**48**(12):4793–9.
33. Bratu S, Landman D, Alam M, Tolentino E, Quale J. Detection of KPC carbapenem-hydrolyzing enzymes in *Enterobacter* spp. from Brooklyn, New York. *Antimicrob Agents Chemother* 2005;**49**(2):776–8.
34. Bratu S, Tolaney P, Karumudi U, Quale J, Mooty M, Nichani S, et al. Carbapenemase-producing *Klebsiella pneumoniae* in Brooklyn, NY: molecular epidemiology and in vitro activity of polymyxin B and other agents. *J Antimicrob Chemother* 2005;**56**(1):128–32.
35. Naas T, Cuzon G, Villegas M-V, Lartigue M-F, Quinn JP, Nordmann P. Genetic structures at the origin of acquisition of the beta-lactamase bla_{KPC} gene. *Antimicrob Agents Chemother* 2008;**52**(4):1257–63.
36. Giakkoupi P, Tzouvelekis LS, Tsakris A, Loukova V, Sofianou D, Tzelepi E. IBC-1, a novel integron-associated class A beta-lactamase with extended-spectrum properties produced by an *Enterobacter cloacae* clinical strain. *Antimicrob Agents Chemother* 2000;**44**(9):2247–53.
37. Poirel L, Le Thomas I, Naas T, Karim A, Nordmann P. Biochemical sequence analyses of GES-1, a novel class A extended-spectrum beta-lactamase, and the class 1 integron In52 from *Klebsiella pneumoniae. Antimicrob Agents Chemother* 2000;**44**(3):622–32.
38. Poirel L, Weldhagen GF, Naas T, De Champs C, Dove MG, Nordmann P. GES-2, a class A beta-lactamase from *Pseudomonas aeruginosa* with increased hydrolysis of imipenem. *Antimicrob Agents Chemother* 2001;**45**(9):2598–603.
39. Moubareck C, Brémont S, Conroy M-C, Courvalin P, Lambert T. GES-11, a novel integron-associated GES variant in *Acinetobacter baumannii. Antimicrob Agents Chemother* 2009;**53**(8):3579–81.
40. Bonnin RA, Rotimi VO, Al Hubail M, Gasiorowski E, Al Sweih N, Nordmann P, et al. Wide dissemination of GES-type carbapenemases in *Acinetobacter baumannii* isolates in Kuwait. *Antimicrob Agents Chemother* 2013;**57**(1):183–8.
41. Kuwabara S, Abraham EP. Some properties of two extracellular beta-lactamases from *Bacillus cereus* 569/H. *Biochem J* 1967;**103**(3):27C–30C.
42. Iaconis JP, Sanders CC. Purification and characterization of inducible beta-lactamases in *Aeromonas* spp. *Antimicrob Agents Chemother* 1990;**34**(1):44–51.
43. Saino Y, Kobayashi F, Inoue M, Mitsuhashi S. Purification and properties of inducible penicillin beta-lactamase isolated from *Pseudomonas maltophilia. Antimicrob Agents Chemother* 1982;**22**(4):564–70.
44. Zhao W-H, Hu Z-Q. IMP-type metallo-β-lactamases in Gram-negative bacilli: distribution, phylogeny, and association with integrons. *Crit Rev Microbiol* 2011;**37**(3):214–26.
45. Lauretti L, Riccio ML, Mazzariol A, Cornaglia G, Amicosante G, Fontana R, et al. Cloning and characterization of bla_{VIM}, a new integron-borne metallo-beta-lactamase gene from a *Pseudomonas aeruginosa* clinical isolate. *Antimicrob Agents Chemother* 1999;**43**(7):1584–90.
46. Poirel L, Naas T, Nicolas D, Collet L, Bellais S, Cavallo JD, et al. Characterization of VIM-2, a carbapenem-hydrolyzing metallo-beta-lactamase and its plasmid- and integron-borne gene from a *Pseudomonas aeruginosa* clinical isolate in France. *Antimicrob Agents Chemother* 2000;**44**(4):891–7.
47. Zhao W-H, Hu Z-Q. Epidemiology and genetics of VIM-type metallo-β-lactamases in Gram-negative bacilli. *Future Microbiol* 2011;**6**(3):317–33.
48. Yong D, Toleman MA, Giske CG, Cho HS, Sundman K, Lee K, et al. Characterization of a new metallo-beta-lactamase gene, *bla*(NDM-1), and a novel erythromycin esterase gene carried on a unique genetic structure in *Klebsiella pneumoniae* sequence type 14 from India. *Antimicrob Agents Chemother* 2009;**53**(12):5046–54.
49. Rafei R, Dabboussi F, Hamze M, Eveillard M, Lemarié C, Mallat H, et al. First report of *bla*NDM-1-producing *Acinetobacter baumannii* isolated in Lebanon from civilians wounded during the Syrian war. *Int J Infect Dis* 2014;**21**:21–3.

50. Walsh TR, Weeks J, Livermore DM, Toleman MA. Dissemination of NDM-1 positive bacteria in the New Delhi environment and its implications for human health: an environmental point prevalence study. *Lancet Infect Dis* 2011;**11**(5):355–62.
51. Nordmann P, Poirel L, Walsh TR, Livermore DM. The emerging NDM carbapenemases. *Trends Microbiol* 2011;**19**(12):588–95.
52. Berçot B, Poirel L, Dortet L, Nordmann P. In vitro evaluation of antibiotic synergy for NDM-1-producing Enterobacteriaceae. *J Antimicrob Chemother* 2011;**66**(10):2295–7.
53. Jacoby GA. AmpC beta-lactamases. *Clin Microbiol Rev* 2009;**22**(1):161–82.
54. Mammeri H, Guillon H, Eb F, Nordmann P. Phenotypic and biochemical comparison of the carbapenem-hydrolyzing activities of five plasmid-borne AmpC β-lactamases. *Antimicrob Agents Chemother* 2010;**54**(11):4556–60.
55. Périchon B, Goussard S, Walewski V, Krizova L, Cerqueira G, Murphy C, et al. Identification of 50 class D β-lactamases and 65 *Acinetobacter*-derived cephalosporinases in *Acinetobacter* spp. *Antimicrob Agents Chemother* 2014;**58**(2):936–49.
56. Upadhyay S, Mishra S, Sen MR, Banerjee T, Bhattacharjee A. Co-existence of Pseudomonas-derived cephalosporinase among plasmid encoded CMY-2 harbouring isolates of *Pseudomonas aeruginosa* in north India. *Indian J Med Microbiol* 2013;**31**(3):257–60.
57. Evans BA, Amyes SGB. OXA β-lactamases. *Clin Microbiol Rev* 2014;**27**(2):241–63.
58. Donald HM, Scaife W, Amyes SG, Young HK. Sequence analysis of ARI-1, a novel OXA beta-lactamase, responsible for imipenem resistance in *Acinetobacter baumannii* 6B92. *Antimicrob Agents Chemother* 2000;**44**(1):196–9.
59. Poirel L, Marqué S, Héritier C, Segonds C, Chabanon G, Nordmann P. OXA-58, a novel class D {beta}-lactamase involved in resistance to carbapenems in *Acinetobacter baumannii*. *Antimicrob Agents Chemother* 2005;**49**(1):202–8.
60. Poirel L, Héritier C, Tolün V, Nordmann P. Emergence of oxacillinase-mediated resistance to imipenem in *Klebsiella pneumoniae*. *Antimicrob Agents Chemother* 2004;**48**(1):15–22.
61. Aubert D, Naas T, Héritier C, Poirel L, Nordmann P. Functional characterization of IS1999, an IS4 family element involved in mobilization and expression of beta-lactam resistance genes. *J Bacteriol* 2006;**188**(18):6506–14.
62. Hammoudi D, Ayoub Moubareck C, Aires J, Adaime A, Barakat A, Fayad N, et al. Countrywide spread of OXA-48 carbapenemase in Lebanon: surveillance and genetic characterization of carbapenem-non-susceptible Enterobacteriaceae in 10 hospitals over a one-year period. *Int J Infect Dis* 2014;**29**:139–44.
63. Doumith M, Ellington MJ, Livermore DM, Woodford N. Molecular mechanisms disrupting porin expression in ertapenem-resistant *Klebsiella* and *Enterobacter* spp. clinical isolates from the UK. *J Antimicrob Chemother* 2009;**63**(4):659–67.
64. Nakae T. Identification of the outer membrane protein of *E. coli* that produces transmembrane channels in reconstituted vesicle membranes. *Biochem Biophys Res Commun* 1976;**71**(3):877–84.
65. Elliott E, Brink AJ, van Greune J, Els Z, Woodford N, Turton J, et al. In vivo development of ertapenem resistance in a patient with pneumonia caused by *Klebsiella pneumoniae* with an extended-spectrum beta-lactamase. *Clin Infect Dis* 2006;**42**(11):e95–8.
66. Lister PD, Wolter DJ, Hanson ND. Antibacterial-resistant *Pseudomonas aeruginosa*: clinical impact and complex regulation of chromosomally encoded resistance mechanisms. *Clin Microbiol Rev* 2009;**22**(4):582–610.
67. Poole K. *Pseudomonas aeruginosa*: resistance to the max. *Front Microbiol* 2011;**2**:65.
68. Abbott I, Cerqueira GM, Bhuiyan S, Peleg AY. Carbapenem resistance in *Acinetobacter baumannii*: laboratory challenges, mechanistic insights and therapeutic strategies. *Expert Rev Anti Infect Ther* 2013;**11**(4):395–409.

69. Catel-Ferreira M, Nehmé R, Molle V, Aranda J, Bouffartigues E, Chevalier S, et al. Deciphering the function of the outer membrane protein OprD homologue of *Acinetobacter baumannii*. *Antimicrob Agents Chemother* 2012;**56**(7):3826–32.
70. Suh B, Bae IK, Kim J, Jeong SH, Yong D, Lee K. Outbreak of meropenem-resistant *Serratia marcescens* comediated by chromosomal AmpC beta-lactamase overproduction and outer membrane protein loss. *Antimicrob Agents Chemother* 2010;**54**(12):5057–61.
71. Gülmez D, Woodford N, Palepou M-FI, Mushtaq S, Metan G, Yakupogullari Y, et al. Carbapenem-resistant *Escherichia coli* and *Klebsiella pneumoniae* isolates from Turkey with OXA-48-like carbapenemases and outer membrane protein loss. *Int J Antimicrob Agents* 2008;**31**(6):523–6.
72. Fernández L, Hancock REW. Adaptive and mutational resistance: role of porins and efflux pumps in drug resistance. *Clin Microbiol Rev* 2012;**25**(4):661–81.
73. Coyne S, Courvalin P, Périchon B. Efflux-mediated antibiotic resistance in *Acinetobacter* spp. *Antimicrob Agents Chemother* 2011;**55**(3):947–53.
74. Marqué S, Poirel L, Héritier C, Brisse S, Blasco MD, Filip R, et al. Regional occurrence of plasmid-mediated carbapenem-hydrolyzing oxacillinase OXA-58 in *Acinetobacter* spp. in Europe. *J Clin Microbiol* 2005;**43**(9):4885–8.
75. Strateva T, Yordanov D. *Pseudomonas aeruginosa*—a phenomenon of bacterial resistance. *J Med Microbiol* 2009;**58**(Pt 9):1133–48.
76. Piddock LJV. Clinically relevant chromosomally encoded multidrug resistance efflux pumps in bacteria. *Clin Microbiol Rev* 2006;**19**(2):382–402.
77. Baroud M, Dandache I, Araj GF, Wakim R, Kanj S, Kanafani Z, et al. Underlying mechanisms of carbapenem resistance in extended-spectrum β-lactamase-producing *Klebsiella pneumoniae* and *Escherichia coli* isolates at a tertiary care centre in Lebanon: role of OXA-48 and NDM-1 carbapenemases. *Int J Antimicrob Agents* 2013;**41**(1):75–9.
78. Yang F-C, Yan J-J, Hung K-H, Wu J-J. Characterization of ertapenem-resistant *Enterobacter cloacae* in a Taiwanese university hospital. *J Clin Microbiol* 2012;**50**(2):223–6.
79. Temime L, Boëlle PY, Courvalin P, Guillemot D. Bacterial resistance to penicillin G by decreased affinity of penicillin-binding proteins: a mathematical model. *Emerg Infect Dis* 2003;**9**(4):411–17.
80. Fernández-Cuenca F, Martínez-Martínez L, Conejo MC, Ayala JA, Perea EJ, Pascual A. Relationship between beta-lactamase production, outer membrane protein and penicillin-binding protein profiles on the activity of carbapenems against clinical isolates of *Acinetobacter baumannii*. *J Antimicrob Chemother* 2003;**51**(3):565–74.
81. Cayô R, Rodríguez M-C, Espinal P, Fernández-Cuenca F, Ocampo-Sosa AA, Pascual A, et al. Analysis of genes encoding penicillin-binding proteins in clinical isolates of *Acinetobacter baumannii*. *Antimicrob Agents Chemother* 2011;**55**(12):5907–13.
82. Neuwirth C, Siébor E, Duez JM, Péchinot A, Kazmierczak A. Imipenem resistance in clinical isolates of *Proteus mirabilis* associated with alterations in penicillin-binding proteins. *J Antimicrob Chemother* 1995;**36**(2):335–42.
83. Yamachika S, Sugihara C, Kamai Y, Yamashita M. Correlation between penicillin-binding protein 2 mutations and carbapenem resistance in *Escherichia coli*. *J Med Microbiol* 2013;**62**(Pt 3):429–36.
84. Farra A, Islam S, Strålfors A, Sörberg M, Wretlind B. Role of outer membrane protein OprD and penicillin-binding proteins in resistance of *Pseudomonas aeruginosa* to imipenem and meropenem. *Int J Antimicrob Agents* 2008;**31**(5):427–33.
85. Giske CG, Buarø L, Sundsfjord A, Wretlind B. Alterations of porin, pumps, and penicillin-binding proteins in carbapenem resistant clinical isolates of *Pseudomonas aeruginosa*. *Microb Drug Resist* 2008;**14**(1):23–30.

86. Hammoudi D, Moubareck CA, Sarkis DK. How to detect carbapenemase producers? A literature review of phenotypic and molecular methods. *J Microbiol Methods* 2014;**107**:106–18.
87. Hrabák J, Chudáčková E, Papagiannitsis CC. Detection of carbapenemases in Enterobacteriaceae: a challenge for diagnostic microbiological laboratories. *Clin Microbiol Infect* 2014;**20**(9):839–53.
88. The European Committee on Antimicrobial Susceptibility Testing. *Breakpoint tables for interpretation of MICs and zone diameters. Version 2.0.* EUCAST; 2012. Available at: http://www.eucast.org [accessed as of 01.01.12]. Report No: 2.0.
89. Clinical Laboratory and Standards Institute. *Performance standards for antimicrobial susceptibility testing. Twenty-fourth informational supplement M 100-S24.* Wayne, PA: CLSI; 2014.
90. Nordmann P, Poirel L. Strategies for identification of carbapenemase-producing Enterobacteriaceae. *J Antimicrob Chemother* 2013;**68**(3):487–9.
91. Vrioni G, Daniil I, Voulgari E, Ranellou K, Koumaki V, Ghirardi S, et al. Comparative evaluation of a prototype chromogenic medium (ChromID CARBA) for detecting carbapenemase-producing Enterobacteriaceae in surveillance rectal swabs. *J Clin Microbiol* 2012;**50**(6):1841–6.
92. Lee K, Chong Y, Shin HB, Kim YA, Yong D, Yum JH. Modified Hodge and EDTA-disk synergy tests to screen metallo-beta-lactamase-producing strains of *Pseudomonas* and *Acinetobacter* species. *Clin Microbiol Infect* 2001;**7**(2):88–91.
93. Cohen Stuart J, Leverstein-Van Hall MA. Dutch working party on the detection of highly resistant microorganisms. Guideline for phenotypic screening and confirmation of carbapenemases in Enterobacteriaceae. *Int J Antimicrob Agents* 2010;**36**(3):205–10.
94. Tijet N, Boyd D, Patel SN, Mulvey MR, Melano RG. Evaluation of the Carba NP test for rapid detection of carbapenemase-producing Enterobacteriaceae and *Pseudomonas aeruginosa*. *Antimicrob Agents Chemother* 2013;**57**(9):4578–80.
95. Bernabeu S, Poirel L, Nordmann P. Spectrophotometry-based detection of carbapenemase producers among Enterobacteriaceae. *Diagn Microbiol Infect Dis* 2012;**74**(1):88–90.
96. Hrabák J, Walková R, Studentová V, Chudácková E, Bergerová T. Carbapenemase activity detection by matrix-assisted laser desorption ionization-time of flight mass spectrometry. *J Clin Microbiol* 2011;**49**(9):3222–7.
97. Sanbongi Y, Shimizu A, Suzuki T, Nagaso H, Ida T, Maebashi K, et al. Classification of OprD sequence and correlation with antimicrobial activity of carbapenem agents in *Pseudomonas aeruginosa* clinical isolates collected in Japan. *Microbiol Immunol* 2009;**53**(7):361–7.
98. Dumas J-L, van Delden C, Perron K, Köhler T. Analysis of antibiotic resistance gene expression in *Pseudomonas aeruginosa* by quantitative real-time-PCR. *FEMS Microbiol Lett* 2006;**254**(2):217–25.
99. Jiang M, Zhang Z, Zhao S. Epidemiological characteristics and drug resistance analysis of multidrug-resistant *Acinetobacter baumannii* in a China hospital at a certain time. *Pol J Microbiol* 2014;**63**(3):275–81.
100. Pournaras S, Maniati M, Spanakis N, Ikonomidis A, Tassios PT, Tsakris A, et al. Spread of efflux pump-overexpressing, non-metallo-beta-lactamase-producing, meropenem-resistant but ceftazidime-susceptible *Pseudomonas aeruginosa* in a region with bla_{VIM} endemicity. *J Antimicrob Chemother* 2005;**56**(4):761–4.
101. El Amin N, Giske CG, Jalal S, Keijser B, Kronvall G, Wretlind B. Carbapenem resistance mechanisms in *Pseudomonas aeruginosa*: alterations of porin OprD and efflux proteins do not fully explain resistance patterns observed in clinical isolates. *APMIS Acta Pathol Microbiol Immunol Scand* 2005;**113**(3):187–96.
102. Daikos GL, Markogiannakis A. Carbapenemase-producing *Klebsiella pneumoniae*: (when) might we still consider treating with carbapenems? *Clin Microbiol Infect* 2011;**17**(8):1135–41.

103. Tängdén T, Giske CG. Global dissemination of extensively drug-resistant carbapenemase-producing Enterobacteriaceae: clinical perspectives on detection, treatment and infection control. *J Intern Med* 2015;**277**(5):501–12.
104. Garonzik SM, Li J, Thamlikitkul V, Paterson DL, Shoham S, Jacob J, et al. Population pharmacokinetics of colistin methanesulfonate and formed colistin in critically ill patients from a multicenter study provide dosing suggestions for various categories of patients. *Antimicrob Agents Chemother* 2011;**55** (7):3284–94.
105. Yahav D, Lador A, Paul M, Leibovici L. Efficacy and safety of tigecycline: a systematic review and meta-analysis. *J Antimicrob Chemother* 2011;**66**(9):1963–71.
106. Karaiskos I, Giamarellou H. Multidrug-resistant and extensively drug-resistant Gram-negative pathogens: current and emerging therapeutic approaches. *Expert Opin Pharmacother* 2014;**15**(10): 1351–70.
107. Souli M, Galani I, Boukovalas S, Gourgoulis MG, Chryssouli Z, Kanellakopoulou K, et al. In vitro interactions of antimicrobial combinations with fosfomycin against KPC-2-producing *Klebsiella pneumoniae* and protection of resistance development. *Antimicrob Agents Chemother* 2011;**55**(5):2395–7.
108. Rahal JJ. Antimicrobial resistance among and therapeutic options against Gram-negative pathogens. *Clin Infect Dis* 2009;**49**(Suppl. 1):S4–10.
109. Pontikis K, Karaiskos I, Bastani S, Dimopoulos G, Kalogirou M, Katsiari M, et al. Outcomes of critically ill intensive care unit patients treated with fosfomycin for infections due to pandrug-resistant and extensively drug-resistant carbapenemase-producing Gram-negative bacteria. *Int J Antimicrob Agents* 2014;**43**(1):52–9.
110. Centers for Disease Control and Prevention (CDC). Guidance for control of infections with carbapenem-resistant or carbapenemase-producing Enterobacteriaceae in acute care facilities. *MMWR Morb Mortal Wkly Rep* 2009;**58**(10):256–60.
111. Gupta N, Limbago BM, Patel JB, Kallen AJ. Carbapenem-resistant Enterobacteriaceae: epidemiology and prevention. *Clin Infect Dis* 2011;**53**(1):60–7.
112. Miriagou V, Cornaglia G, Edelstein M, Galani I, Giske CG, Gniadkowski M, et al. Acquired carbapenemases in Gram-negative bacterial pathogens: detection and surveillance issues. *Clin Microbiol Infect* 2010;**16**(2):112–22.
113. Pan C-Y, Chen J-C, Chen T-L, Wu J-L, Hui C-F, Chen J-Y. Piscidin is highly active against carbapenem-resistant *Acinetobacter baumannii* and NDM-1-producing *Klebsiella pneumoniae* in a systemic Septicaemia infection mouse model. *Mar Drugs* 2015;**13**(4):2287–305.

CHAPTER

ANTIBIOTIC RESISTANCE: CAN NANOPARTICLES TACKLE THE PROBLEM?

6

M. Rai[1], K. Kon[2], A. Gade[1], A. Ingle[1], D. Nagaonkar[1], P. Paralikar[1] and S.S. da Silva[3]

[1]Nanobiotechnology Laboratory, Department of Biotechnology, SGB Amravati University, Amravati, Maharashtra, India [2]Department of Microbiology, Virology, and Immunology, Kharkiv National Medical University, Kharkiv, Ukraine [3]Department of Biotechnology, School of Engineering of Lorena, University of São Paulo, Lorena, São Paulo, Brazil

INTRODUCTION

Antimicrobial agents (particularly antibiotics) have played an indispensable role in human health over the past 50 years. While looking back at the preantibiotic era, the serendipitous discovery of penicillin as a wonder drug by Alexander Fleming in 1928 revolutionized conventional treatment therapies.[1] This was due to its enormous ability to cure or prevent serious infectious diseases, which were once the main cause of death or chronic illness. However, Fleming warned about the consequences of improper and overuse of penicillin.[2] Since the last century, medical science has seen the discovery of more than 100 antimicrobial compounds (including azolides and sulfonamides), out of which none of the novel class of antibiotics have been established since 1987. Due to the extensive use of antibiotics, resistance in microbes has developed, resulting in inactivation of the drug. Moreover, even before the introduction of penicillin as a therapeutic in a global market, members of the penicillin discovery team, Abraham and Chain,[1] identified an enzyme capable of inactivating or hydrolyzing penicillin. So the release of new antibiotics was always followed by the emergence of resistance to them. It seems that evolution is cleverer than even highly evolved organisms like humans. The discovery of newer antibiotics to kill the pathogens has led to the evolution of new ways to counteract them by developing resistance. Moreover, the number of antibiotic-producing genes and resistance-developing genes were found in the same component of natural microbial populations.[3] Hence, in order to prevent the strains from developing resistance, synthetic strategies were designed to modify penicillin chemically to prevent its degradation from penicillinases (β-lactamases). Continuous increases in drug resistance in Gram-negative bacteria (which are extremely drug resistant) has posed a clinical threat due to the development of resistance to first-line antibiotic treatment (eg, ciprofloxacin, levofloxacin, imipenem, amikacin, tazobactam, tobramycin, etc.), as well as second-line antibiotics (eg, polymyxins).[4]

Antibiotic Resistance. DOI: http://dx.doi.org/10.1016/B978-0-12-803642-6.00006-X

Kumarasamy et al.[5] by his discovery of New Delhi metallo-β-lactamase (NDM-1), shocked the medical community worldwide and raised general awareness about superbugs, newly evolved bacteria species consisting of several antibiotic-resistance genes. Now, superbugs have emerged as the most threatening problem to human health in terms of morbidity and mortality worldwide. With reference to the number of infections and health hazards, *Vibrio cholerae* is of the highest rank in the list of superbugs.[6] Currently, the Gram-positive bacterium *Staphylococcus aureus* has become the most iniquitous superbug, closely associated with both nasal and common skin infections such as boils. Methicillin was the first antiresistance antibiotic to be discovered; however, within just 3 years, resistance developed that led to the appearance of the superbug methicillin-resistant *S. aureus* (MRSA), followed by the multidrug-resistant (MDR) *S. aureus*. Another example of severe superbugs is MDR *Mycobacterium tuberculosis* strains, which primarily arise from improper and discontinued treatment.

Understanding of the antibiotic-resistance mechanism is very much needed to overcome this resistance. Many bacteria are equipped with a collection of genes which synthesize proteins required to neutralize antibiotics and their activity.[7] Bacteria employ efflux pumps, which are located in the cell membrane to protect themselves against the influx of antibiotics without causing any harm to cellular components. Although less common, these efflux pumps can be specific for the antibiotics that they pump out of the cell. Ribosomal protection proteins (RPPs) are the mechanism by which bacteria develop resistance to antibiotics. Upon binding with ribosomes, these RPPs change the conformation of ribosomes, which further protect the binding of ribosomes to the antibiotic. This leads to protection of interference of antibiotics with protein synthesis, giving bacteria the chance to overcome the ill-effects of antibiotics.[7] However, these RPPs do not interfere with normal protein synthesis.

Modifying enzymes is another protective mechanism by which bacteria overcome antibiotics. Some bacterial enzymes are known to neutralize antibiotics by the addition of acetyl or phosphate groups to the binding sites of antibiotics.[8,9] Surprisingly, *Streptomyces* spp. have ability to produce all the three genes responsible for antibiotic resistance including efflux pumps, RPP and enzymatic neutralization of antibiotics. It seems that the *Streptomyces* spp. protects itself from their own antibiotics. The two most interesting mechanisms reported in Gram-negative bacteria include *NDM-1*[5] and aminoglycoside 16s r-RNA methylation.[10] *NDM-1* is a bacterial gene that encodes the carbapenemase enzyme, offering the bacteria resistance to all types of antibiotics.[5] Some classes and groups have shown β-lactamase-related resistance, which may disseminate by horizontal gene transfer.[11–14] Further, the transfer of antibiotic-resistance genes by plasmids, integrons, and transposons has been an area to be explored because of its medical implications.[15] Taking all these facts into consideration, we can conclude that antibiotic resistance is present in natural microbial flora, which enables them to battle antibiotics that target them.

Disease-causing Gram-negative pathogens like *M. tuberculosis*, *Klebsiella pneumoniae*, *Salmonella* Typhi, and *Escherichia coli* exhibit a strong correlation between the use of antibiotics treatment therapies and rapid development of resistance to the same.[16] The term, *ESKAPE* has been coined for antibiotic-resistant endemic bacteria including *Enterococcus faecium*,

S. aureus, *K. pneumoniae*, *Acinetobacter baumannii*, *Pseudomonas aeruginosa*, and *Enterobacter* sp.[17] Over the years, a large number of antibiotics (eg, chloramphenicol, tetracyclines, ampicillin, and vancomycin) were incorporated into clinical practices. However, continuous exposure of microorganisms to a variety of antibiotics resulted in the adaptation of microorganisms to these compounds.

With reference to the worldwide management and control of epidemic and infectious diseases, the search for reliable systems for tracking outbreaks of endemic infections is the most pressing current need. The World Health Organization has geared up for heroic efforts toward appropriate remedial action, as well as the development of sustainable antibiotics and drugs for the treatment of life-threatening disease-causing pathogens.

Although it is very tough to fight MDR microbes, there is a concept emerging from studies in the field of nanobiotechnology. It is a well-established fact that the reduction in size of metals (up to the nanoscale) offers more advanced physiochemical properties to the nanostructures than do their bulky counterparts. Owing to their enhanced magnetic and electrochemical properties, nanoparticles can offer accurate and highly sensitive diagnostic tools for detection of even MDR infections.[18–22] Most of the nanoparticles, due to their increased surface area-to-volume ratio, have shown significant antimicrobial potential. There are several reports on the antimicrobial potential of such nanoparticles as silver,[23,24] copper,[25] and zinc,[26] which showed significant activity against many pathogenic microbes, including viruses. Hence, these nanoparticles have shown the potential to tackle the problem of antibiotic resistance.

This chapter emphasizes the prevalence of the MDR problem worldwide. Further, the role of nanotechnology and nanoparticles, individually as well as in combination with antibiotics, in overcoming this problem will be discussed here in detail.

THE MDR PROBLEM

MDR is a condition in which microbes are resistant to more than one antibiotic. The reason behind this is sequential use of the same antibiotics against diseases for years. Antibiotic resistance occurs when bacteria, fungi, viruses, or protozoa cannot be fully inhibited or killed by antibiotics. They become resistant by adopting their structure and function in a way that prevents them from being killed by the antibiotics. These resistant microbes are now able to survive, multiply, spread, and cause infectious diseases in community even in the presence of antibiotics. The patients with drug-resistant bacterial infection are at high risk of dying from infection and may lead to longer and costly hospital treatment. The Centers for Disease Control and Prevention (CDC) estimated that 23,000 deaths were caused by drug-resistant bacteria in the United States alone in 2013.[27] If we estimated the worldwide number of illness and deaths in patients, then the tally is much greater. The bacteria, fungi, and viruses that have become highly resistant to antibiotics are listed in Table 6.1.

Table 6.1 List of the Resistance of Bacteria, Fungi, and Viruses to Different Antibiotics

S. No.	Resistant Organisms	Disease/Infection Caused	Antibiotic Showing Resistance	References
Bacteria				
1	*Clostridium difficile*	Diarrhea	Fluoroquinolones, metronidazole	109,110
2	*Neisseria gonorrhoeae*	Gonorrhea	Cefixime (an oral cephalosporin), ceftriaxone (an injectable cephalosporin), azithromycin, tetracycline	111,112
3	*Acinetobacter baumannii*	Pneumonia	Carbapenem	113,114
4	*Campylobacter*	Diarrhea	Ciprofloxacin, azithromycin	115,116
5	*Escherichia coli*	Intestinal infection	Ampicillin, cephalosporins, fluoroquinolones	117
6	*Enterococcus faecium*	Meningitis, urinary tract infection (UTI), bacteremia (central venous catheter related), endocarditis	Vancomycin, streptomycin, gentamycin, penicillin, ampicillin	118,119
7	Methicillin-resistant *Staphylococcus aureus*	UTI, bacteremia, toxic shock syndrome, pneumonia, osteomyelitis, endocarditis, meningitis, skin infection	β-Lactam antibiotics (eg, oxacillin, penicillin, nafcillin, amoxicillin, and most cephalosporins), erythromycin	120,121
8	*Pseudomonas aeruginosa*	UTIs, pneumonias, skin and soft tissue infections, endocarditis, meningitis	Carbapenems, ceftazidime, ciprofloxacin, imipenem	122,123
9	*Klebsiella pneumoniae*	Pneumonia, bloodstream infections, wound infections, UTIs, and meningitis	Carbapenems, piperacillin, carbenicillin, ofloxacine, ampicillin, co-trimoxazole, chloramphenicol	124–126
10	Multidrug-resistant *Mycobacterium tuberculosis*	Tuberculosis	Isoniazid, rifampicin, pyrazinamide	127–130
11	*Staphylococcus epidermidis*	Bacteremia, catheter, implant, and prostheses related infection (biofilm formations), endocarditis	Penicillin, amoxicillin, tetracycline	131,132
12	*Treponema pallidum*	Treponemal diseases such as syphilis, bejel, pinta, and yaws	Azithromycin, clindamycin, rifampin	133–135
13	*Vibrio cholerae*	Cholerae (watery diarrhea)	Fluoroquinolones, tetracycline	136–139
Fungi				
1	*Candida* spp.	Vaginal infection	Azoles (especially fluconazole), Echinocandins, caspofungin	140,141
2	*Aspergillus fumigatus*	Tuberculosis, atypical tuberculosis, chronic obstructive pulmonary disease (COPD, emphysema), pneumothorax, sarcoidosis, allergic bronchopulmonary aspergillosis, lung cancer	Azoles (voriconazole), echinocandin	142–144
3	*Scopulariopsis* spp.	Amphotericin B, flucytosine, azoles	Onychomycosis	145,146
4	*Cryptococcus neoformans*	Fluconazole	Cryptococcosis	147,148

Table 6.1 List of the Resistance of Bacteria, Fungi, and Viruses to Different Antibiotics *Continued*

S. No.	Resistant Organisms	Disease/Infection Caused	Antibiotic Showing Resistance	References
Viruses				
1	Herpes simplex virus	Herpes	Acyclovir	149,150
2	Varicella-zoster virus	Chickenpox	Acyclovir	151,152
3	Cytomegalovirus (human herpes virus 5)	Congenital infections and birth defect	Ganciclovir, foscarnet	153–155
Protozoan				
1	*Trypanosoma cruzi*	Chagas disease	Benznidazole	156
2	*Leishmania donovani*	Leishmaniasis (or kala-azar)	Amphotericin B, miltefosine	157–159
3	*Plasmodium vivax*	Malaria	Chloroquine	160,161
4	*Toxoplasma gondii*	Artemisinin, atovaquone, sulfadiazine	Toxoplasmosis	162

MECHANISM FOR ANTIMICROBIAL ACTIVITY OF NANOPARTICLES

There are many reports on the mechanism of antimicrobial activity of metallic nanoparticles. Unfortunately, the exact mechanism by which nanoparticle inhibits the growth of organism has not been well elucidated.[28,29]

Silver nanoparticles (AgNPs) are widely used as antimicrobial agents; therefore, they are discussed here as an example (Fig. 6.1). A review of the research literature shows that many reports are available to explain the mechanism of antimicrobial action of AgNPs, but still, there is a pressing need to provide a convincing explanation.[30,80] The AgNPs undergo slow oxidation, as a result of which silver ions are released, which arrests the replication of DNA.[31] Moreover, the expression of ribosomal subunits and inactivation of some cellular proteins and enzymes necessary for adenosine triphosphate (ATP) production are inhibited.[32] According to another hypothesis, silver ions arrest the normal function of membrane-bound respiratory enzymes.[33] The silver ions inhibit the thiol group–containing enzymes like NADH-dehydrogenase II in the respiratory system, which is known as the site of production of reactive oxygen species (ROS) in vivo.[29,34,35]

Many studies support the proposition that the positive charge on the silver ions released from nanoparticles plays an important role in antimicrobial activity through the electrostatic attraction between positively charged nanoparticles and negatively charged cell membranes of microbes.[29,36,37] The nanoparticles affect permeability and respiration.[30,38] In contrast, Sondi and

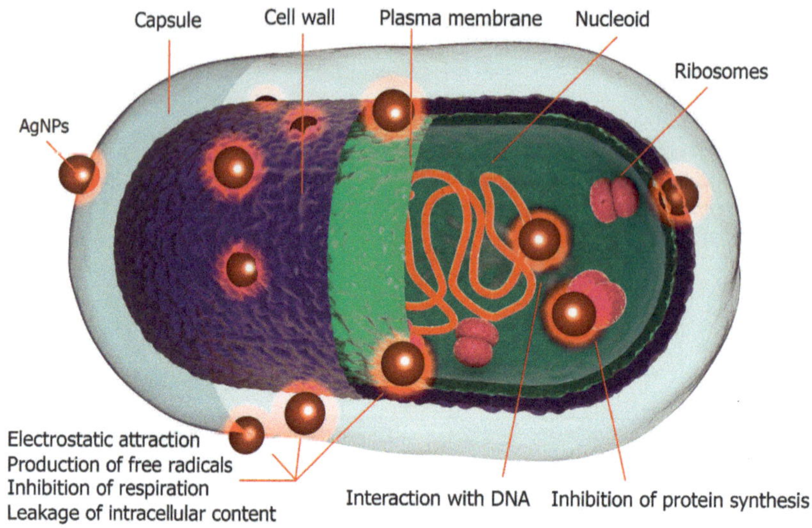

FIGURE 6.1

Mechanism of antibacterial activity of silver nanoparticles.

Salopek-Sondi[39] found that antibacterial action was due to the formation of pits in the Gram-negative bacterial cell wall, leading to cell death. The effect was closely associated with the concentration of AgNPs. Another mechanism suggests generation of free radicals, which affect lipids in membrane, and the eventual formation of pores on the wall.[40,41]

Chatterjee et al.[29] suggested that the mechanisms regarding the activity of metallic nanostructures on bacteria are yet to be elucidated. According to this study, there are three hypothetical mechanisms: (1) accumulation of nanoparticles in the membrane of bacteria, which changes its permeability; consequently, intracellular contents are released; (2) generation of ROS with oxidative deterioration to cell content, and (3) uptake of ions derived from metallic NPs, ATP production, and inhibition of DNA replication.

THE USE OF CeO_2 FOR DETECTION OF DRUG RESISTANCE

Antimicrobial resistance is increasing at an alarming rate. The early detection of antimicrobial resistance is the basic requirement for better treatment strategies. Noll[42] reported that CeO_2 nanoparticles may play a pivotal role in detecting drug resistivity due to their ability to exist in a mixed-valence state and act as either oxidizing agents or reducing agents. The paper reported that the use of CeO_2 NPs for detection of glucose concentrations was useful for the detection of an inhibitory versus noninhibitory concentration of the antibiotic ampicillin on both *E. coli* and *S. aureus*. Interestingly, in this technique, only a spectrophotometer is required for detection. Moreover, the method is simple and does not require acidic pH or other chemicals.

THE ROLE OF NPs TO OVERCOME DRUG RESISTANCE

The development of antibiotics and other chemotherapeutics revolutionized human health, providing a simple cure for a number of dreadful diseases. However, widespread production, misuse, and overuse of antibiotics have contributed to the emergence of MDR infectious organisms, which is considered a next-generation concern for global public health.[43] The infections caused by MDR organisms cannot be cured easily; moreover, the treatment includes the use of multiple broad-spectrum antibiotics, which are less effective, toxic, and more expensive. Considering these facts, modifying existing antimicrobial compounds and developing novel antimicrobial agents with significant antimicrobial potential are urgent needs.[44,45]

Nanotechnology offers tremendous potential for this; it acts as a decent platform to modify the important properties of metal in the form of nanoparticles. Various metal nanoparticles have been explored for promising biomedical applications in the diagnosis, treatment, and therapies of many infectious diseases like tuberculosis, AIDS, and cancer.[46,47] Moreover, metal nanoparticles in general and AgNPs in particular showed significant antimicrobial efficacy against a wide range of organisms, including MDR microbes. Hence, AgNPs are considered as next-generation antimicrobials and nanoweapons for the treatment and prevention of MDR microbes.[45,48,49] Not only AgNPs, but also other metal nanoparticles, are reported to have broad-spectrum antimicrobial activity. Some of the studies carried out in this area are briefly discussed next.

SILVER NANOPARTICLES (AgNPs)

It is widely known that resistance in microbes toward antibiotics is a global issue that is worsening at an alarming rate. MDR has become a huge problem with many infectious diseases, such as tuberculosis, AIDS, and cancer. The problem is not restricted to these diseases, but it was reported that infections due to MDR bacteria are also responsible for infertility in females. Recently, Gopinath et al.[50] reported that MDR bacteria like *Enterobacter* sp., *P. aeruginosa*, *K. pneumoniae*, and *E. coli* are responsible for infertility in females. Further, they demonstrated the activity of AgNPs synthesized from *Fusarium oxysporum* and antibiotics both singly and in combination against these bacteria. The obtained results confirmed that all the resistant bacteria that were normally resistant to antibiotics were found to be susceptible to them in the presence of AgNPs.

Some studies carried out in the past have proved that biologically synthesized metal nanoparticles possess more bioactivity than physically and chemically synthesized nanoparticles.[51,52] In this context, biologically synthesized nanoparticles are becoming the first choice for researchers. Panacek et al. studied the antimicrobial activity of AgNPs against both Gram-positive and Gram-negative bacteria, including MDR strains such as MRSA.[53] It was also reported that the AgNPs exhibited size-dependent bactericidal activity against MRSA and other Gram-positive and Gram-negative bacteria. MRSA is a major cause of nosocomial and related infections at the global level, and its prevalence is increasing due to its strong resistance to antibiotics. Considering this fact, an attempt was made by Haq et al.[54] for the control of three different strains of MRSA. They demonstrated the activity of AgNPs synthesized from five species of mushroom (*Agaricus bisporus*, *Helvella lacunosa*, *Ganoderma applanatum*, *Pleurotus florida*, and *Fomes fomentarius*). It was reported that AgNPs synthesized from *Agaricus bisporus* were found to be more potent than the other four AgNPs. In another study, Behera and Nayak reported the activity of phytosynthesized AgNPs from the extract of *Syzygium cumini* (jamun) against MDR bacteria (MRSA, ampicillin-resistant *E. coli*, and erythromycin-resistant *Streptococcus pyogenes*). Thus, synthesized AgNPs were in the range of 30–100 nm and polydisperse in nature. The authors claimed it was a green and effective synthesis route for obtaining AgNPs that can be used as an alternative antibacterial agent against MDR pathogens.[55] Surface-modified AgNPs like pluronic-coated silver nanoprisms were also found to be effective against two MRSA strains.[56]

Similarly, Agarwal et al.[57] evaluated the activity of biosynthesized AgNPs against *M. tuberculosis* and 26 various clinical isolates, including drug-sensitive, MDR, extreme drug resistance (XDR), and *Mycobacterium* other than *tuberculosis* strains, exhibiting effective bactericidal activity. Ninganagouda et al.[58] reported that MDR strains of *P. aeruginosa*, *E. coli*, and *K. pneumoniae* were found to be susceptible to AgNPs synthesized from *Aspergillus flavus*. Saeb et al.[59] reported the biological synthesis of AgNPs from *Escherichia hermannii*, *Citrobacter sedlakii*, and *Pseudomonas putida* isolated from soil. Further, they tested the antibacterial activity of these AgNPs against *S. aureus*, *Staphylococcus epidermidis*, *P. aeruginosa*, *E. coli*, *K. pneumoniae*, and MRSA. They reported that AgNPs significantly inhibited the growth of MRSA, and also proposed that the activity of AgNPs could be enhanced by using them in combination with antibiotic vancomycin. Some other studies on the use of AgNPs and functionalized AgNPs against various MDR bacteria include AgNPs against MDR *P. aeruginosa*, ampicillin-resistant *E. coli*, and erythromycin-resistant *Staphylococcus pyogenes*,[44] *P. aeruginosa*,[60,61] and MRSA,[62] and against uropathogens such as *P. aeruginosa*, *K. pneumoniae*, and *E. coli*,[63] silver-carbine nanoparticle complex against MRSA, MDR *A. baumannii*,[64] and polymersome nanocarriers embedded with AgNPs against antibiotic-resistant *E. coli*.[65]

GOLD, COPPER, CHITOSAN, AND TITANIUM NANOPARTICLES

Uropathogens are responsible for UTIs, which are very common nowadays in both males and females.[66] Mahitha et al.[67] reported a fast, extracellular, and convenient method for the synthesis of metallic gold nanoparticles (AuNPs) by reducing aqueous $HAuCl_4$ with the help of *Bacopa monnieri* plant extract. Further, they reported their potential against MDR human pathogens such as *E. coli*, *S. aureus*, *Bacillus subtilis*, and *Enterococcus*. Li et al.[68] developed functionalized AuNPs by tuning functional groups on the surface of AuNPs, which were effective against both Gram-negative and Gram-positive uropathogens, including MDR pathogens (*E. coli*, *P. aeruginosa*, and MRSA). Further, it was proposed that the activity of functionalized AuNPs was tailored through surface hydrophobicity.

Among metal nanoparticles, copper nanoparticles (CuNPs) also possess potent antimicrobial activity against wide range of microorganism.[69] However, Ashajyothi et al.[70] proved the potential of CuNPs against MDR bacteria. They demonstrated the activity of biologically synthesized CuNPs using *Enterococcus faecalis* against MDR bacteria such as *E. coli*, *K. pneumoniae*, and MRSA. In another study, Betancourt-Galindo et al.[71] reported remarkable activity of physically synthesized CuNPs against MDR strains of *S. aureus* and *P. aeruginosa*. Similarly, chitosan- and cellulose-based nanocomposites have been found to be effective against MDR pathogens.[72] Moreover, in another study, biologically synthesized titanium nanoparticles using *Lactobacillus crispatus* showed the potential to control biofilm formation. It was reported that synthesized titanium nanoparticles played a potential role in significant higher antimicrobial efficacy against MDR bacteria such as *K. pneumoniae*, *S. aureus*, *A. baumannii*, *E. coli*, and *Morganella morganii*, which are responsible for UTIs.[73]

METAL OXIDE NANOPARTICLES

Not only metal, but also metal oxide nanoparticles such as titanium dioxide (TiO_2), copper oxide (CuO), zinc oxide (ZnO), and manganese oxide (MnO_2), are known to possess good antibacterial properties. One of the common properties they share is photocatalytic activity due to wide band gaps. This photocatalytic activity is mainly involved in its mechanism of action, which occurs through the generation of ROS.[74] Huang et al.[75] synthesized ZnO nanoparticles in the aqueous system through the hydrolysis of ionic Zn^{2+} salts and demonstrated activity against methicillin-resistant bacteria (*S. aureus* and *Streptococcus agalactiae*). On the other hand, Vincent et al.[76] claimed the bactericidal efficacy of ZnO and TiO_2 nanoparticles against MDR organisms. In their study, they determined the activity of ZnO and TiO_2 nanoparticles against MDR *P. aeruginosa* involved in biofilm production. Here, the authors isolated 51 isolates of biofilm-producing *P. aeruginosa* from Bioline Laboratory in Coimbatore, India, out of which 47 were reported to be resistant to many antibiotics. The findings reported here proved significant activity of ZnO and TiO_2 nanoparticles against these isolates. Moreover, Jesline et al.[77] also demonstrated the activity of ZnO and TiO_2 nanoparticles against another biofilm-producing bacterium (MRSA). From these, it can be concluded that these nanoparticles may serve as promising antibacterial agents against MDR and biofilm-producing organisms.

Like metal nanoparticles, modified forms of metal nanoparticles are also used for the control of antibiotic-resistant bacteria. Copper-doped ZnO nanoparticles (Cu:ZnO NPs) are one example of surface-modified nanoparticles that can be effectively used against MDR bacteria like *E. coli* and *S. aureus*.[78] On the other hand, Malka et al.[79] reported the use of zinc-doped CuO nanoparticles against MRSA and *E. coli*. Kunkalekar et al.[81] found that chemically synthesized silver-doped manganese oxide (MnO_2) nanoparticles have antibacterial activity against MRSA. Necula et al.[82] synthesize porous TiO_2-Ag composite coating and evaluated their potential against MRSA. The coating was developed by the process of plasma electrolytic oxidation of a Ti-6Al-7Nb medical alloy in a calcium acetate/calcium glycerophosphate electrolyte having AgNPs. The findings reported showed the complete killing of MRSA within 24 h.

Ansari et al.[83] investigated the interaction of aluminum oxide (Al_2O_3) nanoparticles with clinical isolates of MDR *E. coli* and its biomolecules present on the surface of the envelope. The nanoparticles demonstrated remarkable activity against MDR *E. coli*. The mechanism involved was revealed through scanning electron microscopy and high-resolution transmission electron microscopy analysis. In the analysis, attachment of nanoparticles to the surface of cell membrane and inside the cells was observed. It might be due to the formation of irregularly shaped pits and perforations on the surfaces of bacterial cells. It was hypothesized that the nanoparticles entered in the cell might have interacted with cellular biomolecules and caused adverse effects, eventually triggering cell death. Further, it was proposed that the Al_2O_3 nanoparticles induced structural changes in phospholipids, which lead to the loss of amphiphilic properties, destruction of the membrane, and cell leaking. Similarly, Allahverdiyev et al.[84] reviewed the antibacterial nature of TiO_2 and silver oxide (Ag_2O) nanoparticles against drug-resistant bacteria and *Leishmania* parasites. Table 6.2 summarizes the efficacy of metal nanoparticles against various MDR bacteria.

Table 6.2 Efficacy of Metal Nanoparticles Against Various MDR Bacteria

S. No.	Metal Nanoparticles Used	Test MDR Bacteria	References
1	Silver	*Staphylococcus aureus*	86
2	Silver	MRSA	62
3	Silver	*Pseudomonas aeruginosa*	60,61
4	Silver	*P. aeruginosa*, ampicillin-resistant *Escherichia coli*, erythromycin-resistant *Streptococcus pyogenes*	44
5	Silver	*Mycobacterium tuberculosis*	57
6	Silver	*Enterobacter* sp., *P. aeruginosa*, *Klebsiella pneumoniae*, *E. coli*	50
7	Silver	MRSA	53,54,59
8	Silver	MRSA, ampicillin-resistant *E. coli*, erythromycin-resistant *S. pyogenes*	55
9	Gold	*E. coli*, *P. aeruginosa*, MRSA	68
10	Gold	*E. coli*, *S. aureus*, *Bacillus subtilis*, *Enterococcus sp.*	67
11	Titanium	*K. pneumoniae*, *S. aureus*, *Acinetobacter baumannii*, *E. coli*, *Morganella morganii*	73
12	Zinc oxide	MRSA, *Streptococcus agalactiae*	75
13	Titanium oxide–silver composite coating	MRSA	82
14	Copper-doped zinc oxide nanoparticles	*E. coli*, *S. aureus*	78
15	Manganese oxide	MRSA	81
16	Pluronic-coated silver nanoprisms	MRSA	56
17	Zinc oxide and Titanium oxide	*P. aeruginosa*, MRSA	76,77
18	Aluminum oxide	*E. coli*	83
19	Zinc-doped copper oxide nanocomposite	MRSA, *E. coli*	79

THE COMBINATION OF NANOPARTICLES AND ANTIBIOTICS IS A BETTER OPTION

Combined antibiotic therapy has paved the way toward the improvization of present antimicrobial treatment strategies. In support of these efforts, the Gram-negative resistance summit 2011, proposed a scheme of combination therapy using carbapenem with other antibiotics as a first-line treatment for patients at risk for MDR Gram-negative bacteria. The summit reviewed increased pathogen coverage, pharmacokinetics, and pharmacodynamics as the probable mechanisms by which combination therapy plays an important role in overcoming MDR bacterial infections.[86]

In the field of nanomedicine, the conjugation or combination of antibiotics with antimicrobial nanoparticles has emerged as a novel route to increase the efficacy of both nanoparticles and antibiotics. Chemical compositions of the nanosurface enables prolonged binding, protection from enzymes, and active targeting of surface-functionalized antibiotics at the target site. Hence, attainment of higher concentration of antibiotic within the cell eliminates the requirement of higher dosage, thereby reducing the side effects.[87] Many researchers have demonstrated nanoparticle-antibiotic conjugates as a new class of antimicrobial agents that can actually eliminate MDR problem in pathogens. A large variety of antimicrobial nanoparticles (eg, silver, gold, zinc oxide, titanium oxide, etc.) combined with commercially available antibiotics have been tested against MDR pathogens.

ANTIMICROBIAL EFFECT OF AgNPs WITH ANTIBIOTICS

As discussed in the previous section, although AgNPs have exerted good antimicrobial properties, various attempts have been made toward improving its efficacy by conjugating or combining them with commercial antibiotics. Naqvi et al.[88] investigated the combined activity of AgNPs with ciprofloxacin, imipenem, gentamycin, trimethoprim, and vancomycin against MDR *E. coli*, *P. aeruginosa*, *E. faecalis*, *S. aureus*, *Micrococcus luteus*, *A. baumannii*, *K. pneumoniae*, and *Bacillus* spp. The results of the study showed a 0.2–7 (average 2.8) fold- area increment in antibacterial action of antibiotics when coupled with AgNPs. The most activity was exerted by AgNPs + ciprofloxacin, followed by AgNPs + imipenem, while the lowest activity was shown by AgNPs + vancomycin. The study also revealed that AgNPs were more potent antimicrobial agents against MDR bacteria in comparison with commercial antibiotics.

A similar study carried out by Anitha et al.[89] also demonstrated enhanced efficacy of AgNPs in combination with antibiotics (specifically ampicillin, gentamycin, and ciprofloxacin) against MDR Gram-positive *S. aureus* and Gram-negative *E. coli* strains. Here, the findings also revealed Gram-negative *E. coli* (with a 50-mm zone of inhibition) as more susceptible strains to the nanoparticle-antibiotic combination than Gram-positive *S. aureus* (with a 42-mm zone of inhibition). Kulkarni et al.[90] reported combined activity of AgNPs with antibiotic oxacillin against MRSA (VPG1 to VPG7) strains. The maximum level of synergistic bioactivity (20 mm) was shown against MSRA (VPG7) strain, while the minimum (16 mm) was exerted against MRSA (VPG3 and VPG4). In all seven cases, an increase from 6.66% to 33% was achieved, which clearly revealed the efficacy of nanoparticles-antibiotic conjugate even against MRSA strains.

In another study on growth inhibition of MRSA, Saeb et al.[59] confirmed the enhanced activity of AgNPs combined with vancomycin (with a zone of inhibition of 12.2 ± 1.6 mm) against MRSA, rather than independent activity of both AgNPs (9.6 ± 1.2 mm) and vancomycin (8.7 ± 4.9 mm). Ampicillin-functionalized AgNPs were prepared by Brown et al.[91] and compared with AgNPs alone against drug-resistant *P. aeruginosa*, *E. coli*, *Enterobacter aerogenes*, and *V. cholerae*. Here, the authors reported that within 1 h of treatment, the AgNP-amp complex (4 μg/mL) was able to kill inoculums containing ampicillin-resistant *P. aeruginosa* and *E. coli* at a much faster rate than AgNPs alone. The overall study suggested that combined activity of AgNPs and ampicillin loaded on the surface of the AgNPs was the key factor of enhanced bactericidal activity of the complex.

AgNPs were also reported to enhance the activity of ampicillin when tested against ampicillin-resistant *Enterobacter* sp. ANT 02 [HM803168], *P. aeruginosa* ANT 04 [HM803170],

K. pneumoniae ANT 03 [HM803169], and *E. coli* ANT 01 [HM803167] at a concentration of 20 μg/mL. Gopinath et al. proposed a mechanism for the synergistic effect between ampicillin and AgNPs against ampicillin-resistant, β-lactamase-producing bacteria. In resistant strains, upon treatment with AgNPs-ampicillin complex, AgNPs causes the destruction of cell wall, followed by leakage of intracellular material that led to death of the pathogen. On the other hand, ampicillin makes the entry into the cell through AgNPs mediated damaged cell wall and causes irreparable inhibition of the enzyme transpeptidase, resulting in the ultimate stoppage of cell wall synthesis. This two-step action of the nanoparticle-ampicillin complex clearly represents the possibilities of treating ampicillin-resistant MDR bacterial infections more effectively than with ampicillin alone.[50]

A similar mechanism was explored for the amoxicillin-AgNP complex, in which the growth inhibition of MDR pathogens was attributed to structural changes made on the surface of nanosilver due to the formation of a complete surrounding of amoxicillin molecules. The binding of hydroxyl and amido groups in the amoxicillin molecule with nanosilver by means of chelation ultimately results in the prevention of DNA unwinding followed by DNA damage.[92,93] Li et al.[93] further stated that hydrophobic AgNPs interact more profoundly with a bacterial membrane made up of hydrophobic phospholipids and glycoproteins, facilitating increased transportation of amoxicillin molecules in the vicinity of cell surface. Moreover, Duran et al.[94] confirmed the binding action of nanosilver with sulfur bridges of binding sites of amoxicillin as a responsible mechanism for synergistic activity of the antibiotic-AgNP complex.

ANTIMICROBIAL EFFECT OF AuNPs WITH ANTIBIOTICS

Although the biological inertness of AuNPs does not result in any significant antimicrobial activity, their strong catalytic activity opens up new possibilities in their use in antibiotic-AuNP combination therapies. Hence, efforts have been made by many researchers to analyze whether in synergy, AuNPs can enhance the activity of commercially available antibiotics. In this scenario, Brown et al.[91] demonstrated growth inhibitory activity of ampicillin-functionalized AuNPs against MDR *P. aeruginosa*, *E. aerogenes*, and MRSA. The study showed that AuNPs did not show any bactericidal effect; however, strong bactericidal activity was exhibited only when ampicillin molecules were functionalized on the surface of AuNPs at the concentration 2 and 4 μg/mL.[91] Bresee et al.[95] reported on *p*-mercaptobenzoic acid-capped AuNPs and tested against MDR *E. coli* and *K. pneumoniae*. Fayaz et al.[96] impregnated vancomycin over the surface of biologically synthesized AuNPs by ionic interactions between positively charged amine groups of vancomycin and negatively charged AuNPs. The complex was found to be bactericidal against vancomycin-resistant *S. aureus* at the concentration of 8 μg/mL. Moreover, the vancomycin-loaded AuNPs demonstrated activity against vancomycin-resistant *E. coli* by facilitating the binding of vancomycin molecules to the bacterial cell surface. Zhao et al.[97] studied the growth-inhibition activity of amino-substituted, pyrimidine-capped-AuNPs against MDR clinical isolates. In this study, amino-substituted pyrimidines exerted bactericidal activity in the presence of AuNPs without any energy input. The authors stated that pyrimidine-capped AuNPs exert antibacterial activity by sequestration of magnesium or calcium ions, which further results in the interruption of the bacterial cell membrane, followed by leakage of intracellular content. Moreover, interactions of internalized AuNPs with DNA cause the inhibition of protein synthesis, leading to growth inhibition of MDR pathogen.[97] Significant bactericidal activity of polysiloxane polymers containing methylene blue and AuNPs against MRSA and *E. coli* was demonstrated by Perni et al.,[98] who explained the activity of AuNPs to enhance the light-induced oxidation reaction of methylene blue with subsequent production of ROS as a responsible mechanism for synergistic activity of the antibiotic-AuNP complex.

ANTIMICROBIAL EFFECT OF ZnO NANOPARTICLES WITH ANTIBIOTICS

The US Food and Drug Administration (FDA) has recognized ZnO as one of the five compounds that is generally recognized as safe.[99] Moreover, various researchers have already demonstrated the antibacterial potential of ZnO nanoparticles. Therefore, the attempts have been made toward the use of nanoZnO for preparing the antibiotic-ZnO-NP complex for treatment of MDR pathogenic infections. For instance, ciprofloxacin-conjugated Zn nanoparticles (Zn-CIPs) were prepared by Patra et al.[100] and evaluated for their activity against MDR *E. coli*, *S. aureus*, and *Klebsiella* sp. It was observed that the conjugated complex was successfully able to inhibit the growth of MDR-tested bacterial strains at the minimum concentration of 20 μg/mL. Here, the antibacterial action of the Zn-CIP complex was attributed to the Zn nanoparticles' mediated damage of cell membrane through generation of ROS. This was followed by the entry of ciprofloxacin into the cell, ultimately causing the inhibition of bacterial growth.

Bhande et al.[101] tested time-kill and membrane leakage assays for ZnO nanoparticles in combination with cefotaxime, ampicillin, ceftriaxone, and cefepime against the β-lactamase producer strains of *E. coli*, *K. pneumoniae*, *Sphingomonas paucimobilis*, and *P. aeruginosa*, respectively. The time-kill kinetics of ZnO nanoparticles in combination with β-lactam antibiotics confirmed enhanced antibacterial activity of the complex (50-, 85-, 58-, and 50-fold inhibition, respectively). The post-time-kill effect on cell membranes revealed that synergism between ZnO nanoparticles and β-lactam antibiotics by increasing membrane permeability causes membrane leakage leading to the death of all extended-spectrum β-lactamase-producing strains. The study presented the possibility of using combination therapy consisting of ZnO nanoparticles to enhance the use β-lactam antibiotics for the treatment of UTIs.

ANTIMICROBIAL EFFECT OF OTHER NANOPARTICLES WITH ANTIBIOTICS

Many researchers have reported on the antibacterial activity of TiO_2 nanoparticles by generation of free radical oxides and peroxides, causing cell death.[102] In addition, few of them have also demonstrated the interactions of nano-TiO_2 with commercially available standard antibiotics. In this case, Roy et al.[103] showed the synergistic effect of TiO_2 nanoparticles in combination with a group of antibiotics (β-lactams, cephalosporin, glycopeptides, aminoglycosides, flouroquinolones, azlides, macrolides, lincosamides, and sulfonamides). The interactions between nanoparticles and antibiotics showed improved antibacterial activity in the form of increased zones of inhibition, particularly by combinations of penicillin and amikacilin (10 mm), followed by ampicillin and gentamycin (8 mm). However, more or less similar activity was also shown by TiO_2 nanoparticles in combination with other substances, including oxacillin, cloxacillin, amoxicillin, cephalexin, cefotaxime, ceftazidime, vancomycin, streptomycin, and erythromycin.[103]

Researchers have developed nanoparticles as carriers for effective drug delivery systems to deliver antibiotics with effectiveness against MDR human pathogens. For instance, the use of chitosan nanoparticles as a carrier system for the delivery of antimicrobial peptides and temporin B

(TB-CS-NPs) was developed by Piras et al.[104] to treat *S. epidermidis*. The TB-CS-NPs were prepared by ionic gelation of chitosan oligomers into nanoparticles, and temporin was entrapped within the hollow core with an encapsulation efficiency of 74.7 ± 2.3%. TB functionalized CS-NPs showed strong growth inhibitory effects against tested strains as compared to free TB and blank CS-NPs. Jiang et al.[85] prepared nanohydroxyapatite (nHA) pellets loaded with vancomycin and assayed their effectiveness in the treatment of MRSA-induced chronic osteomyelitis and bone defects in rabbits. The drug release kinetics were studied for vancomycin-loaded nHA pellets both in vitro and in vivo. Biphasic drug release with a rapid initial burst, followed by gradual release of the drug over up to 12 weeks, was achieved. The vancomycin-loaded nHA pellets were found to be effective bone graft materials for reconstruction of bone defects without any recurrence of infection.[85] Zero-valent iron nanoparticles (Fe^0) were combined with standard antibiotic discs (namely, ampicillin, amoxicillin, methicillin, chloramphenicol, tetracycline, amikacin, kanamycin, streptomycin, vancomycin, and erythromycin). Among the tested resistant strains, synergism between zero-valent iron nanoparticles and standard antibiotics showed the greatest increase (29.75%) against *K. pneumoniae*. However, of all the tested combinations, the combination of Fe^0 nanoparticles with streptomycin with a 63.33% increase exhibited the most effective synergism against MDR *P. aeruginosa*.

After this, a 53.33% increase in bactericidal activity was observed for Fe^0 NPs and tetracycline complex against *Bacillus cereus*. The 50% increase in growth inhibition was achieved by Fe^0 nanoparticles in combination with amikacin and methicillin against *S. epidermidis*. The enhancement in the combined effect was ascribed to the generation of oxidative stress by ROS causing damage to DNA and proteins in bacteria.[105] Standard antibiotic tetracycline was entrapped within the silica nanospheres and tested against drug-resistant bacteria. In this case, nanosilica composite exhibited significant bactericidal activity with no mammalian cytotoxicity.[106] In another study, silica nanoparticles were conjugated to aminoglycosides (namely, gentamycin, neomycin, and kanamycin) to combat clinical pathogens and kanamycin-resistant *E. coli*. The drug-resistant *E. coli* was found to be most susceptible to the silica NP-gentamycin and silica NP-neomycin complex at much lower concentrations of 95 and 100 μg/mL, respectively, than the concentrations required for nanoparticles or antibiotics alone.[107] El-Newehy et al.[108] developed a technique of encapsulation of antibiotics within the nanofibers made up of polyvinyl alcohol and polyethylene oxide. The nanofibers showed significant bactericidal activity against MDR *E. coli* and *P. aeruginosa* by enhancing the bioactivity of entrapped antibiotics.[108]

Taking into consideration all the possibilities provided by combination therapy of antibiotics with nanoparticles, the advanced class of nanomaterials can come up with better solutions for combating the drug-resistance problem in human pathogens.

CONCLUSIONS

Antibiotic resistance is increasing at a fast pace in almost all microorganisms, including bacteria, fungi, viruses, and protozoans. Emerging and reemerging diseases pose a great challenge to scientists, which has generated a pressing need to search for new antibiotics or antimicrobial agents to fight the resistant microbes. Various studies have provided evidence that different types of nanoparticles, like silver, copper, zinc oxide, and titanium dioxide, have the potential to combat antimicrobial resistance.

Encouragingly, the existing antimicrobial agents (eg, amphotericin B, oxacillin, cloxacillin, amoxicillin, cephalexin, cefotaxime, ceftazidime, vancomycin, streptomycin, and erythromycin), in combination with different nanoparticles, can be utilized for the treatment of microbial diseases. However, these combinations need extensive animal trials before they can be tried against human infections. The synergistic effects of the antibiotics with nanoparticles may open new vistas in the treatment of MDR pathogens.

ACKNOWLEDGMENTS

We thank the Department of Science and Technology (DST), New Delhi; University Gants Commission, New Delhi, and CNPq Brazil for financial assistance.

REFERENCES

1. Abraham EP, Chain E. An enzyme from bacteria able to destroy penicillin. *Rev Infect Dis* 1940;**10**: 677–8.
2. Blazyk JF. *Cationic, amphipathic beta-sheet peptides and uses thereof.* US Patent Pub. No.2004/0249122 A1, (Filed on February 15, 2001).
3. D'Costa VM, McGrann KM, Hughes DW, Wright GD. Sampling the antibiotic resistome. *Science* 2006;**311**:374–7.
4. Paterson DL, Doi Y. A step closer to extreme drug resistance (XDR) in gram-negative bacilli. *Clin Infect Dis* 2007;**45**:1179–81.
5. Kumarasamy KK, Toleman MA, Walsh TR, Bangaria J, Butt F, Balakrishnan R, et al. Emergence of a new antibiotic resistance mechanism in India, Pakistan, and the UK: a molecular, biological, and epidemiological study. *Lancet Infect Dis* 2010;**10**:597–602.
6. Lipp EK, Huq A, Colwell RR. Effects of global climate on infectious disease: the cholera model. *Clin Microbiol Rev* 2002;**15**:757–70.
7. Chopra I, Roberts M. Tetracycline antibiotics: mode of action, applications, molecular biology, and epidemiology of bacterial resistance. *Microbiol Mol Biol Rev* 2001;**65**(2):232–60.
8. Wright GD. Aminoglycoside-modifying enzymes. *Curr Opin Microbiol* 1999;**2**:499–503.
9. Llano-Sotelo B, Azucena Jr EF, Kotra LP, Mobashery S, Chow CS. Aminoglycosides modified by resistance enzymes display diminished binding to the bacterial ribosomal aminoacyl-tRNA site. *Chem Biol* 2002;**9**:455–63.
10. Doi Y, Arakawa Y. 16S ribosomal RNA methylation: emerging resistance mechanism against aminoglycosides. *Clin Infect Dis* 2007;**45**:88–94.
11. Bush K, Jacoby GA. Updated functional classification of β-lactamases. *Antimicrob Agents Chemother* 2010;**54**:969–76.
12. Poirel L, Naas T, Nordmann P. Diversity, epidemiology, and genetics of class D β-lactamases. *Antimicrob Agents Chemother* 2010;**54**:24–38.
13. Livermore DM, Canton R, Gniadkowski M, Nordmann P, Rossolini GM, Arlet G, et al. CTX-M: changing the face of ESBLs in Europe. *J Antimicrob Chemother* 2007;**59**:165–74.
14. Queenan AM, Bush K. Carbapenemases: the versatile β-lactamases. *Clin Microbiol Rev* 2007;**20**:440–58.
15. Rashid H, Rahman M. Possible transfer of plasmid mediated third generation cephalosporin resistance between *Escherichia coli* and *Shigella sonnei* in the human gut. *Inf Gene Evol* 2015;**30**:15–18.

16. Barker KF. Antibiotic resistance: a current perspective. *Br J Clin Pharmacol* 1999;**48**(2):109–24.
17. Boucher HW, Talbot GH, Bradley JS, et al. Bad bugs, no drugs: no ESKAPE! An update from the Infectious Diseases Society of America. *Clin Infect Dis* 2009;**48**:1–12.
18. Ghosh Dastider S, Barizuddin S, Yuksek NS, Dweik M, Almasri MF. Efficient and rapid detection of *Salmonella* using microfluidic impedance based sensing. *J Sensors* 2015. 2015 Article ID 293461:8 pp., http://dx.doi.org/10.1155/2015/293461.
19. Kashid SB, Tak RD, Raut W. Antibody tagged gold nanoparticles as scattering probes for the pico molar detection of the proteins in blood serum using nanoparticle tracking analyzer. *Colloid Surface B* 2015;**133** (2015):208–13.
20. Fan Z, Senapati D, Khan SA, Singh AK, Hamme A, Yust B, et al. Popcorn-shaped magnetic core–plasmonic shell multifunctional nanoparticles for the targeted magnetic separation and enrichment, label-free SERS imaging and photothermal destruction of multidrug-resistant bacteria. *Chem Eur J* 2013;**19**(12):2839–47.
21. Lam B, Das J, Holmes RD, Live L, Sage A, Sargent EH, et al. Solution-based circuits enable rapid and multiplexed pathogen detection. *Nat Commun* 2013;**4**:2001. Available from: http://dx.doi.org/10.1038/ncomms3001.
22. Nguyen YH, Ma X, Qin L. Rapid identification and drug susceptibility screening of ESAT-6 secreting Mycobacteria by a NanoELIwell assay. *Sci Rep* 2012;**2**:635. Available from: http://dx.doi.org/10.1038/srep00635.
23. Maiti S, Krishnan D, Barman G, Ghosh SK, Laha JK. Antimicrobial activities of silver nanoparticles synthesized from *Lycopersicon esculentum* extract. *J Anal Sci Technol* 2014;**5**:40.
24. Kandi V, Kandi S. Antimicrobial properties of nanomolecules: potential candidates as antibiotics in the era of multi-drug resistance. *Epidemiol Health* 2015;**37**:e2015020. Available from: http://dx.doi.org/10.4178/epih/e2015020.
25. Kruka T, Szczepanowicza K, Stefańskab J, Robert P, Sochaa RP, Warszyński P. Synthesis and antimicrobial activity of monodisperse copper nanoparticles. *Colloid Surface B* 2015;**128**(2015):17–22.
26. Salem W, Leitner DR, Zingl FG, Schratter G, Prassl R, Goessler W, et al. Antibacterial activity of silver and zinc nanoparticles against *Vibrio cholerae* and enterotoxic *Escherichia coli. Int J Med Microbiol* 2015;**305**(1):85–95.
27. *Antibiotic resistance threats in the United States*. Centers for Disease Control and Prevention; 2013.
28. Chaloupka K, Malam Y, Seifalian AM. Nanosilver as a new generation of nanoproduct in biomedical applications. *Trends in Biotechnol* 2010;**28**:580–8.
29. Chatterjee AK, Chakraborty R, Basu T. Mechanism of antibacterial activity of copper nanoparticles. *Nanotechnology* 2014;**25**:1–12.
30. Bawskar MS, Deshmukh SD, Bansod S, Gade AK, Rai MK. Comparative analysis of biosynthesised and chemosynthesised silver nanoparticles with special reference to their antibacterial activity against pathogens. *IET Nanobiotechnol* 2015;**9**(3):107–13. Available from: http://dx.doi.org/10.1049/iet-nbt.2014.0032.
31. Feng QL, Wu J, Chen GQ, Cui FZ, Kim TM, Kim JO. A mechanistic study of the antibacterial effect of silver ions on *Escherichia coli* and *Staphylococcus aureus*. *J Biomed Mater Res* 2000;**52**:662–8.
32. Yamanaka M, Hara K, Kudo J. Bactericidal actions of a silver ion solution on *Escherichia coli*, studied by energy-filtering transmission electron microscopy and proteomic analysis. *Appl Environ Microbiol* 2005;**71**:7589–93.
33. McDonnell G, Russell AD. Antiseptics and disinfectants: activity, action, and resistance. *Clin Microbiol Rev* 1999;**12**:147–79.
34. Matsumura Y, Yoshikata K, Kunisak S, Tsuchido T. Mode of bactericidal action of silver zeolite and its comparison with that of silver nitrate. *Appl Environ Microbiol* 2003;**69**:4278–81.

35. Pal S, Tak YK, Song JM. Does the antibacterial activity of silver nanoparticles depend on the shape of the nanoparticle? A study of the Gram negative bacterium *Escherichia coli*. *Appl Environ Microbiol* 2007;**73**:1712–20.
36. Hamouda T, Myc A, Donovan B, Shih A, Reuter JD, Baker Jr. JR. A novel surfactant nanoemulsion with a unique non-irritant topical antimicrobial activity against bacteria, enveloped viruses and fungi. *Microbiol Res* 2000;**156**:1–7.
37. Dibrov P, Dzioba J, Gosink KK, Hase CC. Chemiosmotic mechanism of antimicrobial activity of Ag(+) in *Vibrio cholerae*. *Antimicrob Agents Chemother* 2000;**46**:2668–70.
38. Singh M, Sing S, Prasad S, Gambhir IS. Nanotechnology in medicine and antibacterial effect of silver nanoparticles. *Digest J Nanomater Biostructures* 2008;**3**:115–22.
39. Sondi I, Salopek-Sondi B. Silver nanoparticles as antimicrobial agent: a case study on *E. coli* as a model for Gram-negative bacteria. *J Colloid Interface Sci* 2004;**275**:177–82.
40. Banerjee M, Mallick S, Paul A, Chattopadhyay A, Ghosh SS. Heightened reactive oxygen species generation in the antimicrobial activity of a three component iodinated chitosan-silver nanoparticle composite. *Langmuir* 2010;**26**:5901–8.
41. Prabhu S, Poulose EK. Silver nanoparticles: mechanism of antimicrobial action, synthesis, medical applications and toxicity effects. *Int Nano Lett* 2012;**2**:32.
42. Noll AJ. *Cerium oxide nanoparticles for the detection of antimicrobial resistance*. A thesis submitted in partial fulfillment of the requirements for the Honors in the Major Program in Molecular Biology and Microbiology in the College of Medicine and in the Burnett Honors College at the University of Central Florida Orlando, FL; 2011. p. 31.
43. Namasivayam SKR, Jayakumar D, Kumar R, Bharani RSA. Antibacterial and anticancerous biocompatible silver nanoparticles synthesized from the cold-tolerant strain of *Spirulina platensis*. *J Coastal Life Med* 2015;**3**(4):265–72.
44. Lara HH, Ayala-Nunez NV, Turrent LCI, Padilla CR. Bactericidal effect of silver nanoparticles against multidrug-resistant bacteria. *World J Microbiol Biotechnol* 2010;**26**:615–21.
45. Rai MK, Deshmukh SD, Ingle AP, Gade AK. Silver nanoparticles: the powerful nanoweapon against multidrug-resistant bacteria. *J Appl Microbiol* 2012;**112**:841–52.
46. Rai M, Ingle AP, Bansod S, Kon K. Tackling the problem of tuberculosis by nanotechnology: disease, diagnosis and drug delivery. In: Rai MK, Kon K, editors. *Nanotechnology in diagnosis, treatment and prophylaxis of infectious diseases*. UK: Elsevier; 2014. p. 133–49.
47. Rai MK, Ingle AP, Birla S, Yadav A, Santo CAD. Strategic role of selected noble metal nanoparticles in medicine. *Crit Rev Microbiol* 2015;1–24. Available from: http://dx.doi.org/10.3109/1040841X.2015.1018131.
48. Rai MK, Yadav AP, Gade AK. Silver nanoparticles as a new generation of antimicrobials. *Biotechnol Adv* 2009;**27**(1):76–82.
49. Rai M, Kon K, Ingle A, Duran N, Galdiero S, Galdiero M. Broad-spectrum bioactivities of silver nanoparticles: the emerging trends and future prospects. *Appl Microbiol Biotechnol* 2014;**98**:1951–61.
50. Gopinath PM, Narchonai G, Dhanasekaran D, Ranjani A, Thajuddin N. Mycosynthesis, characterization and antibacterial properties of AgNPs against multidrug resistant (MDR) bacterial pathogens of female infertility cases. *Asian J Pharma Sci* 2015;**10**:138–45.
51. Antony JJ, Sivalingam P, Siva D, Kamalakkannan S, Anbarasu K, Sukirtha R, et al. Comparative evaluation of antibacterial activity of silver nanoparticles synthesized using *Rhizophora apiculata* and glucose. *Colloids Surf B Biointerfaces* 2011;**88**(1):134–40.
52. Manikprabhu D, Lingappa K. Microwave assisted rapid and green synthesis of silver nanoparticles using a pigment produced by *Streptomyces coelicolor* klmp33. *Bioinorg Chem Appl* 2013. 2013: Article ID 341798, http://dx.doi.org/10.1155/2013/341798.

53. Panacek A, Kvitek L, Prucek R, Kolar M, Vecerova R, Pizurova N, et al. Silver colloid nanoparticles: synthesis, characterization, and their antibacterial activity. *J Phys Chem B* 2006;**110**(33):16248–53.
54. Haq M, Rathod V, Singh D, Singh AK, Ninganagouda S, Hiremath J. Dried mushroom *Agaricus bisporus* mediated synthesis of silver nanoparticles from Bandipora District (Jammu and Kashmir) and their efficacy against *Methicillin resistant Staphylococcus aureus* (MRSA) strains. *Nanosci Nanotechnol Int J* 2015;**5**(1):1–8.
55. Behera S, Nayak PL. In vitro antibacterial activity of green synthesized silver nanoparticles using Jamun extract against multiple drug resistant bacteria. *World J Nano Sci Technol* 2013;**2**(1):62–5.
56. Marta B, Jakab E, Potara M, Simon T, Imre-Lucaci F, Barbu-Tudoran L, et al. Pluronic-coated silver nanoprisms: synthesis, characterization and their antibacterial activity. *Colloids Surf A* 2014;**441**:77–83.
57. Agarwal P, Mehta A, Kachhwaha S, Kothari SL. Green synthesis of silver nanoparticles and their activity against *Mycobacterium tuberculosis*. *Adv Sci Eng Med* 2013;**5**(7):709–14.
58. Ninganagouda S, Rathod V, Jyoti H, Singh D, Prema K, Ul-Haq M. Extracellular biosynthesis of silver nanoparticles using *Aspergillus flavus* and their antimicrobial activity against gram negative MDR strains. *Int J Pharma Bio Sci* 2013;**4**(2):222–9.
59. Saeb ATM, Alshammari AS, Al-Brahim H, Al-Rubeaan KA. Production of silver nanoparticles with strong and stable antimicrobial activity against highly pathogenic and multidrug resistant bacteria. *Sci World J* 2014. 2014: Article ID 704708, http://dx.doi.org/10.1155/2014/704708.
60. Singh K, Panghal M, Kadyan S, Chaudhary U, Yadav JP. Antibacterial activity of synthesized silver nanoparticles from *Tinospora cordifolia* against multidrug resistant strains of *Pseudomonas aeruginosa* isolated from burn patients. *J Nanomed Nanotechnol* 2014;**5**. 1000192 (6 pp.), http://dx.doi.org/10.4172/2157-7439.1000192.
61. Singh K, Panghal M, Kadyan S, Chaudhary U, Yadav JP. Green silver nanoparticles of *Phyllanthus amarus*: as an antibacterial agent against multidrug resistant clinical isolates of *Pseudomonas aeruginosa*. *J Nanobiotechnol* 2014;**12**:40.
62. Chandrakanth RK, Ashajyothi C, Oli AK, Prabhurajeshwar C. Potential bactericidal effect of silver nanoparticles synthesized from *Enterococcus* spp. *Orient J Chem* 2014;**30**(3):1253–62.
63. Upgade A, Prabakaran P. A novel combinatorial herbal drug development using nanotechnology against MDR bacterial uropathogens. *J Pharm Sci Res* 2015;**7**(2):51–4.
64. Leid J, Ditto A, Knapp A, Shah P, Wright B, Blust R, et al. *In vitro* antimicrobial studies of silver carbine complexes: activity of free and nanoparticle carbine formulations against clinical isolates of pathogenic bacteria. *J Antimicrob Chemother* 2012;**67**:138–48.
65. Geilich BM, van de Ven AL, Singleton GL, Sepúlveda LJ, Sridhara S, Webster TJ. Silver nanoparticle-embedded polymersome nanocarriers for the treatment of antibiotic resistant infections. *Nanoscale* 2015;**7**:3511–19.
66. Gatermann SG. Bacterial infections of the urinary tract. In: 10th ed. Borriello P, Murray PR, Finke BG, editors. *Topley & Wilson's microbiology & microbial infections*, vol. III. UK: Hodder Arnold Publishers; 2007. p. 671–83.
67. Mahitha B, Raju BDP, Madhavi T, Durga Maha-lakshmi CHN, Sushma NJ. Evaluation of antibacterial efficacy of phyto-fabricated gold nanoparticles using *Bacopa monnieri* plant extract. *Indian J Adv Chem Sci* 2013;**1**:94–8.
68. Li X, Robinson SM, Gupta A, Saha K, Jiang Z, Moyano DF, et al. Functional gold nanoparticles as potent antimicrobial agents against multi-drug-resistant bacteria. *ACS Nano* 2014;**8**(10):10682–6.
69. Ingle A, Duran N, Rai M. Bioactivity, mechanism of action and cytotoxicity of copper-based nanoparticles: a review. *Appl Microbiol Biotechnol* 2014;**98**:1001–9.
70. Ashajyothi C, Jahanara K, Kelmani CR. Biosynthesis and characterization of copper nanoparticles from *Enterococcus faecalis*. *Int J Pharm Bio Sci* 2014;**5**(4):204–11.

71. Betancourt-Galindo R, Reyes-Rodriguez PY, Puente-Urbina BA, Avila-Orta CA, Rodríguez-Fernández OS, Cadenas-Pliego G, et al. Synthesis of copper nanoparticles by thermal decomposition and their antimicrobial properties. *J Nanomater* 2014. 2014: Article ID 980545, http://dx.doi.org/10.1155/2014/980545.
72. Das S, Das MP, Das J. Fabrication of porous chitosan/silver nanocomposite film and its bactericidal efficacy against multi-drug resistant (MDR) clinical isolates. *J Pharma Res* 2013;**6**:11–15.
73. Ibrahem KH, Salman JAS, Ali FA. Effect of titanium nanoparticles biosynthesis by *Lactobacillus crispatus* on urease, hemolysin & biofilm forming by some bacteria causing recurrent UTI in Iraqi women. *European Sci J* 2014;**10**:324–38.
74. Singh R, Smitha MS, Singh SP. The role of nanotechnology in combating multi-drug resistant bacteria. *J Nanosci Nanotechnol* 2014;**14**:1–12.
75. Huang Z, Zheng X, Yan D, Yin G, Liao X, Kang Y, et al. Toxicological effect of ZnO nanoparticles based on bacteria. *Langmuir* 2008;**24**:4140–4.
76. Vincent MG, John NP, Narayanan PM, Vani C, Murugan S. In vitro study on the efficacy of zinc oxide and titanium dioxide nanoparticles against metallo beta-lactamase and biofilm producing *Pseudomonas aeruginosa. J App Pharma Sci* 2014;**4**:41–6.
77. Jesline A, John NP, Narayanan PM, Vani C, Murugan S. Antimicrobial activity of zinc and titanium dioxide nanoparticles against biofilm-producing methicillin-resistant *Staphylococcus aureus. Appl Nanosci* 2015;**5**:157–62.
78. Kalantar K, Kabir K, Gharibi F, Hatami S, Maleki A. Effect and properties of surface-modified copper doped ZnO nanoparticles (Cu:ZnO NPs) on killing curves of bacterial pathogens. *J Med Bacteriol* 2013;**2** (1):20–6.
79. Malka E, Perelshtein I, Lipovsky A, Shalom Y, Naparstek L, Perkas N, et al. Eradication of multi-drug resistant bacteria by a novel Zn-doped CuO nanocomposite. *Small* 2013;**9**(23):4069–76.
80. Li WR, Xie XB, Shi QS, Duan SS, Ouyang YS, Chen YB. Antibacterial effect of silver nanoparticles on *Staphylococcus aureus. Biometals* 2011;**4**:135–41.
81. Kunkalekar RK, Naik MM, Dubey SK, Salker AV. Antibacterial activity of silver-doped manganese dioxide nanoparticles on multidrug-resistant bacteria. *J Chem Technol Biotechnol* 2013;**88**:873–7.
82. Necula BS, Fratila-Apachitei LE, Zaat SAJ, Apachitei I, Duszczyk J. In vitro antibacterial activity of porous TiO_2–Ag composite layers against methicillin resistant *Staphylococcus aureus. Acta Biomater* 2009;**5**:3573–80.
83. Ansari MA, Khan HM, Khan AA, Cameotra SS, Saquib Q, Musarrat J. Interaction of Al(2)O(3) nanoparticles with *Escherichia coli* and their cell envelope biomolecules. *J Appl Microbiol* 2014;**116**(4):772–83.
84. Allahverdiyev AM, Abamor ES, Bagirova M, Rafailovich M. Antimicrobial effects of TiO_2 and Ag_2O nanoparticles against drug-resistant bacteria and *Leishmania* parasites. *Future Microbiol* 2011;**6**: 933–40.
85. Jiang JL, Li YF, Fang TL, Zhou J, Li XL, Wang YC, et al. Vancomycin loaded nanohydroxyapatite pellets to treat MRSA induced chronic osteomyelitis with bone defects in rabbits. *Inflamm Res* 2012;**61** (3):207–15.
86. Ingle AP, Gade AK, Pierrat S, Sönnichsen C, Rai MK. Mycosynthesis of silver nanoparticles using the fungus *Fusarium acuminatum* and its activity against some human pathogenic bacteria. *Curr Nanosci* 2008;**4**:141–4.
87. Kollef MH, Golan Y, Micek ST, Shorr AF, Restrepo MI. Appraising contemporary strategies to combat multidrug resistant Gram-negative bacterial infections–proceedings and data from the Gram-negative resistance summit. *Clin Infect Dis* 2011;**53**(Suppl. 2):S33–55.
88. Naqvi SZH, Kiran U, Ali MI, Jamal A, Hameed A, Ahmed S, et al. Combined efficacy of biologically synthesized silver nanoparticles and different antibiotics against multidrug-resistant bacteria. *Int J Nanomed* 2013;**8**:3187–95.

89. Anitha M, Karthika J, Arun KJ. Comparison of silver nanoparticles with routinely used antibiotics and its efficacy against gram positive and gram negative organism. *Int J Biol Parma All Sci* 2014;**3**(2):228–34.
90. Kulkarni P, Rathod V, Jyoti H, Patil S, Dattu S, Krishnaveni R. Production of silver nanoparticles by using *Aspergillus terreus* and its antibacterial activity against methicillin resistant *S. aureus* (MRSA). *Int J Latest Res Sci Technol* 2014;**3**(4):144–8.
91. Brown AN, Smith K, Samuels TA, Lu J, Obare SO, Scott ME. Nanoparticles functionalized with ampicillin destroy multiple-antibiotic-resistant isolates of *Pseudomonas aeruginosa* and *Enterobacter aerogenes* and methicillin-resistant *Staphylococcus aureus*. *Appl Environ Microbiol* 2012;**78**(8):2768–74.
92. Fayaz AM, Balaji K, Girilal M, Yadav R, Kalaichelvan PT, Venketesan R. Biogenic synthesis of silver nanoparticles and their synergistic effect with antibiotics: a study against Gram-positive and Gram-negative bacteria. *Nanomedicine* 2009;**6**(1):103–9.
93. Li P, Li J, Wu C, Wu Q, Li J. Synergistic antibacterial effects of b-lactam antibiotic combined with silver nanoparticles. *Nanotechnology* 2005;**16**:1912–17.
94. Duran N, Marcato PD, De Conti R, Alves OL, Costa FTM, Brocchi M. Potential use of silver nanoparticles on pathogenic bacteria, their toxicity and possible mechanisms of action. *J Braz Chem Soc* 2010;**21**:949–59.
95. Bresee J, Bond CM, Worthington RJ, et al. Nanoscale structure-activity relationships, mode of action, and biocompatibility of gold nanoparticle antibiotics. *J Am Chem Soc* 2014;**136**:5295–300.
96. Fayaz MA, Girilal M, Mahdy SA, Somsundar SS, Venkatesan R, Kalaichelvan PT. Vancomycin bound biogenic gold nanoparticles: a different perspective for development of anti VRSA agents. *Process Biochem* 2011;**46**:636–41.
97. Zhao Y, Tian Y, Cui Y, Liu W, Ma W, Jiang X. Small molecule-capped gold nanoparticles as potent antibacterial agents that target Gram-negative bacteria. *J Am Chem Soc* 2010;**132**:12349–56.
98. Perni S, Piccirillo C, Pratten J, Prokopovich P, Chraznowski W, Parkin IP, et al. The antimicrobial properties of light-activated polymers containing methylene blue and gold nanoparticles. *Biomaterials* 2009;**30**:89–93.
99. Mitra S, Chandra S, Laha D, Patra P, Debnath N, Pramanik A, et al. Unique chemical grafting of carbon nanoparticle on fabricated ZnO nanorod: antibacterial and bioimaging property. *Mater Res Bull* 2012;**47**:586–94.
100. Patra P, Mitra S, Debnatha N, Pramanik P, Goswami A. Ciprofloxacin conjugated zinc oxide nanoparticle: a camouflage towards multidrug resistant bacteria. *Bull Mater Sci* 2014;**37**(2):199–206.
101. Bhande RM, Khobragade CM, Mane RS, Bhande S. Enhanced synergism of antibiotics with zinc oxide nanoparticles against extended spectrum b-lactamase producers implicated in urinary tract infections. *J Nanopart Res* 2013;**15**:1413. Available from: http://dx.doi.org/10.1007/s11051-012-1413-4.
102. Muranyi P, Schraml C, Wunderlich J. Antimicrobial efficiency of titanium dioxide-coated surfaces. *J Appl Microbiol* 2010;**108**:1966–73.
103. Roy AS, Parveen A, Koppalkar AR, Ambika Prasad MVN. Effect of nano-titanium dioxide with different antibiotics against methicillin-resistant *Staphylococcus aureus*. *J Biomater Nanobiotechnol* 2010;**1**:37–41.
104. Piras AM, Maisetta G, Sandreschi S, et al. Chitosan nanoparticles loaded with the antimicrobial peptide temporin B exert a long-term antibacterial activity in vitro against clinical isolates of *Staphylococcus epidermidis*. *Front Microbiol* 2015;**6**:372.
105. Murugan S, Paulpandian P. Synergistic antibacterial evaluation of commercial antibiotics combined with nanoiron against human pathogens. *Int J Pharm Sci Rev Res* 2013;**18**(27):183–90.
106. Capeletti LB, de Oliveira LF, Gonçalves Kde A, et al. Tailored silica-antibiotic nanoparticles: overcoming bacterial resistance with low cytotoxicity. *Langmuir* 2014;**30**:7456–64.

107. Agnihotri S, Pathak R, Jha D, et al. Synthesis and antimicrobial activity of aminoglycoside-conjugated silica nanoparticles against clinical and resistant bacteria. *New J Chem* 2015. Available from: http://dx.doi.org/10.1039/C5NJ00007F.
108. El-Newehy MH, Al-Deyab SS, Kenawy ER, Abdel-Megeed A. Fabrication of electrospun antimicrobial nanofibers containing metronidazole using nanospider technology. *Fibers Polym* 2012;**13**(6):709–17.
109. Lynch T, Chong P, Zhang J, Hizon R, Du T, Graham MR, et al. Characterization of a stable, metronidazole-resistant *Clostridium difficile* clinical isolate. *PLoS ONE* 2013;**8**:53757.
110. Tenover FC, Tickler IA, Persing DH. Antimicrobial-resistant strains of *Clostridium difficile* from North America. *Antimicrob Agents Chemother* 2012;**56**(6):2929–32.
111. Kirkcaldy RD, Bolan GA, Wasserheit JN. Cephalosporin-resistant gonorrhea in North America. *JAMA* 2013;**209**:185–7.
112. Unemo M, Golparian D, Stary A, Eigentler A. First *Neisseria gonorrhoeae* strain with resistance to cefixime causing gonorrhoea treatment failure in Austria. *Euro Surveill* 2011;**16**(43):1–3.
113. Aydemir H, Akduman D, Piskin N, Comert F, Horuz E, Terzi A, et al. Colistin vs. the combination of colistin and rifampicin for the treatment of carbapenem-resistant *Acinetobacter baumannii* ventilator-associated pneumonia. *Epidemiol Infect* 2013;**141**(06):1214–22.
114. Batirel A, Balkan II, Karabay O, Agalar C, Akalin S, Alici O, et al. Comparison of colistin–carbapenem, colistin–sulbactam, and colistin plus other antibacterial agents for the treatment of extremely drug-resistant *Acinetobacter baumannii* bloodstream infections. *Eur J Clin Microbiol Infect Dis* 2014;**33**:1311–22.
115. Wieczorek K, Osek J. Antimicrobial resistance mechanisms among *Campylobacter*. *Biomed Res Int* 2013;**2013**:340605.
116. Lehtopolku M, Nakari U-M, Kotilainen P, Huovinen P, Siitonen A, Hakanen AJ. Antimicrobial susceptibilities of multidrug-resistant *Campylobacter jejuni* and *E. coli* strains: *In vitro* activities of 20 antimicrobial agents. *Antimicrob Agents Chemother* 2010;**54**(3):1232–6.
117. Niranjan V, Malini A. Antimicrobial resistance pattern in *Escherichia coli* causing urinary tract infection among inpatients. *Indian J Med Res* 2014;**139**(6):945–8.
118. Landman D, Quale JM. Management of infections due to resistant enterococci: A review of therapeutic options. *J Antimicrob Chemother* 1997;**40**:161–70.
119. Arias CA, Contreras GA, Murray BE. Management of multidrug-resistant enterococcal infections. *Clin Microbiol infect* 2010;**16**(6):555–62.
120. Liu C, Bayer A, Cosgrove SE, Daum RS, Fridkin SK, Gorwitz RJ, et al. Clinical practice guidelines by the Infectious Diseases Society of America for the treatment of methicillin resistant *Staphylococcus aureus* infections in adults and children. *Clin Infect Dis* 2011;**52**(3):18–55.
121. Jahan M, Rahman M, Parvej MS, Chowdhury SMZH, Haque ME, Talukder MAK, et al. Isolation and characterization of *Staphylococcus aureus* from raw cow milk in Bangladesh. *J Adv Vet Anim Res* 2015;**2**(1):49–55.
122. Golshani Z, Sharifzadeh A. Prevalence of blaOxa10 type β-Lactmase gene in carbapenemase producing *Pseudomonas aeruginosa* strains isolated from patients in Isfahan. *J Microbiol* 2013;**6**(5):e9002.
123. Ghamgosha M, Shahreki ZS, Kafilzadeh F, Bameri Z. Metallo-beta lactamase genes Vim-1, Spm-1 and Imp-1 in *Pseudomonas aeruginosa* isolated from zahedan hospitals. *Int J Infect* 2014;**1**(1):e19635.
124. Gupta N, Limbago BM, Patel JB, Kallen AJ. Carbapenem-resistant Enterobacteriaceae: Epidemiology and prevention. *Clin Inf Dis* 2011;**53**(1):60–7.
125. Sikarwar AS, Batra HV. Prevalence of antimicrobial drug resistance of *Klebsiella pneumoniae* in India. *Int J Biosci Biochem Bioinforma* 2011;**1**(3):211–15.

126. Mantzarlis K, Makris D, Manoulakas E, Karvouniaris M, Zakynthinos E. Risk factors for the first episode of *Klebsiella pneumoniae* resistant to carbapenems infection in critically ill patients: A prospective study. *Biomed Res Int* 2013;**2013**:850547.
127. Jawahar MS. Current trends in chemotherapy of tuberculosis. *Indian J Med Res* 2004;**120**(4):398–417.
128. Siu GK, Zhang Y, Lau TC, Lau RW, Ho PL, Yew WW, et al. Mutations outside the rifampicin resistance-determining region associated with rifampicin resistance in *Mycobacterium tuberculosis. J Antimicrob Chemother* 2011;**66**(4):730–3.
129. Alexander DC, Ma JH, Guthrie JL, Blair J, Chedore P, Jamieson FB. Gene sequencing for routine verification of pyrazinamide resistance in *Mycobacterium tuberculosis*: A role for *pncA* but Not *rpsA*. *J Clin Microbiol* 2012;**50**(11):3726–8.
130. Coll F, McNerney R, Preston MD, Guerra-Assunção JA, Warry A, Hill-Cawthorne G, et al. Rapid determination of anti-tuberculosis drug resistance from whole-genome sequences. *Genome Med* 2015;**7**(1):51–61.
131. Fey PD, Olson ME. Current concepts in biofilm formation of *Staphylococcus epidermidis. Future Microbiol* 2010;**5**(6):917–23.
132. Chovanova R, Mezovska J, Vaverkova S, Mikulasova M. The inhibition the Tet(K) efflux pump of tetracycline resistant *Staphylococcus epidermidis* by essential oils from three Salvia species. *Lett Appl Microbiol* 2015;**61**(1):58–62.
133. Katz KA, Klausner JD. Azithromycin resistance in *Treponema pallidum. Curr Opin Infect Dis* 2008;**21**:83–91.
134. Woznicova V, Smajs D, Wechsler D, Matejkova P, Flasarova M. Detection of *Treponema pallidum* subsp. *pallidum* from skin lesions, serum, and cerebrospinal fluid in an infant with congenital syphilis after clindamycin treatment of the mother during pregnancy. *J Clin Microbiol* 2007;**45**:659–61.
135. Huigen E, Stolz E. Letter: Action of rifampicin on *Treponema pallidum. Br J Vener Dis* 1974;**50**(6):465.
136. Roychowdhury A, Pan A, Dutta D, Mukhopadhyay AK, Ramamurthy T, Nandy RK, et al. Emergence of tetracycline-resistant *Vibrio cholerae* O1 serotype Inaba, in Kolkata, India. *Jpn J Infect Dis* 2008;**61**:128–9.
137. Garg P, Sinha S, Chakraborty R, Bhattacharya SK, Nair GB, Ramamurthy T, et al. Emergence of fluoroquinolone resistant strains of *Vibrio cholerae* O1 biotype El Tor among hospitalized patients with cholera in Calcutta, India. *Antimicrob Agents Chemother* 2001;**45**:1605–6.
138. Krishna BV, Patil AB, Chandrasekhar MR. Fluoroquinolone-resistant *Vibrio cholerae* isolated during a cholera outbreak in India. *Trans R Soc Trop Med Hyg* 2006;**100**:224–6.
139. Kitaoka M, Miyata ST, Unterweger D, Pukatzki S. Antibiotic resistance mechanisms of *Vibrio cholera. J Med Microbiol* 2011;**60**:397–407.
140. Spampinato C, Leonardi D. *Candida* infections, causes, targets, and resistance mechanisms: Traditional and alternative antifungal agents. *Biomed Res Int* 2013;**2013**:204237.
141. Pasquale T, Tomada JR, Ghannoun M, Dipersio J, Bonilla H. Emergence of *Candida tropicalis* resistant to caspofungin. *J Antimicrob Chemother* 2008;**61**(1):219–29.
142. van der Linden JW, Camps SM, Kampinga GA, Arends JP, Debets-Ossenkopp YJ, Haas PJ, et al. Aspergillosis due to voriconazole highly resistant *Aspergillus fumigatus* and recovery of genetically related resistant isolates from domiciles. *Clin Infect Dis* 2013;**57**(4):513–20.
143. Howard SJ, Arendrup MC. Acquired antifungal drug resistance in *Aspergillus fumigatus*: epidemiology and detection. *Med Mycol* 2011;**49**:90–5.
144. Bader O, Tünnermann J, Dudakova A, Tangwattanachuleeporn M, Weig M, Grob U. Environmental isolates of azole-resistant *Aspergillus fumigatus* in Germany. *Antimicrob Agents Chemother* 2015;**59**(7):4356–9.

145. Cuenca-Estrella M, Gomez-Lopez A, Buitrago MJ, Mellado E, Garcia-Effron G, Rodriguez-Tudela JL. In vitro activities of 10 combinations of antifungal agents against the multiresistant pathogen *Scopulariopsis brevicaulis*. *Antimicrob Agents Chemother* 2006;**50**(6):2248–50.
146. Rodero L, Mellado E, Rodriguez AC, Salve A, Guelfand L, Cahn P, et al. G484S amino acid substitution in lanosterol 14-α demethylase (ERG11) is related to fluconazole resistance in a recurrent *Cryptococcus neoformans* clinical isolate. *Antimicrob Agents Chemother* 2003;**47**(11):3653–6.
147. Cawcutt K, Baddour LM, Burgess M. A case of *Scopulariopsis brevicaulis* endocarditis with mycotic aneurysm in an immunocompetent host. *Case Reports Med.* 2015;**2015**:872871.
148. Tanwar J, Das S, Fatima Z, Hameed S. Multidrug resistance: An emerging crisis. *Interdiscip Perspect Infect Dis* 2014;**2014**:1–7.
149. Levin M.J., Bacon T.H., Leary J.J. Resistance of herpes simplex virus infections to nucleoside analogues in HIV-infected patients. 2004;39(5):248-57.
150. Morfin F, Thouvenot D. Herpes simplex virus resistance to antiviral drugs. *J Clin Virol* 2003;**6**(1):29–37.
151. Andrei G, Topalis D, Fiten P, Andrei G, Topalis D, Fiten P, et al. *In vitro*-selected drug-resistant varicella-zoster virus mutants in the thymidine kinase and DNA polymerase genes yield novel phenotype-genotype associations and highlight differences between antiherpes virus drugs. *J Virol* 2012;**86**(5):2641–52.
152. Gueudry J, Boutolleau D, Gueudin M, Burrel S, Miri A, Bodaghi B, et al. Acyclovir-resistant varicella-zoster virus keratitis in an immunocompetent patient. *J Clin Virol* 2013;**58**(1):318–20.
153. Ducancelle A, Champier G, Alain S, Petit F, Le Pors M-JS, Mazeron MC. A novel mutation in the UL54 gene of human cytomegalovirus isolates that confers resistance to foscarnet. *Antiviral Ther* 2006;**11**(4):537–40.
154. Gohring K, Hamprecht K, Jahn G. Antiviral Drug- and Multidrug resistance in cytomegalovirus infected SCT patients. *Comput Struct Biotechnol J* 2015;**13**:153–9.
155. Erice A. Resistance of human cytomegalovirus to antiviral drugs. *Clin Microbiol Rev* 1999;**12**(2):286–97.
156. Mejia AM, Hall BS, Taylor MC, Gómez-Palacio A, Wilkinson SR, Triana-Chávez O, et al. Benznidazole-resistance in *Trypanosoma cruzi* is a readily acquired trait that can arise independently in a single population. *J Infect Dis* 2012;**206**:220–8.
157. Maltezou HC. Drug resistance in visceral Leishmaniasis. *J Biomed Biotechnol* 2009;**2010**:617521.
158. Purkait B, Kumar A, Nandi N, Sardar AH, Das S, Kumar S, et al. Mechanism of amphotericin B resistance in clinical isolates of *Leishmania donovani*. *Antimicrob Agents Chemother* 2011;**56**(2):1031–41.
159. Mishra J, Singh S. Miltefosine resistance in *Leishmania donovani* involves suppression of oxidative stress-induced programmed cell death. *Expt Parasitol* 2013;**135**(2):397–406.
160. Baird JK. Chloroquine resistance in *Plasmodium vivax*. *Antimicrob Agents Chemother* 2004;**48**(11):4075–83.
161. Chehuan YF, Costa MRF, Costa JS, Alecrim MGC, Nogueira F, Silveira H, et al. In vitro chloroquine resistance for *Plasmodium vivax* isolates from the Western Brazilian Amazon. *Malaria J* 2013;**12**:226.
162. Doliwa C, Escotte-Binet S, Aubert D, Sauvage V, Velard F, Schmid A, et al. Sulfadiazine resistance in *Toxoplasma gondii*: no involvement of overexpression or polymorphisms in genes of therapeutic targets and ABC transporters. *Parasite* 2013;**20**:19.

CHAPTER

7 NANOARCHITECTONICS USED IN ANTIINFECTIVE THERAPY

V. Grumezescu[1,2], A.M. Holban[2,3], I. Barbu[3], R.C. Popescu[2,4], A.E. Oprea[2], V. Lazar[3], A.M. Grumezescu[2] and M.C. Chifiriuc[3,5]

[1]Lasers Department, National Institute for Lasers, Plasma and Radiation Physics, Bucharest, Romania [2]Department of Science and Engineering of Oxide Materials and Nanomaterials, Faculty of Applied Chemistry and Materials Science, University Politehnica of Bucharest, Bucharest, Romania [3]Microbiology Immunology Department, Faculty of Biology, University of Bucharest, Bucharest, Romania [4]Department of Life and Environmental Physics, Horia Hulubei National Institute of Physics and Nuclear Engineering, Bucharest, Romania [5]Research Institute of the University of Bucharest, Life, Environmental and Earth Sciences, Bucharest, Romania

INTRODUCTION

Since the 1940s, antibiotics have saved millions of lives and reduced the risk from infectious diseases with bacterial etiology. However, resistant bacteria appeared shortly after the introduction into clinical practice of these "miraculous drugs."

Currently, pathogen resistance to antibiotics is one of the top eight most serious and complex public health problems worldwide.[1] It is estimated that in the United States alone, about $30 billion is spent annually to treat infections caused by antibiotic-resistant bacteria. The source of this phenomenon is the inappropriate and unjustified use of antibiotics, as well as the potential for the spread of resistant organisms and the emergence of resistance genes, correlated with a decreased number of new antibiotics that can be active against these superpathogens (superbugs).[2] The spread of these pathogens is facilitated by social factors (globalization and increased international travel), the increasing number of nosocomial infections, and excessive prescription of broad-spectrum antibiotics.

Currently, at least 25,000 people die annually in Europe due to infections caused by only five types of antibiotic-resistant organisms: methicillin-resistant *Staphylococcus aureus*, vancomycin-resistant *Enterococcus* spp., penicillin-resistant *Streptococcus pneumoniae*, different strains of Enterobacteriaceae (eg, *Escherichia coli*, *Klebsiella pneumoniae*) resistant to third-generation cephalosporins, carbapenems and carbapenem-resistant nonenteric gram-negative bacilli (eg, *Pseudomonas aeruginosa*, *Acinetobacter baumannii*).[3] Six pathogenic agents, harboring epidemiologically significant resistance phenotypes, known as ESKAPE (ie, ***E****nterococcus faecium*, ***S****. aureus*, ***K****. pneumoniae*, ***A****. baumannii*, ***P****. aeruginosa*, ***E****nterobacter* sp.), have been found to be responsible of the majority of hospital-acquired infections in the United States,[4,5] where hospital-acquired resistant infections are associated with more than 63,000 deaths per year.[6] Patients with antibiotic-resistant infections have higher treatment costs (ranging from about $6000 to $30,000), which are almost double the costs of hospitalization of patients with infections related to

Antibiotic Resistance. DOI: http://dx.doi.org/10.1016/B978-0-12-803642-6.00007-1

susceptible microorganisms.[7] The rate of such infections is expected to grow in the future, being correlated with the increasing demand for surgery, transplantation, and chemotherapy in elderly populations and increases in the use of neonatal intensive care units.[8,9]

For almost every existing antibiotic, bacteria have developed a resistance factor that protects them from its action; each time this has happened, pharmaceutical companies have responded by developing a "stronger" antibiotic.[10] But bacteria have a generation time of approximately 20–40 min, while a new antibiotic requires at least 10 years to research and develop. Therefore, in the war between bacteria and antibiotics, the balance of victory begins to lean toward these microorganisms. We cannot fight evolution, but we can try to stay one step ahead of it.[11,12] It has been shown that bacteria exposed to subinhibitory concentrations of antibiotics produce increased levels of reactive oxygen species (ROS) that increase the mutation rate and recombination efficiency of the respective bacteria,[13,14] increase the horizontal transmission of mobile genetic elements,[15,16] and induce competence for transformation, leading to increased transfer and distribution of antibiotic-resistant genes.[17]

It also should be underlined that in bacteria, antibiotics have a lot of physiological functions mediated through cell receptors that are implicated in recombination, horizontal gene transfer, mutation, metabolism, gene regulation, and signaling. The negative side effects of antibiotics and their mutagenic or hormone-like effects result from their interactions with a variety of human cell receptors (eg, erythromycin binds to motilin, a peptide that stimulates smooth muscle contraction in the gut).[18]

Because developing new antibiotics requires huge amounts of time and money, the investments made by pharmaceutical companies in research programs to this end have decreased. In addition, the introduction of such new compounds on the market must be preceded by a series of tests that can last up to several years, an additional time investment. Not only that, but novel molecules should be used only when handling cases for which no other solutions exist, so profits made from the sale of the new compound are not very large. Therefore, without new antibiotics capable of annihilating these superpathogens, humanity could face infections which, although considered trivial until recently, could become untreatable or last for an indefinite period of time.[19]

MECHANISMS OF MICROBIAL RESISTANCE: GENETIC SUPPORT AND BIOCHEMICAL PARTICULARITIES

The effect of antibiotics and antimicrobial chemical agents can be countered through several mechanisms. Some of them are intrinsic or natural, resulting from the physiological features of some bacteria groups; others result from mutations of genes encoding target structures (acquired or clinical resistance). Thus, for any antimicrobial substance, some microbial species are sensitive and others are resistant.[20] Chromosomal DNA and bacterial plasmids represent a large reservoir of resistance genes, which encode different resistance mechanisms: the activation of efflux pumps, antibiotics inactivating/modifying enzymes, target-modifying enzymes, and synthesis of additional modified structures.[21]

Natural resistance is a characteristic of all strains from a species or genus, whatever the conditions of isolation, which is always transmitted to offspring (vertical transmission) and is determined by chromosomal genes, corresponding to the "wild" resistance phenotype of different bacterial species to antibiotics.[22] It is the result of three mechanisms: (1) the target of the antibiotic is inaccessible; (2) the antibiotic has a low affinity for the target; and (3) the target is missing. For example, the outer membrane of Gram-negative bacilli is impermeable to certain classes of antibiotics (glycopeptides, macrolides, lincosamides, streptogramins etc.).[23] Target inaccessibility due to the impermeability of structures as the outer membrane of Gram-negative bacteria is a passive resistance mechanism. But it also can be an active process, resulting through the activation of efflux pumps (to cite just one example), an energy-dependent process that eliminates the drug against a concentration gradient, as is the case of low-level resistance of enterobacteria to quinolones.[24] Sometimes antibiotic-inactivating enzymes, such as β-lactamases, could be naturally expressed in some enterobacteria (eg, *Enterobacter cloacae*), causing resistance to aminopenicillins.

One of the mechanisms of *acquired resistance* to antibiotics is the modification of the specific target of the antibiotic. Once accumulated into the bacterial cell at active concentrations, antibiotics act by binding to a specific target, which is usually an essential protein for the bacterium. A large number of strains have become resistant to certain antibiotics by a mutation process, which led to the synthesis of a modified target protein, which in turn has a low affinity to the antibiotic. Substituting even one single amino acid into a protein can lead to changes in its ability to bind antibiotics without influencing other biological functions.[25]

The best example is the modification of target proteins from the penicillin-binding protein (PBP) category, which is involved in cell wall synthesis. Modifications of these proteins can lead to reduced affinity for β-lactam antibiotics. Sometimes the loss of affinity is correlated with the acquisition of a new versions of PBP. For example, antibiotic resistance in staphylococci is the result of the acquisition of a gene encoding an alternative enzyme, called *PBP2a*, with a much lower affinity for antibiotics than the wild-type enzyme.[21,24]

Antibiotic resistance could occur through changes in the bacteria's cytoplasmic membrane. Antibiotic-resistant strains are defective in the synthesis of one or more porins. In *E. coli*, the minimum inhibitory concentration for penicillins and cephalosporins is inversely proportional to the level of expression of OmpC and OmpF porins. OmpF porin loss is associated with a high level of resistance because the protein is participating in the formation of larger pores than with OmpC.[26]

In some cases, a combination of impermeability with the low-level synthesis of β-lactamase could confer resistance to β-lactams, while strains expressing only a basal level of β-lactamase remain susceptible.[27] Resistance by efflux systems depends on the energy and transport systems, which can be of the ATP-binding cassette (ABC) type, which use energy from adenosine triphosphate (ATP) hydrolysis or a secondary energy source using the transmembrane chemical gradient (proton-motive force).[28] Efflux pumps can be specific to a particular antibiotic (eg, TetB from *E. coli* is specific for tetracycline and a narrow range of structural analogs) or they can transport a large structural variety of drugs, providing multidrug-resistant (MDR) phenotypes.[29–31]

Acquired resistance to antibiotics may result from consecutive mutations or transfer of resistance genes between strains of the same species or between different species and genera. From the clinical point of view, the transfer of genes is the most common mechanism of resistance.[32] Resistance genes are usually located on plasmids, but they also can be on mobile genetic elements (transposons, integrons) that can move from a plasmid to another, within the chromosome, in the same cell, or between different cells.[33]

Plasmids (4–400 kbp) are extrachromosomal self-replicating genetic elements consisting of double-stranded DNA molecules that are circular and covalently closed. Plasmids carry resistance or virulence genes that are dispensable under certain conditions.[34] In most cases, the presence of plasmids provides new properties to the bearing cell, resulting in a wide range of functions, such as resistance to one or more groups of antibiotics, resistance to heavy metal cations (Hg^{2+}, Ni, Co, etc.), resistance to anions (tellurite, bromine, etc.), intercalating compounds (acridine, ethidium, bromide etc.), biosynthesis properties (antibiotics, bacteriocins), and metabolic properties.

Some plasmids are conjugative, playing a role in the dissemination of resistance genes.[35] Conjugative plasmids are also known as *male factors, conjugons, transferons*, or *fertility factors* conferring the property of the donor cell's genetic material in relation to a cell that does not have such a plasmid, which acts as a receiver. The neconjugative are not self-transmissible plasmids; however, they can be transferred from one cell to another, either through a phage transduction or by a conjugation process initiated by a conjugative plasmid that exists in the same cell.[21] Currently, there are only 27 plasmid incompatibility groups in the Enterobacteriaceae family, many of which are involved in the transmission of antibiotic-resistant genes.[36]

Plasmid-mediated antibiotic resistance is versatile and able to disseminate both vertically and by horizontal gene transfer among various pathogens or between pathogenic and commensal ones. The level of resistance can be changed in terms of quantity by changing the number of copies of the plasmid. Plasmid resistance changes neither the growth rate nor the virulence of the bacterial strains.[21]

Insertion sequences (ISs) are the simplest transposable elements described in bacteria. They contain about 1000 bp. IS genetic plasticity plays an important role in the host organism, with major impact on the structure and function of the genome.[37] IS are small and compact, consisting of a central region coding for a transposase mediating the transposition function. The central region is flanked on both ends by short terminal sequences (10–40 bp) repeated in sequential or in reverse order. ISs have different effects on gene expression, such as increased virulence and pathogenicity, activation of downstream gene expression, and changed efflux mechanisms, resulting in increased resistance to antibiotics (eg, IS1 and IS10 insertion causes the overexpression of AcrAB-TolC efflux pump *Salmonella enterica*).[38]

Transposons are mobile genetic elements (2100–9300 bp) that include in their composition a series of structural genes, flanked on the ends by IS (1–2 kb), most often repeated in reverse order. In terms of genetic structure, simple transposons are similar to IS, but they differ in that the central sequence genes whose expression is detectable phenotypically confer the bearing cell new properties (eg, resistance genes).

Integrons are mobile genetic elements with specific structure, consisting of two well preserved terminal sequences flanking a central region where resistance genes can be inserted as gene cassettes. The conserved five region of the int gene is encoding a site-specific recombinase capable of taking DNA sequences, including genes for resistance. In integrons, more resistance genes may be present, their expression level being determined by their distance from the promoter. The moving of resistance genes in other plasmid or in chromosome creates the possibility of the linked transfer of gene clusters, which causes the simultaneous acquisition of several genes for resistance to several classes of antibiotics, causing the occurrence of MDR phenotypes.[39]

Biofilm-associated infections represent a particular challenge for the phenomenon of resistance and treatment of chronic infections, which are involved in about 65–80% of all infections in developed countries.[40]

Due to the distinct phenotype of sessile bacteria embedded in biofilms, they become tolerant to high concentrations of antibiotics as a result of multiple mechanisms. This includes the selection of persister cells found in metabolic latency and slow-growing or nongrowing states, which is largely responsible for the inability of antibiotics to eradicate infections.[41]

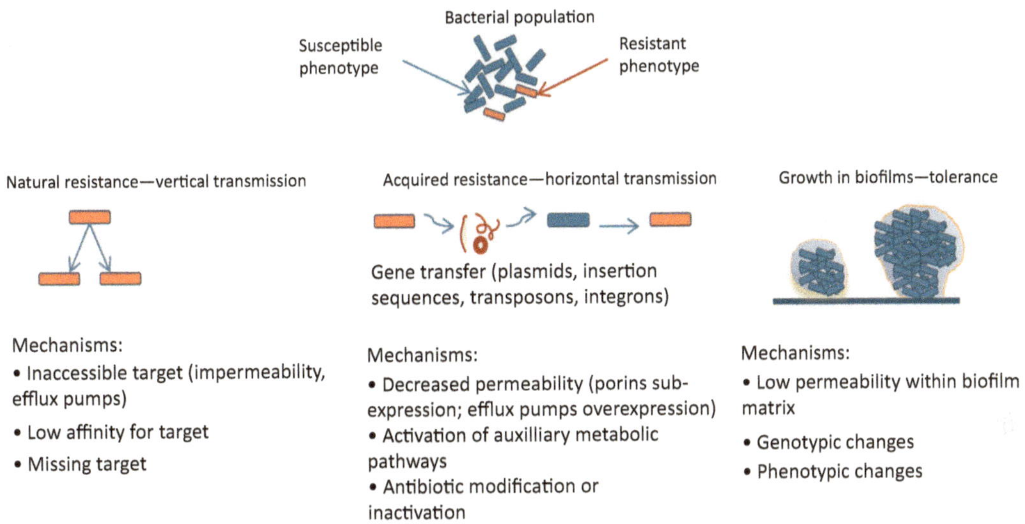

FIGURE 7.1

Comparative representation of the main types of resistance and resistance mechanisms.

The selection of the distinct category persisters seems to be linked to the production of intracellular toxins in bacteria exposed to antibiotics for long periods of time (eg, HipA toxins found in Gram-negative bacteria),[42] in a manner dependent on the SOS gene network.[43]

The main types of resistance and resistance mechanisms are displayed in Fig. 7.1.

CURRENT TRENDS IN ANTIMICROBIAL TREATMENT

In parallel with efforts made for preserving the effectiveness of current antimicrobials by developing evidence-based standards for prudent antimicrobial use for therapeutic and nontherapeutic purposes and by improving the susceptibility testing, the development of novel antimicrobials and the search for alternatives to antimicrobial drugs are urgently needed.

IDENTIFICATION OF NOVEL ANTIBIOTICS AND TARGETS

Spellberg et al. (2004) evaluated the US Food and Drug Administration (FDA) databases of approved/research and development drugs of the most important pharmaceutical and biotechnology industries and found that FDA approval for antibiotics decreased by 56% in 1998–2002 compared to 1983–1987, while new antibiotics represent only 6 of the 506 drugs found in the developmental stage.[44] Possible explanations include the increasing need to find drugs to treat more and more prevalent chronic conditions (eg, hypercholesterolemia, hypertension, mood disorders, dementia, and autoimmune and inflammatory diseases), as well as the fact that antimicrobials are used only for short-course therapies and the rapid selection of resistance limits the efficiency of antibiotics.[45,46]

The increasing rate of antimicrobial resistance in pathogenic agents and the risk of a bioterrorism attack with MDR pathogens are sufficient reasons for the reinforcement of the research and development of antimicrobial drugs (Spellberg et al., 2004). The combined approach of genomic, proteomic, and functional data through databases and bioinformatic tools could provide a promising platform for the identification of novel antimicrobial drugs or targets.[47] *Chemoinformatics* could exploit the existing chemical data provided by high-throughput screening, protein structure libraries, structure-activity relationship studies, ligand-binding affinity, microbial genome projects, and proved biological activity of a huge number of molecules to provide a rational and focused selection of compounds with druglike properties.[48,49]

Another research focus is the development of novel aminoglycosides that are tolerant to aminoglycoside-inactivating enzymes (eg, plazomicin, which is derived from sisomicin) which are frequently expressed, particularly in Gram-negative bacteria.[50,51] The natural or synthetic small size (below 50 kDa) antimicrobial peptides (AMPs) (acting on the cellular prokaryotic membrane and interfering with DNA synthesis, different metabolic pathways, or both) could represent a new class of antimicrobial agents.[52–58] However, many challenges related to their susceptibility to proteolytic degradation, high toxicity, and production costs should be addressed.[59,60]

Nanotechnology could be used successfully to improve the resistance of AMPs to proteolytic degradation, enhance its bioavailability, and decrease its toxicity.[61,62]

NATURAL ANTIMICROBIAL COMPOUNDS

The emergence of antibiotic resistance shifted the research focus toward natural compounds with antimicrobial properties. Since ancient times, plants and bee products have been used empirically to treat various infectious diseases. However, intensive research is needed in order to elucidate the molecular mechanisms of action of these compounds. The volatile oils have been shown to act both at the level of the microbial wall and on intracytoplasmic components. The proposed mechanisms of antimicrobial action of the essential oils include the degradation of cell walls; cytoplasmic membrane damage with increased permeability, leading to loss of cell content; cytoplasm coagulation; and membrane protein damage, thus decreasing the proton-motive force.[63]

Some essential oil components are well known for their antimicrobial properties, such as flavonoids, for which correlations between the intensity of the antimicrobial effect and their chemical structure have been achieved.[64] Phenolic acids were found to induce significant changes in the bacterial cell wall's hydrophobicity, leading to the formation of pores in the cellular membranes of Gram-positive and Gram-negative bacteria[65] that inhibit nucleic acid synthesis and energetic metabolism.[66]

SYNERGIC ASSOCIATIONS

It has been shown that many plant-, bee-, and probiotic-derived compounds could act synergically with antibiotics. The extracts from *Indigofera suffruticosa* leaves enhanced the activity of erythromycin against *S. aureus*.[67] *Melaleuca alternifolia* essential oil exhibited an additive effect with nystatin.[68] The combination of clindamycin with kaempferol or quercetin extracted from *Impatiens balsamina* showed a great synergic effect against *Propionibacterium acnes* demonstrated the potential of these combinations to treat acne.[69] Combined therapy of fungal infections produced by *Candida albicans* with amphotericin B and grape seed extract decreased the therapeutic dose of amphotericin B by more than 75%, demonstrating a clear synergic effect.[70] Garlic showed a significant synergic effect with antibiotics acting on nucleic acid synthesis against Gram-negative diarrheagenic pathogens.[71] *Punica granatum* methanolic extract enhanced the activity of chloramphenicol, gentamicin, ampicillin, tetracycline, and oxacillin against *S. aureus*, offering an alternative way to extend the lifetime of these antibiotics.[72] Myricetin flavonoid exhibited significant synergic activity against extended-spectrum, β-lactamase-producing *K. pneumoniae* in combination with amoxicillin/clavulanate, ampicillin/sulbactam, and cefoxitin.[73]

Taking into account the implication of the SOS response in the selection of persisters, both in bacteria and fungal strains,[74] one of the strategies envisaged for increasing the efficiency of existent antibiotics is their association with molecules that knock out the SOS response. It has been shown that the suppression of the SOS network in *E. coli* with engineered bacteriophage potentiated the bactericidal activity of quinolones in vitro and in vivo against planktonic and biofilm cells, including persisters, acting as a potent adjuvant for antibiotic treatment.[75] It has been shown that AMPs could act synergically with antibiotics, as proved by the combination of penicillin with pediocin and of ampicillin with nisin Z against *Pseudomonas fluorescens*.[76]

ALTERNATIVES TO ANTIMICROBIAL AGENTS

ANTIPATHOGENIC STRATEGIES

The understanding of the mechanism of the intra- and inter-regn crosstalk regulated by the quorum-sensing and response system during the infectious process may contribute to the elaboration of efficient strategies for controlling the severity of clinical symptoms in both acute and biofilm-associated chronic infections, based on the disruption of the coordinated regulation of the expression of virulence factors; this strategy is termed antipathogenic because it is not based on interference with microbial growth, acting only by decreasing the pathogenic potential of the infecting agent.[77] The inhibition of intercellular communication could interfere with different stages of the signaling pathway, such as inhibition of the synthesis of the chemical mediator, signaling mediator inactivation, and blocking signal receptors or transmission.[78] In this strategy, either synthetic substances or natural compounds produced by bacteria or eukaryotic organisms (algae, lichens, molds, plants, insects, etc.) are used.[79–83] Research on fimbrial adhesin inhibitors for the treatment of urinary tract infections produced by *E coli* has been developed based on the selection with hemagglutination inhibition assay.[84,85]

PHAGE THERAPY

Bacteriophages have been used longer than antibiotics (ie, since the early 20th century) for the treatment of bacterial infections, and they have proved to be very efficient, specific, and cost-effective, and having low toxicity and good tolerance.[86,87] They can display a wider spectrum of mechanisms of action than antibiotics and also could be used as drug cocktails to address polymicrobial infections.[88] Due to the emergence of the resistance phenomenon, interest in phage therapy and its products (eg, lysins) has been revived.[89,90] The recent progress in biotechnology, microbiological diagnostic techniques, drug delivery systems, and synthetic biology may help solve the technical difficulties of bacteriophage selection, maintenance, production, and approval.[91]

TARGETING THE HOST COMPONENT IMPLICATED IN THE HOST-PATHOGENIC AGENT RELATIONSHIP

Host-oriented therapeutics, such as vaccination, immunotherapy, immunomodulation, or targeting of genes implicated in the anti-infective response, have the major advantage of having very low probability to select resistance.

Artificial antiinfectious immunity could be acquired actively, either by vaccination or by the passive transfer of exogenous antibodies.[92] The administration of a vaccine is based on a well-defined strategy. The purpose of vaccination is to eradicate (destroy the pathogen consecutive to vaccination), eliminate (remove pathological manifestations of a pathogen, although it might be present in human or animal reservoirs), or limit (control the infectious disease to a level that it is no longer a public health problem) an infectious process.[93]

The development of molecular biology has allowed the development of subunitary vaccines obtained by genetic engineering techniques (recombinant DNA technology).[94]

Immunotherapy is based on the administration of specific antibodies that can neutralize the action of the corresponding antigen. According to the nature of the used antigens, there are three categories of immune sera: (1) antimicrobial immune sera specific for cell-associated antigens, (2) antitoxic sera, and (3) mixed immune sera (antimicrobial and antitoxic).[95]

Passive immunity following the administration of immune sera is installed immediately or in the hours right after intravenous injection. It provides short-term protection (ie, for about 30 days).

The increasing and alarming rate of antimicrobial resistance has refocused research attention on the use of alternative treatments oriented toward the host immune response to infection by the administration of immunomodulators, which interact with the nonspecific and specific immune cells and alter their reactivity to antigens, through the cytokine signaling pathways.

ANTIMICROBIAL NANOPARTICLES

One of the most recently investigated biomedical applications of nanotechnology is the development of new antimicrobial strategies that are effective against planktonic and sessile bacteria, based on the fabrication of new biomaterials or on the improvement of existent ones.[96] The fabrication at nanoscale size and incorporation of antimicrobial substances in nanoparticle shuttles could improve their bioavailability, facilitate accumulation in active concentrations at the site of infection, control the release, and decrease their toxicity by reducing the required therapeutic doses.

Metal oxide nanomaterials (eg, zinc oxide, copper oxide, and magnetite) have been shown to have a broad spectrum of antimicrobial activity,[97,98] mediated by different mechanisms, such as cell wall and cytoplasmic membrane destabilization and production of ROS. The antimicrobial activity of metal oxide nanoparticles is depending on nanoparticle size, form, aggregation state, concentration and exposure time, but also on external factors, such as the susceptibility of microbial strain, growth phase, chemical composition of the local environment, temperature, and pH.

In the next section, we further present the main applications and limitations of magnetite nanoparticles for the development of antimicrobial strategies that are effective against microbial species responsible for difficult-to-treat infections.

MAGNETITE NANOPARTICLES FOR DRUG DELIVERY

Magnetite (Fe_3O_4) nanoparticles have been investigated extensively due to their size-dependent intrinsic magnetic properties, as well as wide usage in the area of biomedical engineering.[99–101] In recent years, several methods to synthesize and functionalize magnetic nanoparticles for biomedical use have been developed.

Magnetite-based nanosystems have been utilized efficiently to deliver many biologically active compounds. These nanoparticles are the preferred shuttles for drug delivery because they increase biocompatibility, ensure the targeted delivery and controlled and prolonged release of many therapeutic compounds, and decrease the amount of the needed therapeutic compound for many biomedical applications, facilitating lower and less toxic drug doses. Moreover, recent studies have demonstrated that magnetite nanoshuttles are able to stabilize and maintain the efficiency of many natural and synthetic compounds utilized in antimicrobial therapy.[102] Because of their low dimensions, magnetite nanoparticles may penetrate biological membranes easily, which are often inaccessible for other delivery systems. These nanosystems may reach intracellular and intercellular active concentrations and proved their efficiency in treating different infections caused by Gram-positive and Gram-negative bacteria and various fungi. Their specific properties are responsible for this enhanced efficiency, which is demonstrated by the reduction of the required active dose and the controlled and prolonged release of the drug. Magnetic nanoparticles with antimicrobial properties may be fabricated by loading a therapeutic agent into a magnetic nanoparticle through encapsulation or adsorption.[103]

Many recent studies reported the synthesis of antimicrobial magnetite nanoparticles using a coprecipitation method. In order to enhance the antimicrobial specificity of the magnetite nanoparticles, they have been functionalized by different methods using several synthetic antimicrobials (antibiotic drugs) and natural antimicrobials (such as essential oils and plant-derived compounds). Table 7.1 presents the most investigated types of magnetite nanoparticles, functionalized with different antimicrobial compounds and the microbial species against which they work.

MAGNETITE-BASED NANOSTRUCTURED ASSEMBLIES

It is well known that the size- and shape-dependent magnetic properties of magnetite nanoparticles are different in different assemblies.[122] Depending on the application and utility of magnetite-based nanomaterials in the biomedical field, multiple types of materials obtained by highly specialized methods and technologies have emerged. Nanostructured assemblies, such as hierarchical structures, layer-by-layer assemblies, zeolites, and nanotubes, were recently developed for use in multiple medical areas, such as tissue engineering, regeneration, prosthesis development, cancer therapy, and infection control.[123]

Multiple stabilization factors of magnetic particles have been proposed recently, the most investigated being carboxylates, phosphates, inorganic materials (ie, silica and gold), polymer stabilizers (such as dextran), polyethylene glycol, polyvinyl alcohol, alginate, chitosan (CS), poly(*N*-2-hydroxyethyl)-D,L-aspartamide-graft-poly(butyl methacrylate), and folic acid–functionalized composite copolymers.[137]

Polymeric nanomaterials

Organized bio-inorganic and hybrid inorganic-organic magnetite-containing nanostructures have been obtained by using different methods, depending on the intended application and use.

Table 7.1 Antimicrobial Magnetite Nanoparticles Functionalized with Different Compounds and Utilized for Drug Delivery

Microbial Species	Type of Antimicrobial Magnetite Nanoparticles	References
Gram-positive bacteria		
Staphylococcus aureus	Magnetite nanoparticles cross-linked with chitosan and grafted with the aminoglycoside antibiotics kanamycin and neomycin	104
	Fe_3O_4 nanoparticles functionalized with eugenol	105
	Fe_3O_4 nanoparticles coated with chitosan-carboxymethylcellulose and functionalized with different antibiotics (penicillins, macrolides, aminoglycosides, rifampicines, and quinolones)	106
	Magnetite nanoparticles stabilized with thioglycerol	107
	Fe_3O_4 nanoparticles functionalized with usnic acid	116
Staphylococcus epidermidis	Fe_3O_4 nanoparticles functionalized with silver and chitosan	108
	66-nm-diameter magnetite nanoparticles	109
	Fe_3O_4 nanoparticles stabilized with thioglycerol	110
Enterococcus faecalis	Magnetite nanoparticles functionalized with vancomycin, penicillin, and streptomycin	111
	Fe_3O_4 nanoparticles functionalized with chitosan and silver	112
	Magnetite nanoparticles coated with dextran and sucrose	113
Bacillus subtilis	Magnetite nanoparticles stabilized with thioglycerol	110
	66-nm-diameter iron oxide magnetite nanoparticles	108
Bacillus licheniformis	66-nm-diameter iron oxide magnetite nanoparticles	116
Bacillus brevis	66-nm-diameter iron oxide magnetite nanoparticles	116
Listeria monocytogenes	Magnetite nanoparticles coated with L-arginine and L-lysine	114
Escherichia coli	Magnetite nanoparticles functionalized with chitosan and cephalosporins	115
	Magnetite nanoparticles functionalized with chitosan and silver	112
	Magnetite nanoparticles doped with polyacrylamide	116
	Magnetite nanoparticles stabilized with thioglycerol	110
	66-nm-diameter iron oxide magnetite nanoparticles	108
	Dextran/sucrose-coated magnetite nanoparticles	113
	Fe_3O_4 nanoparticles functionalized with *Cinnamomum verum* essential oil	117
Pseudomonas aeruginosa	Magnetite nanoparticles coated with chitosan and the aminoglycosides kanamycin and neomycin	104
	Microspheres containing magnetite nanoparticles functionalized with eugenol	106
	Magnetite nanoparticles functionalized with chitosan, carboxymethylcellulose, and antibiotics (penicillins, macrolides, aminoglycosides, rifampicines, and quinolones)	107
	Magnetite nanoparticles functionalized with silver and chitosan	112
	Magnetite nanoparticles stabilized with thioglycerol	110

Table 7.1 Antimicrobial Magnetite Nanoparticles Functionalized with Different Compounds and Utilized for Drug Delivery *Continued*

Microbial Species	Type of Antimicrobial Magnetite Nanoparticles	References
Klebsiella pneumoniae	Magnetite nanoparticles functionalized with silver and chitosan	130
Vibrio cholerae	66 nm of iron oxide magnetite nanoparticles	130
	Yeasts	
Candida albicans	20-nm-diameter Fe_3O_4 nanoparticles functionalized with *Rosmarinus officinalis* essential oil	118
	8-nm-diameter Fe_3O_4 nanoparticles coated with chitosan functionalized with nystatin	119
	7–17-nm-diameter Fe_3O_4 coated with chitosan functionalized with silver nanoparticles	130
Candida tropicalis	20-nm-diameter Fe_3O_4 nanoparticles functionalized with *R. officinalis* essential oil	118
	8-nm-diameter Fe_3O_4 nanoparticles coated with chitosan functionalized with nystatin	135
Candida krusei	Core/shell/coated-shell hybrid magnetite functionalized with *Anethum graveolens* essential oil	120
	Magnetite nanoparticles coated with dextran	113
Candida glabrata	Core/shell/coated-shell hybrid magnetite functionalized with *A. graveolens* essential oil	139
Saccharomyces cerevisiae	Magnetite nanostructures functionalized with fatty acids	121

The most investigated approach to obtain a biomedical product is to introduce nanoparticles into a matrix. For this purpose, several in situ and ex situ methods have been utilized. The most applied ex situ method involves mixing the nanoparticles into a polymer matrix after synthesizing the magnetic nanoparticles.[124] Ex situ methods are utilized in particular circumstances since they generally cause a high agglomeration of magnetite nanoparticles in the polymer. Therefore, recent methods rely on the in situ technologies for the formation of magnetic nanoparticles in different hydrogels. In addition, this method is suitable to develop composite materials that have gained significant advantages for many biomedical applications due to their similarity to natural living tissue and inherent biocompatibility.[124]

In order to increase biocompatibility and obtain tailored nanostructured materials for a particular biomedical application, current research is focusing on involving different polymers in their structure. Scaffolds fabricated from biocompatible and biodegradable polymers play vital roles in the success of any nanostructured material utilized for tissue engineering and regenerative medicine.

Studies have revealed significant interest in the development of smart polymer beads, which are designed to respond to such external stimuli as temperature, pH, solvent composition, and magnetic field.[125] The applications of the particles encapsulated in the polymer depends on the particles' magnetic properties and size.

Poly(lactide-*co*-glycolide)

Among various scaffolding materials, poly(lactide-*co*-glycolide) (PLGA) is a very popular and important biodegradable polyester owing to its tunable degradation rates, good mechanical properties and processibility, and other characteristics. In the latest decades, different fabrication approaches at room temperature were utilized; more appropriate pore structures were designed and achieved; the mechanical properties were investigated for both dry and wet scaffolds; a longtime biodegradation of the PLGA scaffold was observed and a three-stage model was established; and even the effects of pore size and porosity on in vitro biodegradation were revealed. In addition, PLGA scaffolds containing various embedded nanoparticles, including magnetite, have been implanted into animals, and some tissues have been regenerated in vivo after loading cells.[126–128] Organic PLGA nanoparticles and microparticles have been employed extensively in the applications of drug delivery, tissue engineering, and molecular imaging.[129,130] By combining the advantages of PLGA microcapsules and magnetic Fe_3O_4 nanoparticles, organic–inorganic hybrid composite biomaterials could be produced that had broader and more feasible applications than stand-alone applications of either PLGA or magnetic Fe_3O_4 nanoparticles.

Moreover, different scaffolding materials and microspheres and nanospheres composed of polymers and magnetite nanoparticles have been shown to inhibit microbial colonization, biofilm formation, and microbial viability in specific conditions. Nanospheres containing polylactic acid, CS, magnetite, and the antimicrobial natural compound eugenol (PLA-CS-Fe_3O_4@EUG) demonstrated good antimicrobial activity against the Gram-positive *S. aureus* and the Gram-negative *P. aeruginosa*. These nanospheres also inhibited the attachment and biofilm formation of these microbial strains when deposited as a thin film on different medical surfaces by advanced laser techniques.[131] The antibiofilm activity of the magnetite-based polymeric thin film opens a new avenue for the prevention of microbial colonization on medical surfaces.

Another study investigated the in vitro properties of magnetic polylactic-*co*-glycolic acid-polyvinyl alcohol (PLGA-PVA) microspheres containing the active compound usnic acid (UA). UA is a secondary lichen metabolite that possesses great antimicrobial activity against Gram-positive bacteria, including *S. aureus*. This polymeric magnetic nanomaterial proved to inhibit the colonization and biofilm formation of *S. aureus*–tested strains, interfering with all formation stages, starting with attachment, biofilm formation, and maturation.[132]

Polyvinyl alcohol

In 2013, our research group reported on the fabrication of resorbable bioactive wound dressings based on polyvinyl alcohol and magnetite nanoparticles functionalized with the antimicrobial compound UA. These wound dressings were tested in vitro for their antimicrobial effects, and the results revealed that they exhibit enhanced inhibition of attachment and biofilm formation of the pathogen *S. aureus*. Moreover, this nanostructured polymeric assembly produced efficient and prolonged release of the active compound, manifesting its antimicrobial effect in multiple conditions, both when the infection was present at the beginning of the treatment and when the wound dressing was infected after several hours posttreatment.[133]

Polyacrylamide

Polyacrylamide (PAM) is a commercially relevant cationic polymer utilized mainly for water treatment due to its high efficiency and rapid dissolution. Being a cationic polymer, PAM can increase the settling rate of bacterial floc and improve the capture of dispersed bacterial cells, suspended solids, and cell fragments; therefore, one of its largest uses is to flocculate solids in a liquid. Cationic polymers are widely used for removing undesirable organisms from water.[116] Studies revealed that magnetite nanoparticles doped with cationic polymers manifested improved antimicrobial activities in vitro. PAM functionalized magnetite nanoparticles significantly inhibited the growth of *E. coli* strains, and the enhanced antimicrobial activity of the resultant nanopolymer is explained by the activity of the bond between the superoxide radicals of Fe_3O_4 and the NH_2 group of polyacrylamides.[143]

Chitosan

CS and chitin are biopolymers with immense structural possibilities for chemical and mechanical modifications to generate novel properties, functions, and applications that are widely used in the biomedical area. Nanostructured magnetic materials with antifungal properties were obtained by using CS as a scaffolding material and the antifungal drug nystatin. The tetraene diene polyene antibiotic nystatin was loaded into a nanocomposite containing CS-coated magnetic nanoparticles. The study revealed that the nystatin was loaded on the nanomaterial in a proportion of 14.9% and the transmission electron micrographs showed that the diameter of the prepared nanoparticles ranged from 8 to 11 nm. The obtained nanomaterial inhibited the growth of *C. albicans* strains, introducing the idea that nanocomposite materials containing the polymer CS and antimicrobial agents can be utilized to reduce fungal activity for numerous medical applications.[137]

A recent study demonstrated that CS may be used to obtain biocompatible microspheres for coating medical devices. Iordache and coworkers obtained PLGA/CS microspheres containing magnetite nanoparticles functionalized with *Cinnamomi aetheroleum* extract to coat catheters. The obtained modified medical devices proved to have very good biocompatibility and inhibited the attachment and biofilm formation of *S. aureus*.[134]

Other types of microspheres averaging 500 nm in size and made of 8-nm magnetite nanoparticles functionalized with *Melissa officinalis* essential oil, polylactic acid, and CS were prepared by a solvent evaporation method. These microspheres demonstrated enhanced antimicrobial effect in vitro, inhibiting biofilm development of *S. aureus* strains. A recent study revealed that the obtained microspheres can also be utilized as efficient medical device coatings due to their good biocompatibility and enhanced inhibition against microbial colonization.[135]

Lysozyme immobilization

Currently, many studies are investigating suitable methods for the immobilization or adsorption of different enzymes on magnetite nanoparticle-polymer composites.[136–138]

In designing nanostructured magnetite-based assemblies, different antimicrobial compounds and enzymes have been investigated. Lysozyme (*N*-acetylmuramide glycanhydrolase) is a commercially valuable enzyme for food industry and pharmaceutical applications, with potential use for cancer chemotherapy and antimicrobial approaches.[139] Lysozyme can be used for the hydrolysis of the thick cell wall of Gram-positive bacteria and acts by cleaving the 1,4-*b*-glycosidic linkage between *N*-acetylmuramic acid and *N*-acetylglucosamine in the bacterial murein peptidoglycan. Fuertes et al.[140] reported the immobilization of lysozyme in a core/shell composite containing iron oxide ferrite nanoparticles. Shamim et al.[141] reported the adsorption of lysozyme on thermosensitive poly(*N*-isopropylacrylamide)–coated nanomagnetic Fe_3O_4 particles. Recent studies performed by our research group revealed that polymeric microspheres containing magnetite nanoparticles functionalized with lysozyme have an enhanced antimicrobial and antibiofilm effect against several Gram-positive and Gram-negative bacterial species (unpublished results). Despite promising preliminary results regarding the antipathogenic impact of enzyme-functionalized nanostructured materials, many in vitro and in vivo studies are still required to better investigate their potential.

Antibody immobilization

Recent studies are focusing on the development of antibody-coated magnetite nanoparticles that can be utilized in different biomedical fields that depend on biomarker immobilization, including the control of severe infections, especially those with difficult-to-reach localization, which could benefit greatly from nanotechnological applications aimed at pampering drug targeting. Since antibody binding requires special conditions, their adherence is facilitated by coating magnetic nanoparticles with a biocompatible or biodegradable material (eg, natural polymers such as CS, collagen, and folic acid; or synthetic materials like dextran, tetraethyl orthosilicate, polylactic-*co*-glycolic acid, and polyethylene glycol) to modify their surface.[142] Coating materials are providing different functional groups attached to the surface of the nanoparticles, such as aldehyde (–CHO), hydroxyl (–OH), and amine ($-NH_2$) to further facilitate antibody binding. The type of the ligand proved to be very important, as it defines the effectiveness of the nanostructured system. Ongoing research is showing an enhanced interest in alternative antimicrobial therapies, and relevant results are expected to materialize.

CONCLUSIONS

Much progress regarding the biomedical utility of nanostructured materials based on magnetite has been achieved in the latest decade. In the infection control field, magnetic nanoshuttles were primary developed for targeting and controlled delivery of both natural and synthetic drugs with antimicrobial properties. The specific properties of magnetite-based nanoparticles and assembled nanomaterials offer the significant advantages of novel therapeutic approaches related to the use of nanotechnology in the development of future antipathogenic therapies. These nanomaterials proved their efficiency in multiple ways: (1) improving the effect of different antimicrobial compounds and drugs; (2) reducing the toxicity of current antibiotics by reducing the necessary amount of the active drug; (3) modulating key microbial behaviors, such as attachment and biofilm formation, which are important for the persistence and outcome of the infectious process; and (4) by their suitability for use on multiple biomedical applications related to infection control, such as the development of novel therapeutic drugs, the production of tailored medical surfaces, and the development of antimicrobial coatings for prosthetic devices.

Future studies should focus more on fundamental research into the particular effects of different magnetite nanomaterials on the key phenotypic and molecular changes of microbial cells and also in vitro and in vivo cell-material interactions and biodistribution should be addressed. The careful evaluation of the potential positive and adverse effects, as well as the efficient and practical synthesis and modification techniques based on those insights, would be highly beneficial for supporting the great efforts toward various clinical applications.

ACKNOWLEDGEMENTS

This work was supported by a grant of the Romanian National Authority for Scientific Research and Innovation, CNCS – UEFISCDI, project number PN-II-RU-TE-2014-4-2269.

REFERENCES

1. Suk JE, Semenza JC. Future infectious disease threats to Europe. *Am J Public Health* 2011;**101**:2068–79.
2. http://www.ncbi.nlm.nih.gov/books/NBK54257/.
3. http://ecdc.europa.eu/en/publications/surveillance_reports/Pages/index.aspx.
4. Rice LB. Federal funding for the study of antimicrobial resistance in nosocomial pathogens: no ESKAPE. *J Infect Dis* 2008;**197**(8):1079–81.
5. Hidron AI, Edwards JR, Patel J, Horan TC, Sievert DM, Pollock DA, et al. National healthcare safety network team, and participating national healthcare safety network facilities. NHSN annual update: antimicrobial-resistant pathogens associated with health-care-associated infections: annual summary of data reported to the National Healthcare Safety Network at the Centers for Disease Control and Prevention. *Infect Control Hosp Epidemiol* 2008;**29**(11):996–1011.
6. http://www.rff.org/about/media/press-release/new-study-finds-mrsa-rise-hospital-outpatients#sthash.Ot4vIIMa.dpuf.
7. Cosgrove SE. The relationship between antimicrobial resistance and patient outcomes: mortality, length of hospital stay, and health care costs. *Clin Infect Dis* 2006;**42**(Suppl. 2):S82–9.
8. Boucher HW, Talbot GH, Bradley JS, Edwards JE, Gilbert D, Rice LB, et al. Bad bugs, no drugs: no ESKAPE! An update from the infectious diseases society of America. *Clin Infect Dis* 2009;**48**(1):1–12.
9. Chopra I, Schofield C, Everett M, O'Neill A, Miller K, Wilcox M, et al. Treatment of health-care-associated infections caused by gram-negative bacteria: a consensus statement. *Lancet Infect Dis* 2009;**8**:133–9.
10. McKenna M. The enemy within: a new pattern of antibiotic resistance. *Sci Am* 2011;**304**(4):28–33.
11. Livermore DM. Current epidemiology and growing resistance of Gram negative pathogens. *Korean J Intern Med* 2012;**27**(2):128–42.
12. Lederberg J. Infectious history. *Science* 2000;**288**(5464):287–93.
13. Kohanski MA, DePristo MA, Collins JJ. Sublethal antibiotic treatment leads to multidrug resistance via radical-induced mutagenesis. *Mol Cell* 2010;**37**(3):311–20.
14. Kohanski MA, Dwyer DJ, Collins JJ. How antibiotics kill bacteria: from targets to networks. *Nat Rev Microbiol* 2010;**8**(6):423–35.
15. Maiques E, Ubeda C, Campoy S, Salvador N, Lasa I, Novick RP, et al. β-Lactam antibiotics induce the SOS response and horizontal transfer of virulence factors in *Staphylococcus aureus*. *J Bacteriol* 2006;**188**(7):2726–9.
16. Beda C, Maiques E, Knecht E, Lasa I, Novick RP, Penadés JR. Antibiotic-induced SOS response promotes horizontal dissemination of pathogenicity island-encoded virulence factors in staphylococci. *Mol Microbiol* 2005;**56**(3):836–44.
17. Prudhomme M, Attaiech L, Sanchez G, Martin B, Claverys JP. Antibiotic stress induces genetic transformability in the human pathogen *Streptococcus pneumoniae*. *Science* 2006;**313**(5783):89–92.
18. Davies J. *Antibiotic resistance and the future of antibiotics, microbial evolution and co-adaptation*. Washington, DC: IOM, The National Academies Press; 2009. p. 160–72.

19. McKenna AJ, O'Donnell ME, McMullan R, Irwin T. Long-term gastrointestinal outcomes after *Streptococcus bovis* bacteraemia. *Int J Clin Pract* 2011;**65**:1203–5.
20. Andreotti D, Biondi S, Di Modugno E. Beta-lactam antibiotics. In: Abraham DJ, editor. *Burger's medicinal chemistry and drug discovery*. 6th ed. Richmond, VA: Wiley-Interscience, John Wiley and Sons, Inc; 2003. p. 607–736.
21. Mihăescu Gr, Chifiriuc M-C, Diţu L-M. *Microbiologie generală*. Bucuresti: Editura Universităţii din Bucureşti; 2007. p. 552.
22. Jehl F, Chomarat M, Weber M, Gerard A. *De la antibiogramă la prescripţie*. Ed Orizonturi; 2010.
23. Kotra LP, Samama JP, Mobashery S. Beta-Lactamases and resistance to β-lactam antibiotics. *Bacterial Resist Antimicrob* 2002;**2002**:144–80.
24. Levêque D, Jehl F. Molecular determinants of fluoroquinolone antibacterial agents pharmacokinetics. *Curr Clin Pharmacol* 2009;**4**:191–7.
25. Kotra LP, Haddad J, Mobashery S. Aminoglycosides: perspectives on mechanisms of action and resistance and strategies to counter resistance. *Antimicrob Agents Chemother* 2000;**44**:3249–56.
26. Depardieu F, Podglajen I, Leclercq R, Collatz E, Courvalin P. Modes and modulations of antibiotic resistance gene expression. *Clin Microbiol Rev* 2007;**20**(1):79–114.
27. Du S, Kuo H, Cheng C, Fe ACY, Wei H, Chang S. Molecular mechanisms of ceftazidime resistance in *Pseudomonas aeruginosa* isolates from canine and human infections. *Vet Med* 2010;**2010**(4):172–82.
28. Dubois R, Arpin C, Melon M, Melon B, Frigo C, Quentin C, et al. Nosocomial outbreak due to a multiresistant strain of *Pseudomonas aeruginosa* P12: efficacy of cefepime-amikacin therapy and analysis of β-lactam resistance. *J Clin Microbiol* 2001;**39**(6):2072–8.
29. Lister PD, Wolter DJ, Hanson ND. Antibacterial-resistant *Pseudomonas aeruginosa*: clinical impact and complex regulation of chromosomally encoded resistance mechanisms. *Clin Microbiol Rev* 2009;**22**(4):582–610.
30. Liu JH, Pan YS, Yuan L, Wu H, Hu GZ, Chen YX. Genetic variations in the active efflux pump genes acrA/B and tolC in different drug-induced strains of *Escherichia coli*. *Genet Mol Res* 2013;**12**(3):2829–36.
31. Strateva T, Yordanov D. *Pseudomonas aeruginosa*—a phenomenon of bacterial resistance. *J Med Microbiol* 2009;**58**(Pt 9):1133–48.
32. Martínez JL, Baquero F. Emergence and spread of antibiotic resistance: setting a parameter space. *Ups J Med Sci* 2014;**119**(2):68–77.
33. Levy SB. Factors impacting on the problem of antibiotic resistance. *J Antimicrob Chemother* 2002;**49**(1):25–30.
34. Garcillán-Barcia MP, Alvarado A, de la Cruz F. Identification of bacterial plasmids based on mobility and plasmid population biology. *FEMS Microbiol Rev* 2011;**35**(5):936–56.
35. Sandegren L, Linkevicius M, Lytsy B, Melhus Å, Andersson DI. Transfer of an *Escherichia coli* ST131 multiresistance cassette has created a *Klebsiella pneumoniae*-specific plasmid associated with a major nosocomial outbreak. *J Antimicrob Chemother* 2012;**67**(1):74–83.
36. Carattoli A. Resistance plasmid families in Enterobacteriaceae. *Antimicrob Agents Chemother* 2009;**53**(6):2227–38.
37. Siguier P, Gourbeyre E, Chandler M. Bacterial insertion sequences: their genomic impact and diversity. *FEMS Microbiol Rev* 2014;1–28.
38. Siguier P, Gourbeyre E, Chandler M. Bacterial insertion sequences: their genomic impact and divers. *FEMS Microbiol Rev* 2014;**38**:865–91.
39. Andreotti M, Guidotti G, Galluzzo CM, Mancinelli S, Germano P, Pirillo MF, et al. Resistance mutation patterns in plasma and breast milk of HIV-infected women receiving highly-active antiretroviral therapy for mother-to-child transmission prevention. *AIDS* 2007;**12**:2360–2.

40. Lewis K. Persister cells, dormancy and infectious disease. *Nat Rev Microbiol* 2007;**5**(1):48–56.
41. Shah D, Zhang Z, Khodursky A, Kaldalu N, Kurg K, Lewis K. Persisters: a distinct physiological state of *E. coli. BMC Microbiol* 2006;**6**:53.
42. Schumacher MA, Piro KM, Xu W, Hansen S, Lewis K, Brennan RG. Molecular mechanisms of HipA-mediated multidrug tolerance and its neutralization by HipB. *Science* 2009;**23**(5912):396–401.
43. Dörr T, Lewis K, Vulič M. SOS response induces persistence to fluoroquinolones in *Escherichia coli. PLoS Genet* 2009;**5**(12):e1000760.
44. Spellberg B, Powers JH, Brass EP, Miller LG, Edwards Jr. JE. Trends in antimicrobial drug development: implications for the future. *Clin Infect Dis* 2004;**38**(9):1279–86.
45. Projan SJ. Why is big pharma getting out of antibacterial drug discovery? *Curr Opin Microbiol* 2003;**6**:427–30.
46. Powers JH. Development of drugs for antimicrobial-resistant pathogens. *Curr Opin Infect Dis* 2003;**16**:547–51.
47. Hammami R, Fliss I. Current trends in antimicrobial agent research: chemo- and bioinformatics approaches. *Drug Discov Today* 2010;**15**(13–14):540–6.
48. Sardari S, Dezfulian M. Cheminformatics in anti-infective agents discovery. *Mini Rev Med Chem* 2007;**7** (2):181–9.
49. de Azevedo Jr. WF, Diasm R, Timmers LF, Pauli I, Caceres RA, Soares MB. Bioinformatics tools for screening of antiparasitic drugs. *Curr Drug Targets* 2009;**10**(3):232–9.
50. Ramirez MS, Nikolaidis N, Tolmasky ME. Rise and dissemination of aminoglycoside resistance: the aac (6′)-Ib paradigm. *Front Microbiol* 2013;**4**:121.
51. Livermore DM, Mushtaq S, Warner M, Zhang J, Maharjan S, Doumith M, et al. Activity of aminoglycosides, including ACHN-490, against carbapenem-resistant Enterobacteriaceae isolates. *J Antimicrob Chemother* 2011;**66**:48–53.
52. Marr AK, Gooderham WJ, Hancock RE. Antibacterial peptides for therapeutic use: obstacles and realistic outlook. *Curr Opin Pharmacol* 2006;**6**:468–72.
53. Hancock RE, Lehrer R. Cationic peptides: a new source of antibiotics. *Trends Biotechnol* 1998;**16**: 82–8.
54. Brogden KA. Antimicrobial peptides: pore formers or metabolic inhibitors in bacteria? *Nat Rev Microbiol* 2005;**3**:238–50.
55. Collin F, Thompson RE, Jolliffe KA, Payne RJ, Maxwell A. Fragments of the bacterial toxin microcin B17 as gyrase poisons. *PLoS One* 2001;**8**:e61459.
56. Zasloff M. Antimicrobial peptides of multicellular organisms. *Nature* 2002;**415**:389–95.
57. Panyutich A, Shi J, Boutz PL, Zhao C, Ganz T. Porcine polymorphonuclear leukocytes generate extracellular microbicidal activity by elastase-mediated activation of secreted proprotegrins. *Infect Immun* 1997;**65**:978–85.
58. Sader HS, Fedler KA, Rennie RP, Stevens S, Jones RN. Omiganan pentahydrochloride (MBI 226), a topical 12-amino-acid cationic peptide: spectrum of antimicrobial activity and measurements of bactericidal activity. *Antimicrob Agents Chemother* 2004;**48**:3112–18.
59. Peters BM, Shirtliff ME, Jabra-Rizk MA. Antimicrobial peptides: primeval molecules or future drugs? *PLoS Pathog* 2010;**6**:e1001067.
60. http://f1000.com/prime/reports/b/5/51/pdf.
61. Baltzer SA, Brown MH. Antimicrobial peptides: promising alternatives to conventional antibiotics. *J Mol Microbiol Biotechnol* 2011;**20**:228–35.
62. Chifiriuc MC, Grumezescu AM, Lazar V, Bolocan A, Triaridis S, Grigore R, et al. Contribution of antimicrobial peptides to the development of new and efficient antimicrobial strategies. *Curr Proteomics* 2014;**11**(2):98–107.

63. Nazzaro F, Fratianni F, De Martino L, Coppola R, De Feo V. Effect of essential oils on pathogenic bacteria. *Pharmaceuticals* 2013;**6**(12):1451–74.
64. Cushnie TP, Lamb AJ. Antimicrobial activity of flavonoids. *Int J Antimicrob Agents* 2005;**26** (5):343–56.
65. Del Rio D, Rodriguez-Mateos A, Spencer JPE, Tognolini M, Borges G, Crozier A. Dietary (poly)phenolics in human health: structures, bioavailability, and evidence of protective effects against chronic diseases. *Antioxid Redox Signal* 2013;**18**(14):1818–92.
66. Mandal SM, Chakraborty D, Dey S. Phenolic acids act as signaling molecules in plant-microbe symbioses. *Plant Signal Behav* 2010;**5**(4):359–68.
67. Bezerra Dos Santos AT, Araújo TF, Nascimento da Silva LC, da Silva CB, de Oliveira AF, Araújo JM, et al. Organic extracts from *Indigofera suffruticosa* leaves have antimicrobial and synergic actions with erythromycin against *Staphylococcus aureus*. *Front Microbiol* 2015;**6**:13.
68. Rosato A, Vitali C, Piarulli M, Mazzotta M, Argentieri MP, Mallamaci R. In vitro synergic efficacy of the combination of Nystatin with the essential oils of *Origanum vulgare* and *Pelargonium graveolens* against some Candida species. *Phytomedicine* 2009;**16**(10):972–5.
69. Lim YH, Kim IH, Seo JJ. In vitro activity of kaempferol isolated from the *Impatiens balsamina* alone and in combination with erythromycin or clindamycin against *Propionibacterium acnes*. *J Microbiol* 2007;**45**(5):473–7.
70. Han Y. Synergic effect of grape seed extract with amphotericin B against disseminated candidiasis due to *Candida albicans*. *Phytomedicine* 2007;**14**(11):733–8.
71. Eja ME, Asikong BE, Abriba C, Arikpo GE, Anwan EE, Enyi-Idoh KH. A comparative assessment of the antimicrobial effects of garlic (*Allium sativum*) and antibiotics on diarrheagenic organisms. *Southeast Asian J Trop Med Public Health* 2007;**38**(2):343–8.
72. Braga LC, Leite AA, Xavier KG, Takahashi JA, Bemquerer MP, Chartone-Souza E, et al. Synergic interaction between pomegranate extract and antibiotics against *Staphylococcus aureus*. *Can J Microbiol* 2005;**51**(7):541–7.
73. Lin RD, Chin YP, Lee MH. Antimicrobial activity of antibiotics in combination with natural flavonoids against clinical extended-spectrum beta-lactamase (ESBL)-producing *Klebsiella pneumoniae*. *Phytother Res* 2005;**19**(7):612–17.
74. Lafleur MD, Qi Q, Lewis K. Patients with long-term oral carriage harbor high-persister mutants of *C. albicans*. *Antimicrob Agents Chemother* 2010;**54**:39–44.
75. Lu TK, Collins JJ. Engineered bacteriophage targeting gene networks as adjuvants for antibiotic therapy. *Proc Natl Acad Sci USA* 2009;**106**(12):4629–34.
76. Naghmouchi K, le Lay C, Baah J, Drider D. Antibiotic and antimicrobial peptide combinations: synergistic inhibition of Pseu- contribution of antimicrobial peptides to the development of new and efficient antimicrobial strategies current proteomics. *Res Microbiol* 2014;**11**(2):9.
77. Israil AM, Chifiriuc MC. *Bacterial communication: new concepts in the antimicrobial therapy*. Asclepius Publ House; 2009. 110 pp. ISBN: 978-973-88785-7-0579.6.
78. Chifiriuc MC, Grumezescu AM, Lazar V. Quorum sensing inhibitors from the sea: lessons from marine symbiotic relationships. *Curr Org Chem* 2014;**18**:823–39.
79. Rasmussen TB, Bjarnsholt T, Skindersoe ME, Hentzer M, Kristoffersen P, Kote M, et al. Screening for quorum-sensing inhibitors (QSI) by use of a novel genetic system, the QSI selector. *J Bacteriol* 2005;**187**:1799–814.
80. Rasmussen TB, Givskov M. Quorum sensing inhibitors: a bargain of effects. *Microbiology* 2006;**152**:895–904.
81. Stan T, Marutescu L, Chifiriuc CM, Mateescu C, Lazar V. Study of the antimicrobial and antibiofilm activity of romanian propolis. *Biointerface Res Appl Chem* 2013;**3**:541.

82. Cotar A, Saviuc C, Nita A, Bezirtzoglou E, Lazar V, Chifiriuc CM. Anti-pathogenic strategies for fighting *Pseudomonas aeruginosa* infections-probiotic soluble compounds as inhibitors of quorum sensing genes expression. *Curr Org Chem* 2013;**17**(2):155–61.
83. Cotar A, Ionescu B, Pelinescu D, Voidarou C, Lazar V, Bezirtzoglou E, et al. Current solutions for the interception of quorum sensing in *Staphylococcus aureus*. *Curr Org Chem* 2013;**17**(2):97–104.
84. Cusumano CK, Pinkner JS, Han Z, Greene SE, Ford BA, Crowley JR, et al. Treatment and prevention of urinary tract infection with orally active FimH inhibitors. *Sci Transl Med* 2011;**3**:109ra115.
85. Han Z, Pinkner JS, Ford B, Chorell E, Crowley JM, Cusumano CK, et al. Lead optimization studies on FimH antagonists: discovery of potent and orally bioavailable ortho-substituted biphenyl mannosides. *J Med Chem* 2012;**55**:3945–59.
86. Twort FW. An investigation on the nature of ultra-microscopic viruses. *Lancet* 1915;**189**:1241–3.
87. D'Hérelle F. On an invisible microbe antagonistic to dysentery bacilli. *CR Acad Sci Paris* 1917;**165**:373–5.
88. Chan BK, Abedon ST, Loc-Carrillo C. Phage cocktails and the future of phage therapy. *Future Microbiol* 2013;**8**(6):769–83.
89. Burrowes B, Harper DR, Anderson J, McConville M, Enright MC. Bacteriophage therapy: potential uses in the control of antibiotic-resistant pathogens. *Expert Rev Anti Infect Ther* 2011;**9**(9): 775–85.
90. Knoll BM, Mylonakis E. Antibacterial bioagents based on principles of bacteriophage biology: an overview. *Clin Infect Dis* 2014;**58**(4):528–34.
91. Lu TK, Koeris MS. The next generation of bacteriophage therapy. *Curr Opin Microbiol* 2011;**14** (5):524–31.
92. Lazar V, Balotescu C, Cernat R, Bulai D, Stewart-Tull D. *Immunobiology*. Univ of Bucharest Publ House; 2005. 250 pp. ISBN-973-73-7124-0.
93. Mihăescu G, Chifiriuc MC. *The immune system organization in vertebrates. Immunogenetics and oncogenetics*. Romanian Academy Publ. House; 2009ISBN: 9789732718483.
94. Mihaescu G, Chifiriuc C, Lazar V. *Principles and techniques for the immunologic and molecular analysis used in the clinical laboratory*. Univ. of Bucharest Publ. House; 2013. 350 pp. ISBN: 978-606-16-0229-2.
95. Mihaescu G, Chifiriuc C, Ditu LM. *Immunobiology*. Univ. of Bucharest Publ House; 2009. 572 pp. ISBN: 978-973-737-734-0.
96. Pop CS, Hussien MD, Popa M, Mares A, Grumezescu AM, Grigore R, et al. Metallic-based micro and nanostructures with antimicrobial activity. *Curr Top Med Chem* 2015;**15**(16):1577–82.
97. Tran N, Mir A, Mallik D, Sinha A, Nayar S, Webster TJ. Bactericidal effect of iron oxide nanoparticles on *Staphylococcus aureus*. *Int J Nanomedicine* 2010;**5**:277–83.
98. Azam A, Ahmed AS, Oves M, Khan MS, Habib SS, Memic A. Antimicrobial activity of metal oxide nanoparticles against Gram-positive and Gram-negative bacteria: a comparative study. *Int J Nanomedicine* 2012;**7**:6003–9.
99. Lee JH, Kim JW, Levy M, Kao A, Noh SH, Bozovic D, et al. Magnetic nanoparticles for ultrafast mechanical control of inner ear hair cells. *ACS Nano* 2014;**8**:6590–8.
100. Alveroğlu E, Sözeri H, Baykal A, Kurtan U, Şenel M. Fluorescence and magnetic properties of hydrogels containing Fe_3O_4 nanoparticles. *J Mol Struct* 2013;**1037**:361–6.
101. Mahmoudi M, Sant S, Wang B, Laurent S, Sen T. Superparamagnetic iron oxide nanoparticles (SPIONs): development, surface modification and applications in chemotherapy. *Adv Drug Deliv Rev* 2011;**63**:24–46.
102. Liakos I, Grumezescu AM, Holban AM. Magnetite nanostructures as novel strategies for anti-infectious therapy. *Molecules* 2014;**19**:12710–26.

103. Azhar SL, Lotfipour F. Magnetic nanoparticles for antimicrobial drug delivery. *Pharmazie* 2012;**67**(10):817–21.
104. Grumezescu AM, Andronescu E, Holban AM, Ficai A, Ficai D, Voicu G, et al. Water dispersible cross-linked magnetic chitosan beads for increasing the antimicrobial efficiency of aminoglycoside antibiotics. *Int J Pharm* 2013;**454**:233–40.
105. Grumezescu V, Holban AM, Iordache F, Socol G, Mogoşanu GD, Grumezescu AM, et al. MAPLE fabricated magnetite@eugenol and (3-hidroxybutyric acid-*co*-3-hidroxyvaleric acid)-polyvinyl alcohol microspheres coated surfaces with anti-microbial properties. *Appl Surf Sci* 2014;**306**:16–22.
106. Grumezescu AM, Andronescu E, Ficai A, Bleotu C, Mihaiescu DE, Chifiriuc MC. Synthesis, characterization and in vitro assessment of the magnetic chitosan–carboxymethylcellulose biocomposite interactions with the prokaryotic and eukaryotic cells. *Int J Pharm* 2012;**436**:771–7.
107. Ramteke C, Sarangi BK, Chakrabarti T, Mudliar S, Satpute D, Pandey RA. Synthesis of silver nanoparticles from the aqueous extract of leaves of *Ocimum sanctum* for enhanced antibacterial activity. *J Chem* 2012;**2013**. 7 pp. ID 278925.
108. Behera SS, Patra JK, Pramanik K, Panda N, Thatoi H. Characterization and evaluation of antibacterial activities of chemically synthesized iron oxide nanoparticles. *World J Nano Sci Eng* 2012;**2**:196–200.
109. Inbaraj BS, Kao TH, Tsai TY, Chiu CP, Kumar R, Chen BH. The synthesis and characterization of poly (γ-glutamic acid)-coated magnetite nanoparticles and their effects on antibacterial activity and cytotoxicity. *Nanotechnology* 2011;**22**:1–9.
110. Ramteke C, Sarangi BK, Chakrabarti T, Mudliar S, Satpute D, Pandey RA. Synthesis and broad spectrum antibacterial activity of magnetite ferrofluid. *Curr Nanosci* 2010;**6**:587–91.
111. Chifiriuc MC, Grumezescu AM, Andronescu E, Ficai A, Cotar AI, Grumezescu V, et al. Water dispersible magnetite nanoparticles influence the efficacy of antibiotics against planktonic and biofilm embedded *Enterococcus faecalis* cells. *Anaerobe* 2013;**22**:14–19.
112. Markova Z, Siskova K, Filip J, Safarova K, Prucek R, Panacek A, et al. Chitosan-based synthesis of magnetically-driven nanocomposites with biogenic magnetite core, controlled silver size, and high antimicrobial activity. *Green Chem* 2012;**14**:2550–8.
113. Iconaru SL, Prodan AM, Motelica-Heino M, Sizaret S, Predoi D. Synthesis and characterization of polysaccharide-maghemite composite nanoparticles and their antibacterial properties. *Nanoscale Res Lett* 2012;**7**:576.
114. Ebrahiminezhad A, Davaran S, Rasoul-Amini S, Barar J, Moghadam M, Ghasemi Y. Synthesis, characterization and anti-listeria monocytogenes effect of amino acid coated magnetite nanoparticles. *Curr Nanosci* 2012;**8**:868–74.
115. Chifiriuc C, Grumezescu V, Grumezescu AM, Saviuc C, Lazăr V, Andronescu E. Hybrid magnetite nanoparticles/*Rosmarinus officinalis* essential oil nanobiosystem with antibiofilm activity. *Nanoscale Res Lett* 2012;**7**:209.
116. Mukherje M. In vitro antimicrobial activity of polyacrylamide doped magnetic iron oxide nanoparticles. *Int J Mater, Mech Manuf* 2014;**2**(1):64–6.
117. Anghel AG, Grumezescu AM, Chirea M, Grumezescu V, Socol G, Iordache F, et al. MAPLE Fabricated Fe_3O_4@*Cinnamomum verum* antimicrobial surfaces for improved gastrostomy tubes. *Molecules* 2014;**19**:8981–94.
118. Chifiriuc CM, Grumezescu AM, Saviuc C, Croitoru C, Mihaiescu DE, Lazar V. Improved antibacterial activity of cephalosporins loaded in magnetic chitosan microspheres. *Int J Pharm* 2012;**436**:201–5.
119. Hussein-Al-Ali SH, El Zowalaty ME, Kura AU, Geilich B, Fakurazi S, Webster TJ, et al. Antimicrobial and controlled release studies of a novel nystatin conjugated iron oxide nanocomposite. *BioMed Res Int* 2014;**2014**:651831.

120. Saviuc C, Grumezescu AM, Chifiriuc CM, Mihaiescu DE, Hristu R, Stanciu G, et al. Hybrid nanosystem for stabilizing essential oils in biomedical applications. *Dig J Nanomater Biostruct* 2011;**6**:1657–66.
121. Saviuc C, Grumezescu AM, Chifiriuc MC, Bleotu C, Stanciu G, Hristu R, et al. In vitro methods for the study of microbial biofilms. *Biointerface Res Appl Chem* 2011;**1**:31–40.
122. Bennet M, Bertinetti L, Neely RK, Schertel A, Körnig A, Flors C, et al. Biologically controlled synthesis and assembly of magnetite nanoparticles. *Faraday Discuss* 2015;**181**:71–83.
123. Podsiadlo P, Paternel S, Rouillard J-M, Zhang Z, Lee J, Lee J-W, et al. Layer-by-layer assembly of Nacre-like nanostructured composites with antimicrobial properties. *Langmuir* 2005;**21**(25):11915–21.
124. Kalia S, Kango S, Kumar A, Haldorai Y, Kumari B, Kumar R. Magnetic polymer nanocomposites for environmental and biomedical applications. *Colloid Polym Sci* 2014;**292**:2025–52.
125. Philippovaa O, Barabanovab A, Molchanova V, Khokhlova A. Magnetic polymer beads: recent trends and developments in synthetic design and applications. *Eur Polym J* 2011;**47**(4):542–59.
126. Langer R, Vacanti JP. Tissue engineering. *Science* 1993;**260**:920–6.
127. Yuan HP, Kurashina K, de Bruijn JD, Li YB, de Groot K, Zhang XD. A preliminary study on osteoinduction of two kinds of calcium phosphate ceramics. *Biomaterials* 1999;**20**:1799–806.
128. Yu L, Ding J. Injectable hydrogels as unique biomedical materials. *Chem Soc Rev* 2008;**37**:1473–81.
129. Acharya S, Sahoo SK. PLGA nanoparticles containing various anticancer agents and tumour delivery by EPR effect. *Adv Drug Deliv Rev* 2011;**63**:170–83.
130. Xu JS, Huang J, Qin R, Hinkle GH, Povoski SP, Martin EW, et al. Synthesizing and binding dual-mode poly (lactic-*co*-glycolic acid) (PLGA) nanobubbles for cancer targeting and imaging. *Biomaterials* 2010;**31**:1716–22.
131. Holban AM, Grumezescu V, Grumezescu AM, Vasile BŞ, Truşcă R, Cristescu R, et al. Antimicrobial nanospheres thin coatings prepared by advanced pulsed laser technique. *Beilstein J Nanotechnol* 2014;**5**:872–80.
132. Grumezescu V, Holban AM, Grumezescu AM, Socol G, Ficai A, Vasile BS, et al. Usnic acid-loaded biocompatible magnetic PLGA-PVA microsphere thin films fabricated by MAPLE with increased resistance to staphylococcal colonization. *Biofabrication* 2014;**6**:035002.
133. Holban AM, Grumezescu AM, Andronescu E, Grumezescu V, Chifiriuc CM, Lazar V, et al. Magnetite-usnic acid nanostructured bioactive material with antimicrobial activity. *Rom J Mater* 2013;**43** (4):402–7.
134. Iordache F, Oprea AE, Grumezescu V, Andronescu E, Socol G, Grumezescu AM, et al. Poly(lactic-*co*-glycolic) acid/chitosan microsphere thin films functionalized with *Cinnamomi aetheroleum* and magnetite nanoparticles for preventing the microbial colonization of medical surfaces. *J Sol-Gel Sci Technol* 2015;**73**(3):679–86.
135. Grumezescu AM, Andronescu E, Oprea AE, Holban AM, Socol G, Grumezescu V, et al. MAPLE fabricated magnetite@*Melissa officinalis* and poly lactic acid: chitosan coated surfaces with anti-staphylococcal properties. *J Sol-Gel Sci Technol* 2015;**73**(3):612–19.
136. Park S, Lee Y, Kim DN, Park S, Jang E, Koh WG. Entrapment of enzyme-linked magnetic nanoparticles within poly(ethylene glycol) hydrogel microparticles prepared by photopatterning. *React Funct Polym* 2009;**69**:293–9.
137. Kondo A, Fukuda H. Preparation of thermo-sensitive magnetic hydrogel microspheres and application to enzyme immobilization. *J Ferment Bioeng* 1997;**84**:337–41.
138. An X, Su Z, Zeng H. Preparation of highly magnetic chitosan particles and their use for affinity purification of enzymes. *J Chem Technol Biotechnol* 2003;**78**:596–600.
139. Başar N, Uzun L, Güner A, Denizli A. Lysozyme purification with dye-affinity beads under magnetic field. *Int J Biol Macromol* 2007;**41**:234–42.

140. Fuertes AB, Sevilla M, Valdes-Solis T, Tartaj P. Templated synthesis of mesoporous superparamagnetic polymers. *Chem Mater* 2007;**19**:5418–23.
141. Shamim N, Liang H, Hidajat K, Udin MS. Adsorption, desorption, and conformational changes of lysozyme from thermosensitive nanomagnetic particles. *J Colloid Interface Sci* 2008;**320**:15–21.
142. Reza RT, Pérez CM, Martínez AM, Baques DB, García-Casillas P. Study of the particle size effect on the magnetic separation of bovine serum albumin (BSA). *Sens Lett* 2010;**8**(3):476–81.
143. Atta AM, Al-Lohedan HA, Al-Hussain SA. Functionalization of Magnetite Nanoparticles as Oil Spill Collector. *Int J Mol Sci* 2015;**16**(4):6911–31. Available from: http://dx.doi.org/10.3390/ijms16046911.

CHAPTER

NANOBIOCOMPOSITES OF METALS AS ANTIMICROBIAL AGENTS

8

I.A. Shurygina[1,2], M.G. Shurygin[1,2] and B.G. Sukhov[3]

[1]*Irkutsk Scientific Centre of Surgery and Traumatology, Irkutsk, Russia*

[2]*Irkutsk Scientific Centre SB RAS, Irkutsk, Russia*

[3]*A. E. Favorsky Irkutsk Institute of Chemistry SB RAS, Irkutsk, Russia*

INTRODUCTION

Metal can be extremely toxic to most bacteria and yeast at exceptionally low concentrations.[1] Nanocomposites, which consist of a metal core and a polymer matrix, are a special group of hybrid composite materials. This polymer nanostabilization matrix plays an important role in shaping the structure of the nanocomposite. Therefore, one of the main tasks is to prepare storage-stable ultrafine particles with desired properties.[2]

Nanocomposites combine the properties of the core and the polymer matrix. The matrix gives them additional unique properties, such as solubility and biocompatibility.

The interest in composites containing silver nanoparticles (AgNPs) is due to the prospects for their wide use as antibacterial and antiviral agents[3]; functional nonlinear optical, plasmonic, and chiroplasmonic materials[3–13]; and catalytic systems for organic synthesis, including its asymmetric and plasmon-enhanced types.[14–21] In most cases, to obtain AgNPs using chemical recovery of silver from a salt thereof with a reducing compound such as sodium borohydride, citrate, glucose, hydrazine, ascorbate irradiation reduction,[3–5,22–25] thermolysis,[5] photolysis, and radiation chemistry,[26] various materials are used to stabilize NPs, such as solid zeolite matrices, aqueous organic emulsions, and solutions of macromolecules,[3–5,25,26] organisms, and plant extract reduction.[27–33] Using strong reducing leads to the formation of small, monodisperse particles.

The possibility of using polysaccharide matrices as a stabilizing macromolecular component of nanocomposites has attracted the attention of researchers.[34–40]

NANOCOMPOSITES AS ANTIMICROBIAL AGENTS

Nanocomposites containing various metals and metal oxides are widely used as antibacterial agents. We have extensive experience working with nanocomposites based on noble metals such as silver, gold,[34,37,41–44] palladium,[45] platinum,[46] and other multipurpose metal and elements such as iron,[36,47–49] cobalt, selenium,[50–55] and some prebiotic multielement nanominerals[56] (Fig. 8.1).

Antibiotic Resistance. DOI: http://dx.doi.org/10.1016/B978-0-12-803642-6.00008-3

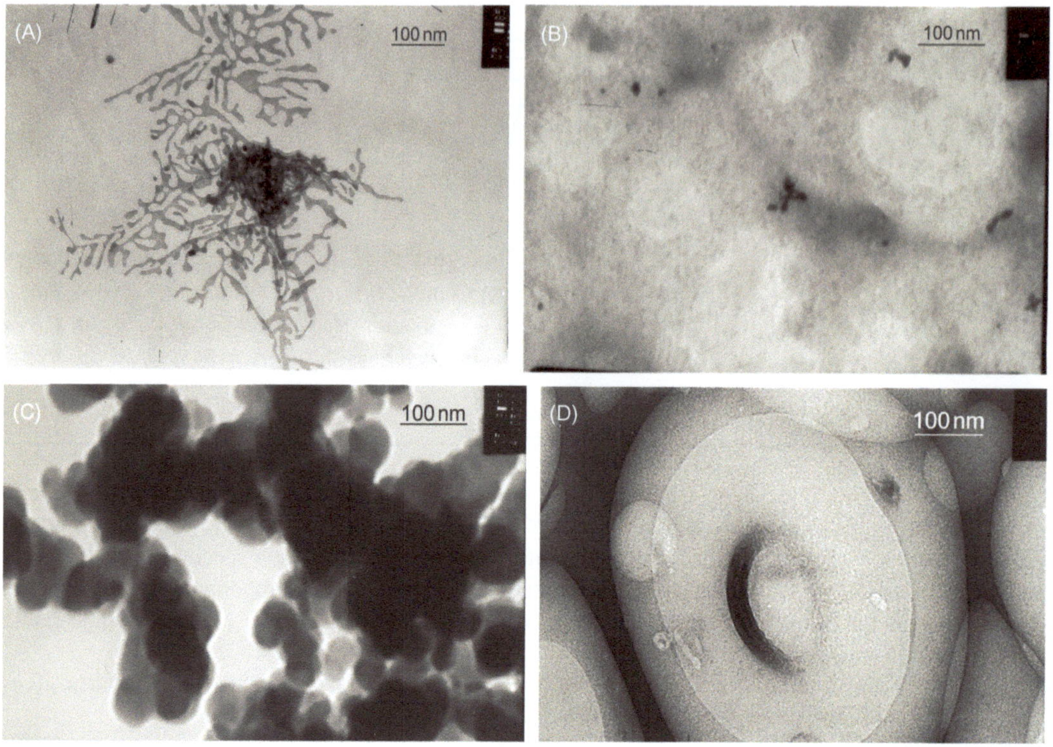

FIGURE 8.1

Samples of nanocomposites: (A) $Ag^{(0)}$ galactomannan; (B) $Au^{(0)}$ galactomannan; (C) $Au^{(0)}$ arabinogalactan; and (D) $Se^{(0)}$ arabinogalactan.

To stabilize the NPs, mostly natural polysaccharide, as well as biogeopolymer (humic substances)[57–60] and synthetic polymer matrix,[61–63] are used.

We conducted a study on the antimicrobial activity of various nanocomposites. In particular, we showed that composites containing nanoplatinum and polysaccharide arabinogalactan do not experience antibacterial activity. Selenium–arabinogalactan nanocomposites have general antimicrobial activity against phytopathogens.[50–52] The minimum bactericidal concentration (MBC) for nanocomposites containing gold was 2000 mg/mL. The highest activity was found in nanocomposites containing silver (MBC 1–100 μg/mL).

At the same time, the highest antibacterial activity is found in silver nanocomposites. For example, Hernández-Sierra et al.[64] compared the bactericidal and bacteriostatic effects of AgNPs, zinc oxide, and gold against *Streptococcus mutans*. For AgNPs, the minimum inhibitory concentration (MIC) was 4.86 ± 2.71 μg/mL, MBC 6.25 μg/mL; for zinc, MIC 500 ± 306.18 μg/mL and MBC 500 μg/mL; gold NPs have the bacteriostatic effect observed at concentrations above 197 μg/mL. Thus, the authors found that the highest antimicrobial activity against *S. mutans* was with AgNPs.

Gold NPs impart unusual biological properties to nanocomposites, which are widely used in medicine for the transportation of substances into a cell by endocytosis, the delivery of genetic material into cell nucleus by ballistic transfection, and the target delivery of pharmaceuticals. The use of gold NPs allows the diagnosis of cancer based on binding of the conjugates of specific antibodies and gold NPs to the surface of cancer cells, with healthy cells intact.

Currently, antimicrobial properties of silver-containing drugs are attracting an ever-growing research interest due to the antibiotic resistance of modern pathogenic and conditional pathogenic microorganisms. The term *silver nanomaterials* denotes any material containing silver in nanosized form. Commercial products containing metallic AgNPs 5–50 nm in size are termed *nanosilver*. AgNPs are clusters of atoms of silver metal, most often designed as antimicrobial agents.

Nanocomposite materials containing AgNPs in a polymer matrix have unique properties and are candidates for medicine, optoelectronics, nanophotonics, and catalytic systems. AgNPs exhibit high antimicrobial activity and can be used for producing efficient bactericide preparations and drugs, with stable aqueous dispersions having special significance. For enhancement of the stability of nanosized silver particles, of considerable importance is the nature of the nanostabilizing matrix.

The unflagging interest of researchers in AgNPs and their composites comes from the ever-increasing application of these multipurpose nanomaterials in modern medicine and related fields. AgNPs are attractive because they are nontoxic to the human body at low concentrations and have broad-spectrum antibacterial actions.[65] Nanosilver is an effective killing agent against a broad spectrum of Gram-positive (Table 8.1) and Gram-negative bacteria (Table 8.2), including antibiotic-resistant strains and fungi (Table 8.3).

Table 8.1 Antimicrobial Activities of Nanocomposites Against Gram-Positive Bacteria

Bacterial Strain	Nanobiocomposite	Me	Size of NP	MIC	MBC	References
Streptococcus mitis ATCC 6249	Chitlac-nAg (lactose-substituted chitosan and AgNPs)	Ag	NA	0.05 (% v/v)	0.1 (% v/v)	Di Giulio et al. (2013)
Streptococcus mutans ATCC 25175	Chitlac-nAg	Ag	NA	0.1 (% v/v)	0.1 (% v/v)	Di Giulio et al. (2013)
Streptococcus oralis ATCC 9811	Chitlac-nAg	Ag	NA	0.1 (% v/v)	0.1 (% v/v)	Di Giulio et al. (2013)
Streptococcus pyogenes	Se NPs synthesized by *Ralstonia eutropha*	Se	NA	100 μg/mL	NA	Srivastava, Mukhopadhyay (2015)
Staphylococcus aureus CCM 3953	Oleic acid–coated maghemite NPs embedded in a 1,4-phenylenediamine/phosphotriazine matrix and AgNPs	Ag/Fe	Ag 10–20 nm	0.078 g/L	NA	Dallas et al. (2010)
S. aureus CCM 3953	2-Hydroxyethyl cellulose-Ag	Ag	9 nm	2.5 mg/L	NA	Suchomel et al. (2015)
S. aureus CCM 3953	2-Hydroxyethyl cellulose-AgBr	AgBr	70 nm	13.5 mg/L	NA	Suchomel et al. (2015)

(*Continued*)

Table 8.1 Antimicrobial Activities of Nanocomposites Against Gram-Positive Bacteria *Continued*

Bacterial Strain	Nanobiocomposite	Me	Size of NP	MIC	MBC	References
S. aureus (MRSA)	Oleic acid–coated maghemite NPs embedded in a 1,4-phenylenediamine/ phosphotriazine matrix and AgNPs	Ag/Fe	Ag 10–20 nm	0.078 g/L	NA	Dallas et al. (2010)
S. aureus (MRSA)	2-Hydroxyethyl cellulose-Ag	Ag	9 nm	2.5 mg/L	NA	Suchomel et al. (2015)
S. aureus (MRSA)	2-Hydroxyethyl cellulose-AgBr	AgBr	70 nm	13.5 mg/L	NA	Suchomel et al. (2015)
Staphylococcus epidermidis	Oleic acid–coated maghemite NPs embedded in a 1,4-phenylenediamine/ phosphotriazine matrix and AgNPs	Ag/Fe	Ag 10–20 nm	0.15 g/L	NA	Dallas et al. (2010)
S. epidermidis	2-Hydroxyethyl cellulose-Ag	Ag	9 nm	1.3 mg/L	NA	Suchomel et al. (2015)
S. epidermidis	2-Hydroxyethyl cellulose-AgBr	AgBr	70 nm	3.4 mg/L	NA	Suchomel et al. (2015)
Enterococcus faecalis CCM 4224	Oleic acid–coated maghemite NPs embedded in a 1,4-phenylenediamine/ phosphotriazine matrix and AgNPs	Ag/Fe	Ag 10–20 nm	0.15 g/L	NA	Dallas et al. (2010)
E. faecalis CCM 4224	2-Hydroxyethyl cellulose-Ag	Ag	9 nm	5.1 mg/L	NA	Suchomel et al. (2015)
E. faecalis CCM 4224	2-Hydroxyethyl cellulose-AgBr	AgBr	70 nm	13.5 mg/L	NA	Suchomel et al. (2015)
Enterococcus faecium (VRE)	2-Hydroxyethyl cellulose-Ag	Ag	9 nm	5.1 mg/L	NA	Suchomel et al. (2015)
E. faecium (VRE)	2-Hydroxyethyl cellulose-AgBr	AgBr	70 nm	6.7 mg/L	NA	Suchomel et al. (2015)
Bacillus subtilis	Ag/Cu chitosan	Ag/Cu	NA	0.054 mg/L	NA	Zain et al. (2014)

Data from:

Di Giulio M, Di Bartolomeo S, Di Campli E, Sancilio S, Marsich E, Travan A, Cataldi A, Cellini L. The effect of a silver nanoparticle polysaccharide system on streptococcal and saliva-derived biofilms. Int J Mol Sci *2013;**14**(7):13615–25. doi:10.3390/ijms140713615.*

Srivastava N, Mukhopadhyay M. Green synthesis and structural characterization of selenium nanoparticles and assessment of their antimicrobial property. Bioprocess Biosyst Eng *2015 Sep;**38**(9):1723–30. doi:10.1007/s00449-015-1413-8.*

Dallas P, Tucek J, Jancik D, Kolar M, Panacek A, Zboril R. Magnetically controllable silver nanocomposite with multifunctional phosphotriazine matrix and high antimicrobial activity. Adv Funct Mater *2010;**20**(14):2347–554. doi:10.1002/adfm.200902370.*

Suchomel P, Kvitek L, Panacek A, Prucek R, Hrbac J, Vecerova R, Zboril R. Comparative study of antimicrobial activity of AgBr and Ag nanoparticles (NPs). PLoS One *2015;**10**(3):e0119202. doi:10.1371/journal.pone.0119202.*

Zain NM, Stapley AG, Shama G. Green synthesis of silver and copper nanoparticles using ascorbic acid and chitosan for antimicrobial applications. Carbohydr Polym *2014;**112**:195–202. doi:10.1016/j.carbpol.2014.05.081.*

Table 8.2 Antimicrobial Activities of Nanocomposites Against Gram-Negative Bacteria

Bacterial Strain	Nanobiocomposite	Me	Size of NP	MIC	MBC	References
Escherichia coli	Extracellular biogenic synthetic AgNPs by *Trichoderma viride* fungi	Ag	5–40 nm	30 μg/mL	NA	Fayaz et al. (2010)
E. coli	Ag/Cu chitosan	Ag/Cu	NA	0.076 mg/L	NA	Zain et al. (2014)
E. coli	Iodine-stabilized Cu NP chitosan composite	Cu	8 ± 4 nm	130.8 μg/mL	NA	Mallick et al. (2012)
E. coli CCM 3954	Oleic acid–coated maghemite NPs embedded in a 1,4-phenylenediamine/phosphotriazine matrix and AgNPs	Ag/Fe	Ag 10–20 nm	0.078 g/L	NA	Dallas et al. (2010)
E. coli CCM 3954	2-Hydroxyethyl cellulose-Ag	Ag	9 nm	2.5 mg/L	NA	Suchomel et al. (2015)
E. coli CCM 3954	2-Hydroxyethyl cellulose-AgBr	AgBr	70 nm	6.7 mg/L	NA	Suchomel et al. (2015)
E. coli	Se NPs synthesized by *Ralstonia eutropha*	Se	NA	250 μg/mL	NA	Srivastava, Mukhopadhyay (2015)
Pseudomonas aeruginosa	Dextrose reduced gelatin-capped Ag-NPs	Ag	NA	12.5 μg/mL	NA	Mohan et al. (2014)
P. aeruginosa CCM 3955	2-Hydroxyethyl cellulose-AgBr	AgBr	70 nm	0.8 mg/L	NA	Suchomel et al. (2015)
P. aeruginosa CCM 3955	Oleic acid–coated maghemite NPs embedded in a 1,4-phenylenediamine/phosphotriazine matrix and AgNPs	Ag/Fe	Ag 10–20 nm	0.15 g/L	NA	Dallas et al. (2010)
P. aeruginosa CCM 3955	2-Hydroxyethyl cellulose-Ag	Ag	9 nm	2.5 mg/L	NA	Suchomel et al. (2015)
P. aeruginosa	Se NPs synthesized by *Ralstonia eutropha*	Se	NA	100 μg/mL	NA	Srivastava, Mukhopadhyay (2015)
Pseudomonas fluorescens	AgNPs with cashew nut shell liquid	Ag	5–50 nm	180 ± 14.6 μg/mL	183 ± 14.6 μg/mL	Velmurugan et al. (2014)
P. fluorescens	Gold NPs with cashew nut shell liquid	Au	5–50 nm	363 ± 16.2 μg/mL	386 ± 12.7 μg/mL	Velmurugan et al. (2014)
Edwardsiella tarda	AgNPs with cashew nut shell liquid	Ag	5–50 nm	124 ± 9.1 μg/mL	128 ± 11.0 μg/mL	Velmurugan et al. (2014)
Aeromonas bestiarum	AgNPs with cashew nut shell liquid	Ag	5–50 nm	205 ± 10.5 μg/mL	312 ± 9.4 μg/mL	Velmurugan et al. (2014)

(Continued)

Table 8.2 Antimicrobial Activities of Nanocomposites Against Gram-Negative Bacteria ***Continued***

Bacterial Strain	Nanobiocomposite	Me	Size of NP	MIC	MBC	References
Aeromonas hydrophila	AgNPs with cashew nut shell liquid	Ag	5–50 nm	250 ± 10.2 μg/mL	322 ± 9.2 μg/mL	Velmurugan et al. (2014)
A. bestiarum	Gold NPs with cashew nut shell liquid	Au	5–50 nm	294 ± 12.8 μg/mL	294 ± 9.42 μg/mL	Velmurugan et al. (2014)
Klebsiella pneumoniae (ESBL)	2-Hydroxyethyl cellulose-Ag	Ag	9 nm	5.1 mg/L	NA	Suchomel et al. (2015)
K. pneumoniae (ESBL)	2-Hydroxyethyl cellulose-AgBr	AgBr	70 nm	3.4 mg/L	NA	Suchomel et al. (2015)
Helicobacter pylori	Green synthesis of AgNPs through reduction with *Solanum xanthocarpum* L. berry extract	Ag	5–25 nm	2–8 μg/mL	NA	Amin et al. (2012)

Data from:

Srivastava N, Mukhopadhyay M. Green synthesis and structural characterization of selenium nanoparticles and assessment of their antimicrobial property. Bioprocess Biosyst Eng *2015Sep;**38**(9):1723–30. doi:10.1007/s00449-015-1413-8.*

Dallas P, Tucek J, Jancik D, Kolar M, Panacek A, Zboril R. Magnetically controllable silver nanocomposite with multifunctional phosphotriazine matrix and high antimicrobial activity. Adv Funct Mater *2010;**20**(14):2347–554. doi:10.1002/adfm.200902370.*

Suchomel P, Kvitek L, Panacek A, Prucek R, Hrbac J, Vecerova R, Zboril R. Comparative study of antimicrobial activity of AgBr and Ag nanoparticles (NPs). PLoS One *2015;**10**(3):e0119202. doi:10.1371/journal.pone.0119202.*

Zain NM, Stapley AG, Shama G. Green synthesis of silver and copper nanoparticles using ascorbic acid and chitosan for antimicrobial applications. Carbohydr Polym *2014;**112**:195–202. doi:10.1016/j.carbpol.2014.05.081.*

Fayaz AM, Balaji K, Girilal M, Yadav R, Kalaichelvan PT, Venketesan R. Biogenic synthesis of silver nanoparticles and their synergistic effect with antibiotics: a study against gram-positive and gram-negative bacteria. Nanomedicine *2010;**6**(1):103–9. doi:10.1016/j.nano.2009.04.006.*

Mallick S, Sharma S, Banerjee M, Ghosh SS, Chattopadhyay A, Paul A. Iodine-stabilized Cu nanoparticle chitosan composite for antibacterial applications. ACS Appl Mater Interfaces *2012;**4**(3):1313–23. doi:10.1021/am201586w.*

Mohan S, Oluwafemi OS, George SC, Jayachandran VP, Lewu FB, Songca SP, Kalarikkal N, Thomas S. Completely green synthesis of dextrose reduced silver nanoparticles, its antimicrobial and sensing properties. Carbohydr Polym *2014;**106**:469–74. doi:10.1016/j.carbpol.2014.01.008.*

Velmurugan P, Iydroose M, Lee SM, Cho M, Park JH, Balachandar V, Oh BT. Synthesis of silver and gold nanoparticles using cashew nut shell liquid and its antibacterial activity against fish pathogens. Indian J Microbiol *2014;**54**(2):196–202. doi:10.1007/s12088-013-0437-5.*

Amin M, Anwar F, Janjua MR, Iqbal MA, Rashid U. Green synthesis of silver nanoparticles through reduction with Solanum xanthocarpum *L. Berry extract: characterization, antimicrobial and urease inhibitory activities against Helicobacter pylori.* Int J Mol Sci *2012;**13**(8):9923–41. doi:10.3390/ijms13089923.*

Table 8.3 Antifungal Activities of Nanocomposites

Species of Fungi	Nanobiocomposite	Me	Size of NP	MIC	MBC	References
Candida albicans	Oleic acid–coated maghemite NPs embedded in a 1,4-phenylenediamine/phosphotriazine matrix and AgNPs	Ag/Fe	Ag 10–20 nm	0.078 g/L	NA	Dallas et al. (2010)
C. albicans	2-Hydroxyethyl cellulose-Ag	Ag	9 nm	1.3 mg/L	NA	Suchomel et al. (2015)
C. albicans	2-Hydroxyethyl cellulose-AgBr	AgBr	70 nm	1.7 mg/L	NA	Suchomel et al. (2015)
Candida tropicalis	Oleic acid–coated maghemite NPs embedded in a 1,4-phenylenediamine/phosphotriazine matrix and AgNPs	Ag/Fe	Ag 10–20 nm	0.078 g/L	NA	Dallas et al. (2010)
C. tropicalis	2-Hydroxyethyl cellulose-Ag	Ag	9 nm	1.3 mg/L	NA	Suchomel et al. (2015)
C. tropicalis	2-Hydroxyethyl cellulose-AgBr	AgBr	70 nm	1.7 mg/L	NA	Suchomel et al. (2015)
Candida parapsilosis	Oleic acid–coated maghemite NPs embedded in a 1,4-phenylenediamine/phosphotriazine matrix and AgNPs	Ag/Fe	Ag 10–20 nm	0.078 g/L	NA	Dallas et al. (2010)
C. parapsilosis	2-Hydroxyethyl cellulose-Ag	Ag	9 nm	1.3 mg/L	NA	Suchomel et al. (2015)
C. parapsilosis	2-Hydroxyethyl cellulose-AgBr	AgBr	70 nm	1.7 mg/L	NA	Suchomel et al. (2015)
Aspergillus clavatus	Se NPs synthesized by *Ralstonia eutropha*	Se	NA	500 μg/mL	NA	Srivastava, Mukhopadhyay (2015)

Data from:
Dallas P, Tucek J, Jancik D, Kolar M, Panacek A, Zboril R. Magnetically controllable silver nanocomposite with multifunctional phosphotriazine matrix and high antimicrobial activity. Adv Funct Mater *2010;**20**(14):2347–554. doi:10.1002/adfm.200902370.*
Suchomel P, Kvitek L, Panacek A, Prucek R, Hrbac J, Vecerova R, Zboril R. Comparative study of antimicrobial activity of AgBr and Ag nanoparticles (NPs). PLoS One *2015;**10**(3):e0119202. doi:10.1371/journal.pone.0119202.*
Srivastava N, Mukhopadhyay M. Green synthesis and structural characterization of selenium nanoparticles and assessment of their antimicrobial property. Bioprocess Biosyst Eng *2015 Sep;**38**(9):1723–30. doi:10.1007/s00449-015-1413-8.*

Li et al.[66] showed that Catechin-Cu NPs at concentrations of 10 and 20 ppm provided rapid and effective killing of up to 90% and 85% of *S. aureus* and *Escherichia coli* within 3 h, respectively. After treatment with 20- and 40-ppm NPs, the bacteria are killed completely. Furthermore, on the basis of assessing the antibacterial effects by scanning electron microscopy, transmission electron microscopy (TEM), and atomic force microscopy, it was found that the cell membrane damage of the bacteria was caused by direct contact of the bacteria with the NPs, and it was the effective mechanism in the bacterial inactivation.

Chemically synthesized ZnO NPs with positive surface potential show very high antimicrobial propensity with MIC of 50 and 100 μg/mL for Gram-negative and Gram-positive bacterium, respectively. On the other hand, ZnO NPs of the same size, but with negative surface potential, show insignificant antimicrobial propensity against the studied bacteria. Unlike the positively charged NPs, neither a $Zn^{(2+)}$ ion nor negatively charged ZnO NP shows any significant inhibition in the growth or morphology of the bacterium.[67]

Hsueh et al.[68] showed that *Bacillus subtilis* growth is inhibited by high concentrations of ZnO NPs (≥50 ppm), with cells exhibiting a prolonged lag phase. RedoxSensor and Phag-GFP fluorescence data further showed that at ZnO NP concentrations above 50 ppm, *B. subtilis* reductase activity, membrane stability, and protein expression decreased. ZnO NPs negatively affect exopolysaccharide production. Moreover, it was found that *B. subtilis* biofilm surface structures became smooth under ZnO NP concentrations of only 5–10 ppm, with concentrations ≤25 ppm significantly reducing biofilm formation activity. ZnO NPs can affect *B. subtilis* viability through the inhibition of cell growth, cytosolic protein expression, and biofilm formation, which suggests that future ZnO NP waste management strategies would do well to mitigate the potential environmental hazard caused by the disposal of these NPs.[68]

Selenium NPs demonstrated excellent antimicrobial activity. It was found that 100-, 100-, 250-, and 100-μg/mL selenium NPs inhibited 99% growth of *Pseudomonas aeruginosa*, *S. aureus*, *E. coli*, and *Streptococcus pyogenes*, respectively. Selenium NPs were biologically synthesized by the nonpathogenic bacterium *Ralstonia eutropha*. Similarly, 500-μg/mL selenium NPs was found to inhibit the growth of the pathogenic fungus *Aspergillus clavatus*. The antimicrobial efficacy of selenium NPs was comparable with the commercially available antibiotic drug ampicillin.[69]

It is proved that antimicrobial activity depends on the size and shape of AgNPs. For example, smaller particle sizes tend to enhance antibacterial properties, and triangular particles tend to have strong antibacterial properties.

Zille et al.[70] studied the surface characteristics, antimicrobial activity, and aging effect of plasma-pretreated polyamide 6.6 fabrics (PA66) coated with AgNPs, with the aim of identifying the optimum size of nanosilver exhibiting antibacterial properties suitable for manufacturing of hospital textiles. The release of bactericidal Ag^+ ions from the 10-, 20-, 40-, 60-, and 100-nm AgNP-coated PA66 surface was a function of the particle size, number, and age. Plasma pretreatment promoted both ionic and covalent interactions between AgNPs and the formed oxygen species on the fibers, favoring the deposition of AgNPs smaller in diameter. These findings consequently showed better immediate and durable antimicrobial effects against Gram-negative *E. coli* and Gram-positive *S. aureus* bacteria. Surprisingly, after 30 days of aging, a comparable bacterial growth inhibition was achieved for all the fibers treated with AgNPs of <100 nm in size. The Ag^+ in the coatings also favored the electrostatic stabilization of the plasma-induced functional groups on the PA66 surface, thereby retarding the aging process. At the same time, the size-related ratio Ag^+/Ag^0 of the AgNPs between 40 and 60 nm allowed for the controlled release of Ag^+ rather than bulk silver. Overall, the results suggest that instead of reducing the AgNP size, which is associated with higher toxicity, similar long-term effects can be achieved with larger NPs (40–60 nm), even in lower concentrations. Since the antimicrobial efficiency of AgNPs larger than 30 nm is mainly ruled by the release of Ag^+ over time and not by the size and number of the AgNPs, this parameter is crucial for the development of efficient antimicrobial coatings on plasma-treated surfaces and the contribution to the safety and durability of clothing used in clinical settings.[70]

Pal et al.[71] investigated the antibacterial properties of differently shaped AgNPs against the Gram-negative bacterium *E. coli*. Energy-filtering TEM images revealed considerable changes in the cell membrane upon treatment, resulting in cell death. Truncated triangular silver nanoplates with a {111} lattice plane as the basal plane displayed the strongest biocidal action, compared with spherical and rod-shaped NPs and with $Ag^{(+)}$ (in the form of $AgNO_3$). It is proposed that nanoscale size and the presence of a {111} plane combine to promote this biocidal property. The size of metallic NPs ensures that a significantly large surface area of the particles is in contact with the bacterial effluent. Considering a hypothetical case with spherical particles of uniform size, a reduction in the particle size from approximately 10 μm to 10 nm with the same mass of matter will increase the contact surface area by $10.^9$ Such a large contact surface is expected to enhance the extent of bacterial elimination. However, smallness in itself is not the goal.[71]

The diverse unique properties of nanocomposites are largely due to the size of their particles. In addition, nanobiocomposites (nanosized particles stabilized by natural polymers) can exhibit a synergism of the properties that is peculiar both to the stabilizing natural matrix and to the central core materials. Therefore, methods of synthesis and properties of such materials are extensively studied.[72]

The continuing attention of researchers devoted to noble metal NPs and composites is due to their widespread application in biomedical technologies. One promising field is the production of high-efficiency antimicrobial preparations based on argentic NPs and stabilized by various polymers.[73] Of great importance in creating argentic nanocomposites is the nature and nanostabilizing efficacy of the matrix, which is usually a natural substance.[38,74,75] Investigating nanomaterials with bactericidal and fungicidal properties is now vital because of the increased resistance of bacteria and fungi to most antimicrobial and antimycotic preparations.[73] It is actually possible to prepare nanomaterials that vary widely in the size, shape, composition, and structure of their NPs and their aggregates, as well as their nanostabilizing polymeric matrices. In addition, the wide variability in the properties of nanomaterials is determined by the distribution of metallic NPs in a polymer matrix and the nature of the interaction between the polymer and NPs.[76,77] All of these characteristics can affect the interaction with prokaryotic cells. We therefore studied the relationship between the structures of argentic nanocomposites and their antimicrobial activities. Dealing with this problem is of considerable theoretical and practical importance since it could allow the production of highly active antimicrobial nanomaterials with low silver content, reducing the dosage and thus the inevitable side effects.

Sulfated arabinogalactan argentic nanocomposites were synthesized according to techniques briefly described by Shurygina et al.[78] Synthesis was based on the redox reaction between argentic hydroxide and polysaccharide in an aqueous alkaline medium. The oxidant was Ag^+, and sulfated arabinogalactan served as the reducing agent.[2,34,35] The silver content of the samples varied from 5.3% to 9.2%. The structure and properties of nanocomposites were determined via TEM using a Philips TEM-410 system and ultraviolet (UV) spectroscopy on a Shimadzu 1800 apparatus. Four samples were studied.

The antimicrobial activity of sulfated arabinogalactan argentic nanocomposites was explored via batch cultivation in a liquid medium (for bacteria beef extract broth and fungi Sabouraud broth) to determine the MIC. We used the standard cultures of bacteria as test cultures for determining the antimicrobial action of preparations, along with freshly extracted clinical test cultures of Gram-positive and Gram-negative microorganisms and fungi. Antibacterial activity was tested for *E. coli* (ATCC 25922), *E. coli* (ESBL1224), *P. seudomonas aeruginosa* (ATCC 27853), *S. aureus* (ATCC 29213), and *S. aureus* (MRSA34R), as well as for *E. coli* and *Proteus mirabilis* hospital cultures. Our samples were typical in their cultural, morphological, and biochemical properties. Antimicrobial activity was studied via nanocomposite cultivation within the limits of 0.15–500 μg/mL at finite concentrations of a microorganism that were equal to 5×10^5 colony-forming units (CFU)/mL. Free nanocomposite test tubes served as our standard.

TEM and UV spectroscopy revealed the structural nanocomposite properties of our synthesized samples, allowing us to explore their main physicochemical characteristics. The silver content in the samples were 9.2% in sample no. 1 (Fig. 8.2A); 5.3% in no. 2 (Fig. 8.2B); 6.3% in no. 3 (Fig. 8.2C); and 8.1% in no. 4 (Fig. 8.2D). UV spectroscopy revealed that plasmon peaks attributable to AgNPs arose at 422 nm for sample no. 1; at 431 nm for sample no. 2; at 417 nm for sample no. 3; and at 411 nm for sample no. 4. According to TEM, samples 1 and 4 were predominantly round AgNPs integrated into the polymeric matrix in numbers ranging from 1 to 4. In sample no. 3, round AgNPs were incorporated into a single polymeric shell in groups of 50–200. Sample no. 2 was the most polymorphous in its structure.

In addition to round NPs, triangular, trapezoidal, and irregular polygon–shaped particles were observed, each of those were incorporated into isolated polymeric matrices. In our first series of experiments, the antibacterial activity of sample no. 4 was tested for *E. coli* (ATCC 25922), *E. coli* (ESBL1224), *P. aeruginosa* (ATCC 27853), *S. aureus* (ATCC 29213), and *S. aureus* (MRSA34R) cultures and for *E. coli* and *P. mirabilis* hospital strains. According to our results, the bacteriostatic concentration of sulfated arabinogalactan argentic nanocomposite varied from 2.5 to 30 μg/mL; and the bactericide concentration ranged from 3 to 100 μg/mL (Table 8.4).

The next stage of our studies was to analyze antimicrobial nanocomposite activity at different silver content. The activity of the AgNP was studied against Gram-negative microorganisms of *E. coli* (ATCC 25922) (Table 8.5). Our study of the bactericidal action of AgNPs revealed no direct relationship between silver content in nanomaterial and the intensity of the antibacterial effect. The highest bactericidal efficacy was found for the sample with the lowest silver concentration (5.3%). The studied argentic nanocomposites exhibited a wide spectrum of bactericidal activity at the minimal silver concentrations that met the safety requirements for medical productions. It was found that antimicrobial activity depended on composite morphology rather than on the silver content of a sample. Our results thus can be used to create safe and effective new medical and veterinary preparations and to lower the cost of argentic preparations and make them available for the treatment of many infectious diseases and for surgery.[79]

The properties of the NPs are defined as the stabilizing polymer matrix and conditions of formation of NPs in the composite. And the development of new nanostructured materials based on different polymeric matrices is one of the fundamental problems of material science. The examination of regularities in the NP formation, the investigation of their specific properties arising from size effects provides firsthand information on the nanosized state of matter. Polymer nanocomposites play a key role in current materials science, as they enable suitable properties of polymeric matrices to be combined with those of NPs that are embedded in them, are attached on their surface, or both.[80]

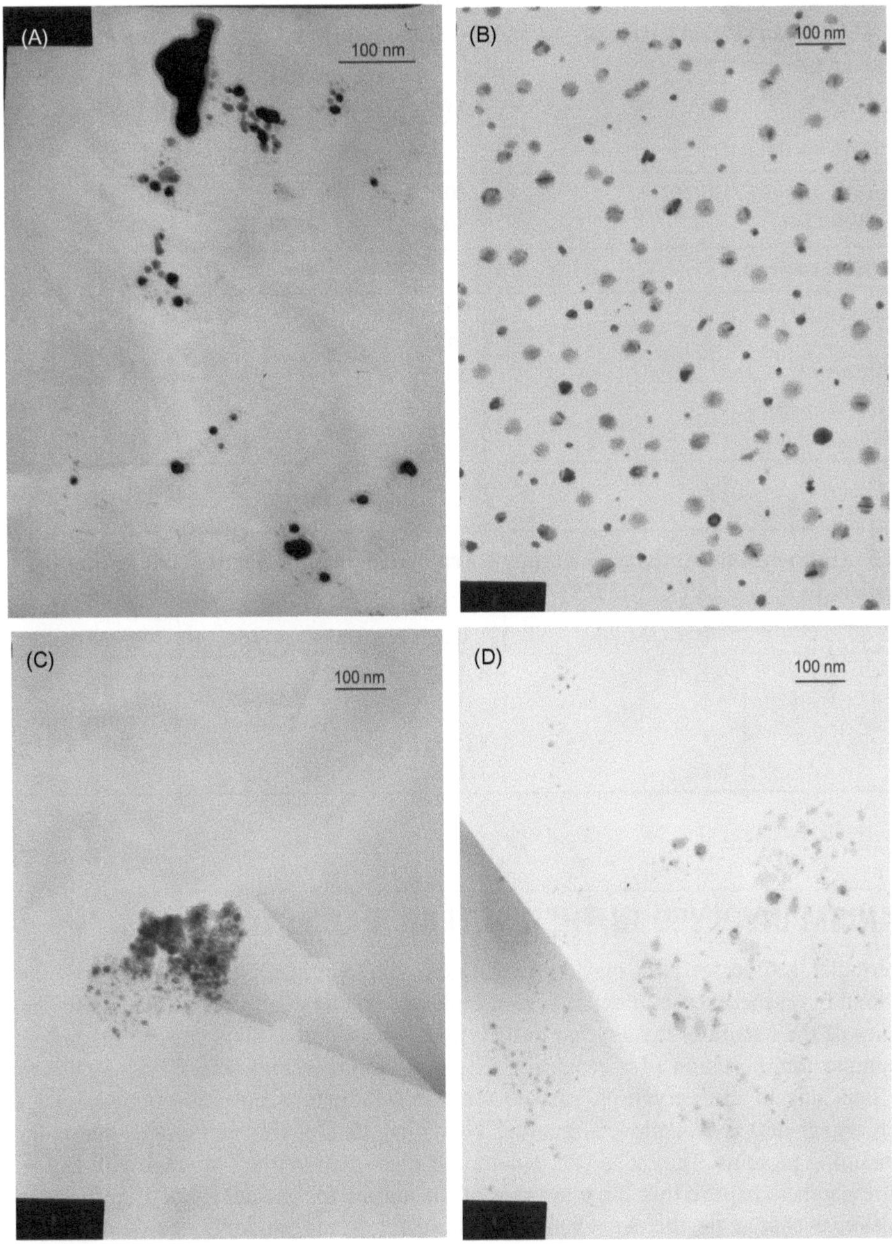

FIGURE 8.2

Samples of $Ag^{(0)}$ sulfated arabinogalactan: (A) sample no. 1, 9.2% Ag; (B) sample no. 2, 5.3% Ag; (C) sample no. 3, 6.3% Ag; and (D) sample no. 4, 8.1% Ag.

Table 8.4 Antimicrobial Activity of Sulfated Arabinogalactan Argentic Nanocomposite (8.1% Silver Content)

Bacterium	MIC (μg/mL)	MBC (μg/mL)
Gram-positive bacteria		
Staphylococcus aureus ATCC 29213	20	20
S. aureus MRSA34R	30	50
Gram-negative bacteria		
Escherichia coli ATCC 25922	2.5	3
E. coli	3	5
E. coli ESBL1224	10	20
Pseudomonas aeruginosa ATCC 27853	3	20
Proteus mirabilis	20	100

Table 8.5 Antimicrobial Activity of Sulfated Arabinogalactan Argentic Nanocomposite with Respect to *E. coli* (ATCC 25922)

Sample	Silver (%)	UV Spectroscopy Peaks (nm)	MIC (μg/mL)
1	9.2	422	5
2	5.3	431	0.15
3	6.3	417	0.6
4	8.1	411	2.5

MECHANISM INVOLVED IN THE BACTERICIDAL ACTIVITY OF NPs

The bactericidal activity of NPs may be associated with multiple mechanisms. First, the AgNPs may act directly on the microbial cells. Second, microbial cells may affect silver ions. They inhibit the enzymes of the respiratory chain due to the effects on oxidative phosphorylation and violate the transmembrane transportation of electrons.[81] The high antimicrobial activity of AgNPs compared to silver salts due to the very high surface area of NPs, which ensures good contact of silver with microorganisms. It is believed that the NPs bind to the cell membrane, particularly with sulfur-containing proteins. They also can penetrate bacteria, where they interact with DNA.[82]

Some researchers believe that the main role of the antimicrobial activity of silver nanocomposite plays a positive charge on the $Ag+$ ion. And the electrostatic attraction between the negatively charged cell membrane of the microorganism and the positively charged NPs determines the effect of NPs on the cell membrane.[83]

Sondi and Salopek-Sondi[84] noted that the antimicrobial action of AgNPs on Gram-negative bacteria depends on the concentration of AgNPs and is closely related to the formation of holes in the cell wall of bacteria. The authors believe that AgNPs accumulate in the bacterial membrane and disrupt its permeability, resulting in cell death.[84] However, this data are not sufficient to explain the mechanism of antimicrobial activity of AgNPs.[85]

Thus, many researchers believe that nanosilver can interact directly with the cell membrane, resulting in toxic effects.[86–89] However, the mechanism of interaction between the components of bacterial membranes and AgnPs is still not clear. Danilczuk et al.[90] believe that antimicrobial activity is responsible for the Ag-generated free radicals on the surface of AgNPs. Grigor'eva et al.[91] found that AgNPs are adsorbed on the outer membrane of gram-negative (*Salmonella* Typhimurium), and the cell wall of Gram-positive bacteria (*S. aureus*). They can penetrate and accumulate in cells of bacteria without aggregation and damage the cytoplasm. At the same time, Lara et al.[92] have suggested a different mechanism of bactericidal action of AgNPs. They believe that AgNPs inhibit the synthesis of cell wall components, with protein synthesis mediated by the 30S subunit of the ribosome, as well as nucleic acid synthesis. Wigginton et al.[93] found that nanosilver inhibits the activity of bacterial enzymes by binding to proteins. Khan et al.[88,89] reported that the adsorption of nanosilver on the bacterial surface depends on several factors: the concentration of NaCl, pH, and zeta potential.

We propose not only that the size of NPs influences the antibacterial activity of nanomaterial, but also that the matrix is very important for nanomaterial quality. One goal of special importance is the designing of antimicrobial drugs based on AgNPs stabilized by natural and synthetic polymers since including those provides additional functions. In this connection, the key point remains the complex interaction of microbial culture and silver nanocomposite as a whole, taking into account the nature of a specific, functional nanostabilizing polymer matrix.

The aim of our study was to investigate the mechanisms of interaction of microbial cells with silver nanocomposites depending on the properties of their polymer matrices. Experimental investigation was carried out on Gram-negative bacteria, Gram-positive bacteria, and fungi. The test strains were typical regarding their cultural, morphological, and biochemical properties. The used strains were resistant to multiple antibiotics.

The nanocomposites used in our study were as follows:

- Original silver$^{(0)}$-containing nanocomposites on the basis of acidic arabinogalactan sulfate were prepared. The matrix for this nanocomposite was low-toxic antithrombotic and antiatherogenic arabinogalactan sulfate ($LD_{50} > 6000$ mg/kg).
- Silver$^{(0)}$-containing nanocomposites on the basis of acidic arabinogalactan have good antibacterial properties, but the potassium salts of sulfated arabinogalactan did not have antibacterial effects.
- Silver$^{(0)}$-containing nanocomposite possessed antimicrobial activity. MIC with respective *E. coli* ranged from 3 to 10 μg/mL, for *S. aureus*, it was 20 μg/mL. MBC for *E. coli* ranged from 3 to 20 μg/mL, for *S. aureus*, it was 20 μg/mL.

Along with antimicrobial activity, our group studied the nanocomposite-retained antithrombotic properties of the matrix arabinogalactan sulfate. Ganenko et al.[39] investigated the effects of silver and sulfated arabinogalactan nanocomposites on platelet aggregation of healthy donors under the action of adenosine diphosphate aggregation inducer. They found that the nanocomposite exhibits antiplatelet therapeutic effects at a concentration of 0.5 mg/mL. The study showed a reduction in platelet aggregation under the action of inductors by 50% compared with the control. Thus, the nanocomposite silver sulfated arabinogalactan has pronounced antithrombotic activity.

Our study showed that the nanocomposite is promising in the designing of AgNP-based biocompatible antimicrobial drugs, as well as special antimicrobial and athrombogenic pharmaceuticals. In addition, the nanocomposite showed antiinflammatory and wound-healing activity in an experimental model. The use of silver nanocomposite sulfated arabinogalactan in the form of 1% gel showed good antiinflammatory effects in an experimental model of skin-muscle injury. It significantly reduced neutrophil infiltration in the wound area and decreased edema as compared to controls. It reduced the period of wound epithelization. Epithelization was observed on the third day, compared to the seventh day in the control group of laboratory animals.

Nanocomposite silver sulfated arabinogalactan in the form of 1% gel showed good results when tested on a model of a burn wound. The drug accelerated wound epithelialization compared with controls and the use of a comparator. All experimental studies were performed in a certified vivarium with Wistar rats and approved by the local ethics committee.

Three series of runs were conducted to study the interaction of the nanocomposite with *E. coli* using TEM. In each series of runs, the phosphate buffer (0.9 mL, pH = 7.6) was added to a 0.5% solution (0.1 mL) of silver nanocomposite and silver-free comparison preparation to reach the final concentration of the nanocomposite (0.05%; 40 μg/mL calculated by silver).

The microbial suspension of ATCC 25922 *E. coli* (0.1 mL) was introduced into the nanocomposite (or silver-free comparison preparation) solution. The seeds were incubated at 37°C for 1, 2, and 24 h. After the corresponding incubation was completed, the quantitative seeding on solid growth media (5% blood agar) was performed.

Simultaneously, the samples for electron microscopy were prepared. After the incubation, the samples were fixed by a 6.5% solution of glutaraldehyde (Fluka) on the phosphate-salt buffer (pH = 7.6) for 2 h. Then the samples were absorbed on nickel grids covered with evaporated carbon film.

TEM studies have shown that the nanocomposite represents as the conglomerates of sulfated arabinogalactan with average size of about 200 nm. The nanocomposite included 10–20 particles of nanosized silver (10–25 nm). The silver particles were not observed out of the conglomerates. About 5% of sulfated arabinogalactan particles incorporating no silver nanoclusters were found. The investigated samples exhibited high adhesive ability toward the coal substrate, probably caused by the properties of the polysaccharide matrix.

It was found that neither the sulfated arabinogalactan particles with incorporated AgNPs nor those with silver-free ones were detected in extracellular space, but they were entirely localized on the microorganism's surface 1 h after the incubation of *E. coli* with the nanocomposite. This can also be attributed to adhesive and surface-active properties of the polysaccharide matrix with respect to the microbial cell membrane. The microscopic study revealed that free AgNPs were not only fixed on the cellular membrane, but also penetrated the membrane and were introduced into cytoplasm space.

Homogenization of a part of the bacterial content and destruction of the membrane was observed. In some cases, the bacterial content fluxed into the external medium.[78] By 2 h after the incubation, no individual particles of the nanocomposite were found. The polysaccharide also was not detected (likely, due to its utilization by bacteria). Most of the bacteria had high homogenous electron-contrast, probably owing to the accumulation of ionic or elemental silver forming the black center of the cell.

By 24 h after the incubation began, the preserved bacteria became nonviable while also being in different stages of destruction. There were many traces of dead bacteria, empty shells. The dying bacteria were surrounded by homogeneous electron-contrast silver like a cloud. Simultaneously, AgNPs were formed again on the surface of the surviving bacteria, as well as on the remains of the bacteria as on a matrix. It is pertinent to note that the newly generated AgNPs in all the cases were 2–10 times smaller than those in the starting nanocomposite.

Their morphology differed from the initial spherical one (eg, triangular nanocrystals appeared). Potentially, the antimicrobial activity of these NPs is 10 times higher than that of the spherical NPs.[78]

The incubation of *E. coli* (in 10^5 CFU/mL concentrations) on the solid growth media led to complete inhibition of the microorganism's growth 1 h after the incubation with the nanocomposite studied. When *E. coli* was used in 10^7 CFU/mL concentrations, the concentration of viable bacteria decreased to 10^3 CFU/mL 1 h after incubation with the nanocomposite.[78]

The bacteriological study proved the bactericidal action of AgNPs occurring in 2 and 24 h (there was no microorganism growth on solid media in samples with *E. coli* concentrations of 10^5 and 10^7 CFU/mL). Thus, at the first stage, owing to the high adhesion of sulfated arabinogalactan to the microbial cell, absorption of the silver-containing NPs on the membranes occurs. Consequently, the primary fixed approach of the AgNPs located in the matrix to the microbe membrane takes place (transportation and concentration of antimicrobial silver on the microbe target).

In the next stage, the formation of silver ions and the formation of metallic NPs are in dynamic equilibrium. Further, some silver atoms (owing to their high surface energy) are stabilized to afford metal clusters—new NPs of silver, whose sizes and morphology are determined by specific conditions of their synthesis involving the microorganism's membrane. Therefore, in the experiment, the AgNPs that formed on membranes and fragments of the destructed bacteria have other morphology (including triangular) and smaller sizes compared to the starting nanocomposite that additionally enhances antimicrobial activity of such NPs.[78]

A tentative mechanism of the AgNP evolution from polysaccharide matrix of sulfated arabinogalactan into the components of microbial cells seems to be irreversible. This could be explained by both more effective coordination of silver by these components and the utilization of the initial nanostabilizing matrix (namely, sulfated arabinogalactan).

Nanocomposites combine the properties of the metal constituting the basis of the nanocomposite and a matrix. In the implementation of the properties of nanocomposites, the properties of the matrix are very important. Multifunctional composites are promising as catheter locks, implants, and other applications.

CONCLUSIONS

Metal-based nanocomposites are a promising group of agents with antibacterial activity. The greatest activity was in nanocomposites based on silver. The significant influence on the properties of the composite has the morphological characteristics of NPs. The characteristics of the polymer matrix play an important role in the properties of the nanocomposites. The matrix is not a passive carrier of NPs. The nanocomposites imparts additional properties, such as water solubility, biocompatibility, and ability to digest the bacteria. The combination of these properties determines the unique characteristics of nanocomposite materials. All of the abovementioned elements make this group promising in the area of developing new commercial products, especially in the growing resistance of bacteria to antimicrobial agents.

ACKNOWLEDGMENT

This work was supported by the Russian Academy of Medicine under Contracts 0120965321 and 01201280990; by the Federal Agency of Scientific Organizations under State Scientific Research Assignment for the Irkutsk Institute of Chemistry SB RAS V.46.4.1; by "Development of Methodology for Production of Functional Biopolymers and Their Hybrid Nano-Biocomposites Based on Renewable Vegetable Raw Materials: The Design of New Pharmacologically Active Substances and Innovative Materials for Key Technologies" (State Registration Number 01201281996); and by the Russian Foundation for Basic Research under Grant 14-03-00859_a. The main results were obtained using the material and technical base of the Baikal Analytical Center for collective uses of SB RAS.

To the best of our knowledge, no conflict of interest, financial or otherwise, exists.

REFERENCES

1. Lemire JA, Harrison JJ, Turner RJ. Antimicrobial activity of metals: mechanisms, molecular targets and applications. *Nat Rev Microbiol* 2013;**11**:371–84.
2. Sukhov BG, Aleksandrova GP, Grishchenko LA, Feoktistova LP, Sapozhnikov AN, Proidakova OA, et al. Nanobiocomposites of noble metals based on arabinogalactan: preparation and properties. *J Struct Chem* 2007;**48**(5):922–7.
3. Rao CNR, Muller A, Cheetham AK, editors. *The chemistry of nanomaterials: synthesis, properties and applications*. Weinheim: Wiley-VCH Verlag GmbH & Co. KGaA; 2004.
4. Katz E, Willner I. Integrated nanoparticle–biomolecule hybrid systems: synthesis, properties, and applications. *Angew Chem Int Ed Engl* 2004;**43**(45):6042–108.
5. Pomogailo AD, Rozenberg AS, Uflyand IE. *Nanochastitsy metallov v polimerakh (nanoparticles of metals in polymers)*. Moscow: Khimiya; 2000 [in Russian].
6. Aleksandrova GP, Grishchenko LA, Sukhov BG, Trofimov BA, Klimenkov IV. Spectral characteristics of biocompatible nanoparticles of noble metals. In: Borisenko VE, Gurin VS, Gaponenko SV, editors. *Physics, chemistry and application of nanostructures—proceedings of the international conference*. Minsk: World Scientific Pub Co Inc; 2009. p. 357–60.
7. Aizpurua J, Hillenbrand R. Localized surface plasmons: basics and applications in field-enhanced spectroscopy. In: Enoch S, Bonod N, editors. *"Plasmonics", from basics to advanced topics*. Berlin: Springer; 2012. p. 151–76.
8. Droulias S, Yannopapas V. Broad-band giant circular dichroism in metamaterials of twisted chains of metallic nanoparticles. *J Phys Chem C* 2013;**117**:1130–5.
9. Maoz BM, Chaikin Y, Tesler AB, Elli OB, Fan Z, Govorov AO, et al. Amplification of chiroptical activity of chiral biomolecules by surface plasmons. *Nano Lett* 2013;**13**(3):1203–9.
10. Valev VK, Baumberg JJ, Sibilia C, Verbiest T. Chirality and chiroptical effects in plasmonic nanostructures: fundamentals, recent progress, and outlook. *Adv Mater* 2013;**25**(18):2517–34.

11. Ben-Moshe A, Maoz BM, Govorov AO, Markovich G. Chirality and chiroptical effects in inorganic nanocrystal systems with plasmon and exciton resonances. *Chem Soc Rev* 2013;**42**(16):7028–41.
12. Dahl JA, Maddux BL, Hutchison JE. Toward greener nanosynthesis. *Chem Rev* 2007;**107**(6):2228–69.
13. Panda T, Deepa K. Biosynthesis of gold nanoparticles. *J Nanosci Nanotech* 2011;**11**:10279–94.
14. Trofimov BA, Sukhov BG, Nosyreva VV, Mal'kina AG, Aleksandrova GP, Grishchenko LA. Pd(0)-arabinogalactan nanocomposites as catalysts for dimerization of acetylenic compounds. *Doklady Chem* 2007;**417**(1):261–3.
15. Yasukawa T, Miyamura H, Kobayashi S. Chiral metal nanoparticle-catalyzed asymmetric C–C bond formation reactions. *Chem Soc Rev* 2014;**43**:1450–61.
16. Xiao M, Jiang R, Wang F, Fang C, Wang J, Yu JC. Plasmon-enhanced chemical reactions. *J Mater Chem A* 2013;**1**:5790–805.
17. Hou W, Cronin SB. A review of surface plasmon resonance-enhanced photocatalysis. *Adv Func Mater* 2013;**23**(13):1612–19.
18. Sarina S, Waclawik ER, Zhu H. Photocatalysis on supported gold and silver nanoparticles under ultraviolet and visible light irradiation. *Green Chem* 2013;**15**(7):1814–33.
19. Kale MJ, Avanesian T, Christopher P. Direct photocatalysis by plasmonic nanostructures. *ACS Catal* 2014;**4**(1):116–28.
20. Zhang X, Chen YL, Liu RS, Tsai DP. Plasmonic photocatalysis. *Rep Prog Phys* 2013;**76**(4):046401.
21. Lindquist NC, Nagpal P, McPeak KM, Norris DJ, Oh S-H. Engineering metallic nanostructures for plasmonics and nanophotonics. *Rep Prog Phys* 2012;**75**:036501.
22. Long D, Wu G, Chen S. Preparation of oligochitosan stabilized silver nanoparticles by gamma irradiation. *Rad Phys Chem* 2007;**76**:1126–31.
23. Chen P, Song L, Liu Y, Fang Y. Synthesis of silver nanoparticles by γ-ray irradiation in acetic water solution containing chitosan. *Radiat Phys Chem* 2007;**76**:1165–8.
24. Wang Z, Liu J, Chen X, Wan J, Qian Y. A simple hydrothermal route to large-scale synthesis of uniform silver nanowires. *Chem Eur J* 2004;**11**(1):160–3.
25. Lee GJ, Shin SI, Oh SG. Preparation of silver dendritic nanoparticles using sodium polyacrylate in aqueous solution. *Chem Lett* 2004;**33**(2):118–19.
26. Tsuji M, Hashimoto M, Nishizawa Y, Kubokawa M, Tsuji T. Microwave-assisted synthesis of metallic nanostructures in solution. *Chem Eur J* 2005;**11**(2):440–52.
27. Vigneshwaran N, Kathe AA, Varadarajan PV, Nachane RP, Balasubramanya RH. Biomimetics of silver nanoparticles by white rot fungus, *Phaenerochaete chrysosporium. Colloids Surf B* 2006;**53**:55–9.
28. Li S, Shen Y, Xie A, Yu X, Qiu L, Zhang L, et al. Green synthesis of silver nanoparticles using *Capsicum annuum* L. extract. *Green Chem* 2007;**8**:852–8.
29. Vigneshwaran N, Ashtaputre NM, Varadarajan PV, Nachane RP, Paralikar KM, Balasubramanya RH. Biological synthesis of silver nanoparticles using the fungus *Aspergillus flavus. Mater Lett* 2007; **61**:1413–18.
30. Basavaraja S, Balaji SD, Lagashetty A, Rajasab AH, Venkataraman A. Extracellular biosynthesis of silver nanoparticles using the fungus fusarium semitectum. *Mater Res Bull* 2008;**43**:1164–70.
31. Kalimuthu K, Babu RS, Venkataraman D, Bilal M, Gurunathan S. Biosynthesis of silver nanocrystals by *Bacillus licheniformis. Colloid Surf B* 2008;**65**:150–3.
32. Mohanpuria P, Rana NK, Yadav SK. Biosynthesis of nanoparticles: technological concepts and future applications. *J Nanopart Res* 2008;**10**:507–17.
33. Mukherjee P, Roy M, Mandal BP, Dey GK, Mukherjee PK, Ghatak J, et al. Green synthesis of highly stabilized nanocrystalline silver particles by a nonpathogenic and agriculturally important fungus *T. asperellum. Nanotechnology* 2008;**19**:075103.
34. Grishchenko LA, Medvedeva SA, Aleksandrova GP, Feoktistova LP, Sapozhnikov AN, Sukhov BG, et al. Redox reactions of arabinogalactan with silver ions and formation of nanocomposites. *Russ J Gen Chem* 2006;**76**(7):1111–16.

35. Trofimov BA, Sukhov BG, Aleksandrova GP, Medvedeva SA, Grishchenko LA, Malkina AG, et al. Nanocomposites with magnetic, optical, catalytic, and biologically active properties based on arabinogalactan. *Doklady Chem* 2003;**393**(4–6):287–8.
36. Aleksandrova GP, Medvedeva SA, Grishchenko LA, Sukhov BG, Trofimov BA. Metal and metal oxide nanoparticle producing method. Patent RU 2260500; 2004.
37. Aleksandrova GP, Grishchenko LA, Fadeeva TV, Sukhov BG, Trofimov BA. Features of formation of silver and gold nanobiocomposites with antimicrobial activity. *Nanotekhnika* 2010;**23**:34–42 [in Russian].
38. Lesnichaya MV, Aleksandrova GP, Sukhov BG, Trofimov BA, Feoktistova LP, Sapozhnikov AN, et al. Silver-containing nanocomposites based on galactomannan and carrageenan: synthesis, structure, and antimicrobial properties. *Russ Chem Bull* 2010;**59**(12):2323–8.
39. Ganenko TV, Kostyro JA, Sukhov BG, Trofimov BA, Fadeeva TV, Vereshchagina SA, et al. Silver nanocomposite of sulphated arabinogalactan exhibiting antimicrobial and antithrombotic activity and method for preparing it. Patent RU 2462254; 2010.
40. Kostyro JA, Alekseev KV, Petrova EN, Gumennikova EN, Romanko TV, Romanko VG, et al. Agent for burn and wound healing. Patent RU 2513186; 2012.
41. Aleksandrova GP, Grishchenko LA, Fadeeva TV, Medvedeva SA, Sukhov BA, Trofimov BA. Agent possessing antibacterial activity. Patent RU 2278669; 2004.
42. Prozorova GF, Korzhova SA, Kon'kova TV, Ermakova TG, Pozdnyakov AS, Sukhov BG, et al. Specific features of formation of silver nanoparticles in the polymer matrix. *Doklady Chem* 2011;**437**(1):47–9.
43. Gasilova E, Aleksandrova G, Sukhov B, Trofimov B. Colloids of gold nanoparticles protected from aggregation with arabinogalactan. *Macromol Symp* 2012;**317–318**(1):1–6.
44. Myachina GF, Kon'kova TV, Korzhova SA, Ermakova TG, Pozdnyakov AS, Sukhov BG, et al. Gold nanoparticles stabilized with water-soluble biocompatible poly(1-vinyl-1,2,4-triazole). *Doklady Chem* 2010;**431**(1):63–4.
45. Gasilova ER, Matveeva GN, Aleksandrova GP, Sukhov BG, Trofimov BA. Colloidal aggregates of Pd nanoparticles supported by larch arabinogalactan. *J Phys Chem B* 2013;**117**(7):2134–41.
46. Morozkin ES, Zaporozhchenko IA, Kharkova MV, Cherepanova AV, Laktionov PP, Vlasov VV, et al. Cytotoxic and immunomodulating properties of silver and platinum nanocomposites. *Chem Sustainable Dev* 2013;**21**:147–54.
47. Aleksandrova GP, Grischenko LA, Medvedeva SA, T'kov AV, Feoktistova LP, Sapozhnikov AN, et al. The synthesis of nanoparticles possessing magnetic properties for biomedicine purposes. *Phys Mesomech* 2004;**7**:139–42 [in Russian].
48. Aleksandrova GP, Grishchenko LA, Bogomyakov AS, Sukhov BG, Ovcharenko VI, Trofimov BA. Magnetic activity of nanostructured biopolymeric nanomagnets. *Russ Chem Bull* 2010;**59**(12):2318–22.
49. Petrova MV, Kiryutin AS, Savelov AA, Lukzen NN, Vieth H-M, Yurkovskaya AV, et al. A theoretical and experimental study of NMR contrasting properties of nanocomposites based on ferric oxides stabilized by arabinogalactan matrix. *Appl Magn Reson* 2011;**41**(2–4):525–36.
50. Graskova IA, Zhivet'yev MA, Borovskii GB, Sukhov BG. Bactericide impact of polymer-stabilized multi-functional nano-composites. *J Stress Physiol Biochem* 2012;**8**(3):33.
51. Papkina AV, Perfileva AI, Zhivetev MA, Borovskiy GB, Graskova IA, Lesnichaya MV, et al. Effect of selenium and arabinogalactan nanocomposite on viability of the phytopathogen *Clavibacter michiganensis* subsp. sepedonicus. *Dokl Biol Sci* 2015;**461**(1):89–91.
52. Papkina AV, Perfileva AI, Zhivet'yev MA, Borovskii GB, Graskova IA, Klimenkov IV, et al. Complex effects of selenium–arabinogalactan nanocomposite on both phytopathogen *Clavibacter michiganensis* subsp. sepedonicus and potato plants. *Nanotechnol Russ* 2015;**10**(5–6):484–91.
53. Kolesnikova LI, Karpova EA, Vlasov BY, Sukhov BG, Trofimov BA. Lipid peroxidation–antioxidant defense system during toxic liver damage and its correction with a nanocomposite substance containing selenium and arabinogalactan. *Bull Exp Biol Med* 2015;**159**(2):183–7.

54. Shurygina IA, Rodionova LV, Shurygin MG, Sukhov BG, Kuznetsov SV, Popova LG, et al. Using confocal microscopy to study the effect of an original pro-enzyme Se/arabinogalactan nanocomposite on tissue regeneration in a skeletal system. *Bull Russ Acad Sci Phys* 2015;**79**(2):256–8.
55. Rodionova LV, Shurygina IA, Sukhov BG, Popova LG, Shurygin MG, Artem'ev AV, et al. Nanobiocomposite based on selenium and arabinogalactan: synthesis, structure, and application. *Russ J Genl Chem* 2015;**85**(2):485–7.
56. Lesnichaya MV, Sukhov BG, Sapozhnikov AN, Safronova LA, Evseenko OV, Ilyash VM, et al. New nanobiocomposites of ammonium magnesium phosphate and carrageenan as efficient prebiotics. *Doklady Chem* 2014;**457**(2):144–7.
57. Aleksandrova GP, Dolmaa G, Tserenpil S, Grishenko LA, Sukhov BG, Regdel D, et al. A new humic acid preparation with addition of silver nanoparticles. In: Xu J, Wu J, He Y, editors. *Functions of natural organic matter in changing environment*. Netherlands, Dordrecht: Springer; 2013. p. 783–8.
58. Aleksandrova GP, Dolmaa G, Enkhbadral U, Grishenko GL, Tserenpil S, Sukhov BG, et al. A new humic acid remedy with addition of silver nanoparticles. *Mongolian J Chem* 2012; **13**(39):7–11.
59. Lesnichaya MV, Aleksandrova GP, Dolmaa G, Sapozhnikov AN, Sukhov BG, Regdel D, et al. Synthesis of silver-containing nanocomposites based on humic substances of brown coal and their antioxidant activity. *Doklady Chem* 2014;**456**(1):72–5.
60. Lesnichaya MV, Aleksandrova GP, Dolmaa G, Sapozhnikov AN, Nomintsetseg B, Sukhov BG, et al. Features of gold nanoparticle formation in matrices of humic substances of different origin. *Doklady Chem* 2015;**460**(1):13–16.
61. Myachina GF, Korzhova SA, Ermakova TG, Sukhov BG, Trofimov BA. Silver-poly(1-vinyl-1,2,4-triazole) nanocomposites. *Doklady Chem* 2008;**420**(1):123–4.
62. Myachina GF, Korzhova SA, Ermakova TG, Kon'kova TV, Pozdnyakov AS, Sukhov BG, et al. Nanocomposites of silver and 1-vinyl-1,2,4-triazole copolymer with sodium methacrylate. *Doklady Chem* 2009;**427**(2):199–201.
63. Prozorova GF, Korzhova SA, Kon'kova TV, Ermakova TG, Pozdnyakov AS, Sapozhnikov AN, et al. Synthesis and properties of silver and gold nanocomposites in poly-1-vinyl-1,2,4-triazole matrix. *J Struct Chem* 2010;**51**(1):105–8.
64. Hernández-Sierra JF, Ruiz F, Pena DC, Martínez-Gutiérrez F, Martínez AE, Guillén Ade J, et al. The antimicrobial sensitivity of *Streptococcus mutans* to nanoparticles of silver, zinc oxide, and gold. *Nanomedicine* 2008;**4**(3):237–40.
65. Baker C, Pradhan A, Pakstis L, Pochan DJ, Shah SI. Synthesis and antibacterial properties of silver nanoparticles. *J Nanosci Nanotechnol* 2005;**5**:244–9.
66. Li H, Chen Q, Zhao J, Urmila K. Enhancing the antimicrobial activity of natural extraction using the synthetic ultrasmall metal nanoparticles. *Sci Rep* 2015;**5**:11033.
67. Arakha M, Saleem M, Mallick BC, Jha S. The effects of interfacial potential on antimicrobial propensity of ZnO nanoparticle. *Sci Rep* 2015;**5**:9578.
68. Hsueh YH, Ke WJ, Hsieh CT, Lin KS, Tzou DY, Chiang CL. ZnO nanoparticles affect *Bacillus subtilis* cell growth and biofilm formation. *PLoS One* 2015;**10**(6):e0128457.
69. Srivastava N, Mukhopadhyay M. Green synthesis and structural characterization of selenium nanoparticles and assessment of their antimicrobial property. *Bioprocess Biosyst Eng* 2015;**38**(9):1723–30.
70. Zille A, Fernandes MM, Francesko A, Tzanov T, Fernandes M, Oliveira FR, et al. Size and aging effects on antimicrobial efficiency of silver nanoparticles coated on polyamide fabrics activated by atmospheric DBD plasma. *ACS Appl Mater Interf* 2015;**7**(25):13731–44.
71. Pal S, Tak YK, Song JM. Does the antibacterial activity of silver nanoparticles depend on the shape of the nanoparticle? A study of the Gram-negative bacterium *Escherichia coli*. *Appl Environ Microbiol* 2007;**73**(6):1712–20.

72. Pomogailo AD. Hybrid polymer–inorganic nanocomposites. *Russ Chem Rev* 2000;**69**(1):53–80.
73. Marambio-Jones C, Hoek EMV. A review of the antibacterial effects of silver nanomaterials and potential implications for human health and the environment. *J Nanopart Res* 2012;**12**(5):1531–51.
74. Barud HS, Regiani T, Marques RFC, Lustri WR, Messaddeq Y, Ribeiro SJL. Antimicrobial bacterial cellulose-silver nanoparticles composite membranes. *J Nanomater* 2011;**2011** ID 721631.
75. Darroudi M, Ahmad MB, Abdullah AH, Ibrahim NA. Green synthesis and characterization of gelatin-based and sugar-reduced silver nanoparticles. *Int J Nanomed* 2011;**6**:569–74.
76. Kvítek L, Panásek A, Soukupová J, Kolár M, Vecerová R, Prucek R, et al. Effect of surfactants and polymers on stability and antibacterial activity of silver nanoparticles (Nps). *J Phys Chem* 2008;**112**(15):5825–34.
77. Dabrowiak JC. Nanomedicine, Chapter 9. In: Dabrowiak JC, editor. *Metals in medicine*. Chichester, UK: John Wiley & Sons; 2009.
78. Shurygina IA, Sukhov BG, Fadeeva TV, Umanets VA, Shurygin MG, Ganenko TV, et al. Bactericidal action of Ag(0)-antithrombotic sulfated arabinogalactan nanocomposite: coevolution of initial nanocomposite and living microbial cell to a novel nonliving nanocomposite. *Nanomedicine* 2011;**7**(6):827–33.
79. Fadeeva TV, Shurygina IA, Sukhov BG, Rai MK, Shurygin MG, Umanets VA, et al. Relationship between the structures and antimicrobial activities of argentic nanocomposites. *Bull Russian Acad Sci Phys* 2015;**79**(2):273–5.
80. Dallas P, Tucek J, Jancik D, Kolar M, Panacek A, Zboril R. Magnetically controllable silver nanocomposite with multifunctional phosphotriazine matrix and high antimicrobial activity. *Adv Funct Mater* 2010;**20**(14):2347–554.
81. Rajeshkumar S, Malarkodi C. In vitro antibacterial activity and mechanism of silver nanoparticles against foodborne pathogens. *Bioinorg Chem Appl* 2014;**2014** ID 581890.
82. Rai M, Yadav A, Gade A. Silver nanoparticles as a new generation of antimicrobials. *Biotechnol Adv* 2009;**27**(1):76–83.
83. Kim JS, Kuk E, Yu KN, Kim JH, Park SJ, Lee HJ, et al. Antimicrobial effects of silver nanoparticles. *Nanomedicine* 2007;**3**(1):95–101.
84. Sondi I, Salopek-Sondi B. Silver nanoparticles as antimicrobial agent: a case study on *E. coli* as a model for Gram-negative bacteria. *J Colloid Interface Sci* 2004;**275**:177–82.
85. Lara HH, Garza-Treviño EN, Ixtepan-Turrent L, Singh DK. Silver nanoparticles are broad-spectrum bactericidal and virucidal compounds. *J Nanobiotechnol* 2011;**9**:30.
86. McShan D, Ray PC, Yu H. Molecular toxicity mechanism of nanosilver. *J Food Drug Anal* 2014; **22**(1):116–27.
87. El Badawy AM, Silva RG, Morris B, Scheckel KG, Suidan MT, Tolaymat TM. Surface charge-dependent toxicity of silver nanoparticles. *Environ Sci Technol* 2011;**45**:283–7.
88. Khan SS, Mukherjee A, Chandrasekaran N. Studies on interaction of colloidal silver nanoparticles (SNPs) with five different bacterial species. *Coll Surf B Biointerfaces* 2011;**87**:129–38.
89. Khan SS, Srivatsan P, Vaishnavi N, Mukherjee A, Chandrasekaran N. Interaction of silver nanoparticles (SNPs) with bacterial extracellular proteins (ECPs) and its adsorption isotherms and kinetics. *J Hazard Mater* 2011;**192**:299–306.
90. Danilczuk M, Lund A, Sadlo J, Yamada H, Michalik J. Conduction electron spin resonance of small silver particles. *Spectrochim Acta A Mol Biomol Spectrosc* 2006;**63**:189–91.
91. Grigor'eva A, Saranina I, Tikunova N, Safonov A, Timoshenko N, Rebrov A, et al. Fine mechanisms of the interaction of silver nanoparticles with the cells of *Salmonella* Typhimurium and *Staphylococcus aureus*. *BioMetals* 2013;**26**:479–88.
92. Lara HH, Ayala-Nuñez NV, Ixtepan-Turrent L, Rodriguez-Padilla C. Bactericidal effect of silver nanoparticles against multidrug-resistant bacteria. *World J Microbiol Biotechnol* 2010;**26**:615–21.
93. Wigginton NS, de Titta A, Piccapietra F, Dobias J, Nesatyy VJ, Suter MJF, et al. Binding of silver nanoparticles to bacterial proteins depends on surface modifications and inhibits enzymatic activity. *Environ Sci Technol* 2010;**44**:2163–8.

CHAPTER

CAN NITRIC OXIDE OVERCOME BACTERIAL RESISTANCE TO ANTIBIOTICS?

9

A.B. Seabra, M.T. Pelegrino and P.S. Haddad

Exact and Earth Sciences Department, Universidade Federal de São Paulo, Diadema, SP, Brazil

ANTIBIOTIC RESISTANCE AND NITRIC OXIDE

The emergence of resistant bacterial strains is a consequence of the inappropriate use of antibiotics, decreasing the chances of treatment success.[1] The number of cases of disease related to resistance to antimicrobial therapeutics has increased, and this issue has become widespread, representing a significant threat to public health.[2] As a consequence, it is necessary to increase antibiotic doses and implement alternative treatments with drugs that are more toxic and cause longer hospital stays and increased mortality.[2] Some bacteria, such as *Pseudomonas aeruginosa*, can cause severe damage because of their ability to provoke human morbidity and mortality.[3] In addition, these bacteria have the ability to develop resistance to new therapies through several mechanisms.[3] In fact, aerobic, Gram-negative *P. aeruginosa* is intrinsically resistant to several traditional antibacterials, including the macrolides, many β-lactams, cotrimoxazole, tetracylines, and most fluoroquinolones,[3,4] and it can develop resistance to new antimicrobials. Overall, antibiotic resistance involves the alteration of target structures, enzymatic degradation, active efflux from the cell, and blockage of entry.[3] Because of the increasing occurrence of bacterial resistance to traditional antibiotics, the development of novel approaches to combat microbes has become essential.[1,5] Nitric oxide (NO) represents a new avenue. Since the early 1990s, NO has been considered as an attractive and potent antibiotic against a wide range of Gram-positive and Gram-negative bacteria.[6–8]

NO is a powerful antimicrobial agent that is endogenously synthesized by NO synthase enzymes, which have three isoforms: endothelial (eNOS), neuronal (nNOS), and inducible (iNOS).[9,10] Nitrogen oxide species (NOS) catalyzes the oxidation of L-arginine to L-citrulline, yielding free NO. The calcium-independent iNOS isoform produces relatively high amounts of NO (in the micromolar range) for long periods of time as part of the host's antimicrobial defense.[9,10] iNOS is found primarily in immune cells (most importantly in macrophages), and it produces cytotoxic amounts of NO to defend the host against pathogens.[3] NO is a small gaseous molecule that is devoid of charge and relatively lipophilic, and thus it is able to cross biological barriers such as bacterial lipid membranes, causing oxidative and nitrosative effects and killing the bacteria.[3] NO and its autoxidation species (NOx) act through multiple pathways and have many targets. Examples of the harmful reactive oxygen species (ROS) generated by NO that is

Antibiotic Resistance. DOI: http://dx.doi.org/10.1016/B978-0-12-803642-6.00009-5

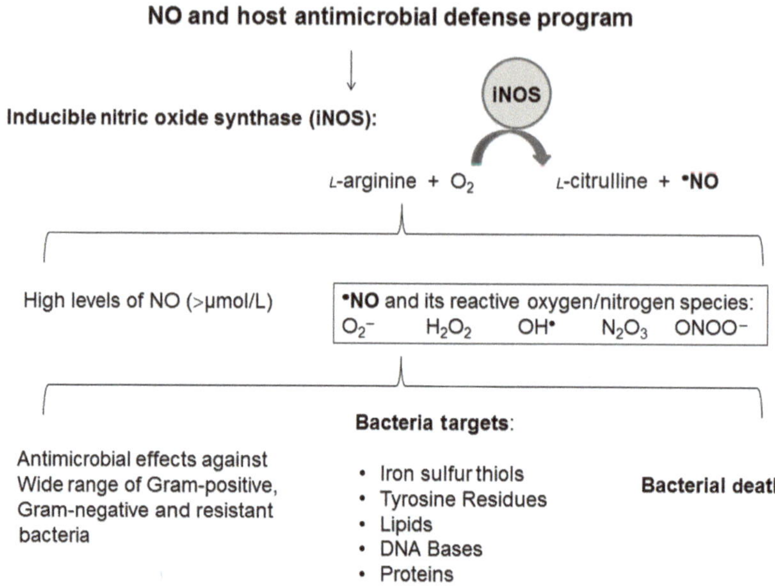

FIGURE 9.1

Schematic representation of NO synthesized by inducible nitric oxide synthase (iNOS) in host cell systems, the formation of reactive oxygen species (ROS) and reactive nitrogen species (RNS), and the multimodal antimicrobial activity of these species in bacteria.

produced by iNOS are superoxide (O_2^-), hydroxyl radicals ($OH^{\bullet}$), and hydrogen peroxide (H_2O_2).[10] Examples of reactive nitrogen species (RNS) generated by NO are dinitrogen trioxide (N_2O_3) and peroxynitrite ($OONO^-$). NO itself, ROS, and RNS interact with a wide variety of bacterial targets, including lipids, tyrosine residues, DNA bases, and vital iron-sulfur thiols, leading to cytotoxic effects.[9,10] Fig. 9.1 shows a schematic representation of the effects of NO synthesized by iNOS in host cell systems, the formation of ROS and RNS, and the multimodal antimicrobial activity of these species in bacteria. With this knowledge in mind, this chapter highlights recent progress in the use of NO against resistant bacteria, with a special focus on the combination of nanotechnology and NO donors to fight bacterial resistance. Despite great advances in this area, more studies are necessary to design clinically efficient, NO-releasing nanomaterials. The hope is that this chapter will inspire new avenues in this promising field.

ANTIBACTERIAL EFFECTS OF GASEOUS NO

The medical-grade gaseous form of NO (gNO) is the simplest way to administer NO directly to the target site by inhalation or topical application. The US Food and Drug Administration (FDA) has approved the use of gNO as an inhaled vasodilator for treating pulmonary hypertension.[11] The Ghaffari group demonstrated the in vitro antibacterial effects of gNO toward a range of pathogens, including methicillin-resistant *Staphylococcus aureus* (MRSA), *S. aureus*, *Escherichia coli*, Group B *Streptococcus*, *P. aeruginosa*, and *Candida albicans*.[11] Cell cultures were exposed to a gNO chamber containing 160 parts per million (ppm) of the gas for 24 h, and the data show a 90% reduction in bacterial growth. These effects were observed to be dose-dependent, as the bactericidal or bacteriostatic effects were limited at lower gNO concentrations.[11] Similarly, Miller and collaborators evaluated the effects of intermittent gNO inhalation at a concentration of 40 ppm in rats that were infected with *P. aeruginosa*.[12] The results show a 90% reduction in the bacterial load, and the data suggest that there is a dual effect depending on whether higher or lower concentrations of NO are administered. Higher concentrations of NO have a direct bactericidal effect, whereas lower concentrations produce an indirect effect by stimulating the host immune system.[12,13] An interesting paper from 2012 focused on investigating the safe use of gNO for antimicrobial purposes.[14] A potential side effect of NO might arise from the binding of NO with host hemoglobin to form methemoglobin, which can decrease oxygen transport. Fortunately, an experiment in which 160 ppm gNO was administered to the airways of 10 healthy adult individuals did not show significant side effects, and this experiment showed acceptable arterial hemoglobin oxygen saturation.[14]

In the clinic, it is common to observe the persistence of bacterial burdens in injuries, and conventional treatments for wound infection are becoming less effective because of antibiotic-resistant strains and the presence of a complex network of bacterial biofilms. Therefore, another important application for gNO is in the dermatological area, where it is used to treat infected wounds.[11] It is important to highlight the role of NO in wound healing, where there are two important aims: reducing the bacterial burden of the wound and enhancing wound healing.[11,15–18]

Although the administration of gNO as an antimicrobial agent was effective against several bacterial strains, including MRSA, clinical experience with gNO shows that it is a viable option only in a hospital setting, and only when the patient is immobile and has access to specialized equipment. Another important limitation of gNO administration is the cost of treatment, because 1 h of treatment costs an average of $125, which can be impractical in some settings.[19] Therefore, the use of NO donors may circumvent the drawbacks of administering gNO as an antimicrobial agent.

NO DONORS

As a free radical, NO is very reactive and has a short half-life in biological systems (approximately 0.5 s); therefore, there are practical problems and limitations involved in the direct use of gNO for biomedical applications.[20] Clinically and in experimental research, NO is primarily administered through NO donors, which have the ability to mimic continuous endogenous NO production.[21] The most important classes of low-molecular-weight NO donors are organic nitrates, *N*-diazeniumdiolates (NONOates), *S*-nitrosothiols (RSNOs), metal-NO-complexes, NO-drug hybrids, and sydonomines.[22] Organic nitrates are the most frequently used NO donors, and isosorbide monitrate (ISMN) is a clinically approved NO donor for diverse biomedical applications.[1] Organic nitrates and nitrites have been available on the market since 1882, and they include nitroglycerin, isosorbide dinitrate, isosorbide 5-mononitrate, and sodium nitroprusside. These drugs remain useful for the treatment of acute ischemic stroke and heart failure.[23,24] Another class of NO donors of particular interest is the NONOates.[21,25] NONOates are obtained when NO gas reacts with secondary amines under basic conditions, and they are able to undergo proton-initiated decomposition. Under physiological conditions, NONOates release 2 mols of NO per mol of NO donor.[25] Another important class of NO donors is RSNOs.[3] RSNOs can react directly with the sulfhydryl (SH) groups of biomolecules, transferring NO^+ to the SH. A transnitrosation reaction between RSNOs and SH-containing biomolecules allows NO to mediate toxicity against microbes.[3] Among the RSNOs, *S*-nitrosoglutathione (GSNO) has the ability to internalize into microbial systems, where inside the pathogen, it modifies vital thiols and thereby alters protein functions, which can lead to cytotoxicity.[3] In addition, RSNOs act as spontaneous NO donors in biological media through S—N bond cleavage, with the release of free NO, according to Eq. (9.1) as follows:

$$2\ \text{RSNO} \rightarrow 2\ \text{NO} + \text{RS-SR} \tag{9.1}$$

The subproduct of RSNO decomposition is a dimer with disulfide bonds, and in biological systems, this dimer can be reduced to thiols through the action of reducing agents.[26] The next section will focus on recent progress in the antibacterial application of low-molecular-weight NO donors.

ANTIBACTERIAL ACTIONS OF NO DONORS AND NO PRODRUGS

Recently, a combination of NO donors with traditional antibiotics has been shown to be an attractive approach for combating resistant bacteria. For instance, Bang and collaborators[9] reported the in vitro antibacterial effects of the NO donor diethylenetriamine (DETA)/NO, which belongs to the NONOate class, against resistant uropathogenic bacteria. That study was performed with clinical isolates of multidrug-resistant (MDR), extended-spectrum β-lactamase (ESBL)–producing uropathogenic *E. coli*. The bacterial enzyme flavohemoglobin has been demonstrated to have NO dioxygenase activity, and thus it is related to the protection of bacterial walls from nitrosative/oxidative stresses.[9,27] As a consequence, the pharmacological inhibition of flavohemoglobin may help to combat host infections. The antimicrobial actions of DETA/NO against ESBL-producing uropathogenic *E. coli* were significantly enhanced by the coadministration of the NO donor with miconazole and polymyxin B nonapeptide.[9] This work demonstrates that the coadministration of NO donors with traditional antibiotics can have synergistic antibacterial effects. The NO donor DETA/NO (1 mmol/L) was used against a *Mycobacterium tuberculosis* strain.[28] The study included patients in whom pulmonary tuberculosis (TB) had been confirmed by sputum smear and culture. The authors reported a significant relation between a reduced susceptibility to NO and resistance against isoniazid, an anti-TB drug. Therefore, NO contributes to the antibacterial action of this anti-TB drug.[28]

Burkholderia pseudomallei can infect the respiratory tract in mammalian hosts.[29] Conventional antibiotics are inefficient for combating these infections because *B. pseudomallei* is extremely resistant to conventional antibiotics. The bacteria are able to downregulate the synthesis of NO through iNOS in activated macrophages. This NO downregulation in the host confers a significant advantage to the bacteria during infection. Hence, the administration of exogenous NO donors can replace the endogenous NO production that is lacking in host macrophages.

In fact, the NONOates spermine NONOate and DETA NONOate exhibited potent antibacterial activity (99.9%) against the bacteria after 2.5 h of incubation with a 25-μmol/L NO donor, and the minimum inhibitory concentration (MIC) and minimum bactericidal concentrations (MBCs) values confirmed the bacteriostatic effect of the NO donors.[29]

With respect to RSNO NO donors, Kishikawa and collaborators demonstrated the antibacterial effect of *S*-nitrosocysteine (CysNO) against the bacterial species present in urinary tract infections.[30] The authors observed that all tested bacterial species were inhibited at a urinary pH of 5.5, with bactericidal activity against *P. aeruginosa*, *Staphylococcus saprophyticus*, and *Enterobacter cloacae*.[30] Cariello and colleagues showed the bactericidal effect of *S*-nitrosoglutathione (GSNO) and *S*-nitroso-*N*-acetylcysteine (SNAC) against clinical bacterial strains that were isolated from patients with infectious keratitis.[31] In the case of *S. aureus*, the MIC mean values for SNAC and GSNO were 2.1 ± 1.3 and 4.6 ± 3.2 mmol/L, respectively. Overall, both GSNO and SNAC had bactericidal effects; however, SNAC showed greater antimicrobial activity than GSNO against all bacteria. These data indicated that these types of NO donors might be useful for the topical treatment of ocular infections.[31] Among various derivatives of NO, the bacteriostatic effects of RSNOs are 100-fold stronger than those of gNO. However, RSNOs form disulfide bonds in the presence of water and heat; thus, most RSNOs must be kept refrigerated and in the form of a dry powder until they are mixed and administered.[19,32]

NO donors such as organic nitrates have limited antibacterial and biofilm-disrupting properties and are unlikely to be used alone as antibacterial agents, given their potent cardiovascular effects.[19,23] Thus, there is a tendency to cocombine organic nitrates with antibiotics that are already used in clinical practice, and this approach might provide additive or synergistic antibacterial effects.[33]

In addition to NO donors, NO can be generated for antimicrobial purposes. Acidified nitrite (NO_2^-) was devised as a novel method for generating NO for antibiotic therapy.[34] In this approach, nitrite (NO_2^-) is coapplied with an acid, such as citric acid or ascorbic acid, in a cream formulation, and NO is successfully generated, according to Eqs. (9.2)–(9.4),[34] as follows:

$$NO_2^- + H^+ \rightarrow HNO_2 \qquad (9.2)$$

$$2\,HNO_2 \leftrightarrow N_2O_3 + H_2O \qquad (9.3)$$

$$N_2O_3 \leftrightarrow NO + NO_2 \qquad (9.4)$$

In brief, nitrite and H^+ form nitrous acid (HNO_2), which is converted to dinitrogen trioxide (N_2O_3). The dinitrogen trioxide decomposes, generating free NO (Eq. (9.4)). The antimicrobial effects of acidified nitrite are concentration-related; parts per billion levels of NO have a bacteriostatic effect, whereas parts per million levels exhibit a strong bactericidal effect.[34]

Ormerod and colleagues demonstrated the efficacy of NO that was generated from an acidified nitrite formulation as shown in Eqs. (9.2)–(9.4).[35] That study involved the recruitment of hospitalized patients with MRSA-positive wounds. Those patients were treated with 4.5% citric acid in aqueous cream and 3% sodium nitrite twice daily for 5 days. The data showed that 60% of the infected wounds were cleared of MRSA colonization, and in vitro experiments corroborated these results.[35] The efficacy of the acidified nitrite treatment depends on the pKA and the type of acid used. It is also very important that the biological effects could rely on the range of nitrogen oxides produced in addition to the NO. For clinical use, this method shows some limitations, such as the potential to cause irritation to airways and mucosal surfaces and inflammation.[36] Recent studies have mentioned this nitrite/acidified formulation as an alternative approach for food storage[37] and wastewater treatment systems.[38] Making progress in new approaches to combat bacterial resistance in applications such as food storage and wastewater treatment is important because the extensive use of biocides at sublethal concentrations is a source of resistant microorganisms such as *Salmonella enterica*.

Another strategy for generating NO for antibacterial applications is the use of probiotic patches. The use of probiotics in medical clinics has grown during the last two decades, with highlighted use in diet supplements and combating bacterial infections. By definition, probiotics are live microorganisms that may confer health benefits to the host upon administration in adequate amounts. The probiotic patch is intended to influence the metabolic activity of the microorganism.[39,40] NO-generating probiotic patches are filled with glucose and nitrite salt (ie, sodium nitrite, $NaNO_2$). The microorganism metabolizes the glucose by fermentation and produces lactic acid, and this process generates protons and causes the dismutation of nitrite into gNO, according to Eqs. (9.5) and (9.6)[40]:

$$\text{Glucose} \rightarrow \text{Lactic Acid} \tag{9.5}$$

$$3\,NO_2^- + 2\,H^+ \rightarrow 2\,NO + H_2O + NO_3^- \tag{9.6}$$

According to this process, the control of gNO production relies on the metabolism of the bacteria, and therefore it can be sustained for longer durations. The metabolism of the bacteria also releases other products, such as lipopolysaccharides and lipoteichoic acid, which can help to increase the NO concentration in the host. Several studies have shown that these products induce iNOS expression, which is responsible for increased rates of NO production.[40] Jones and colleagues demonstrated in vitro analysis of gNO-producing patches that used *Lactobacillus fermentum*.[40] These patches were able to generate 250 ppm of gNO over 8 h, which inhibited the growth of *E. coli*, *S. aureus*, and *P. aeruginosa* colonies.[40] The same group demonstrated the antimicrobial activity of gNO-generating patches on MRSA-infected wounds in a New Zealand white rabbit model.[41] The authors also monitored the healing process in vertical incisions that were made approximately 1 cm from the bases of the ears and were infected with MRSA. During the MRSA wound treatment, 200 ppm NO by volume was generated for 24 h, and the patches were replaced daily for 20–21 days. By measuring the wound area during the experiment, the authors showed that the healing processes in infected wounds were significantly accelerated in rabbits treated with the gNO-producing patches compared with the control group. Indeed, after 10 days of treatment, the wound area of animals treated with gNO was about 5 mm^2; by contrast, the wound area of the control animal group was 20 mm^2.

Similar results can be observed for wounds that were not infected with MRSA.[24,41] Although gNO-generating patches might have important applications for treating infected wounds, a limitation of this method is that the rate of gNO production depends on the metabolic activity in each patch, which introduces variability into the NOg synthesis among patches.[24,41] Taken together, the results demonstrate that the therapeutic generation of gNO from patches might find important applications in hospitals for treating MRSA-infected wounds. It should be noted that in addition to its potent antibacterial effect, NO is an important mediator of wound repair; hence, the administration of gNO to infected wounds has two biomedical aims: to combat bacterial infection and to promote and accelerate wound closure.[16–18,42]

NO-RELEASING NANOMATERIALS

Although significant advances can be achieved by administering NO donors to treat resistant bacteria, low-molecular-weight NO donors such as NONOates and RSNOs are relatively unstable.[21,43] Free NO donors can be triggered to decompose by several factors, such as heat, light, copper ions, enzymes, and acid environments, leading to uncontrolled NO release.[20] To increase the thermal and photochemical stability of NO donors, these molecules have been trapped in innovative formulations made of nanomaterial scaffolds. Several advantages have been reported for loading NO donors into different nanomaterials, such as polymeric nanoparticles, liposomes, dendrimers, nanoemulsions, micelles, metallic nanoparticles, and carbon nanotubes.[22,44–49] Moreover, nanotechnology has emerged as a potent tool for combating bacterial resistance.[2] The combination of NO donors and nanomaterials represents a promising strategy to overcome microbial resistance. It should be noted that NO-releasing nanomaterials can be combined with the simultaneous delivery of therapeutic agents to enhance and optimize antibacterial effects. In this context, this section highlights recent and important advances in the use of NO-releasing nanomaterials to combat resistant bacteria.

Potential bacterial resistance to NO-releasing silica nanoparticles was evaluated by using spontaneous and serial passage mutagenesis assays.[50] MRSA, *S. aureus*, *E. coli*, *Staphylococcus epidermidis*, and *P. aeruginosa* were systematically exposed to NO-releasing silica nanoparticles at or below MIC levels. The bacterial strains were unable to developed resistance toward NO-releasing silica nanoparticles. In the serial-passage mutagenesis assays, no increase in MIC values was reported upon bacterial exposure to subinhibitory NO concentrations for 20 days. Similarly, no increase in the MIC values was observed for bacteria that survived exposure to lethal NO concentrations, as shown by spontaneous mutagenesis assays.[50] These results showed the inability of several bacterial strains to develop resistance to NO.

Polymeric nanoparticles composed of the oligosaccharide chitosan have been used extensively in antimicrobial applications because of their biocompatibility and biodegradability.[51,52] Chitosan is a polymer of *N*-acetylglucosamine and glucosamine. The NH-groups of chitosan promote the polymer's association with the negatively charged microbial membrane and cell wall. Moreover, chitosan also enhances and accelerates wound healing. These properties contribute to the multiple antimicrobial effects of chitosan.[53] With this knowledge, Cardozo and collaborators[47] investigated the antimicrobial effects of NO-releasing alginate/chitosan and chitosan/tripolyphosphate nanoparticles against MDR strains of *S. aureus* and *E. coli* isolated from bovine mastitis. The low-molecular-weight NO donor *S*-nitroso-mercaptosuccinic acid (*S*-nitroso-MSA) was encapsulated into chitosan-based nanoparticles. The resulting nanoparticles showed average hydrodynamic sizes ranging from 277 to 377 nm and positive zeta potentials (+17 mV), indicating good colloidal stability under physiological conditions. The NO release from these nanoparticles revealed an initial burst during the first 4 h, followed by slower delivery over several hours. The initial burst of NO that was released from the polymeric systems showed potent antibacterial activity. The MIC for NO-releasing, chitosan-based nanoparticles against *S. aureus* ranged from 125 to 250 mg/mL. This result indicates that there is a potential use for NO-releasing polymeric nanoparticles to combat and treat resistant bacteria that cause bovine mastitis.[47] In a similar approach, NO-releasing (NONOate) chitosan and poly-γ-glutamic acid (γ-PGA) composite microparticles demonstrated potential activity against some opportunistic pathogens that cause chronic infections.[54]

The NO released from a thermosensitive conjugate composed of Pluronic F68 and branched polyethylenimine (BPEI) containing a NONOate (as the NO donor) showed an antibacterial effect against Gram-positive *S. aureus*, Gram-negative *E. coli*, and MRSA.[55] NO release from F68-BPEI-NONOate showed an initial burst followed by a sustained release (half-life of NO release of about 30 min). The initial burst is an effective NO release profile for antibacterial activity because a high NO flux is reported to have toxic effects. In fact, the NO-releasing F68-BPEI-NONOate was found to have potent antibacterial action because of the destruction of the bacterial membrane.[55] Moreover, the ability of NO-releasing polymeric nanoparticles to show an antibacterial effect by destroying the membrane was recently reported by Priya and collaborators.[56] RSNOs derived from cysteine were grafted into a polymeric chain of poly(vinyl methyl ether-*co*-maleic anhydride) (PVMMA) through a solvent displacement method, forming an NO-releasing hexagonal nanomaterial. The antibacterial activity of the NO-releasing polymeric hexagons was confirmed against *S. aureus*, *E. coli*, and *P. aeruginosa* with a MIC of 1.9 mg, and it inhibited the visible growth of microorganisms after 24 h of incubation. Bacterial cell death was revealed to occur via membrane destruction, as shown by confocal microscopy.[56] After exposure to the Cys-SNO conjugated polymeric hexagons, morphological changes were observed during the incubation period (1, 2, or 5 h). The observed morphological variations in microbes are caused by the nitrosation of cell surface proteins and intracellular proteins such as enzymes. Because NO is lipophilic in nature, it can penetrate the cell membrane. A high level of NO release close to the cells causes a high intracellular NO concentration. This increased dose of NO causes the accumulation of NO in membranes, accelerating the oxidation of NO to N_2O_3 and creating greater nitrosative stress within and near the bacterial membrane.[56]

In another approach, NO-releasing polymeric nanofibers were prepared by electrospinning from thiol-derivatized poly(lactic-*co*-glycolic-*co*-hydroxymethyl propionic acid) (PLGH) that was functionalized to incorporate NO into the polymer backbone for antibacterial activity.[57] NO release from the polymeric fibers was observed over a period of days in aqueous solution at physiological pH and temperature. The total NO release was found to range from 0.026 to 0.28 mmol/g. Scanning electron microscopy (SEM) revealed the porous morphology of the fibers, with interconnected pores distributed throughout the fibrous structure. The NO-releasing polymeric fibers showed antibacterial activity against *Acinetobacter baumannii*. The results showed that the 0.0035-mmol NO released from the fibers reduced bacterial survival by 96%.[57]

A series of NO-releasing poly(propylene imine) (PPI) dendrimers was also evaluated by Sun et al.[58] against resistant and pathogenic bacteria such as *S. aureus*. Dendrimers have a repetitively branched architecture characterized by a high level of precision in their synthesis and structure.[59] Secondary amine- and NONOate NO donor-functionalized G2 ($n = 8$) and G5 ($n = 64$) PPI dendrimers were synthesized. The dendrimer exterior of G2 or G5 PPI-NH_2 was modified with propylene oxide (PO) or styrene oxide (SO) via a ring-opening reaction or with poly-(ethylene glycol) methyl ether acrylate (PEG) by conjugate addition to yield secondary amine-functionalized dendrimers. These compounds were subsequently functionalized with NONOates, generating NONOate-PPI dendrimers. The results showed that these NO donors released total NO ranging from 1.1 to 3.3 μmol/mg and killed 99.9% of the tested bacterial strains, including MRSA. In addition, no cytotoxic effect was observed in healthy cells.[58] Similarly, NONOate- dendrimers containing quaternary ammonium (QA)- and poly(amidoamine) (PAMAM) with antibacterial effects against Gram-negative *P. aeruginosa* and Gram-positive *S. aureus* bacteria have also been reported.[60] The authors investigated the influence of the alkyl chain length on NO release and antibacterial activity, and for this purpose, PAMAM dendrimers were functionalized with QA moieties containing a series of alkyl chains. NONOate moieties were obtained by modifying secondary amines to form NO-donor QA-PMMA dendrimers. The results showed that, depending on the alkyl chain of the QA, these materials delivered about 0.75 μmol/mg NO with a half-life between 1.9 and 4.4 h. The NO-release from these materials induced damage to resistant bacteria. The bactericidal activity was found to be dependent on dendrimer generation, QA alkyl chain length, and bacterial Gram status. In brief, shorter alkyl chains showed increased bactericidal activity against *P. aeruginosa* (MBC 200–300 μg/mL) and *S. aureus* (MBC 300 μg/mL).[60]

Han and colleagues[61] showed the efficacy of an NO-releasing nanomaterial made from a mixture of hydrogel/glass composite nanoparticles against MRSA abscesses in mice. The hydrogel/glass composite was obtained by mixing tetramethyl orthosilicate, chitosan, polyethylene glycol, sodium nitrate, and glucose. NO is generated upon nitrite reduction. When the nanomaterial is exposed to aqueous medium, trapped NO is released. The antimicrobial activity of the NO-generating nanocomposite was evaluated against MRSA in vitro and in abscesses in vivo. The in vivo experiments showed that the NO-generating nanomaterial significantly reduced bacterial growth after 16 h of treatment in comparison with the control group.[61] Similar results have been reported by other authors who treated infected wounds with NO-releasing materials, confirming the ability of NO to accelerate wound closure and decrease the bacterial burden in a MRSA-infected murine wound model.[2,62,63]

Zeolites are known to be convenient materials for NO storage and release. They are nanoporous materials composed of a metal-organic framework. Their large surface areas and metallic components allow zeolites to bind large amounts of NO relative to their size. Although they are stable when dry, once they are exposed to moisture, zeolites are forced by water molecules to release NO. The rate of NO release can be adjusted by altering the composition of the metal-organic framework.[19,64] Zeolites have been shown to have antibacterial properties against *P. aeruginosa*, MRSA, *Clostridium difficile, E. coli*, and *Bacillus subtilis*.[64] Neidrauer and colleagues[65] demonstrated the antibacterial efficacy of an ointment containing NO-loaded, zinc-exchanged zeolite A, which releases NO upon contact with water. MBC assays were performed by using five common wound pathogens, including Gram-negative bacteria (*E. coli* and *A. baumannii*), Gram-positive bacteria (*S. epidermidis* and MRSA), and a fungus (*C. albicans*). The time dependence of the antimicrobial activity was characterized in vitro by performing log-reduction assays at four time points after 1–8 h of cream exposure in bacterial medium. The cytotoxicity of the cream after 24 h was also tested by using cultured 3T3 fibroblast cells. The minimum microbicidal concentrations (MMCs) ranged from 50 to 100 mg cream/mL media, and furthermore, the MMC for *C. albicans* was 50 mg cream/mL media. Five to eight log reductions in bacterial viability and three log reductions in fungal viability were observed after 8 h of exposure to NO-zeolite cream compared with untreated bacteria. The fibroblasts remained viable after 24 h of exposure to the same concentration of NO-zeolite cream as the concentration that was used in the antimicrobial tests.[65] Moreover, in vivo studies demonstrated the beneficial effects of NO-zeolite ointment on the acceleration of the wound-healing process.[65]

NO-generating nanoparticles made of hydrogel/glass obtained from tetramethyl orthosilicate, chitosan, glucose, and polyethylene glycol, and sodium nitrite in 0.5 mol/L sodium phosphate buffer were used to treat wounds infected with *P. aeruginosa.*[3] In this nanocomposite, NO was generated from nitrite reduction, and 5 mg/mL of the nanomaterial released 15 mmol/L of NO. The antibacterial effect of this NO-releasing nanomaterial was evaluated in vitro and in vivo in murine excisional wounds infected with an MDR clinical isolate of *P. aeruginosa*. The addition of glutathione (GSH) to the NO-releasing nanomaterial enhanced its antibacterial efficacy. In fact, the in vitro experiments revealed that the NO-releasing nanomaterial inhibited bacterial growth for up to 8 h, and the GSH + NO-releasing nanomaterial completely inhibited bacterial growth for 24 h. The in vivo experiments showed the ability of the GSH + NO-releasing nanomaterial to significantly accelerate the closure of infected wounds and to decrease the bacterial burden, in comparison with control groups. After 3 days of treatment, the bacterial counts in animals with infected wounds treated with the nanomaterial were lower than those of the control animals. Therefore, this NO-releasing nanomaterial can be applied to lower the bacterial burden and also to promote and accelerate wound healing.[3]

The use of visible light can be used to trigger NO release from nanomaterials for antibacterial effects. For example, light can induce the photorelease of NO from $[Mn(PaPy_3)(NO)](ClO_4)$ ({Mn-NO}) loaded into a mesoporous nanomaterial.[5] The photorelease of NO (50–80 μmol/L) was observed with 5 mg/mL of the nanomaterial within 5 min upon irradiation with 10–100 mW/cm^2. The photorelease of NO from the nanomaterial successfully eradicated both drug-susceptible and drug-resistant *A. baumannii* in a soft-tissue infection model.[5] Thus, the combination of light and NO-releasing nanomaterials can be used to combat resistant bacteria.

The combination of NO with chemotherapeutic drugs against *M. tuberculosis* has been reported.[66] The NO donor DETA/NO was incorporated into microparticles of polylactide-*co*-glycolide (PLGA). PLGA microparticles containing 10% w/w of DETA/NO alone or in combination with 25% rifabutin (RFB) and 25% isoniazid (INH) were obtained by spray-drying. In an animal model, daily inhalation of NO-releasing microparticles reduced the $\log_{10}$ CFU in the lungs from 6.1 to 4.4 in 1 month. The simultaneous inhalation of NO-releasing microparticles and RFB and INH significantly decreased the CFU in the lungs and spleen by 4 log.[66] Therefore, the coadministration of NO-releasing materials with other drugs might lead to important antibacterial applications.

The antibacterial effect of NO-releasing, thermally hydrocarbonized, porous silicon nanoparticles (NO/THCPSi NPs) was assessed for planktonic *E. coli* and *P. aeruginosa*, as well as *S. aureus* and *S. epidermis* biofilms.[8] The NO release from NO/THCPSi NPs showed an initial burst during the first 2 h. The incubation of *E. coli*, *P. aeruginosa*, and *S. aureus* cultures with 0.1 mg/mL NO/THCPSi NPs for 24 h reduced the bacterial counts by 1 log in comparison with the control groups. SEM images and energy-dispersive X-ray (EDX) spectroscopy analysis revealed the presence of NO/THCSi nanoparticles in *E. coli*. Fig. 9.2 shows the SEM images and EDX spectra of *E. coli* incubated with NO/THCPSi NPs compared with an untreated control. As shown, the nanoparticles are in the bacteria and also on the background surface (Fig. 9.2A). Untreated *E. coli* showed no nanoparticles (Fig. 9.2B). Moreover, the presence of NO/THCPSi NPs on the *E. coli* cell membrane surfaces was confirmed by EDX analysis showing the presence of the characteristic Si peak (Fig. 9.2C). As expected, no Si peak was observed for untreated *E. coli* (Fig. 9.2D).[8] The antibacterial effect of NO-releasing nanoparticles can be attributed to the direct interaction of the nanomaterial with the bacterial cell membrane.

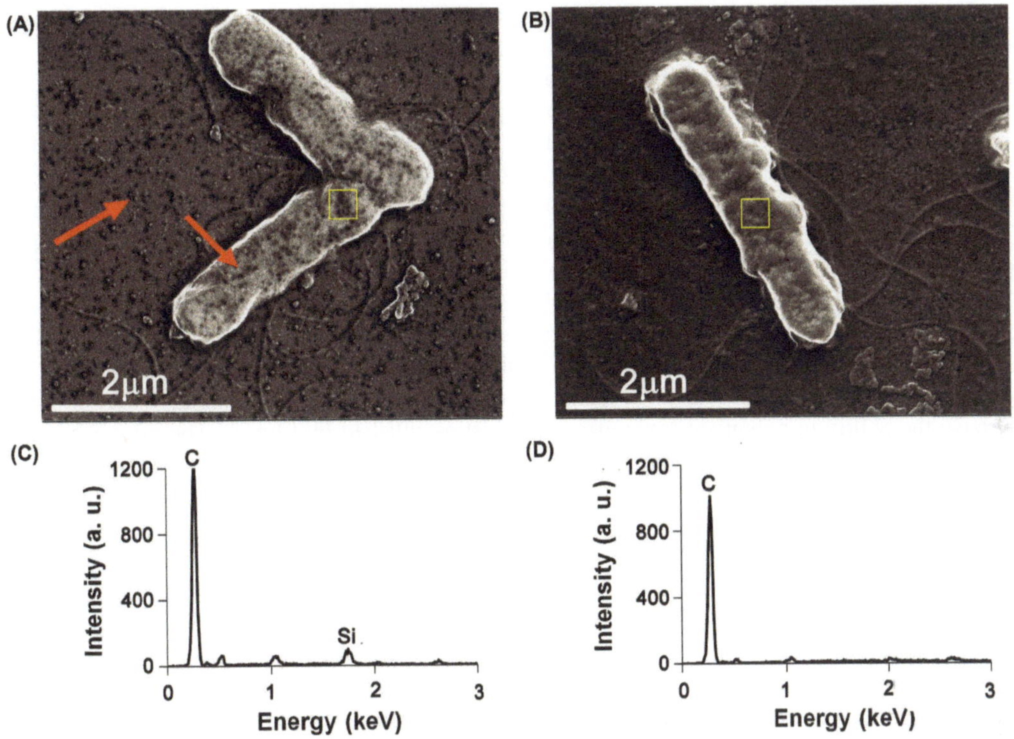

FIGURE 9.2

Scanning electron microscopy (SEM) images and energy-dispersive X-ray (EDX) spectroscopy of NO/THCPSi nanoparticles incubated with *E. coli*; (A) an SEM image of NO/THCPSi nanoparticle-treated *E. coli*, (B) an SEM image of untreated *E. coli*, (C) an EDX spectrum of NO/THCPSi nanoparticle-treated *E. coli*, and (D) an EDX spectrum of untreated *E. coli* (control). The EDX analysis was performed on the bacterial surface (*yellow overlay*). The nanoparticles on the bacterial surface and those settled on the background are indicated by *red arrows*.

Reproduced with permission from Kafshgari, et al. Nitric oxide-releasing porous silicon nanoparticles. Nanoscale Res Lett 2014;9:333. http://www.nanoscalereslett.com/content/9/1/333.

BIOFILMS AND NO

In nature, bacteria can exist as free-floating planktonic forms or in biofilms.[67] Biofilms are communities of encased bacteria in a self-produced extracellular matrix, and they are able to survive in adverse environments.[10] Bacterial biofilms are reported to be up to 10,000 times more resistant to antibiotics than free planktonic bacteria, leading to chronic infections even after antibiotic treatment.[2,68,69] Bacterial infections and biofilms can exist either in living tissues, including nonhealing wounds (diabetic wounds or cystic fibrosis wounds), or on abiotic surfaces such as medical implant devices and artificial prosthetics, causing serious complications.[10,20,67,70] Because bacterial biofilms cannot be combated with conventional antibiotics, the scientific community is actively seeking new approaches to disperse biofilms efficiently. Understanding the events that regulate biofilm dispersal can contribute to a strategy to promote bacterial detachment.[20]

As a signaling molecule, NO was found to play a key role in the regulation of dispersal events in biofilms.[69] Recently, NO was shown to mediate a signaling pathway that results in the dispersion of established biofilms during the late stage of mature biofilms.[20,71] It should be noted that because NO promotes biofilm dispersion via a signaling mechanism rather than by causing a cytotoxic effect, the use of NO to induce the transition from biofilm to planktonic bacteria would not be expected to generate resistant strains, as observed in the case of antibiotics. Interestingly, low doses of NO (picomolar to nanomolar ranges), which are not toxic to mammalian cells, can successfully induce biofilm dispersal in several bacterial strains that are reported to be resistant to antibiotics.[20] Therefore, the use of NO donors represents a potent approach to disperse antibiotic-resistant biofilm-forming bacteria.[71] Moreover, the combination of NO donors with conventional antibiotics may have the ability to potentiate toxic effects against established biofilms.[68] The establishment of biofilms can cause complications in chronic wounds, impairing the healing process, spreading infection to other tissues or to the bloodstream, or contaminating other patients in hospitals.[10] The situation can deteriorate with the increasing prevalence of resistant bacteria, such as MRSA, leading to fatal cases in hospitals.[8]

Barraud and collaborators[72] demonstrated effective disruption of *P. aeruginosa* biofilms treated with the low-molecular-weight NO donors sodium nitroprusside (SNP), *S*-nitrosoglutathione (GSNO), and *S*-nitroso-*N*-acetyl-penicillane (SNAP). Low concentrations of the NO donors (in the nanomolar-to-micromolar range) decreased the biofilm mass and consequently increased the number of planktonic cells. Indeed, 500 nmol/L SNP caused a tenfold decrease in the ratio of biofilm-to-planktonic cells.[72]

In a similar approach, three NO-releasing silica nanoparticles (14, 50, and 150 nm) were used against biofilms of Gram-negative *P. aeruginosa* and Gram-positive *S. aureus*.[73] The total NO release from the three sizes of silica nanoparticles was about 0.3 μmol/mg. The results showed that smaller nanoparticles (14 nm) had enhanced antibacterial activity. For *P. aeruginosa*, the MBCs were found to be 0.5 and 6 mg/mL for planktonic and biofilm bacteria, respectively, upon treatment with NO-releasing silica nanoparticles (14 nm). Similarly, for *S. aureus*, the MBCs for *S. aureus* were found to be 2 and 10 mg/mL for planktonic and biofilm bacteria, respectively. In addition, the bactericidal NO doses were found to be 1.44 ± 0.06 and 2.40 ± 0.10 μmol/mL for *P. aeruginosa* and *S. aureus* biofilms, respectively.

The antibiofilm efficacy of the NO donor ISMN encapsulated into liposomes was evaluated against *S. aureus* biofilms, which are known to be associated with recalcitrant chronic rhinosinusitis with serious resistance to conventional antibiotics.[1] In vitro *S. aureus* biofilms were grown with a minimum biofilm eradication concentration device, and they were treated with 3- and 60-mg/mL ISMN-containing anionic liposomes. ISMN-containing unilamellar liposomes (size 384 nm, 25 mmol/L, encapsulation efficiency of 6.3%) at 3 and 60 mg/mL concentrations showed significant antibiofilm activity after 24 h of exposure in comparison with the control group. ISMN-containing multilamellar liposomes (536 nm and 25 mmol/L with an encapsulation efficiency of 10.7%) at a concentration of 60 mg/mL resulted in lower biofilm biomass ($0.003\ \mu m^3/\mu m^2$) compared with the control group ($4.4\ \mu m^3/\mu m^2$). The charge interactions of liposomes with the bacterial wall are thought to enhance the formulation's penetration into the biofilm.[1]

Similarly, NO-releasing cross-linked star polymers were obtained by reverse addition-fragmentation chain transfer polymerization and were successfully used to disperse *P. aeruginosa* biofilm.[20] The NO release from the star polymers was evaluated to be about 355 nmol/L per hour for 400 μg/mL polymer. The NO release profile displayed a rapid release during the first hour, followed by a sustained NO release for up to 70 h. *P. aeruginosa* biofilms were treated with 100- and 400-μg/mL NO-releasing polymers. The results showed a strong inhibition of biofilm formation, with a reduction of 90% and 95% in the biofilm biomass after 7.5 h of treatment with 100- and 400-μg/mL NO-releasing polymers, respectively. As expected, the number of planktonic cells in the culture medium increased as a consequence of the suspension of the biomass.[20]

The ability of a gNO-releasing dressing to combat biofilms and prevent biofilm formation was characterized against MRSA, *P. aeruginosa* and *A. baumannii* biofilms.[74] The initial rate of gNO release from the dressing was found to be 73 nmol/cm^2 h. After 6 h of treatment, the gNO-releasing dressing could induce complete bacterial death and prevent biofilm formation. Moreover, the gNO-releasing dressing showed higher antibiofilm activity than current antimicrobial treatments, including silver dressings, povidone-iodine dressings, and the antibiotics vancomycin, gentamicin, and chlorhexidine.[74]

HOW DOES NO KILL BACTERIA? WHY ARE BACTERIA UNLIKELY TO DEVELOP RESISTANCE TOWARD NO DONORS?

As demonstrated in this chapter, administration of exogenous NO donors or NO-releasing nanomaterials can kill resistant bacterial strains. This process can be understood by considering the multiple targets against which NO has cytotoxic pathways (Fig. 9.3). Bacteria employ only a few mechanisms to defend themselves against multiple NO actions. Hence, the inability of bacteria to defend themselves against multiple NO pathways explains the rarity of bacterial resistance to NO.[5]

As a free radical, NO readily reacts with superoxide ($O_2^{-\bullet}$), the subproduct of bacterial respiration, forming the powerful oxidant peroxynitrite ($OONO^-$),[2,70] as follows:

$$^{\bullet}NO + O_2^{-\bullet} \rightarrow OONO^- \tag{9.7}$$

At high concentrations (a micromolar-millimolar range), NO exerts its antibacterial effect through the formation of reactive NOS, which includes peroxynitrite, dinitrogen trioxide (N_2O_3), and nitrogen dioxide (NO_2). The antibacterial effect is attributed to the reaction of reactive NOS with bacterial targets, including (1) nitrosative damage to DNA; (2) the inhibition of DNA repair enzymes; and (3) the modification of amino acid residues (such as cys, phe, trp, met, and tyr) in bacterial proteins (eg, the nitrosation of cysteine and the nitration of tyrosine residues in vital bacterial proteins, which lead to the dysfunction of these proteins); (4) lipid peroxidation; (5) the inactivation of important zinc metalloproteins, leading to the inhibition of bacterial cellular respiration; and (6) the inactivation of prosthetic groups of proteins, including heme and Fe-S clusters, leading to the removal of heme from protein and Fe depletion in bacteria.[2,3,70] In addition, NO reportedly increases the susceptibility of *Listeria monocytogenes* to toll-like, receptor-activated macrophages.[75] This stimulation is necessary for host cells to resist intracellular pathogens.

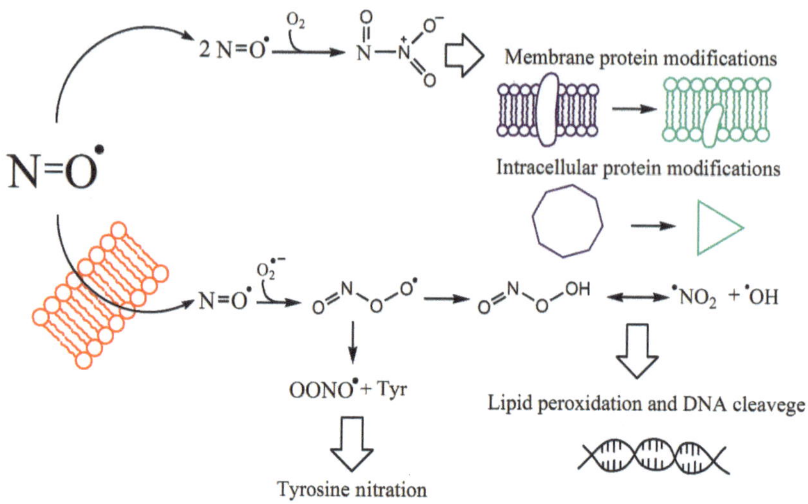

FIGURE 9.3

Schematic representation of the multiple pathways of toxicity activity for NO and its oxidation products in bacterial cells.

The diverse pathways through which NO exerts its antibacterial effects would require multiple and simultaneous mutations to occur in order for the bacteria to survive, which hinders the development of resistance.[50] As a consequence, there is no evidence of bacterial resistance to exogenous NO donors at present. However, it would be naïve to assume that bacteria absolutely cannot develop resistance to NO donors.

FINAL REMARKS

Conventional antibiotics have been used to kill or inhibit bacteria for many years. However, the effectiveness of antibiotics has been significantly reduced because of increased bacterial resistance to these drugs, leading to a growing global problem. In this context, there is concern about the use of antibiotics, and the development of new drugs or strategies to combat bacteria has wide appeal. Although the combination of antibiotics with nanomaterials was shown to be an interesting approach in the fight against resistant bacteria in the first stage, this approach has been hampered by growing bacterial resistance to conventional antibiotics as the active agent in nanomaterials. To circumvent this limitation, a promising, emerging strategy is the use of NO as a potent antibacterial agent. NO is synthesized endogenously in activated macrophages through the action of iNOs, and it plays a key role in host defense against pathogens, including antibiotic-resistant bacteria. According to the literature under review, gNO, NO donors and NO-releasing nanomaterials have been used successfully in recent years to combat antibiotic-resistant bacteria, and also to prevent and/or disperse biofilms. Combinations of NO donors with nanomaterials have shown the ability to potentiate the powerful antibacterial effects of NO against resistant bacteria. Moreover, in the case of biofilm-infected wounds, NO-releasing nanomaterials are able to decrease the wound burden and also promote and accelerate wound closure. NO appears to overcome common limitations of antibiotics—in particular, the prevention of biofilm formation and the capacity to disperse established biofilms. Because NO is known to display numerous simultaneous antibacterial mechanisms, the possibility that bacteria develop resistance to NO is low. It would be naïve to assume that bacteria will not develop resistance toward NO. However, the use of NO donors represents at least a promising strategy to combat resistant bacteria. Moreover, the combination of NO donors with antimicrobial agents should be further explored in the near future. In this context, there is hope that the findings reported in this chapter will inspire new and promising research in this field.

ACKNOWLEDGMENTS

Support was received from the FAPESP, CNPq, INOMAT (MCTI/CNPq), NanoBioss (MCTI), and the Brazilian Network on Nanotoxicology. The authors would like to acknowledge Nature Publishing Group Language Editing for revising the English-language usage in this manuscript.

REFERENCES

1. Jardeleza C, Rao S, Thierry B, Gjjar P, Vreugde S, Prestidge CA, et al. Liposome-encapsulated ISMN: a novel nitric oxide-based therapeutic agent against *Staphylococcus aureus* biofilms. *Plos One* 2014;**9**: e92117. Available from: http://dx.doi.org/10.1371/journal.pone.0092117.
2. Pelfrift RY, Friedman AJ. Nanotechnology as a therapeutic tool to combat microbial resistance. *Adv Drug Deliv Rev* 2013;**65**:1803–15.
3. Chouake J, Schairer D, Kutner A, Sanchez DA, Makdisi J, Blecher-Paz K, et al. Nitrosoglutathione generating nitric oxide nanoparticles as an improved strategy for combating *Pseudomonas aeruginosa*-infected wounds. *J Drugs Dermatol* 2012;**11**:1471–7.
4. Haddadin RN, Saleh SA, Mahmoud RA, Shebabi AA. Multiple drug resistance and strength of attachment to surfaces in *Pseudomonas aeruginosa* isolates. *Lett Appl Microbiol* 2010;**51**:48–53.
5. Heilman BJ, John JS, Oliver SRJ, Mascharak PK. Light-triggered eradication of *Acinetobacter baumannii* by means of NO delivery from a porous material with an entrapped metal nitrosyl. *J Am Chem Soc* 2012;**134**:11573–82.
6. Witte M, Barbul A. Role of nitric oxide in wound repair. *Am J Surg* 2002;**183**:406–12.
7. Friedman A, Friedman J. New biomaterials for the sustained release of nitric oxide: past, present and future. *Expert Opin Drug Deliv* 2009;**6**:1113–22.
8. Kafshgari MH, Cavallaro A, Delalat B, Harding FJ, McInnes SJP, Makila E, et al. Nitric oxide-releasing porous silicon nanoparticles. *Nanoscale Res Lett* 2014;**9**:333 http://www.nanoscalereslett.com/content/9/1/333
9. Bang CS, Kinnunen A, Karlsson M, Onnberg A, Soderquist B, Persson K. The antibacterial effect of nitric oxide against ESBL-producing uropathogenic *E. coli* is improved by combination with miconazole and polymyxin B nonapeptide. *BMC Microbiol* 2014;**14**:65 http://www.biomedcentral.com/1471-2180/14/65
10. Brisboi EJ, Bayliss J, Wu J, Major TC, Xi C, Wang SC, et al. Optimized polymeric film-based nitric oxide delivery inhibits bacterial growth in a mouse burn wound model. *Acta Biomater* 2014;**10**:4136–42.
11. Ghaffari A, Miller CC, McMullin B, Ghahary A. Potential application of gaseous nitric oxide as a topical antimicrobial agent. *Nitric Oxide* 2006;**14**:21–9.

12. Miller CC, Hergott CA, Rohan M, Arsenault-Mehta K, Döring G, Mehta S. Inhaled nitric oxide decreases the bacterial load in a rat model of. *J Cyst Fibros* 2013;**12**:817–20.
13. Yossef AG, Miller CC, Greenberg D. Inhalation of gaseous nitric oxide for treating respiratory diseases. 2015, US 20150044305.
14. Miller C, Miller M, McMullin B, Regev G, Serghides L, Kain K, et al. A phase I clinical study of inhaled nitric oxide in healthy adults. *J Cyst Fibros* 2012;**11**:324–31.
15. Ghaffari A, Jalili R, Ghaffari M, Miller C, Ghahary A. Efficacy of gaseous nitric oxide in the treatment of skin and soft tissue infections. *Wound Repair Regen* 2007;**15**:368–77.
16. Amadeu TP, Seabra AB, de Oliveira MG, Costa AMA. *S*-nitrosoglutathione-containing hydrogel accelerates rat cutaneous wound repair. *J Eur Acad Dermatol Venereol* 2007;**21**:629–37.
17. Amadeu TP, Seabra AB, de Oliveira MG, Costa AMA. Nitric Oxide donor improves healing if applied on inflammatory and proliferative phases. *J Surg Res* 2008;**149**:84–93.
18. Georgii JL, Amadeu TP, Seabra AB, de Oliveira MG, Costa AMA. Topical *S*-nitrosoglutathione-releasing hydrogel improves healing of rat ischaemic wounds. *J Tissue Eng Regen Med* 2011;**5**:612–19.
19. Schairer DO, Chouake JS, Nosanchuk JD, Friedman AJ. The potencial of nitric oxide releasing therapies as antimicrobial agentes. *Virulence* 2012;**3**:271–9.
20. Duong HTT, Jung K, Kutty SK, Agustina S, Adnan NNM, Basuki JS, et al. Nanoparticle (star polymer) delivery of nitric oxide effectively negates *Pseudomonas aeruginosa* biofilm formation. *Biomacromolecules* 2014;**15**:2583–9.
21. Miller MR, Megson IL. Recent developments in nitric oxide donor drugs. *Brit J Pharmacol* 2007;**151**:305–21.
22. Seabra AB, de Lima R, Calderon M. Nitric oxide releasing nanomaterials for cancer treatment: current status and perspectives. *Curr Top Med Chem* 2015;**15**:298–308.
23. Münzel T, Daiber A, Gori T. More answers to the still unresolved question of nitrate tolerance. *Eur Heart J* 2013;**34**:2666–73.
24. Kutner AJ, Friedman AJ. Use of nitric oxide nanoparticulate platform for the treatment of skin and soft tissue infections. *WIREs Nanomed Nanobi* 2013;**5**:502–14.
25. Carpenter AW, Slomberg DL, Rao KS, Schoenfisch MH. Influence of scaffold size on bactericidal activity of nitric oxide-releasing silica nanoparticles. *ACS Nano* 2011;**5**:7235–44.
26. Seabra AB, de Oliveira MG. Poly(vinyl alcohol) and poly(vinyl pyrrolidone) blended films for local nitric oxide release. *Biomaterials* 2004;**17**:3773–82.
27. Gardner PR, Gardner AM, Martin LA, Salzman AL. Nitric oxide dioxygenase: an enzymic function for flavohemoglobin. *Proc Natl Acad Sci USA* 1998;**95**:10378–83.
28. Idh J, Mekonnen M, Abate E, Wedajo W, Werngren J, Angeby KA, et al. Resistance to first-line anti-TB drugs is associated with reduced nitric oxide susceptibility in *Mycobacterium tuberculosis*. *PLoS ONE* 2012;**7**:e39891. Available from: http://dx.doi.org/10.1371/journal.pone.0039891
29. Jones-Carson J, Laughlin JR, Stewart AL, Voskuil MI, Vázquez-Torres A. Nitric oxide-dependent killing of aerobic, anaerobic and persistent *Burkholderia pseudomallei*. *Nitric Oxide* 2012;**27**:25–31.
30. Kishikawa H, Ebberyd A, Römling U, Brauner A, Lüthje P, Lundberg JP, et al. Control of pathogen growth and biofilm formation using a urinary catheter that release antimicrobial nitrogen oxides. *Free Radic Biol Med* 2013;**65**:1257–64.
31. Cariello AJ, Bispo PJM, de Souza PFG, Pignatari ACC, de Oliveira MG, Hofling-Lima AL. Bactericidal effect of *S*-nitrosothiols against clinical isolates from keratitis. *Clin Ophthalmol* 2012;**6**:107–14.
32. Watanabe K, Ishima Y, Akaike T, Sawa T, Kuroda T, Ogawa W, et al. *S*-nitrosated α-1-acid glycoprotein kills drug-resistant bacteria and aids survival in sepsis. *FASEB J* 2013;**27**:391–8.
33. Bátai I, Kerényi M, Tekeres M. The growth of bacteria in intravenous glyceryl trinitrate and sodium nitroprusside. *Anesth Analg* 1999;**89**:1570–2.
34. Gao S, Zhiguo L, Bond P. The concentration-determined and population-specific antimicrobial effects of free nitrous acid on *Pseudomonas aeruginosa* PAO1. *Appl Microbiol Biotechnol* 2015;**99**:2305–13.

35. Ormerod AD, Shah AAJ, Li H, Benjamin NB, Ferguson GP, Leifert C. An observational prospective study of topical acidified nitrite for killing methicillin-resistant *Staphylococcus aureus* (MRSA) in contaminated wounds. *BMC Res Notes* 2011;**4**:458–65.
36. Weller RB. Nitric oxide–containing nanoparticles as an antimicrobial agent and enhancer of wound healing. *J Invest Dermatol* 2009;**129**:2335–7.
37. Molina-González D, Alonso-Calleja C, Alonso-Hernando A, Capita R. Effect of sub-lethal concentrations of biocides on the susceptibility to antibiotics of multi-drug resistant *Salmonella enterica* strains. *Food Control* 2014;**40**:329–34.
38. Pijuan M, Ye L, Yuan Z. Free nitrous acid inhibition on the aerobic metabolism of poly-phosphate accumulating organisms. *Water Res* 2010;**44**:6063–72.
39. Nole KLB, Yim E, Keri JE. Probiotics and prebiotics in dermatology. *J Am Acad Dermatol* 2014;**71**:814–21.
40. Jones ML, Ganopolsky JG, Labbé A, Prakash S. A novel nitric oxide producing probiotic patch and its antimicrobial efficacy: preparation and in vitro analysis. *Appl Microbiol Biotechnol* 2010;**87**:509–16.
41. Jones ML, Ganopolsky JG, Labbé A, Gilardino M, Wahl C, Martoni C, et al. Novel nitric oxide producing probiotic wound healing patch. *Int Wound J* 2012;**9**:330–43.
42. Seabra AB, Martins D, Simões MMSG, da Silva R, Brocchi M, de Oliveira MG. Antibacterial nitric oxide-releasing polyester for the coating of blood-contacting artificial materials. *Artif Organs* 2010;**34**:E204–14.
43. de Oliveira MG, Shishido SM, Seabra AB, Morgon NH. Thermal stability of primary *S*-nitrosothiols: roles of autocatalysis and structural effects on the rate of nitric oxide release. *J Phys Chem A* 2002;**106**:8963–70.
44. Seabra AB, Durán N. Nitric oxide-releasing vehicles for biomedical applications. *J Mater Chem* 2010;**20**:1624–37.
45. Seabra AB, Marcato PD, de Paula LB, Duran N. New strategy for controlled release of nitric oxide. *J Nano Res* 2012;**20**:61–7.
46. Seabra AB, Duran N. Nanotechnology allied to nitric oxide release materials for dermatological applications. *Curr Nanosci* 2012;**8**:520–5.
47. Cardozo VF, Lancheros CAC, Narciso AM, Valereto ECS, Kobayashi RKT, Seabra AB, et al. Evaluation of antibacterial activity of nitric oxide-releasing polymeric particles against *Staphylococcus aureus* from bovine mastites. *Int J Pharm* 2014;**473**:20–9.
48. Seabra AB, Pasquoto T, Ferrarini ACF, Cruz M, Haddad PS, de Lima R. Preparation, characterization, cytotoxicity and genotoxicity evaluations of thiolated- and *S*-nitrosated superparamagnetic iron oxide nanoparticles: implications for cancer treatment. *Chem Res Toxicol* 2014;**27**:1207–18.
49. Seabra AB, Justo GZ, Haddad PS. State of the art, challenges and perspectives in the design of nitric oxide-releasing polymeric nanomaterials for biomedical applications. *Biotechnol Adv* 2015;**33**:1370–9.
50. Privett BJ, Broadnax AD, Bauman SJ, Riccio DA, Schoenfisch MH. Examination of bacterial resistance to exogenous nitric oxide. *Nitric Oxide* 2012;**26**:169–73.
51. Berezin AS, Lomkova EA, Shorik YA. Chitosan conjugates with biologically active compounds: design strategies, properties, and targeted drug delivery. *Russ Chem Bull* 2012;**61**:781–95.
52. Han F, Dong Y, Song AH, Yin R, Li SM. Alginate/chitosan based bi-layer composite membrane as potential sustained-release wound dressing containing ciprofloxacin hydrochloride. *Appl Surf Sci* 2014;**311**:626–34.
53. Huh AJ, Kwon YJ. Nanoantibiotics: a new paradigm for treating infectious diseases using nanomaterials in the antibiotics resistant era. *J Control Release* 2011;**156**:128–45.
54. Sun Y, Liu Y, Liu W, Lu C, Wang L. Chitosan microparticles ionically cross-linked with poly(γ-glutamic acid) as antimicrobial peptides and nitric oxide delivery systems. *Biochem Eng J* 2015;**95**:78–85.
55. Park J, Kim J, Singha K, Han D-K, Park H, Kim WJ. Nitric oxide integrated polyethylenimine-based tri-block copolymer for efficient antibacterial activity. *Biomaterials* 2013;**34**:8766–75.

56. Priya S, Nithya R, Berchmans S. *S*-Nitrosothiol tethered polymer hexagons: synthesis, characterisation and antibacterial effect. *J Mater Sci Mater Med* 2014;**25**:1–10.
57. Wold KA, Damadaran VB, Suazo LA, Bowen RA, Reynolds MM. Fabrication of biodegradable polymeric nanofibers with covalently attached NO donors. *ACS Appl Mater Interfaces* 2012;**4**:3022–30.
58. Sun B, Slomberg DL, Chudasama SL, Yuan L, Schoenfisch MH. Nitric oxide-releasing dendrimers as antibacterial agents. *Biomacromolecules* 2012;**13**:3343–54.
59. Howard MD, Hood ED, Zern B, Shuvaev VV, Grosser T, Muzykantov VR. Nanocarriers for vascular delivery of anti-inflammatory agents. *Annu Rev Pharmacol Toxicol* 2014;**54**:205–26.
60. Worley BV, Slomberg DL, Schoenfisch MH. Nitric oxide-releasing quaternary ammonium-modified poly (amidoamine) dendrimers as dual action antibacterial agents. *Bioconjug Chem* 2014;**25**:918–27.
61. Han G, Martinez LR, Mihu MR, Friedman AJ, Friedman JM, Nosanchuk JD. Nitric oxide releasing nanoparticles are therapeutic for *Staphylococcus aureus* abscesses in a murine model of infection. *Plos One* 2009;**4**:1–7.
62. Cabrales P, Zanini GM, Meays D, Frangos JA, Carvalho LJM. Murine cerebral malaria is associated with a vasospasm-like microcirculatory dysfunction, and survival upon rescue treatment is markedly increased by nimodipine. *Am J Pathol* 2010;**176**:1306–15.
63. Friedman AJ, Han G, Navati MS, Chacko M, Gunther L, Alfieri A, et al. Sustained release nitric oxide releasing nanoparticles: characterization of a novel delivery platform based on nitrite containing hydrogel/glass composites. *Nitric Oxide* 2008;**19**:12–20.
64. Fox S, Wilkinson TS, Wheatley PS, Xiao B, Morris RE, Sutherland A, et al. NO-loaded Zn^{2+}-exchanged zeolite materials: a potential bifunctional anti-bacterial strategy. *Acta Biomater* 2010;**6**:1515–21.
65. Neidrauer M, Ercan UK, Bhattacharyya A, Samuels J, Sedlak J, Trikha R, et al. Antimicrobial efficacy and wound-healing property of a topical ointment containing nitric-oxide-loaded zeolites. *J Med Microbiol* 2014;**63**:203–9.
66. Verma RK, Agrawal AK, Sing AK, Mohan M, Gupta A, Gupta P, et al. Inhalable microparticles of nitric oxide donors induce phagosome maturation and kill *Mycobacterium tuberculosis*. *Tuberculosis* 2013;**93**:412–17.
67. Lu Y, Slomberg DL, Schoenfisch MH. Nitric oxide-releasing chitosan oligosaccharides as antibacterial agents. *Biomaterials* 2014;**35**:1716–24.
68. Yepuri NR, Barraud N, Mohammadi NS, Kardak BG, Kjelleberg S, Rice SA, et al. Synthesis of cephalosporin-30-diazeniumdiolates: biofilm dispersing NO-donor prodrugs activated by b-lactamase. *Chem Commun* 2013;**49**:4791–3.
69. Barraud N, Storey MV, Moore ZP, Webb JS, Rice SA, Kjelleberg S. Nitric oxide-mediated dispersal in single- and multi-species biofilms of clinically and industrially relevant microorganisms. *Microbial Biotechnol* 2009;**2**:370–8.
70. Cai W, Wu J, Xi C, Meyerhoff ME. Diazeniumdiolate-doped poly(lactic-*co*-glycolic acid)-based nitric oxide releasing films as antibiofilm coatings. *Biomaterials* 2012;**33**:7933–44.
71. Romling U, Kjelleberg S, Normark S, Nyman L, Uhlin BE, Akerlund B. Microbial biofilm formation: a need to act. *J Intern Med* 2014;**276**:98–110.
72. Barraud N, Hasset DJ, Hwang S, Rice SA, Kjelleberg S, Webb JS. Involvement of nitric oxide in biofilm dispersal of *Pseudomonas aeruginosa*. *J Bacteriol* 2006;**188**:7344–53.
73. Slomberg DL, Lu Y, Broadnax AD, Hunter RA, Carpenter AW, Schoenfisch MH. Role of size and shape on biofilm eradication for nitric oxide-releasing silica Nanoparticles. *Appl Mater Interfaces* 2013;**5**:9322–9.
74. Sulemankhil I, Ganopolsky JG, Dieni CA, Dan AF, Jones ML, Prakash S. Prevention and treatment of virulent bacterial biofilms with an enzymatic nitric oxide-releasing dressing. *Antimicrob Agents Chemother* 2012;**56**:6095–103.
75. Cole C, Thomas S, Filak H, Henson PM, Lenz LL. Nitric oxide increases susceptibility of toll-like receptor-activated macrophages to spreading *Listeria monocytogenes*. *Immunity* 2012;**36**:807–20.

CHAPTER

CONTROLLING BACTERIAL ANTIBIOTIC RESISTANCE USING PLANT-DERIVED ANTIMICROBIALS

10

A. Upadhyay, D.P. Karumathil, I. Upadhyaya, V. Bhattaram and K. Venkitanarayanan
Department of Animal Science, University of Connecticut, Storrs, CT, United States

BACKGROUND

The discovery of antibiotics was one of the greatest medical victories in the 20th century that significantly reduced mortalities associated with infections. Since then, a number of novel antimicrobial compounds have been discovered, and many synthetic and semisynthetic antimicrobial drugs have been developed for controlling bacterial and fungal infections in humans and animals. However, the excessive and inappropriate use of antimicrobials in human medicine and animal operations has led to the emergence of antimicrobial-resistant bacteria. Today, the treatment of bacterial infections has again become a formidable challenge due to the rapid development of multidrug resistance (MDR) in microbes.[1–4] The situation is further exacerbated by the fact that development of effective antimicrobials is a very time-consuming, scientifically challenging, and costly endeavor. More than 20 novel classes of antibiotics were discovered between 1930 and 1962[5,6]; since then, however, only two new classes of antibiotics have been approved.[7–9] Analogous development of drug variants has kept pace with drug resistance for almost 20 years, but this resource is also fast dwindling, leading to significant public health concerns around the globe.[10]

The development and spread of antimicrobial resistance determinants in microbes is a complex problem that is fueled by many human practices in different fields. A significant correlation has been demonstrated between the emergence of antibiotic resistance in microbes and the increased use of antimicrobials in human and veterinary medicine, animal husbandry, agriculture, community environment, and food industry.[11,4] Therefore, there is a need for a coordinated multidisciplinary strategy for mitigating microbial antibiotic resistance.[12,13] The World Health Organization has published comprehensive recommendations to curb the emergence and spread of antibiotic-resistant bacteria by promoting the prudent use of antimicrobials in humans, agriculture, and food animals.[14] In addition, novel classes of compounds have to be discovered and screened for antimicrobial efficacy against pathogens to develop new antibiotics. Historically, natural compounds have been the source of most antibiotics. In the past two decades, research efforts have intensified to search for antimicrobials in various ecological niches such as deep ocean flora,[15] soil metagenome,[16] reptiles,[17] mammals,[18] and plants.[19,20]

Antibiotic Resistance. DOI: http://dx.doi.org/10.1016/B978-0-12-803642-6.00010-1

The aim of this chapter is to highlight the potential of plant-derived compounds as novel antimicrobials and discuss their efficacy in controlling the development of antibiotic resistance in microbes. The various fields where the application of plant compounds could lead to a reduction in bacterial antibiotic resistance are discussed. In addition, studies highlighting the diverse effects exerted by plant compounds on various factors that facilitate antibiotic resistance development are presented.

PLANT-DERIVED ANTIMICROBIALS

Plant-derived compounds represent an untapped source of safe, effective, and environmentally friendly antimicrobials. Plant extracts have traditionally being used as food preservatives, flavor enhancers, and dietary supplements to prevent food spoilage and improve health. In addition, plant compounds have been employed in complementary and herbal medicine to treat various diseases. The antimicrobial activity of several plant-derived compounds has been documented, and their antimicrobial components have been characterized.[20–22] The primary advantage of using plant-derived antimicrobials for medicinal purpose is that they do not exhibit any side effects often associated with synthetic chemicals.[23] Moreover, no reports of antimicrobial resistance toward these phytochemicals have been documented so far, probably due to their multiple mechanisms of action, which prevents the development of resistant microbes. Due to their nontoxic nature, low cost, and significant antimicrobial efficacy, several plant compounds are used as feed supplements in livestock and food animals, sanitizers and disinfectants in the food industry, antiseptics in veterinary medicine, and as sources of new drugs.

CHEMICAL CHARACTERISTICS

The majority of plant compounds are secondary metabolites and are produced as a result of reciprocal interactions between plants and microbes, where they contribute to the enhanced survival of plants in local environments and fending off pathogenic microorganisms.[24–26] The major groups of plant-derived antimicrobials include polyphenols, alkaloids, flavonoids, terpenoids, lectins, tannins, and saponins.[21,27,28]

MECHANISM OF ACTION

Plant-derived antimicrobials exert antimicrobial action against both Gram-positive and Gram-negative microorganisms. At bactericidal concentrations, plant compounds act primarily by disintegrating bacterial cell walls and cell membranes, leading to the leakage of cellular content, disruption of proton motive force, and efflux pump and enzyme dysfunction, eventually causing cell death.[20,29] The major antimicrobial effect of plant-derived antimicrobials comes from their hydrophobic components, which disrupt cell membrane proteins and associated structures and make them more permeable.[30,31] Also, some plant compounds may stimulate the growth of pseudomycelia, affecting the synthesis of bacterial structural components.[32]

Recent research has revealed that plant-derived antimicrobials at low or sublethal concentrations affect microbial virulence by modulating the expression of critical genes and proteins.[33–39]

In addition, plant compounds disrupt efflux pumps[40–43] and quorum-sensing signals that contribute to virulence in many organisms.[44–48] Since quorum sensing partially contributes to antibiotic-resistant phenotype in bacteria,[49,50] plant-derived antimicrobials that chemically inhibit or reduce quorum sensing represent a viable strategy to control the development of antibiotic resistance in microorganisms. However, this application of plant compounds is largely unexplored except for a few studies discussed in this chapter.

INCREASING THE SENSITIVITY OF DRUG-RESISTANT MICROBES

Plant-derived antimicrobials have been investigated for their efficacy in reversing drug-resistant phenotypes in pathogenic microbes and increasing their sensitivity to currently used antimicrobials.[51]

Enterobacteriaceae

Brehm-Stecher and Johnson[52] investigated the efficacy of several plant compounds in increasing bacterial sensitivity to antibiotics. These researchers observed that low concentrations of sesquiterpene (nerolidol, apritone, and bisabolol) increased the sensitivity of drug-resistant *Escherichia coli* to ciprofloxacin, clindamycin, vancomycin and tetracycline. Similarly, Dickson et al.[53] found that extracts from *Mezoneuron benthamianum*, *Securinega virosa*, and *Microglossa pyrifolia* increased the sensitivity of drug-resistant *Salmonella* sp., *Klebsiella* sp., and *Pseudomonas aeruginosa* to norfloxacin. In another study, geraniol, obtained from the *Helichrysum italicum* extract, improved the antimicrobial efficacy of quinolones, chloramphenicol, and β-lactams against MDR pathogens.[54] A similar synergistic effect was observed between antibiotics and compounds obtained from *Camellia sinensis*,[55] *Caesalpinia spinosa*,[56] *Scutellaria baicalensis*,[57] and the oil of *Croton zehntneri*.[58]

Acinetobacter baumannii

Among non-Enterobacteriaceae organisms, *Acinetobacter baumannii* has emerged as a major challenge to the modern health-care system. It is a Gram-negative, nosocomial pathogen that is resistant to the majority of current antibiotics.[59] The role of environmental persistence of *Acinetobacter* in the transmission of hospital-acquired infections is well recognized.[60–63] The pathogen is able to survive at a wide range of temperature, pH, and moisture levels, either as free-living planktonic cells or as biofilms.[64] Moreover, it utilizes a variety of carbon and energy sources for survival. These attributes contribute to *A. baumannii's* persistence and spread in the hospital setting.[65]

The mechanism of antimicrobial resistance in *A. baumannii* falls into three broad categories: (1) enzymatic inactivation of antimicrobials, (2) modification of antimicrobial targets, and (3) reduced access to antimicrobial targets due to membrane modifications.[66] Chemicals that can alter the aforementioned antimicrobial resistance mechanisms could modulate drug resistance in this pathogen. Recent studies have investigated the efficacy of plant-derived compounds in controlling various MDR pathogens, including *Acinetobacter* sp. The essential oil from *Achillea millefolium* subsp. *Millefolium* Afan (Asteraceae) was found to exert significant antimicrobial effect against several pathogens, including *Acinetobacter lwoffii*.[67] The antimicrobial activity was primarily attributed to eucalyptol, camphor, and borneol components in the oil. Similarly, methanol-based extract of the flower *Peltophorum pterocarpum* resulted in a complete inhibition of *A. baumannii* in vitro.[68] Polyphenols, carboxylic acids, coumarins, flavonoids, and tannins were the active antimicrobial chemicals in the extract. Miyasaki et al.[69] investigated the antimicrobial efficacy of herbal extracts against MDR *A. baumannii* using a direct inoculation method. They identified 18 herbal extracts that inhibited the growth of the pathogen, with the most potent compounds being norwogonin in *Scutellaria baicalensis*; ellagic acid in *Rosa rugosa*, chebulagic acid, chebulinic acid, corilagin, and terchebulin in *Terminalia chebula*. A similar antimicrobial efficacy against multidrug-resistant *A. baumannii* has been reported with other plant compounds, including extracts from *Origanum vulgare*[70] and green tea.[71] Gas chromatography mass spectrophotometric analysis of these compounds revealed that *O. vulgare* extract contained pulegone (68.59%), piperitone (7.8%), piperitenone (7.8%), 1, 8-cineole (1.3%), and carvacrol (1.6%) as the active antimicrobial components, whereas green tea extract primarily consisted of epigallocatechin-3-gallate. Follow-up mechanistic research showed that in general, epigallocatechin and polyphenols acted by forming complexes with proteins and polysaccharides, thereby inhibiting the action of important enzymes that contribute to the antimicrobial characteristics in the microbes.

In addition to testing the antimicrobial effect of plant compounds on MDR *A. baumannii*, recent studies have investigated the efficacy of plant compounds in modulating antibiotic sensitivity of *A. baumannii*. Chusri et al.[72] studied the effect of ethanol extracts of *Holarrhena antidysenterica* on the antibiotic sensitivity of both MDR and extensively drug-resistant *A. baumannii*. These researchers observed that the ethanol extracts were effective in increasing the sensitivity of the pathogens against novobiocin, potentially by weakening the outer membrane of the bacteria. In addition to their effect on bacterial membranes, plant compounds were found to modulate the transcription of antibiotic resistance genes. Karumathil et al.[73] investigated the effect of two plant compounds, *trans*-cinnamaldehyde and eugenol, in increasing the sensitivity of two clinical isolates of MDR *A. baumannii* to seven β-lactam antibiotics. The plant compounds significantly increased the sensitivity of the *A. baumannii* isolates to the antibiotics. Follow-up gene expression analysis revealed that both compounds were effective in downregulating the expression of genes coding for efflux pumps in *A. baumannii*.

Clostridium difficile

Clostridium difficile is another nosocomial pathogen that causes a toxin-mediated enteric disease in humans with high case fatalities. In excess of 300,000 cases of *C. difficile*–associated disease are reported annually in the United States, resulting in approximately $3 billion in health-care costs.[74,75] Most of the *C. difficile* outbreak isolates from North America have been reported to exhibit widespread resistance against clindamycin and moxifloxacin. Moreover, the antimicrobial resistance in North American isolates parallels rates of resistance reported from Europe and the Far East.[76] Very few studies have investigated the effect of plant compounds on *C. difficile* antibiotic resistance or virulence attributes. In a recent study, carvacrol and *trans*-cinnamaldehyde were found to reduce *C. difficile* toxin production and cytotoxicity in vitro.[77] Interestingly, cinnamon oil has also been reported to modulate the antibiotic resistance phenotype of *C. difficile* and enhance the activity of clindamycin to *C. difficile* in vitro.[78]

Methicillin-resistant Staphylococcus aureus

Methicillin-resistant *Staphylococcus aureus* (MRSA) is one of the most common and virulent strains of antibiotic-resistant *S. aureus*. In medical facilities, MRSA causes life-threatening bloodstream infections, pneumonia, and surgical site infections.[79] Several studies have tested the efficacy of plant compounds in inactivating MRSA. Nitta et al.[80] tested the antimicrobial activity of extracts from 181 species of tropical and subtropical plants against MRSA. The active extracts obtained from the barks of *Shorea hemsleyana* and roots of *Cyphostemma bainesii* significantly reduced MRSA counts in vitro. The active compounds were all identified as stilbene derivatives.

Dias et al.[79] investigated the efficacy of isothiocyanates from cruciferous plants in inactivating MRSA and observed a dose-dependent antibacterial effect against the pathogen. Benzyl-isothiocyanate was the most effective compound. In another study, Chovanova et al.[81] investigated the antibacterial and antibiotic resistance modifying the activity of crude extracts from Asteraceae and Lamiaceae plants and essential oils from *Salvia officinalis* and *Salvia sclarea* against methicillin-resistant *Staphylococcus*. The essential oils in combination with oxacillin were found to exert synergistic and additive effects against methicillin-resistant *Staphylococcus epidermidis*.

EFFECT ON DRUG-RESISTANCE MECHANISMS IN MICROBES

The efficacy of plant compounds in increasing the sensitivity of microbes to antibiotics could be due to their efficacy in modulating the drug-resistance strategies of microbes. Inhibition of efflux pumps,[82] modification of antibiotic target sites,[83] and enzymatic degradation of antimicrobials[84] are the major strategies employed by microbes to resist the action of antibiotics. Efflux pumps reduce cytoplasmic drug accumulation by pumping out compounds out of the inner membrane to the periplasmic space or directly to the external environment.[85] In Gram-positive microbes, the main efflux pumps belong to the multidrug and toxic compound extrusion family.[86] In contrast, in Gram-negative microbes, the resistance-nodulation-cell-division family, which forms an efficient channel between inner membrane, periplasmic space, and outer membrane, along with AcrAb-TolC pumps, helps in extruding antimicrobial compounds.[85,86]

Several studies have highlighted the efficacy of plant-derived antimicrobials in reducing the efflux pump activity of Gram-positive microbes and Gram-negative microbes.[40,41,42,87] In a recent study, five plant-derived compounds (namely, *trans*-cinnamaldehyde, carvacrol, thymol, eugenol, and β-resorcylic) acid were found to increase the sensitivity of *Salmonella* Typhimurium DT104 to five antibiotics.[88] In addition, *trans*-cinnamaldehyde and thymol at their subinhibitory concentrations downregulated the expression of genes encoding resistance to ampicillin (*blaPSE*), chloramphenicol (*floR*), streptomycin (*aadA2*), sulfamethoxazole (*sul1*), tetracycline (*tetG*), and invasion genes (*hilA, hilD, invF, fliZ,* and *ygiX*).[89] Since the mechanism of antimicrobial resistance in *Salmonella* Typhimurium DT104 is primarily mediated by interaction between specific antibiotics and the AcrAB-TolC efflux pump, the aforementioned plant compounds could be acting through modulation of these efflux pumps to increase the sensitivity of the pathogen to antibiotics.[43]

β-lactamase is an enzyme produced by drug-resistant microbes that imparts resistance against β-lactam antibiotics.[90] Several plant compounds that exert an inhibitory effect on β-lactamases have been identified.[91] *Garcinia lucida* and *Bridelia micrantha* are medical plants found in Cameroon, and extracts from these plants have been shown to exert significant anti-β-lactamase activity.[92] Similarly, antipenicillinase activity has been observed in epigallocatechin gallate, which was found to increase the sensitivity of *S. aureus* to penicillin and augment the antimicrobial efficacy of ampicillin and sulbactam against MRSA.[93]

APPLICATIONS OF PLANT-DERIVED COMPOUNDS IN COMBATING ANTIBIOTIC RESISTANCE IN MICROBES

Antibiotic resistance is a multifaceted problem primarily driven by the excessive use of antimicrobials in medicine, agriculture, and the food industry. Finding alternatives to antibiotics in these fields could reduce the selection pressure for resistance phenotypes in microbes, thereby controlling resistance development. The following discussion highlights the representative studies that prove the efficacy of plant compounds as potential alternatives for antibiotics.

HUMAN AND VETERINARY MEDICINE

Human and veterinary medicine are heavily dependent on antimicrobial chemotherapy for the treatment of microbial infections. However, with the increasing resistance of microbes, the use of antibiotics for effective treatment is becoming challenging by the day. There is a need to identify antimicrobial alternatives that can complement, reduce, or replace currently used antibiotics without affecting the health of the host. Plant-derived antimicrobials are a viable strategy due to their diverse pharmacological effects. Several plant compounds have shown efficacy in treating human and animal diseases. Tea tree oil, particularly its active component terpinen-4-ol, has been tested for treating bovine mastitis.[94] Moreover, due of its potent antifungal properties, tea tree oil high in terpinen-4-ol content has been developed for treatment of thrush in horses.[95] In addition to its antimicrobial effect, terpinen-4-ol promotes antiinflammatory cytokine production while inhibiting proinflammatory cytokine expression, thereby contributing to tissue healing.[94] Similar antiinflammatory properties have been observed with human monocytes[96,97] and macrophages.[98]

Shorea robusta resin is used for treating infected wounds and burns by tribes in India since ages. Several recent scientific studies have characterized the efficacy of this ethnomedicine using modern approaches. Yaseen et al.[99] observed that the extract of robusta resin healed incision and excision wounds faster than plain ointment and framycetin in rats. Histopathology revealed complete epithelization and new blood vessel formation in treatment groups. The active components of the resin were found to be bergenin and ursolic acid, and these compounds also exhibited significant wound-healing properties.[100] A similar wound-healing efficacy has been reported in extracts from *Pedilanthus tithymaloides* leaves,[101] the *Grewia tiliaefolia* tree,[102] and *Kalanchoe petitiana* leaves.[103]

Curcumin (diferuloylmethane) is another well-characterized, naturally occurring polyphenol that occurs as a primary constituent in turmeric isolated from rhizomes of *Curcuma longa*.[104] More than two decades of research has highlighted the pleiotropic nature of the biological effects of this compound, including antioxidant, antiinflammatory, hypoglycemic, anticancer, and antimicrobial properties. These attributes are the result of the modulatory effects on transcriptional factors, chemokines, tumor suppressor genes, and microRNAs.[104,105] Recent research has also highlighted the efficacy of curcumin in modulating antibiotic resistance in pathogens. Curcumin was found to reverse antibiotic resistance in MRSA[106] and *P. aeruginosa*[107] by enhancing membrane permeability, ATPase inhibition, and efflux pump inhibition, respectively.

APPLICATIONS IN COMMUNITY PRACTICES

Excessive use of antibiotics outside of the hospital setting is another important factor that drives the development of antimicrobial resistance in pathogens. MDR *S. aureus*, which was until recently considered a nosocomial pathogen, is now recognized as a common community-associated pathogen as well.[108,109] Similarly, the last decade has seen a rise in penicillin-resistant *Streptococcus pneumoniae*, largely due to the excessive use of antibiotics for respiratory tract infections outside the hospital setting. The use of synthetic, broad-spectrum antimicrobials such as triclosan in consumer products, including hand soap, toothpaste, cosmetics, and plastics has further accelerated antimicrobial resistance development in microbes through cross-resistance mechanisms.[110,111] In addition, these synthetic chemicals exert their own harmful effects in humans that include cancer.[112]

Plant extracts potentially represent a safer and natural alternative for antibiotics in the community environment. Several plant compounds are well known for their efficacy in preventing common infections for which the general population opts to use over-the-counter drugs or products with synthetic antimicrobials or antibiotics. A summary of such ailments, along with the efficacy of plant compounds, is discussed next.

Skin infections

Skin care is an essential component of good hygiene and infection control. It commonly involves washing using plain or antiseptic soap, application of alcohol-based products, and creams.[113] The common skin ailments include bacterial infections (eg, impetigo, folliculitis, ecthyma, and cellulitis), fungal and yeast infections (eg, dermatophytosis, candidiasis, and pityriasis), and viral infections (eg, herpes simplex and herpes zoster). Several scientific studies have shown the efficacy of essential oils and other plant extracts in treating skin infections. Tea tree oil is an example of essential oil obtained from a Australian native plant, *Melaleuca alternifolia*, which is extensively used in herbal medicine for treating infections of the skin, including acne vulgaris, seborrheic dermatitis, impetigo, psoriasis, and dermatophytosis.[114,115]

Anecdotal evidence exists for extracts obtained from olive, garlic, coconut, and honey for the treatment of impetigo, dermatophytosis, and scientific studies are currently validating the efficacy of these natural therapies.[115] Carmo et al.[116] tested the efficacy of essential oil of *Cymbopogon citratus* (DC) in treating pityriasis versicolor infection in humans. The essential oil was found to be significantly effective in treating the condition in clinical trials. Eucalyptus oil is another example of a plant extract that has broad-spectrum antibacterial, antifungal, and antiviral properties.[117] Application of such plant extracts for the treatment of common skin infections could help in replacing or reducing antimicrobial usage and control the development of antimicrobial resistance in related microbes.

Oral hygiene

Oral hygiene is another essential component of overall health and wellbeing. Two major oral diseases (namely, dental caries, and periodontal disease) are widespread in the human population across the globe.[118,119] Plant-derived compounds have been investigated for their efficacy in improving oral hygiene. Oral mouthwashes containing essential oils have shown to reduce halitosis,[120] gingivitis,[121,122] and plaque development[123] in humans. Forrer et al.[124] tested the efficacy of alpha-bisabolol and tea tree oil against *Solobacterium moorei*, a Gram-positive bacterium associated with halitosis. The two plant compounds were found to be effective in reducing the bacterium. Tea tree oil has been found to be effective against other halitosis causing bacteria, including *Streptococcus* sp. *Actinomyces* sp., and *Prevotella* sp.[125] In another study, 0.5% tea and 2% neem mouthwashes were found to exhibit antiplaque and antigingivitis activity over a 3-week experimental period.[126] Cinnamon oil and clove oil also exhibit significant antimicrobial activity against dental carie–causing organisms, including *Streptococcus mutans*.[127] These results suggest that plant-derived compounds with proven efficacy against pathogenic oral microorganisms could be incorporated into dental health programs to reduce the usage of antibiotics, as combination therapies, and to improve oral hygiene.

ROLE IN AGRICULTURE

Food animals

The past two decades have seen an unprecedented growth in the global demand for animal protein.[128] At present, the global biomass of animals raised for food exceeds the total biomass of humans.[129] To meet this demand, the food industry has adopted highly intensive, vertically integrated, and cost-efficient livestock production systems. Because these systems exert tremendous stress on animals, antibiotics are extensively used in food animals reared in such production systems to prevent disease, maintain overall health, and promote growth. Currently, global antimicrobial usage by animals is approximately double that in humans.[130] In the United States, the use of antimicrobials in the food animal industry accounts for approximately 80% of annual antimicrobial usage.[131] The majority of antimicrobials are administered prophylactically at subtherapeutic levels, thereby creating a conducive environment for the development and spread of antimicrobial resistance microbes in food animals.[132] A substantial fraction of the antimicrobials used in food animals overlap with the antimicrobials used in human medicine during surgeries, organ transplantation, and treatment of infections,[133] potentially leading to major public health implications.

Studies have shown a close association between the prevalence of antibiotic resistance microbes and administration of antibiotics in animals.[134,135] A recent study from European countries, including Norway, Sweden, Denmark, Austria, Switzerland, the Netherlands, and Belgium, revealed a significant correlation between the consumption of eight classes of antimicrobials (ie, fluoroquinolones, streptomycin, amphenicols, gentamicin, tetracycline, sulfonamides, aminopenicillins, and cephalosporins) and the prevalence of resistant commensal *E. coli* in poultry, pigs, and cattle.[136]

Plant-derived antimicrobials are being tested as alternatives for antibiotics for both growth promotion and disease prevention in food animals. Several studies have shown the efficacy of in-feed supplementation of plant compounds such as *trans*-cinnamaldehyde, eugenol, carvacrol, and thymol in reducing the colonization of foodborne pathogens, including *Salmonella* Enteritidis and *Campylobacter* sp. in layer and broiler chicken.[137–141] Similar results have been obtained with other foodborne pathogens. For example, purple prairie clover (*Dalea purpurea* Vent.) was effective in reducing the fecal shedding of *E. coli* O157:H7 in lambs.[142] Similarly, the essential oils thymol, carvacrol, *trans*-cinnamaldehyde, and capsaicin were found to be effective in decreasing the colonization of *Clostridium perfringens* and pathogenic *E. coli* in broilers.[143–145]

Salmonella enterica serovar Heidelberg is one of the most commonly detected serovars in swine and poultry. Strains resistant to multiple antibiotics, including streptomycin, kanamycin, ceftriaxone, ciprofloxacin, and other β-lactam antibiotics, have been isolated from poultry meat, thereby raising concerns on poultry products transferring MDR *Salmonella* to humans. Plant compounds have been found effective in reducing colonization of drug-resistant strains of *Salmonella* Heidelberg. The administration of blends of essential oil (carvacrol, thymol, eucalyptol, and lemon) in drinking water at (0.05%) was found to reduce the cecal colonization of *Salmonella* Heidelberg in broilers.[146] Moreover, the essential oil blend increased weight gain and improved feed conversion ratio in the birds. Similar results were obtained with blend of thymol and cinnamaldehyde.[147] Interestingly, these researchers also observed a decrease in horizontal transfer of the pathogen between birds, which plays a crucial role in spread and survival of pathogens.

In addition to their efficacy in reducing pathogen colonization, survival, or both in food animals, plant compounds such as β-resorcylic acid, caprylic acid, sodium caprylate, and *trans*-cinnamaldehyde have shown efficacy in reducing pathogens on cattle hide,[148] in cattle drinking water,[149–151] and on beef.[152] In another study, Wells et al.[153] reported that the addition of 0.5% *trans*-cinnamic acid and 0.5% para-coumaric acid reduced the survival of *E. coli* O157:H7 in bovine feces. These studies suggest that plant compounds can be used to control pathogens in food animals and the farm environment.

Fresh produce

Antibiotics have been used for more than five decades to control bacterial diseases of fruits, vegetables, and ornamental plants.[154] Streptomycin is commonly used in several countries for this purpose, whereas other drugs such as oxytetracycline, oxolinic acid, and gentamicin are allowed in only a few countries. The majority of antibiotics are used prophylactically as springtime sprays to suppress pathogen growth on the surfaces of flowers and leaves.[155] In the United States, this amounts to approximately 0.5–1% of the total antibiotic usage.[155]

The role of antibiotic usage on plants in the development of antibiotic-resistant microbes is still a subject of debate due to lack of conclusive data. Although there have been reports of emergence of streptomycin-resistant strains in plant pathogens such as *Erwinia amylovora*, *Pseudomonas* spp., and *Xanthomonas campestris*,[154] recent research suggests that the use of antibiotics such as streptomycin,[156] gentamicin, and oxytetracyline[157] on field-grown plants did not affect the abundance of resistant bacteria, resistance genes, and other mobile resistant determinants in the microbial community. This could potentially be due to the effect of biotic and abiotic factors in inactivating the applied biocides in soil and other environmental conditions before it could significantly contribute toward resistant development in resident flora.[156,157] On the other hand, marine sediment bacteria [158,159] and bacteria from fresh water aquaculture effluents[160] have been found to harbor antibiotic resistance genes similar to those observed in MDR human pathogens, thereby raising concerns for the aquaculture industry. Nevertheless, since similar biological cause-and-effect rules apply to antimicrobial resistance development across ecological niches, conclusive evidence highlighting the role of antibiotics use on plants contributing to resistance could emerge in the future.

To decrease or reverse the risks associated with the development of MDR in plantborne pathogens, applications of alternative antimicrobials are warranted either as replacements or as complements to currently used drugs. Ample research evidence exists highlighting the efficacy of plant compounds in inactivating pathogens on plants and fresh produce. Plant-derived compounds such as cinnamon leaf oil, cinnamaldehyde, carvacrol, thymol, and lemongrass oil have been found to inactivate foodborne pathogens, including *Salmonella*, *Listeria monocytogenes,* and *E. coli* O157:H7 on organic leafy greens, lettuce, and cantaloupes when used as antimicrobial spray or dip treatments.[161–164]

Recent research has also addressed the application challenges that arise due the hydrophobic nature of essential oils. Bhargava et al.[165] investigated the efficacy of oregano oil nanoemulsions prepared from food-grade emulsifiers for controlling foodborne bacteria on fresh produce. Dipping in 0.05% and 0.1% nanoemulsions for 1 min reduced pathogen counts (*Salmonella* Typhimurium, *E. coli* O157:H7, and *L. monocytogenes*) by at least 2 and 3 log colony-forming units (CFU)/g, respectively. Antimicrobial coating and films containing essential oils have also proven to be an effective strategy in inactivating foodborne pathogens and improving the shelf life of fresh produce, such as strawberries,[166,167] cantaloupes,[164] and mangoes.[168,169]

Applications in agriculture soil

Agriculture ecosystems such as soil are known to transmit bacterial pathogens such as *Salmonella* and *L. monocytogenes* to the food chain through contamination of crops and fresh produce. Farmers traditionally rely on good agriculture practices and conventional soil amendment strategies to reduce pathogenic microbes in soil. The judicious application of livestock manure has the potential to increase the levels of plant nutrients and enhance soil microbial biomass, activity, and diversity, along with the inactivation of pathogens.[170–172] However, recent metagenomic explorations have identified manure as a reservoir of resistant bacteria and resistant genes on mobile genetic elements such as broad-host range plasmids and integrons.[173,174] Gao et al.[175] isolated extended spectrum β-lactamase producing *E. coli* from pig manure, compost, and soil samples. Similar MDR strains have been isolated by other researchers from pigs,[176,177] poultry,[178] and dairy farm soil.[179]

Plant extracts are potential candidates for natural amendment of soil and manure. Plant compounds have been used for inactivating fungal pathogens in soil[180,181]; however, only a few studies have investigated their efficacy as soil/manure amendments to inactivate bacterial pathogens for future application in the agricultural fields. Yossa et al.[182] tested the efficacy of cinnamaldehyde and eugenol in inactivating *Salmonella* in organic soil. They observed that the plant compounds at concentrations ranging from 0.5% to 2% were very effective in inactivating *Salmonella* serovars in soil as early as the first day of application. In a recent study, Wells et al.[183] evaluated the efficacy of the essential oils thymol, geraniol, glydox, linalool, pine oil, plinol, and terpineol for controlling pathogen levels (ie, coliforms, *E. coli* O157:H7, and *Salmonella*) in manure slurry. Thymol, linalool, pine oil, plinol, and terpineol were found to be the most effective compounds; they reduced the pathogen levels in slurry by at least 2 log CFU compared to control samples. Moreover, these compounds were effective in reducing the prevalence of *L. monocytogenes*, *E. coli* O157:H7, and *Campylobacter* in feedlot surfaces to various degrees.

REDUCING SPREAD OF ANTIBIOTIC RESISTANCE DETERMINANTS IN THE ENVIRONMENT

Antimicrobial resistance in human pathogens is linked to resistance reservoirs present in farm animals and in the wider environment. Human pathogens represent only a small fraction of the vast diversity of bacteria that are exposed to antibiotics.[184] These include the microbiota of farm animals, as well as a large number present in environments such as soil[185] activated sludge,[186] wastewater effluent plants,[187] and lakes.[188] While significant research is required to test the application of plant compounds in these niches, adequate evidence exists concerning their efficacy against bacterial biofilms in the environment. Bacterial biofilms provide microbes with better survival opportunities through enhanced resistance toward environmental stresses and antimicrobials.[189] Therefore, this mode of microbial life deserves attention in the process of developing novel strategies to combat drug-resistant microbes.

A majority of MDR organisms, including MRSA,[190] *A. baumannii*,[191] and *Klebsiella pneumoniae*[192] are strong biofilm formers. The biofilms facilitate the persistence of these pathogens in the hospital environment. Interestingly, extracellular DNA is a crucial component in the biofilm matrix of many pathogens that drives the formation of biofilms in pathogens such as *L. monocytogenes*,[193] *Vibrio cholerae*,[194] *Mycobacterium*,[195] and *A. baumannii*.[196] Since extracellular DNA represents an important mechanism of horizontal gene transfer in bacteria,[197,198] biofilms represent a potential hot spot for the genetic exchange of antimicrobial resistance genes. Thus, controlling pathogen biofilms could lead to potential reduction in antibiotic resistance development. Plant compounds have been shown to exhibit significant antibiofilm efficacy against major nosocomial pathogens, including *S. aureus*,[199–201] *P. aeruginosa*,[202] *K. pneumoniae*,[203–205] *A. baumannii*,[206] and foodborne pathogens such as *Cronobacter sakazakii*,[207] *E. coli* O157:H7,[208–210] and *L. monocytogenes*.[36] Thus, plant compounds have the potential to be used as eco-friendly biosanitizers for controlling pathogens in high-risk environments such as hospitals, farms, or food-processing areas, thereby eliminating infection risk and the development of antibiotic resistance.

CONCLUSION AND FUTURE RESEARCH DIRECTIONS

The studies discussed in this chapter collectively highlight the potential of using plant-derived antimicrobials as a promising strategy for controlling antimicrobial resistance development in microorganisms. Although their safety, pleiotropic biological effects, and suitability for diverse applications favor the usage of plant compounds, low water solubility and bioavailability, as well as a lack of characterization, limit their usage as alternatives for antibiotics. Further studies are needed to better characterize the pharmacodynamics and pharmacokinetic properties of plant compounds for developing effective, natural, and safe antimicrobials.

REFERENCES

1. Cantas L, Shah SQ, Cavaco LM, Manaia CM, Walsh F, Popowska M, et al. A brief multi-disciplinary review on antimicrobial resistance in medicine and its linkage to the global environmental microbiota. *Front Microbiol* 2013;**4**:96.
2. Pallett A, Hand K. Complicated urinary tract infections: practical solutions for the treatment of multiresistant Gram-negative bacteria. *J Antimicrob Chemother* 2010;**65**(Suppl. 3):iii25–33.
3. Van Boeckel TP, Gandra S, Ashok A, Caudron Q, Grenfell BT, Levin SA, et al. Global antibiotic consumption 2000 to 2010: an analysis of national pharmaceutical sales data. *Lancet Infect Dis* 2014;**14**:742–50.

4. Van Boeckel TP, Brower C, Gilbert M, Grenfell BT, Levin SA, Robinson TP, et al. Global trends in antimicrobial use in food animals. *Proc Natl Acad Sci USA* 2015;**112**:5649–54.
5. Coates A, Hu Y, Bax R, Page C. The future challenges facing the development of new antimicrobial drugs. *Nat Rev Drug Discov* 2002;**1**:895–910.
6. Powers JH. Antimicrobial drug development–the past, the present, and the future. *Clin Microbiol Infect* 2004;**10**(Suppl. 4):23–31.
7. Butler MS, Buss AD. Natural products-the future scaffolds for novel antibiotics? *Biochem Pharmacol* 2006;**71**:919–29.
8. Hair PI, Keam SJ. Daptomycin: a review of its use in the management of complicated skin and soft-tissue infections and *Staphylococcus aureus* bacteraemia. *Drugs* 2007;**67**:1483–512.
9. Zappia G, Menendez P, Monache GD, Misiti D, Nevola L, Botta B. The contribution of oxazolidinone frame to the biological activity of pharmaceutical drugs and natural products. *Mini Rev Med Chem* 2007;**7**:389–409.
10. Coates AR, Halls G, Hu Y. Novel classes of antibiotics or more of the same? *Br J Pharmacol* 2011;**163**:184–94.
11. Cheng AC, Turnidge J, Collignon P, Looke D, Barton M, Gottlieb T. Control of fluoroquinolone resistance through successful regulation, Australia. *Emerg Infect Dis* 2012;**18**:1453–60.
12. Serrano PH. Responsible use of antibiotics in aquaculture. In: *FAO Fisheries technical paper 469*. Rome: United Nations; 2005. Available from <ftp://ftp.fao.org/docrep/fao/009/a0282e/a0282e00.pdf>.
13. Smith PR, Le Breton A, Horsberg TE, Corsin F. Guidelines for antimicrobial use in aquaculture. In: Guardabassi L, Jensen LB, Kruse H, editors. *Guide to antimicrobial use in animals*. Oxford: Blackwell Publishing Ltd; 2009. p. 207–18.
14. World Health Organization. *The evolving threat of antimicrobial resistance: options for action*. Geneva: WHO; 2012.
15. Riedlinger J, Reicke A, Zahner H, Krismer B, Bull AT, Maldonado LA, et al. Abyssomicins, inhibitors of the para-aminobenzoic acid pathway produced by the marine verrucosispora strain AB-18-032. *J Antibiot (Tokyo)* 2004;**57**:271–9.
16. Daniel R. The soil metagenome—a rich resource for the discovery of novel natural products. *Curr Opin Biotechnol* 2004;**15**:199–204.
17. Wang Y, Hong J, Liu X, Yang H, Liu R, Wu J, et al. Snake cathelicidin from *Bungarus fasciatus* is a potent peptide antibiotics. *PLoS One* 2008;**3**:e3217.
18. Flores-Villasenor H, Canizalez-Roman A, Reyes-Lopez M, Nazmi K, de la Garza M, Zazueta-Beltran J, et al. Bactericidal effect of bovine lactoferrin, LFcin, LFampin and LFchimera on antibiotic-resistant *Staphylococcus aureus* and *Escherichia coli*. *Biometals* 2010;**23**:569–78.
19. Mitscher LA, Drake S, Gollapudi SR, Okwute SK. A modern look at folkloric use of anti-infective agents. *J Nat Prod* 1987;**50**:1025–40.
20. Burt S. Essential oils: their antibacterial properties and potential applications in foods—a review. *Int J Food Microbiol* 2004;**94**:223–53.
21. Cowan MM. Plant products as antimicrobial agents. *Clin Microbiol Rev* 1999;**12**:564–82.
22. Savoia D. Plant-derived antimicrobial compounds: alternatives to antibiotics. *Future Microbiol* 2012;**7**:979–90.
23. Van Wyk BE, Gericke N. *People's plants*. Pretoria: Briza Publications; 2000.
24. Reichling J. Plant-Microbe Interactions and Secondary Metabolites with Antibacterial, Antifungal and Antiviral Properties. In: Wink M, editor. *Annual Plant Reviews Volume 39: Functions and Biotechnology of Plant Secondary Metabolites*. 2nd ed. Oxford, UK: Wiley-Blackwell; 2010. Available from: http://dx.doi.org/10.1002/9781444318876.ch4.
25. Harborne JB. *Introduction to ecological biochemistry*. 4th ed. London: Academic press; 1993.

26. Kennedy DO, Wightman EL. Herbal extracts and phytochemicals: plant secondary metabolites and the enhancement of human brain function. *Adv Nutr* 2011;**2**:32–50.
27. Geissman TA. Flavonoid compounds, tannins, lignins and related compounds. In: Florkin M, Stotz EH, editors. *Pyrrole pigments, isoprenoid compounds and phenolic plant constituents*. New York, NY: Elsevier; 1963. p. 265.
28. Das K, Tiwari R, Shrivastava D. Techniques for evaluation of medicinal plant products as antimicrobial agent: current methods and future trends. *J Med Plants Res* 2010;**4**:104–11.
28. Lewis K, Ausubel FM. Prospects for plant-derived antibacterials. *Nat Biotechnol* 2006;**24**:1504–7.
30. Sikkema J, de Bont JA, Poolman B. Interactions of cyclic hydrocarbons with biological membranes. *J Biol Chem* 1994;**269**:8022–8.
31. Knobloch K, Pauli A, Iberl B, Weigand H, Weis N. Antibacterial and antifungal properties of essential oil components. *J Essent Oil Res* 1989;**1**:119–28.
32. Conner DE, Beuchat LR. Effects of essential oils from plants on growth of food spoilage yeasts. *J Food Sci* 1984;**49**:429–34.
33. Goh EB, Yim G, Tsui W, McClure J, Surette MG, Davies J. Transcriptional modulation of bacterial gene expression by subinhibitory concentrations of antibiotics. *Proc Natl Acad Sci USA* 2002; **99**:17025–30.
34. Tsui WH, Yim G, Wang HH, McClure JE, Surette MG, Davies J. Dual effects of MLS antibiotics: transcriptional modulation and interactions on the ribosome. *Chem Biol* 2004;**11**:1307–16.
35. Upadhyay A, Johny AK, Amalaradjou MA, Ananda Baskaran S, Kim KS, Venkitanarayanan K. Plant-derived antimicrobials reduce *Listeria monocytogenes* virulence factors in vitro, and down-regulate expression of virulence genes. *Int J Food Microbiol* 2012;**157**:88–94.
36. Upadhyay A, Upadhyaya I, Kollanoor-Johny A, Venkitanarayanan K. Antibiofilm effect of plant derived antimicrobials on *Listeria monocytogenes*. *Food Microbiol* 2013;**36**:79–89.
37. Azizkhani M, Misaghi A, Basti AA, Gandomi H, Hosseini H. Effects of *Zataria multiflora boiss*. essential oil on growth and gene expression of enterotoxins A, C and E in *Staphylococcus aureus* ATCC 29213. *Int J Food Microbiol* 2013;**163**:159–65.
38. Qiu J, Wang D, Xiang H, et al. Subinhibitory concentrations of thymol reduce enterotoxins A and B and alpha-hemolysin production in *Staphylococcus aureus* isolates. *PLoS One* 2010;**5**:e9736.
39. Qiu J, Feng H, Lu J, Xiang H, Wang D, Dong J, et al. Eugenol reduces the expression of virulence-related exoproteins in *Staphylococcus aureus*. *Appl Environ Microbiol* 2010;**76**:5846–51.
40. Cherigo L, Pereda-Miranda R, Fragoso-Serrano M, Jacobo-Herrera N, Kaatz GW, Gibbons S. Inhibitors of bacterial multidrug efflux pumps from the resin glycosides of ipomoea murucoides. *J Nat Prod* 2008;**71**:1037–45.
41. Kumar A, Khan IA, Koul S, Koul JL, Taneja SC, Ali L, et al. Novel structural analogues of piperine as inhibitors of the NorA efflux pump of *Staphylococcus aureus*. *J Antimicrob Chemother* 2008;**61**:1270–6.
42. Holler JG, Christensen SB, Slotved HC, Rasmussen HB, Guzman A, Olsen CE, et al. Novel inhibitory activity of the *Staphylococcus aureus* NorA efflux pump by a kaempferol rhamnoside isolated from *Persea lingue* nees. *J Antimicrob Chemother* 2012;**67**:1138–44.
43. Quinn T, O'Mahony R, Baird AW, Drudy D, Whyte P, Fanning S. Multi-drug resistance in *Salmonella enterica*: efflux mechanisms and their relationships with the development of chromosomal resistance gene clusters. *Curr Drug Targets* 2006;**7**:849–60.
44. Koh CL, Sam CK, Yin WF, Tan LY, Krishnan T, Chong YM, et al. Plant-derived natural products as sources of anti-quorum sensing compounds. *Sensors (Basel)* 2013;**13**:6217–28.
45. Persson T, Hansen TH, Rasmussen TB, Skinderso ME, Givskov M, Nielsen J. Rational design and synthesis of new quorum-sensing inhibitors derived from acylated homoserine lactones and natural products from garlic. *Org Biomol Chem* 2005;**3**:253–62.

46. Rasmussen TB, Bjarnsholt T, Skindersoe ME, Hentzer M, Kristoffersen P, Kote M, et al. Screening for quorum-sensing inhibitors (QSI) by use of a novel genetic system, the QSI selector. *J Bacteriol* 2005; **187**:1799–814.
47. Rasmussen TB, Skindersoe ME, Bjarnsholt T, Phipps Rk, Chritensen KB, Jensen PO, et al. Identity and effects of quorum-sensing inhibitors produced by penicillium species. *Microbiol* 2005;**151**:1325–40.
48. Welsh MA, Eibergen NR, Moore JD, Blackwell HE. Small molecule disruption of quorum sensing cross-regulation in *Pseudomonas aeruginosa* causes major and unexpected alterations to virulence phenotypes. *J Am Chem Soc* 2015;**137**:1510–19.
49. Gerdt JP, Blackwell HE. Competition studies confirm two major barriers that can preclude the spread of resistance to quorum-sensing inhibitors in bacteria. *ACS Chem Biol* 2014;**9**:2291–9.
50. Marshall J. Quorum sensing. *Proc Natl Acad Sci USA* 2013;**110**:2690.
51. Aiyegoro O, Okoh A. Use of bioactive plant products in combination with standard antibiotics: implications in antimicrobial chemotherapy. *J Med Plants Res* 2009;**3**:1147–52.
52. Brehm-Stecher BF, Johnson EA. Sensitization of *Staphylococcus aureus* and *Escherichia coli* to antibiotics by the sesquiterpenoids nerolidol, farnesol, bisabolol, and apritone. *Antimicrob Agents Chemother* 2003;**47**:3357–60.
53. Dickson RA, Houghton PJ, Hylands PJ, Gibbons S. Antimicrobial, resistance-modifying effects, antioxidant and free radical scavenging activities of *Mezoneuron benthamianum* baill., *Securinega virosa* roxb. &Wlld. and *Microglossa pyrifolia* lam. *Phytother Res* 2006;**20**:41–5.
54. Lorenzi V, Muselli A, Bernardini AF, Berti L, Pages JM, Amarai L, et al. Geraniol restores antibiotic activities against multidrug-resistant isolates from Gram-negative species. *Antimicrob Agents Chemother* 2009;**53**:2209–11.
55. Aqil F, Khan MS, Owais M, Ahmad I. Effect of certain bioactive plant extracts on clinical isolates of beta-lactamase producing methicillin resistant *Staphylococcus aureus*. *J Basic Microbiol* 2005;**45**:106–14.
56. Kondo K, Takaishi Y, Shibata H, Higuti T. ILSMRs (intensifier of beta-lactam-susceptibility in methicillin-resistant *Staphylococcus aureus*) from tara [*Caesalpinia spinosa* (molina) kuntze]. *Phytomedicine* 2006;**13**:209–12.
57. Chan BC, Ip M, Lau CB, Lui SL, Jolivalt C, Ganem-Elbaz C, et al. Synergistic effects of baicalein with ciprofloxacin against NorA over-expressed methicillin-resistant *Staphylococcus aureus* (MRSA) and inhibition of MRSA pyruvate kinase. *J Ethnopharmacol* 2011;**137**:767–73.
58. Rodrigues FF, Costa JG, Coutinho HD. Synergy effects of the antibiotics gentamicin and the essential oil of *Croton zehntneri*. *Phytomedicine* 2009;**16**:1052–5.
59. Sacha P, Wieczorek P, Ojdana D, Czaban S, Klosowska W, Jurczak A, et al. Susceptibility, phenotypes of resistance, and extended-spectrum β-lactamases in *Acinetobacter baumannii* strains. *Folia Histochem Cytobiol* 2012;**50**:46–51.
60. Fournier PE, Richet H. The epidemiology and control of *Acinetobacter baumannii* in health care facilities. *Clin Infect Dis* 2006;**42**:692–9.
61. Van Looveren M, Goossens H. Antimicrobial resistance of *Acinetobacter* spp. in Europe. *Clin Microbiol Infect* 2004;**10**:684–704.
62. Raka L, Mulliqi-Osmani G, Begolli L, Kurti A, Lila G, Bajrami R, et al. Acinetobacter. In: Basak S, editor. *Infection control*. In Tech, ISBN: 978-953-51-1145-0, http://dx.doi.org/10.5772/55618. Available from: <http://www.intechopen.com/books/infection-control/acinetobacter>.
63. Aygun G, Demirkiran O, Utku T, Mete B, Urkmez S, Yilmaz M, et al. Environmental contamination during a carbapenem-resistant *Acinetobacter baumannii* outbreak in an intensive care unit. *J Hosp Infect* 2002;**52**:259–62.
64. Espinal P, Marti S, Vila J. Effect of biofilm formation on the survival of *Acinetobacter baumannii* on dry surfaces. *J Hosp Infect* 2012;**80**:56–60.

65. Abbo A, Navon-Venezia S, Hammer-Muntz O, Krichali T, Siegman-Igra Y, Carmeli Y. Multidrug-resistant *Acinetobacter baumannii. Emerg Infect Dis* 2005;**11**:22–9.
66. Manchanda V, Sanchaita S, Singh N. Multidrug resistant *Acinetobacter. J Glob Infect Dis* 2010;**2**:291–304.
67. Candan F, Unlu M, Tepe B, Daferera D, Polissiou M, Sokmen A, et al. Antioxidant and antimicrobial activity of the essential oil and methanol extracts of *Achillea millefolium* subsp. millefolium Afan. (Asteraceae). *J Ethnopharmacol* 2003;**87**:215–20.
68. Sukumaran S, Kiruba S, Mahesh M, Nisha SR, Miller PZ, Ben CP, et al. Phytochemical constituents and antibacterial efficacy of the flowers of *Peltophorum pterocarpum* (DC.) baker ex heyne. *Asian Pac J Trop Med* 2011;**4**:735–8.
69. Miyasaki Y, Rabenstein JD, Rhea J, Crouch ML, Mocek UM, Kittell PE, et al. Isolation and characterization of antimicrobial compounds in plant extracts against multidrug-resistant *Acinetobacter baumannii. PLoS One* 2013;**8**:e61594.
70. Saghi H, Bahador A, Khaledi A, Kachoei RA, Dastjerdi FA, Esmaeili D. Antibacterial effects of *Origanum vulgare* essence against multidrug-resistant *Acinetobacter baumannii* isolated from selected hospitals of Tehran, Iran. *J Clin Microbiol Infect* 2015;**2**:e22982.
71. Osterburg A, Gardner J, Hyon S, Neely A, Babcock G. Highly antibiotic-resistant *Acinetobacter baumannii* clinical isolates are killed by the green tea polyphenol (−)-epigallocatechin-3-gallate (EGCG). *Clin Microbiol Infect* 2009;**15**:341–6.
72. Chusri S, Na-Phatthalung P, Siriyong T, Paosen S, Voravuthikunchai SP. *Holarrhena antidysenterica* as a resistance modifying agent against *Acinetobacter baumannii*: its effects on bacterial outer membrane permeability and efflux pumps. *Microbiol Res* 2014;**169**:417–24.
73. Karumathil DP, Kollanoor-Johny A, Venkitanarayanan K. Effect of Trans-cinnamaldehyde and eugenol in reducing the resistance of MDR *Acinetobacter baumannii* to beta-lactam antibiotics, In: *114th ASM general meeting*; May 2014.
74. Ghose C, Kalsy A, Sheikh A, Rollenhagen J, John M, Young J, et al. Transcutaneous immunization with *Clostridium difficile* toxoid A induces systemic and mucosal immune responses and toxin A-neutralizing antibodies in mice. *Infect Immun* 2007;**75**:2826–32.
75. Wilkins TD, Lyerly DM. *Clostridium difficile* testing: after 20 years, still challenging. *J Clin Microbiol* 2003;**41**:531–4.
76. Tenover FC, Tickler IA, Persing DH. Antimicrobial-resistant strains of *Clostridium difficile* from North America. *Antimicrob Agents Chemother* 2012;**56**:2929–32.
77. Mooyottu S, Kollanoor-Johny A, Flock G, Bouillaut L, Upadhyay A. Sonenshein et al. Carvacrol and trans-cinnamaldehyde reduce *Clostridium difficile* toxin production and cytotoxicity *in vitro. Int J Mol Sci* 2014;**15**:4415–30.
78. Shahverdi AR, Monsef-Esfahani HR, Tavasoli F, Zaheri A, Mirjani R. Trans-cinnamaldehyde from *Cinnamomum zeylanicum* bark essential oil reduces the clindamycin resistance of *Clostridium difficile* in vitro. *J Food Sci* 2007;**72**:S055–8.
79. Dias C, Aires A, Saavedra MJ. Antimicrobial activity of isothiocyanates from cruciferous plants against methicillin-resistant *Staphylococcus aureus* (MRSA). *Int J Mol Sci* 2014;**15**:19552–61.
80. Nitta T, Arai T, Takamatsu H, Inatomi Y, Murata H, Iinuma M, et al. Antibacterial activity of extracts prepared from tropical and subtropical plants on methicillin-resistant *Staphylococcus aureus. J Health Sci* 2002;**48**:273–6.
81. Chovanová R, Mikulášová M, Vaverková Š. In vitro antibacterial and antibiotic resistance modifying effect of bioactive plant extracts on methicillin-resistant *Staphylococcus epidermidis. Int J Microbiol* 2013;**2013**.760969.
82. Nikaido H. Prevention of drug access to bacterial targets: permeability barriers and active efflux. *Science* 1994;**264**:382–8.

83. Spratt BG. Resistance to antibiotics mediated by target alterations. *Science* 1994;**264**:388–93.
84. Davies J. Inactivation of antibiotics and the dissemination of resistance genes. *Science* 1994;**264**:375–82.
85. Horiyama T, Yamaguchi A, Nishino K. TolC dependency of multidrug efflux systems in *Salmonella enterica* serovar Typhimurium. *J Antimicrob Chemother* 2010;**65**(7):1372–6.
86. Alvarez-Ortega C, Olivares J, Martinez JL. RND multidrug efflux pumps: what are they good for? *Front Microbiol* 2013;**4**:7.
87. Piddock LJ. Multidrug-resistance efflux pumps—not just for resistance. *Nat Rev Microbiol* 2006;**4**:629–36.
88. Johny AK, Hoagland T, Venkitanarayanan K. Effect of subinhibitory concentrations of plant-derived molecules in increasing the sensitivity of multidrug-resistant *Salmonella enterica* serovar Typhimurium DT104 to antibiotics. *Foodborne Pathog Dis* 2010;**7**(10):1165–70.
89. Kollanoor-Johny A., Venkitanarayanan K. Trans-cinnamaldehyde and thymol down-regulate antibiotic resistance gene expression in multidrug resistant *Salmonella* Typhimurium DT104 and reduce bacterial invasion of porcine intestinal epithelial cells, In: *IFT annual meeting*, June 11–14, New Orleans, LA; 2011. <http://www.ift.org/Meetings-and-Events/Past-MeetingResources/Technical%20Abstract%20Search%20Details.aspx?id=52309>.
90. Frere JM. Beta-lactamases and bacterial resistance to antibiotics. *Mol Microbiol* 1995;**16**:385–95.
91. Jimenez-Valera M, Ruiz-Bravo A, Ramos-Cormenzana A. Inhibition of beta-lactamases from *Yersinia enterocolitica* by plant extracts. *J Antimicrob Chemother* 1987;**19**:31–7.
92. Pieboji J, Baurin S, Frere JM, Ngassam P, Ngamenni B, Azebaze A, et al. Screening of some medicinal plants from Cameroon for beta-lactamase inhibitory activity. *Phytother Res* 2007;**21**:284–7.
93. Zhao WH, Hu ZQ, Hara Y, Shimamura T. Inhibition of penicillinase by epigallocatechin gallate resulting in restoration of antibacterial activity of penicillin against penicillinase-producing *Staphylococcus aureus*. *Antimicrob Agents Chemother* 2002;**46**:2266–8.
94. Taga I, Lan CQ, Altosaar I. Plant essential oils and mastitis disease: their potential inhibitory effects on pro-inflammatory cytokine production in response to bacteria related inflammation. *Nat Prod Commun* 2012;**7**:675–82.
95. Williams LR. Clonal production of tea tree oil high in terpinen-4-ol for use in formulations for the treatment of thrush. *Complement Ther Nurs Midwifery* 1998;**4**:133–6.
96. Hart PH, Brand C, Carson CF, Riley TV, Prager RH, Finlay-Jones JJ. Terpinen-4-ol, the main component of the essential oil of *Melaleuca alternifolia* (tea tree oil), suppresses inflammatory mediator production by activated human monocytes. *Inflamm Res* 2000;**49**:619–26.
97. Brand C, Ferrante A, Prager RH, Riley TV, Carson CF, Finlay-Jones JJ, et al. The water-soluble components of the essential oil of *Melaleuca alternifolia* (tea tree oil) suppress the production of superoxide by human monocytes, but not neutrophils, activated *in vitro*. *Inflamm Res* 2001;**50**:213–19.
98. Nogueira MNM, Aquino SG. Junior CR, Spolidorio DMP. Terpinen-4-ol and alpha-terpineol (tea tree oil components) inhibit the production of IL-1β, IL-6 and IL-10 on human macrophages. *Inflamm Res* 2014;**63**:769–78.
99. Yaseen Khan M, Ali SA, Pundarikakshudu K. Wound healing activity of extracts derived from *Shorea robusta* resin. *Pharm Biol* 2015;**54**(3):1–7.
100. Mukherjee H, Ojha D, Bharitkar YP, Ghosh S, Mondal S, Kaity S, et al. Evaluation of the wound healing activity of *Shorea robusta*, an Indian ethnomedicine, and its isolated constituent (s) in topical formulation. *J Ethnopharmacol* 2013;**149**:335–43.
101. Ghosh S, Samanta A, Mandal NB, Bannerjee S, Chattopadhyay D. Evaluation of the wound healing activity of methanol extract of *Pedilanthus tithymaloides* (L.) Poit leaf and its isolated active constituents in topical formulation. *J Ethnopharmacol* 2012;**142**:714–22.

102. Khadeer AB, Krishna V, Malleshappa KH. In vivo wound healing activity of the methanolic extract and its isolated constituent, gulonic acid gamma-lactone, obtained from *Grewia tiliaefolia*. *Planta Med* 2009;**75**:478–82.
103. Mekonnen A, Sidamo T, Asres K, Engidawork E. In vivo wound healing activity and phytochemical screening of the crude extract and various fractions of *Kalanchoe petitiana A. Rich* (Crassulaceae) leaves in mice. *J Ethnopharmacol* 2013;**145**:638–46.
104. Ghosh S, Banerjee S, Sil PC. The beneficial role of curcumin on inflammation, diabetes and neurodegenerative disease: a recent update. *Food Chem Toxicol* 2015;**83**:111–24.
105. Zhao XC, Zhang L, Yu HX, Sun Z, Lin XF, Tan C. Curcumin protects mouse neuroblastoma Neuro-2A cells against hydrogen-peroxide-induced oxidative stress. *Food Chem* 2011;**129**:387–94.
106. Mun SH, Kang OH, Joung DK, Kim SB, Choi JG, Shin DW, et al. In vitro anti-MRSA activity of carvone with gentamicin. *Exp Ther Med* 2014;**7**:891–6.
107. Negi N, Prakash P, Gupta ML, Mohapatra TM. Possible role of Curcumin as an efflux pump inhibitor in multi drug resistant clinical isolates of *Pseudomonas aeruginosa*. *J Clin Diagn Res* 2014;**8**:04.
108. Chambers HF. The changing epidemiology of *Staphylococcus aureus*? *Emerg Infect Dis* 2001;**7**:178.
109. Deresinski S. Methicillin-resistant *Staphylococcus aureus*: an evolutionary, epidemiologic, and therapeutic odyssey. *Clin Infect Dis* 2005;**40**:562–73.
110. Aiello AE, Larson EL, Levy SB. Consumer antibacterial soaps: effective or just risky? *Clin Infect Dis* 2007;**45**(Suppl. 2):S137–47.
111. Giuliano CA, Rybak MJ. Efficacy of triclosan as an antimicrobial hand soap and its potential impact on antimicrobial resistance: a focused review. *Pharmacotherapy* 2015;**35**:328–36.
112. Yueh MF, Taniguchi K, Chen S, Evans RM, Hammock BD, Karin M, et al. The commonly used antimicrobial additive triclosan is a liver tumor promoter. *Proc Nat Acad Sci USA* 2014;**111**:17200–5.
113. Bissett L. Skin care: an essential component of hand hygiene and infection control. *Br J Nurs* 2007;**16**:976–81.
114. Pazyar N, Feily A, Yaghoobi R. An overview of interleukin-1 receptor antagonist, anakinra, in the treatment of cutaneous diseases. *Curr Clin Pharmacol* 2012;**7**(4):271–5.
115. Hartman-Adams H, Banvard C, Juckett G. Impetigo: diagnosis and treatment. *Am Fam Physician* 2014;**90**(4):229–35.
116. Carmo ES, Pereira FDO, Moreira ACP, Brito LL, Gayoso CW, Costa JGMD, et al. Essential oil from *Cymbopogon citratus* DC Stapf: a promising natural product against *Malassezia* spp. *Rev Inst Adolfo Lutz* 2012;**71**:386–91.
117. Elaissi A, Rouis Z, Salem NAB, Mabrouk S, ben Salem Y, Salah K, et al. Chemical composition of 8 eucalyptus species' essential oils and the evaluation of their antibacterial, antifungal and antiviral activities. *BMC Complement Altern Med* 2012;**12**(1):81.
118. Chachra S, Dhawan P, Kaur T, Sharma AK. The most effective and essential way of improving the oral health status education. *J Indian Soc Pedod Prev Dent* 2011;**29**(3):216.
119. Lam A. Elements in oral health programs. *NY State Dent J* 2014;**80**(2):26–30.
120. Rostoka D, Kroiča J, Iriste V, Reinis A, Kuznetsova V, Teibe U. Treatment of halitosis with mouth rinsing agents containing essential oils. *Stomatologiia* 2011;**91**(3):27–34.
121. Bauroth K, Charles CH, Mankodi SM, Simmons K, Zhao Q, Kumar LD. The efficacy of an essential oil antiseptic mouthrinse vs. dental floss in controlling interproximal gingivitis: a comparative study. *J Am Dent Assoc* 2003;**134**(3):359–65.
122. Sharma NC, Galustians HJ, Qaqish J, Charles CH, Vincent JW, McGuire JA. Antiplaque and antigingivitis effectiveness of a hexetidine mouthwash. *J Clin Periodontol* 2003;**30**(7):590–4.

123. Cosyn J, Princen K, Miremadi R, Decat E, Vaneechoutte M, Bruyn H. A double-blind randomized placebo-controlled study on the clinical and microbial effects of an essential oil mouth rinse used by patients in supportive periodontal care. *Int J Dent Hyg* 2013;**11**(1):53–61.
124. Forrer M, Kulik EM, Filippi A, Waltimo T. The antimicrobial activity of alpha-bisabolol and tea tree oil against *Solobacterium moorei*, a Gram-positive bacterium associated with halitosis. *Arch Oral Biol* 2013;**58**(1):10–16.
125. Hammer KA, Dry L, Johnson M, Michalak EM, Carson CF, Riley TV. Susceptibility of oral bacteria to *Melaleuca alternifolia* (tea tree) oil in vitro. *Oral Microbiol Immunol* 2003;**18**(6):389–92.
126. Balappanavar AY, Sardana V, Singh M. Comparison of the effectiveness of 0.5% tea, 2% neem and 0.2% chlorhexidine mouthwashes on oral health: a randomized control trial. *Indian J Dent Res* 2013;**24**(1):26.
127. Gupta C, Kumari A, Garg AP. Comparative study of cinnamon oil and clove oil in some oral microbiota. *Acta Biomed* 2012;**82**(3):197–9.
128. Tilman D, Balzer C, Hill J, Befort BL. Global food demand and the sustainable intensification of agriculture. *Proc Natl Acad Sci USA* 2011;**108**(50):20260–4.
129. FAOSTAT. *FAO food balance data.*<http://faostat3.fao.org/faostat-gateway/go/to/browse/FB/FBS/E>; [accessed 26.03.15].
130. Aarestrup F. Sustainable farming: get pigs off antibiotics. *Nature* 2012;**486**(7404):465–6.
131. FDA. 2010. <http://www.fda.gov/NewsEvents/Testimony/ucm219015.htm>.
132. You Y, Silbergeld EK. Learning from agriculture: understanding low-dose antimicrobials as drivers of resistome expansion. *Front Microbiol* 2014;**5**:284.
133. Laxminarayan R, Duse A, Wattal C, Zaidi AK, Wertheim HF, Sumpradit N, et al. Antibiotic resistance-the need for global solutions. *Lancet Infect Dis* 2013;**13**(12):1057–98.
134. Vieira AR, Collignon P, Aarestrup FM, McEwen SA, Hendriksen RS, Hald T, et al. Association between antimicrobial resistance in *Escherichia coli* isolates from food animals and blood stream isolates from humans in Europe: an ecological study. *Foodborne Pathog Dis* 2011;**8**(12):1295–301.
135. Aarestrup FM. Veterinary drug usage and antimicrobial resistance in bacteria of animal origin. *Basic Clin Pharmacol Toxicol* 2005;**96**(4):271–81.
136. Chantziaras I, Boyen F, Callens B, Dewulf J. Correlation between veterinary antimicrobial use and antimicrobial resistance in food-producing animals: a report on seven countries. *J Antimicrob Chemother* 2014;**69**(3):827–34.
137. Upadhyaya I, Upadhyay A, Kollanoor-Johny A, Mooyottu S, Baskaran SA, Yin HB, et al. In-feed supplementation of trans-cinnamaldehyde reduces layer-chicken egg-borne transmission of *Salmonella enterica* serovar enteritidis. *Appl Environ Microbiol* 2015;**81**(9):2985–94.
138. Upadhyaya I, Upadhyay A, Yin HB, Nair MS, Bhattaram VK, Karumathil D, et al. Reducing colonization and eggborne transmission of *Salmonella* Enteritidis in layer chickens by in-feed supplementation of caprylic acid. *Foodborne Pathog Dis* 2015;**12**(7):591–7.
139. Kollanoor-Johny A, Upadhyay A, Baskaran SA, Upadhyaya I, Mooyottu S, Mishra N, et al. Effect of therapeutic supplementation of the plant compounds trans-cinnamaldehyde and eugenol on *Salmonella enterica* serovar enteritidis colonization in market-age broiler chickens. *J Appl Poult Res* 2012;**21**(4):816–22.
140. Kollanoor-Johny A, Mattson T, Baskaran SA, Amalaradjou MA, Babapoor S, March B, et al. Reduction of *Salmonella enterica* serovar enteritidis colonization in 20-day-old broiler chickens by the plant-derived compounds trans-cinnamaldehyde and eugenol. *Appl Environ Microbiol* 2012;**78**(8):2981–7.
141. Arsi K, Donoghue AM, Venkitanarayanan K, Kollanoor-Johny A, Fanatico AC, Blore PJ, et al. The efficacy of the natural plant extracts, thymol and carvacrol against *Campylobacter* colonization in broiler chickens. *J Food Saf* 2014;**34**(4):321–5.

142. Huang QQ, Jin L, Xu Z, Barbieri R, Acharya S, Hu TM, et al. Effects of purple prairie clover (*Dalea purpurea* Vent.) on feed intake, nutrient digestibility and faecal shedding of *Escherichia coli* O157:H7 in lambs. *Anim Feed Sci Technol* 2015;**207**:51–61.
143. Jamroz D, Orda I, Kamel C, Wiliczkiewicz A, Wertelecki T, Skorupinska I. The influence of phytogenic extracts on performance, nutrient digestibility, carcass characteristics, and gut microbial status in broiler chickens. *J Anim Feed Sci* 2003;**12**:583–96.
144. Jamroz D, Wiliczkiewicz A, Wertelecki T, Orda J, Skorupinska J. Use of active substances of plant origin in chicken diets based on maize and locally grown cereals. *Br Poult Sci* 2005;**46**:485–93.
145. Mitsch P, Zitterl-Eglseer K, Kohler B, Gabler C, Losa R, Zimpernik I. The effect of two different blends of essential oil components on the proliferation of *Clostridium perfringens* in the intestines of broiler chickens. *Poult Sci* 2006;**83**:669–75.
146. Alali WQ, Hofacre CL, Mathis GF, Faltys G, Ricke SC, Doyle MP. Effect of non-pharmaceutical compounds on shedding and colonization of *Salmonella enterica* serovar Heidelberg in broilers. *Food Control* 2013;**31**(1):125–8.
147. Amerah AM, Mathis G, Hofacre CL. Effect of xylanase and a blend of essential oils on performance and *Salmonella* colonization of broiler chickens challenged with *Salmonella* Heidelberg. *Poult Sci* 2012;**91**:943–7.
148. Baskaran SA, Bhattaram V, Upadhyaya I, Upadhyay A, Kollanoor-Johny A, Schreiber Jr D, et al. Inactivation of *Escherichia coli* O157:H7 on cattle hides by caprylic acid and β-Resorcylic acid. *J Food Prot* 2013;**76**:318–22.
149. Amalaradjou MAR, Annamalai T, Marek P, Rezamand P, Schreiber D, Hoagland T, et al. Inactivation of *Escherichia coli* O157:H7 in cattle drinking water by sodium caprylate. *J Food Prot* 2006;**69**:2248–52.
150. Charles AS, Baskaran SA, Murcott C, Schreiber D, Hoagland T, Venkitanarayanan K. Reduction of *Escherichia coli* O157:H7 in cattle drinking-water by *trans*-cinnamaldehyde. *Foodborne Pathog Dis* 2008;**5**:763–71.
151. Zhao T, Zhao P, West JW, Bernard JK, Cross HG, Doyle MP. Inactivation of enterohemorrhagic *Escherichia coli* in rumen content-or feces-contaminated drinking water for cattle. *Appl Environ Microbiol* 2006;**72**:3268–73.
152. Ahn J, Grün IU, Mustapha A. Effects of plant extracts on microbial growth, color change, and lipid oxidation in cooked beef. *Food Microbiol* 2007;**24**:7–14.
153. Wells JE, Berry ED, Varel VH. Effects of common forage phenolic acids on *Escherichia coli* O157:H7 viability in bovine feces. *Appl Environ Microbiol* 2005;**71**:7974–9.
154. McManus PS, Stockwell VO, Sundin GW, Jones AL. Antibiotic use in plant agriculture. *Annu Rev Phytopathol* 2002;**40**:443–65.
155. Stockwell VO, Duffy B. Use of antibiotics in plant agriculture. *Rev Sci Tech* 2012;**31**:199–210.
156. Duffy B, Holliger E, Walsh F. Streptomycin use in apple orchards did not increase abundance of mobile resistance genes. *FEMS Microbiol Lett* 2014;**350**:180–9.
157. Rodríguez-Sánchez C, Altendorf K, Smalla K, Lipski A. Spraying of oxytetracycline and gentamicin onto field-grown coriander did not affect the abundance of resistant bacteria, resistance genes, and broad host range plasmids detected in tropical soil bacteria. *Biol Fertil Soils* 2008;**44**:589–96.
158. Yang J, Wang C, Shu C, Liu L, Geng J, Hu S, et al. Marine sediment bacteria harbor antibiotic resistance genes highly similar to those found in human pathogens. *Microb Ecol* 2013;**65**(4): 975–81.
159. Shah SQ, Cabello FC, L'Abée-Lund TM, Tomova A, Godfrey HP, Buschmann AH, et al. Antimicrobial resistance and antimicrobial resistance genes in marine bacteria from salmon aquaculture and non-aquaculture sites. *Environ Microbiol* 2014;**16**(5):1310–20.

160. Lim SJ, Jang E, Lee SH, Yoo BH, Kim SK, Kim TH. Antibiotic resistance in bacteria isolated from freshwater aquacultures and prediction of the persistence and toxicity of antimicrobials in the aquatic environment. *J Environ Sci Health B* 2013;**48**(6):495–504.
161. Todd J, Friedman M, Patel J, Jaroni D, Ravishankar S. The antimicrobial effects of cinnamon leaf oil against multi-drug resistant *Salmonella* Newport on organic leafy greens. *Int J Food Microbiol* 2013;**166**(1):193–9.
162. Yossa N, Patel J, Millner P, Ravishankar S, Lo YM. Antimicrobial activity of plant essential oils against *Escherichia coli* O157:H7 and Salmonella on lettuce. *Foodborne Pathog Dis* 2013;**10**(1):87–96.
163. Moore-Neibel K, Gerber C, Patel J, Friedman M, Ravishankar S. Antimicrobial activity of lemongrass oil against *Salmonella enterica* on organic leafy greens. *J Appl Microbiol* 2012;**112**(3):485–92.
164. Upadhyay A, Upadhyaya I, Mooyottu S, Kollanoor-Johny A, Venkitanarayanan K. Efficacy of plant-derived compounds combined with hydrogen peroxide as antimicrobial wash and coating treatment for reducing *Listeria monocytogenes* on cantaloupes. *Food Microbiol* 2014;**44**:47–53.
165. Bhargava K, Conti DS, da Rocha SR, Zhang Y. Application of an oregano oil nanoemulsion to the control of foodborne bacteria on fresh lettuce. *Food Microbiol* 2015;**47**:69–73.
166. Perdones A, Sánchez-González L, Chiralt A, Vargas M. Effect of chitosan–lemon essential oil coatings on storage-keeping quality of strawberry. *Postharvest Biol Technol* 2012;**70**:32–41.
167. Amal SHA, El-Mogy MM, Aboul-Anean HE, Alsanius BW. Improving strawberry fruit storability by edible coating as a carrier of thymol or calcium chloride. *J Hortic Sci Ornamental Plants* 2010;**2**:88–97.
168. Chien PJ, Sheu F, Yang FH. Effects of edible chitosan coating on quality and shelf life of sliced mango fruit. *J Food Eng* 2007;**78**:225–9.
169. Alikhani M. Enhancing safety and shelf life of fresh-cut mango by application of edible coatings and microencapsulation technique. *Food Sci Nutr* 2014;**2**:210–17.
170. Mandal A, Patra AK, Singh D, Swarup A, Masto RE. Effect of long-term application of manure and fertilizer on biological and biochemical activities in soil during crop development stages. *Bioresour Technol* 2007;**98**:3585–92.
171. Wang Q, Bai Y, Gao H, He J, Chen H, Chesney RC, et al. Soil chemical properties and microbial biomass after 16 years of no-tillage farming on the Loess Plateau, China. *Geoderma* 2008;**144**:502–8.
172. Erickson MC, Habteselassie MY, Liao J, Webb CC, Mantripragada V, Davey LE, et al. Examination of factors for use as potential predictors of human enteric pathogen survival in soil. *J Appl Microbiol* 2014;**116**:335–49.
173. Monier JM, Demanèche S, Delmont TO, Mathieu A, Vogel TM, Simonet P. Metagenomic exploration of antibiotic resistance in soil. *Curr Opin Microbiol* 2011;**14**:229–35.
174. Heuer H, Solehati Q, Zimmerling U, Kleineidam K, Schloter M, Müller T. Accumulation of sulfonamide resistance genes in arable soils due to repeated application of manure containing sulfadiazine. *Appl Environ Microbiol* 2011;**77**:2527–30.
175. Gao L, Hu J, Zhang X, Wei L, Li S, Miao Z. Application of swine manure on agricultural fields contributes to extended-spectrum β-lactamase-producing *Escherichia coli* spread in Tai'an, China. *Front Microbiol* 2015;**6**:313.
176. Gao L, Tan Y, Zhang X, Hu J, Miao Z, Wei L, et al. Emissions of *Escherichia coli* carrying extended-spectrum β-Lactamase resistance from pig farms to the surrounding environment. *Int J Environ Res Public Health* 2015;**12**:4203–13.
177. Tian GB, Wang HN, Zou LK, Tang JN, Zhao YW, Ye MY, et al. Detection of CTX-M-15, CTX-M-22, and SHV-2 extended-spectrum β-lactamases (ESBLs) in *Escherichia coli* fecal-sample isolates from pig farms in China. *Foodborne Pathog Dis* 2009;**6**:297–304.

178. Blaak H, Hamidjaja RA, van Hoek AH, de Heer L, de Roda Husman AM, Schets FM. Detection of extended-spectrum beta-lactamase (ESBL)-producing *Escherichia coli* on flies at poultry farms. *Appl Environ Microbiol* 2014;**80**:239–46.
179. Burgos JM, Ellington BA, Varela MF. Presence of multidrug-resistant enteric bacteria in dairy farm topsoil. *J Dairy Sci* 2005;**88**(4):1391–8.
180. Sealy R, Evans MR, Rothrock C. The effect of a garlic extract and root substrate on soilborne fungal pathogens. *Hort Technology* 2007;**17**:169–73.
181. Soylu S, Yigitbas H, Soylu EM, Kurt Ş. Antifungal effects of essential oils from oregano and fennel on *Sclerotinia sclerotiorum. J Appl Microbiol* 2007;**103**:1021–30.
182. Yossa N, Patel J, Millner P, Lo M. Inactivation of *Salmonella* in organic soil by cinnamaldehyde, eugenol, ecotrol, and sporan. *Foodborne Pathog Dis* 2011;**8**:311–17.
183. Wells JE, Berry ED, Guerini MN, Varel VH. Evaluation of essential oils in beef cattle manure slurries and applications of select compounds to beef feedlot surfaces to control zoonotic pathogens. *J Appl Microbiol* 2015;**118**:295–304.
184. Woolhouse M, Ward M, Van Bunnik B, Farrar J. Antimicrobial resistance in humans, livestock and the wider environment. *Philos Trans R Soc Lond B Biol Sci* 2015;**370**:20140083.
185. Hollowell AC, Gano KA, Lopez G, Shahin K, Regus JU, Gleason N, et al. Native California soils are selective reservoirs for multidrug-resistant bacteria. *Environ Microbiol Rep* 2015;**7**(3):442–9.
186. Guo MT, Yuan QB, Yang J. Insights into the amplification of bacterial resistance to erythromycin in activated sludge. *Chemosphere* 2015;**136**:79–85.
187. Iweriebor BC, Gaqavu S, Obi LC, Nwodo UU, Okoh AI. Antibiotic susceptibilities of *Enterococcus* species isolated from hospital and domestic wastewater effluents in Alice, Eastern Cape province of South Africa. *Int J Environ Res Public Health* 2015;**12**:4231–46.
188. Czekalski N, Sigdel R, Birtel J, Matthews B, Bürgmann H. Does human activity impact the natural antibiotic resistance background? Abundance of antibiotic resistance genes in 21 Swiss lakes. *Environ Int* 2015;**81**:45–55.
189. Penesyan A, Gillings M, Paulsen IT. Antibiotic discovery: combatting bacterial resistance in cells and in biofilm communities. *Molecules* 2015;**20**:5286–98.
190. Ando E, Monden K, Mitsuhata R, Kariyama R, Kumon H. Biofilm formation among methicillin-resistant *Staphylococcus aureus* isolates from patients with urinary tract infection. *Acta Med Okayama* 2004;**58**:207–14.
191. Longo F, Vuotto C, Donelli G. Biofilm formation in *Acinetobacter baumannii. New Microbiol* 2014;**37** (2):119–27.
192. Donelli G, Vuotto C. Biofilm-based infections in long-term care facilities. *Future Microbiol* 2014;**9** (2):175–88.
193. Harmsen M, Lappann M, Knøchel S, Molin S. Role of extracellular DNA during biofilm formation by *Listeria monocytogenes. Appl Environ Microbiol* 2010;**76**(7):2271–9.
194. Seper A, Fengler VH, Roier S, Wolinski H, Kohlwein SD, Bishop AL, et al. Extracellular nucleases and extracellular DNA play important roles in *Vibrio cholerae* biofilm formation. *Mol Microbiol* 2011;**82** (4):1015–37.
195. Rose SJ, Babrak LM, Bermudez LE. *Mycobacterium avium* possesses extracellular DNA that contributes to biofilm formation, structural integrity, and tolerance to antibiotics. *PLoS One* 2015;**10**: e0128772.
196. Sahu PK, Iyer PS, Oak AM, Pardesi KR, Chopade BA. Characterization of eDNA from the clinical strain *Acinetobacter baumannii* AIIMS 7 and its role in biofilm formation. *ScientificWorldJournal* 2012;**2012**.973436

197. Montanaro L, Poggi A, Visai L, Ravaioli S, Campoccia D, Speziale P, et al. Extracellular DNA in biofilms. *Int J Artif Organs* 2011;**34**:824–31.
198. Stewart FJ. Where the genes flow. *Nat Geosci* 2013;**6**:688–90.
199. Al-Bakri AG, Othman G, Afifi FU. Determination of the antibiofilm, antiadhesive, and anti-MRSA activities of seven *Salvia* species. *Pharmacogn Mag* 2010;**6**(24):264.
200. Nostro A, Roccaro AS, Bisignano G, Marino A, Cannatelli MA, Pizzimenti FC, et al. Effects of oregano, carvacrol and thymol on *Staphylococcus aureus* and *Staphylococcus epidermidis* biofilms. *J Med Microbiol* 2007;**56**(4):519–23.
201. Walencka E, Rozalska S, Wysokinska H, Rozalski M, Kuzma L, Rozalska B. Salvipisone and aethiopinone from *Salvia sclarea* hairy roots modulate staphylococcal antibiotic resistance and express antibiofilm activity. *Planta Med* 2007;**73**(6):545–51.
202. Wu DQ, Ye J, Ou HY, Wei X, Huang X, He YW, et al. Genomic analysis and temperature-dependent transcriptome profiles of the rhizosphere originating strain *Pseudomonas aeruginosa* M18. *BMC Genomics* 2011;**12**(1):438.
203. Derakhshan S, Sattari M, Bigdeli M. Effect of cumin (*Cuminum cyminum*) seed essential oil on biofilm formation and plasmid Integrity of *Klebsiella pneumoniae*. *Pharmacogn Mag* 2010;**6**(21):57.
204. Magesh H, Kumar A, Alam A, Sekar U, Sumantran VN, Vaidyanathan R. Identification of natural compounds which inhibit biofilm formation in clinical isolates of *Klebsiella pneumoniae*. *Indian J Exp Biol* 2013;**51**(9):764–72.
205. Selim SA, Adam ME, Hassan SM, Albalawi AR. Chemical composition, antimicrobial and antibiofilm activity of the essential oil and methanol extract of the Mediterranean cypress (*Cupressus sempervirens* L.). *BMC Complement Altern Med* 2014;**14**(1):179.
206. Pelletier RP. Effect of Plant-Derived Molecules on *Acinetobacter baumannii* Biofilm on Abiotic Surfaces. Honors Scholar Theses. Paper 257; 2012. Available from: http://digitalcommons.uconn.edu/srhonors_theses/257.
207. Amalaradjou MAR, Venkitanarayanan K. Effect of trans-cinnamaldehyde on inhibition and inactivation of *Cronobacter sakazakii* biofilm on abiotic surfaces. *J Food Prot* 2011;**74**:200–8.
208. Lee JH, Cho MH, Lee J. 3-Indolylacetonitrile decreases *Escherichia coli* O157:H7 biofilm formation and *Pseudomonas aeruginosa* virulence. *Environ Microbiol* 2013;**13**:62–73.
209. Lee JH, Cho HS, Joo SW, Chandra Regmi S, Kim JA, Ryu CM, et al. Diverse plant extracts and *trans*-resveratrol inhibit biofilm formation and swarming of *Escherichia coli* O157:H7. *Biofouling* 2013;**29**:1189–203.
210. Lee JH, Kim YG, Cho HS, Ryu SY, Cho MH, Lee J. Coumarins reduce biofilm formation and the virulence of *Escherichia coli* O157:H7. *Phytomedicine* 2014;**21**:1037–42.

CHAPTER

11 ESSENTIAL OILS: A NATURAL ALTERNATIVE TO COMBAT ANTIBIOTICS RESISTANCE

M.L. Chávez-González, R. Rodríguez-Herrera and C.N. Aguilar

Food Research Department, School of Chemistry, Universidad Autónoma de Coahuila, Saltillo, Coahuila, CP, Mexico

INTRODUCTION

Antibiotics are probably the drug most used in human medicine, and over the last few years, abuse in the use of these drugs has created multidrug resistance (MDR), which puts at serious risk the effective treatment of a growing number of infections caused by pathogenic microorganisms.[1] MDR is the result of a number of convergent factors, among which are the inappropriate use of antibiotics caused by suspension of treatment, use of inadequate doses, and genetic improvement of microorganisms, among others.[2]

Currently, bacterial infections represent a serious risk around the world,[3] especially bacterial infections that are resistant to drugs and affect a large number of patients; these diseases are usually more severe compared with the same infections caused by microorganisms (nonresistant) and thus are more difficult to be treated and eventually be cured.[1,2] This is a problem that not only confounded medical specialists, but also many governments because the costs and impact are high. It has been estimated that resistance to antibiotics is responsible for about 50,000 deaths per year only in the United States and Europe[4]; there is no reliable record in other countries, but this issue is estimated to cause large numbers of deaths, especially in developing countries where control over the sale of antibiotics is poor.

This situation requires some effort to help combat such problems. One possible solution is to search for alternative therapies to control these diseases. An alternative is the use of essential oils to achieve control over antibiotic-resistant microorganisms.[5,6]

Essential oils are liquid, volatile, natural, and complex mixtures of low-molecular-weight compounds and are formed by aromatic plants as secondary metabolites, which are naturally synthesized by plants in response to attacks by insects, herbivores, and other organisms.[5] Essential oils are characterized by a strong odor. They are usually extracted by steam or hydrodistillation or solvent extraction.[7,8] This type of oil can be produced by all plant organs (ie, seeds, flowers, leaves, buds, stems, fruits, roots, wood, or bark), and are stored in oil ducts, resin ducts, glands, or trichomes of the plants.[7,8] They are commonly used as flavoring agents in food products, drinks, perfumeries, pharmaceuticals, and cosmetics. The essential oil alone or in combination possesses significant medicinal properties; hence, it may be used for chemotherapy of infectious and noninfectious diseases.[5] The presence of different compounds in the oil makes it possible to use it as an antimicrobial agent that offers a low risk of microbial resistance development.[6]

Antibiotic Resistance. DOI: http://dx.doi.org/10.1016/B978-0-12-803642-6.00011-3

In this chapter, the essential oils as a natural alternative to fight MDR pathogens are presented, natural sources of essential oils, predominant chemical composition in these fractions and the mechanisms of action of essential oil are showed. Finally some of the most commonly used essential oils against multidrug-resistant microorganisms such as citrus, rosemary, oregano, basil, and mentha, are presented.

NATURAL SOURCES OF ESSENTIAL OILS AND ITS COMPOSITION

There are around 3000 known essential oils, of which only 300 are commercially important for the pharmaceutical, agronomic, food, sanitary, cosmetic, and perfume industries.[5,8,9] Plants producing essential oils belong to various genera in about 60 families, including Alliaceae, Apiaceae, Asteraceae, Lamiaceae, Myrtaceae, Poaceae, and Rutaceae.[5]

Essential oils are very complex natural mixtures; the concentration of each component can vary.[8] The essential oil components include terpenoids and phenylpropanoids, and other components like aromatic and aliphatic constituents also may be present. Within the terpenes group are found monoterpenes, sesquiterpenes, and oxygenated derivatives, all of which are characterized by low molecular weight. Major components are responsible for the biological properties of essential oils,[8] but sometimes a combination of molecules modifies their activity to significant effect.[5,9]

The antimicrobial and other biological activities of essential oils are directly correlated with the presence of bioactive volatile components. The composition of the essential oil of any particular plant depends on the plant part used, whether it be flowers, green parts (ie, leaves and stems), bark, wood, whole fruits, pericarp, seed, or roots.[7] Terpenoids, discussed in the next section, are the major constituents of essential oils; other important compounds are aromatic and aliphatic constituents.

TERPENES

Terpenoids are naturally occurring hydrocarbons produced by a wide variety of plants. Terpenes are grouped in different types according to their structure and function; they are classified based on five-carbon (isoprene) units. The main terpenes are monoterpenes (C10) and sesquiterpenes (C15), but hemiterpene (C5), diterpenes (C20), triterpenes (C30), and tetraterpenes (C40) are also important, and they represent around 90% of essential oils.[8] More than 55,000 terpene molecules have been discovered to date. The bioactivities of a particular essential oil is based on these molecules most of the time, but sometimes biological activity cannot be attributed to only one compound[5,9] (Fig. 11.1).

AROMATIC COMPOUNDS

Aromatic compounds present in essential oils are usually derived from phenylpropane, and they are in lower concentration than terpenes. Aromatic compounds comprise aldehyde, alcohol, phenols, methoxy derivatives, and methylene dioxy compounds. The syntheses of phenylpropanoic derivatives and terpenes generally are separated in plants but may coexist in some cases.[3]

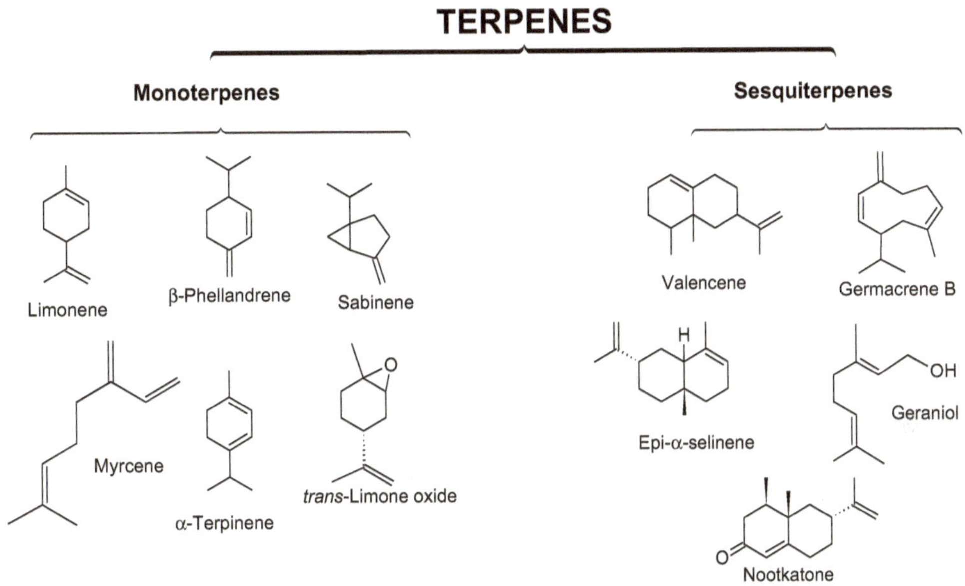

FIGURE 11.1

Examples of some terpenes (monoterpenes and sesquiterpenes) found in essential oils.

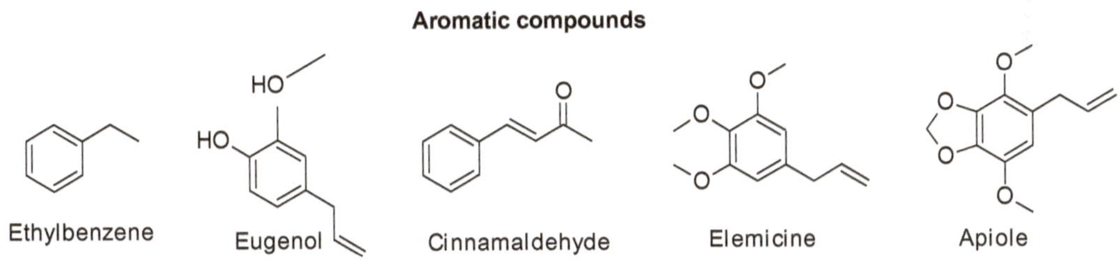

FIGURE 11.2

Examples of some aromatic compounds found in essential oils.

Some examples of aromatic compounds include cinnamaldehyde, cinnamic alcohol, chavicol, eugenol, anethole, elemicine, estragole, methyleugenols, apiole, myristicine, and safrole[8] (Fig. 11.2).

MECHANISM OF ACTION OF ESSENTIAL OILS

The antimicrobial effect of essential oils has been demonstrated with a wide range of microorganisms. Numerous mechanisms of action have been proposed, but none have been completely understood.[10,11] In most of the mechanisms, only the action of chemical compounds present in essential oils is explained, and antibacterial activity involves different mechanisms used to attack pathogenic bacteria.[12]

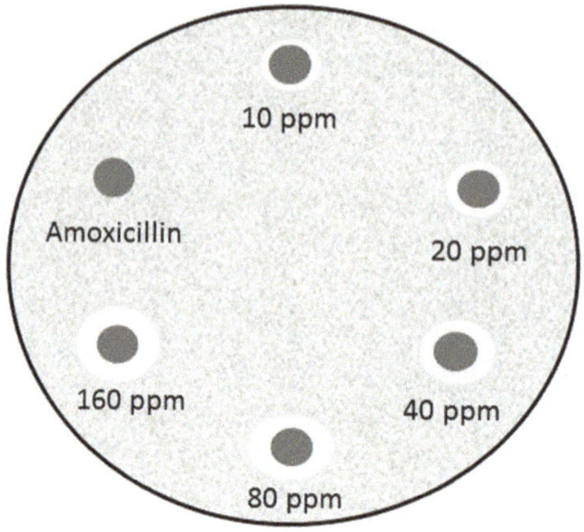

FIGURE 11.3

Schematic effect on the activity of basil essential oils against MDR bacteria.

The mechanism of antimicrobial action of terpenes is associated with their high affinity for lipids because of their hydrophobic nature. Their antibacterial properties are evidently associated with this lipophilic character and by the outer microorganism structures.[13] This allows monoterpenes to penetrate membrane structures inside the cell, which increases membrane fluidity and permeability, changes the topology of membrane proteins, and induce disturbances in the respiration chain.[14,15]

Phenolic compounds present in essential oils disrupt the cell membrane, resulting in the inhibition of cell functional properties and eventually causing leakage of internal cell content.[16] It has been reported that the phenolics thymol and carvacrol also inhibited growth of Gram-negative bacteria by disrupting the outer cell membrane.[17] Other processes associated with the cell membrane include electron transport, ions, protein translocation, phosphorylation, and other enzyme-dependent reactions. The disrupted permeability of the cytoplasmic membrane can result in cell death.[2] Interaction of essential oils with microbial cell membranes results in growth inhibition of some Gram-positive and Gram-negative bacteria.[7] It has been reported that Gram-positive bacteria appear to be more susceptible to the antibacterial properties of essential oil compounds than Gram-negative bacteria. This is expected, as Gram-negative bacteria have an outer layer surrounding their cell wall, limiting the access of hydrophobic compounds (Fig. 11.3).

DIFFERENT ESSENTIAL OILS USED AGAINST MDR

CITRUS ESSENTIAL OIL

The genus *Citrus* has different species, including orange (*Citrus sinensis*), mandarin (*Citrus reticulata*), lime (*Citrus aurantifolia*), grapefruit (*Citrus paradisi*) among others. These fruits have different chemical components, and within the fruit peel, essential oils can be found.

Citrus essential oils contain an extremely wide variety of compounds.[18] Approximately 400 compounds have been identified in citrus oils.[7] The analysis, extraction, and content of the components depend on the seasonal variation, ripeness, and geographical region, the specific citrus type, and even the separation and extraction methods used.[7,10] Around 85–99% are volatile compounds; the remaining ones are nonvolatile compounds.[18] The major components are monoterpenes (representing around 97% of the citrus essential oil), and 1.8–2% represent alcohols, aldehydes, and esters.[10]

Essential oils derived from the citrus industry have been screened for their antimicrobial properties against diverse foodborne pathogens such as *Escherichia coli* O157:H7, *Salmonella* Typhimurium, *Staphylococcus aureus, Listeria monocytogenes, Campylobacter* sp., and others. Several have shown to possess antimicrobial activity.[19,20] The major component of citrus oils is limonene (ranging from 32% to 98% depending on the citrus type), with sweet orange containing 69–98% and lemon containing 45–76%.[10] The chemical, physical, and biological properties of limonene have an important impact on the biological properties of the major components of essential oil.[18]

Limonene has shown to be effective against *S. aureus, L. monocytogenes, Salmonella enterica*, and *Saccharomyces bayanus*, as well as other microorganisms.[7] It was found to be effective against strains of *E. coli*,[10] *Klebsiella pneumoniae*,[21] *Mycoplasma pneumoniae*,[7] and *Staphylococcus epidermidis*.[21] This component can exert potent, broad-spectrum antimicrobial activity.[7] However, limonene is also susceptible to oxidative degradation, which causes a reduction of activity.[7] Linalool exhibits antimicrobial properties against *Shigella sonnei, Salmonella flexneri*,[10] *Staphylococcus epidermidis*,[21] *Arcobacter butzleri*,[22] *Campylobacter jejuni, E. coli* O157, and *L. monocytogenes*.[23] Citrulline exerts potent, broad-spectrum antimicrobial activity.[7] α-Terpineol has an important effect on *E. coli, Salmonella* Typhimurium, *L. monocytogenes*, and *S. aureus*.[24] Citral was reported with power against *A. butzleri*,[22] *C. jejuni, E. coli, Salmonella* Typhimurium, *L. monocytogenes*, and *S. aureus*.[23] Another important terpene that can be used in antimicrobial therapy is (4R)-(+)-carvone (monoterpene), which was effective against *L. monocytogenes* and showed activity against *Enterococcus faecium* and *E. coli*.[13]

In addition to the use of citrus essential oil components against pathogenic microorganisms, complete essential oils have been tested. Pathan et al.[25] indicated that *Citrus aurantifolia* showed high activity against *K. pneumoniae* and *S. aureus*. Mandarin essential oil has been reported to have antimicrobial effects against *E. coli, K. pneumoniae*, and *Salmonella enterica*.[26] Lemon oil was reported to have properties against *Lactobacillus plantarum* and *L. monocytogenes*, which showed inhibition percentages around 99.9%[26,27] and against *Candida albicans, Bacillus subtilis, E. coli, K. pneumoniae*, and *Salmonella enterica*.[26,28,29] In addition, some studies demonstrated inhibition and reduction of the numbers of foodborne pathogens such as *Salmonella* spp., *E. coli* O157:H7, and *L. monocytogenes* by citrus essential oils.[26] It has been observed that the activity of the essential oil used also depends on its concentration. Recently, it has been found that in lemon peel, the maximum concentration of volatile compounds occurs when the fruit is at the intermediate maturation stage.[18]

ROSEMARY ESSENTIAL OIL

Rosemary (*Rosmarinus officinalis* L.) is a herb that belongs to the mint family. It is an evergreen aromatic shrub[30] grown in many parts of the world, and it has been cultivated for a long time for use in folk medicine, cosmetics, and phytocosmetics.[31] Recent studies on rosemary essential oil have focused on its antimicrobial activity against several pathogenic microorganisms.[1]

Chemical analysis of rosemary essential oil revealed the presence of terpenes and terpenoids. The major constituents of this essential oil include α-pinene, myrcene, 1,8-cineole, camphor, camphene, α-terpineol, and borneol.[1,31] There have been some reports on the antimicrobial activity of this essential oil. For instance, Jiang et al.[30] indicated that rosemary essential oil was very active against *Staphylococcus epidermidis*, *S. aureus*, *B. subtilis*, *Proteus vulgaris*, *Pseudomonas aeruginosa*, and *E. coli*. The minimum inhibitory concentration (MIC) was found to range from 0.03% (v/v) to 1.0 (v/v) for all the test microorganisms. Other investigations reported that rosemary essential oil had a strong antimicrobial activity against *S. aureus*,[32] and detailed the antimicrobial efficiency of the main rosemary essential oil components. For example, among α-pinene, myrcene, 1,8-cineole, camphor, and borneol compounds, only α-pinene was able to inhibit *S. aureus* ATCC 25923, *Enterococcus faecalis* ATCC 29212, *E. coli* ATCC 35218, and *K. pneumoniae*, with the MIC ranging from 0.8 to 8 μL/mL.[31] It also has been discussed that the antimicrobial activities of rosemary essential oil were superior to α-pinene and 1,8-cineole.[30] It is difficult to attribute the antimicrobial effect of an essential oil to one active compound or a few, because extracts are a mixture of different chemical compounds with diverse biological properties.

At present, there is a growing interest in enhancing the antibiotic activity of bioactive plant products, and numerous efforts have been made to increase its effectiveness against microorganisms. Studies about the effect of rosemary essential oil with tetracycline, gentamycin, sulfazotrim, chloramphenicol, cefepime, and tetracylcline have showed a synergistic effect against *S. aureus* and *E. coli*.[1]

OREGANO OIL

The *Origanum* (Lamiaceae) genus is an annual, perennial and shrubby herb that is predominantly distributed around the Mediterranean, Euro-Siberian, and Iran-Siberian regions.[33] Its use in traditional medicine has given rise to numerous studies that confirmed the benefits of oregano for human health, and its oil has been used to treat respiratory and gastrointestinal disorders, as well as an oral antiseptic and in dermatological applications.[34] *Origanum* has around 39 species, of which *Origanum vulgare* L. is the most studied. Its essential oil has been reported to have several biological properties, such as antioxidant, antimicrobial, and antimutagenic.[35]

Origanum vulgare essential oil is composed of different compounds. Most of them are thymol and carvacrol, but other compounds include ρ-cymene, thymoquinone, and γ-terpinene.[35,36] It has been reported that oregano essential oils have a powerful antimicrobial action against bacteria (Gram-positive and Gram-negative), yeast and fungi.[34] Oregano (*Origanum vulgare* subsp. vulgare) essential oil showed activity against *Sarcina lutea*, *S. aureus*, *C. albicans*, *E. faecalis*, and *Bacillus cereus*, resulting in inhibition halos of 34.67, 26.67, 24.67, 22.33, and 20.33 mm, respectively.[35] Lv et al.[37] reported that oregano essential oil has high activity against *S. aureus* (27.4 ± 0.5 mm), *B. subtilis* (27.4 ± 0.7 mm), *E. coli* (18.2 ± 0.8 mm), and *Saccharomyces cerevisiae* (27.2 ± 0.6 mm). In the same study, the MIC was determined for all tested microorganisms, they found that MIC was the lowest with 0.625 μL/mL, in all bacterial strains.[37] Hammer et al.[29] reported similar results, MIC values of oregano oil against *E. coli* and *S. aureus* were from 0.5 to 1.2 μL/mL. Actually, oregano has been used against pathogenic bacterial strains such as *E. coli* O157:H7, and when it was in direct contact with *Salmonella* Typhimurium DT104, this microorganism was inactivated.[38]

Oregano essential oil has been tested for use as a cleaning agent in the treatment of antibiotic-resistant bacteria on organic leafy greens; an investigation demonstrated that with a treatment of 0.5% oregano oil, the greatest microorganism population reductions (up to 4.9-log) was seen on all leafy greens, and even biological activity increased over time in storage.[39] In this context, it was investigated the disinfection efficacy of oregano oil on *Salmonella* Typhimurium inoculated into iceberg lettuce. Washing lettuce leaves with oregano oil led to a significant reduction in numbers of *Salmonella* Typhimurium as compared with chlorinated water. The best result occurred at a 75-ppm concentration, with a reduction of 1−92 log CFU/g. The authors stated that oregano oil could be used as a natural alternative to chlorine without affecting sensory properties.[34]

Some studies have reported the effect of oregano oil in combination with other essential oils like marjoram, where the reduction in the maximum specific microbial growth rate achieved was approximately threefold higher than that with the oregano oil alone.[36] Investigations have further examined the effect of oregano essential oil on lag phase duration; when *E. coli* was exposed to oregano in combination with basil, the results indicated that the time of the lag phase increased by 7.44 h with respect to the increase with oregano alone.[36]

BASIL ESSENTIAL OIL

Common basil (*Ocimum basilicum* L.) belongs to the Lamiaceae family, and it has been considered an important herb traditionally used worldwide. More than 150 species of this genus have been recognized, and basil is the most commercially important of these around the world.[40,41] It has been used as a food ingredient for flavoring, in cosmetics, and in traditional medicine for treating coughs, inflammations, and pains.[41,42] Basil essential oil has been indicated to possess high antioxidant, antimicrobial, antihypertensive, anticancer, and antiinflammatory activities.[40] Different authors have described the antibacterial properties of this essential oil as well.

The chemical composition of *O. basilicum* essential oil differs according to the season. These essential oils have oxygenated monoterpenes (60.7−68.9%), followed by sesquiterpene hydrocarbons (16.0−24.3%) and oxygenated sesquiterpenes (12.0−14.4%).[41] Around 29 compounds representing 98.0−99.7% of the oil composition have been reported by Hussain et al.[41] Linalool was the main constituent of essential oils (56.7−60.6%), followed by epi-α-cadinol (8.6−1.4%), α-bergamotene (7.4−9.2%), γ-cadinene (3.3−5.4%), germacrene D (1.1−3.3%), and camphor (1.1−3.1%). In addition, components like methylchavikol, methylcinnamat, linolen, eugenol, camphor, *cis*-geraniol, 1,8-cineole, α-bergamotene, β-caryophyllene, germacrene D, γ-cadinene, epi-α-cadinol, and viridiflorol have been reported as important components.[41,42]

The essential oil obtained from *O. basilicum* has showed strong antimicrobial activity against a wide number of microorganisms. This strong effect has been demonstrated using *O. basilicum* in different seasons. *S. aureus* and *B. subtilis* were two of the tested microorganisms, and these bacteria showed significant sensitivity in the presence of basil essential oil, with low MIC values for *S. aureus* (0.9−1.5 mg/mL) and *B. subtilis* (0.8−1.4 mg/mL). These findings showed that basil essential oil is a strong antimicrobial agent.[41]

Opalchenova and Obreshkova[42] reported on the antimicrobial activity of basil against drug-resistant bacterial strains from *Staphylococcus*, *Pseudomonas*, and *Enterococcus* genera. Other investigations reported that *L. monocytogenes* and *B. cereus* were sensitive to basil essential oil. The viability of these two bacteria treated with 0.1% basil essential oil was 0.16% and 0.08%, respectively. In the same study, *Vibrio* spp. and *Aerobacter hydrophila* showed high sensitivity too; with 0.01% basil essential oil concentration, the viability varied from 0.014% to 3.64%.[43]

A number of studies have discovered the effect of basil essential oil on drugs such as amoxicillin and flumequine. It has demonstrated antimicrobial activity comparable to these drugs, presenting a strong effect against Gram-positive and Gram-negative bacteria. This effect increased when the essential oil is produced in the autumn and winter, when linalool content is in higher concentrations.[41] Also, essential oil from *O. basilicum* L. has been shown to possess an inhibitory effect on fungi like *Aspergillus ochraceus*.[42]

MENTHA ESSENTIAL OIL

Mentha is a well-known genus (in the Lamiaceae family) that has medicinal and aromatic value. The *Mentha* genus includes 25–30 species that are widely grown in temperate areas around the world, particularly in Europe, North America, North Africa, Asia Minor, Northern parts of Iran and near East (Syria, Ethiopia), but nowadays, it is cultivated throughout all regions of the world.[44–46] *Mentha piperita* is a perennial plant that is 50–90 cm high, normally quadrangular, and a prototypical member of the mint family.[44]

Mentha spp. is a plant that exhibits important biological activities. For that reason, it has been used through the years as a remedy for respiratory diseases like bronchitis, sinusitis, tuberculosis, and the common cold. The plant acts as an excellent expectorant.[46] These bioactivities are due to the essential oil extracted from different parts of the plant, like aerial parts of the flowering plant, dried leaves, and fresh flowering.[44]

The composition of the Mentha essential oil directly affects the effectiveness of its antimicrobial activity,[45] which have displayed differences in its constituents depending on the growing area. The chemistry of mentha oil is very complex and highly variable. Analysis by gas chromatography–mass spectrometry (GC-MS) revealed that the prominent components are menthol, isomenthone, limonene, iso-menthanol, menthol acetate, carvone, β-pinene, α-pinene, 1,8-cineole, α-terpineol, isopulegol, pulegone, piperiton, piperitone oxide, and β-phellandrene.[45–47] Both the in vivo and in vitro antimicrobial power of these components on *Streptococcus mutans* and *Streptococcus pyogenes* have been evaluated, showing outstanding activity.[45] In addition, the oil exhibited bactericidal effects against *S. aureus*, *Staphylococcus epidermidis*, *B. cereus*, and *E. coli* 4.7. The MIC of the *M. piperita* essential oil was determined against various bacterial strains and varied from 1.13 to 2.25 mg/mL; the MIC of Gram-positive bacteria (*B. subtilis* and *S. aureus*) was lower than that of Gram-negative bacteria (*E. coli*, *P. aeruginosa*, and *Pseudomonas fluorescens*).[45]

The antimicrobial activity of Mentha oil against different microorganisms in a disk diffusion assay showed significant effects against *S. aureus*, *Staphylococcus epidermidis*, *B. cereus*, *C. albicans*, and *Vibrio cholerae*, with an inhibition zone in a range of 13–21 mm.[46] On the other hand, Singh et al.[44] reported that *S. aureus* and *Staphylococcus pyogenes* were sensitive to this essential oil, with an inhibition zone of 17.2 and 13.1 mm, respectively, which is greater than that of gentamycin.

CONCLUSIONS

The antimicrobial activity of essential oils are of great interest in the food, cosmetic, and pharmaceutical industries, since their possible use as natural additives emerged from the tendency to replace antibiotics due to the high resistance shown by a number of different pathogens. Although essential oils appear to be the answer to the problem of antibiotic resistance, additional investigations need to be performed to confirm the safety of each essential oil and evaluate the appropriate concentration of each one for human consumption.

One of the most important points to be consider for the use of essential oils in fighting pathogenic bacteria resistant to antibiotics is to perform studies to achieve greater yields of oils. Although essential oils are widely distributed by nature, they appear only in low concentrations. Future studies should focus on extraction methodologies that are technically and economically viable in order to improve recovery yields, and also ensure that the methodologies do not harm the bioactive components of the essential oils. Furthermore, it is important to make efforts to evaluate the synergistic effect of the combination of essential oils and pH values to enhance antimicrobial activity and to study the effect of combining water activity and essential oils in order to develop formulations that allow the essential oils to be used to replace specific antibiotics. Studies in vivo and clinical trials are needed to evaluate the potential of these essential oils as antibiotic substitutes or as interveners in this type of treatment.

REFERENCES

1. Barreto H, Silva-Filho EC, Lima E de O, Coutinho HDM, Morais-Braga MFB, et al. Chemical composition and possible use as adjuvant of the antibiotic therapy of the essential oil of *Rosmarinus officinalis* L. *Ind Crops Prod* 2014;**59**:290–4.
2. Nithyanand P, Shafreen RMB, Muthamil S, Murugan R, Pandian SK. Essential oils from commercial and wild Patchouli modulate Group A Streptococcal biofilms. *Ind Crops Prod* 2015;**69**:1806.
3. Suzuki S, Horinouchi T, Furusawa C. Suppression of antibiotic resistance acquisition by combined use of antibiotics. *J Biosci Bioeng* 2015;**120**(4):467–9.
4. Rönnerstrand B, Andersson SK. Trust, reciprocity and collective action to fight antibiotic resistance. An experimental approach. *Soc Sci Med* 2015;**142**:249–55.
5. Raut JS, Karuppayil SM. A status review on the medicinal properties of essential oils. *Ind Crops Prod* 2014;**62**:250–64.
6. Veras HNH, Rodrigues FFG, Colares AV, Menezes IRA, Coutinho HDM, Botelho MA, et al. Synergistic antibiotic activity of volatile compounds from the essential oil of *Lippia sidoides* and thymol. *Fitoterapia* 2012;**83**:508–12.
7. Rivera Calo J, Crandall PG, O'Bryan CA, Ricke S. Essential oils as antimicrobials in food systems—a review. *Food Control* 2015;**54**:111–19.
8. Bakkali F, Averbeck S, Averbeck D, Idaomar M. Biological effects of essential oils—a review. *Food Chem Toxicol* 2008;**46**:446–75.
9. Carson CF, Hammer KA, Riley TV. *Melaleuca alternifolia* (tea tree) oil: a review of antimicrobial and other medicinal properties. *Clin Microbiol Rev* 2006;**19**:50–62.
10. Fisher K, Phillips C. Potential antimicrobial uses of essential oils in food: is citrus the answer? *Trends Food Sci Technol* 2008;**19**:156–64.
11. Holley RA, Patel D. Improvement in shelf-life and safety of perishable foods by plant essential oils and smoke antimicrobials. *Food Microbiol* 2005;**22**(4):273–92.
12. Benchaar C, Calsamiglia S, Chaves AV, Fraser GR, Colombatto D, McAllister TA, et al. A review of plant-derived essential oils in ruminant nutrition and production. *Anim Feed Sci Technol* 2008;**145**:209–28.

13. Paduch R, Kandefer-Szerszen M, Trytek M, Fiedurek J. Terpenes: substances useful in human healthcare. *Arch Immunol Ther Exp* 2007;**55**:315–27.
14. Bajpai VK, Baek KH, Kang SC. Control of *Salmonella* in food by using essential oils: a review. *Food Res Int* 2012;**45**:722–34.
15. Friedly EC, Crandall PG, Ricke SC, Roman M, O'Bryan C, Chalova VI. In vitro antilisterial effect of citrus oil fractions in combination with organic acids. *J Food Sci* 2009;**74**:M67–72.
16. Trombetta D, Castelli F, Sarpietro MG, Venuti V, Cristani M, Daniele C, et al. Mechanisms of antibacterial action of three monoterpenes. *Antimicrob Agents Chemother* 2005;**49**:2474–8.
17. Helander IM, Alakomi HL, Latva-Kala K, Mattila-Sandholm T, Pol L, Smid EJ, et al. Characterization of the action of selected essential oil components on gram negative bacteria. *J Agric Food Chem* 1998; **46**:3590–5.
18. Ruiz B, Flotats X. Citrus essential oils and their influence on the anaerobic digestion process: an overview. *Waste Manag* 2014;**34**:2063–79.
19. Kollanoor-Johny A, Mattson T, Baskaran SA, Amalaradjou MA, Babapoor S, March B, et al. Reduction of *Salmonella enterica* serovar enteritidis colonization in 20-day-old broiler chickens by the plant-derived compounds trans-cinnamaldehyde and eugenol. *Appl Environ Microbiol* 2012;**78**:2981–7.
20. Muthaiyan A, Martin EM, Natesan S, Crandarll PG, Wilkinson BJ, Rickie SC. Antimicrobial effect and mode of action of terpeneless cold pressed Valencia orange essential oil on methicillin-resistant *Staphylococcus aureus* cell lysis. *J Appl Microbiol* 2012;**112**:1020–33.
21. Sonboli A, Eftekhar F, Yousefzadi M, Kanani MR. Antibacterial activity and chemical composition of the essential oil of *Grammosciadium platycarpum* Boiss. from Iran. *Z Naturforsch C* 2005;**60**:30–4.
22. Fisher K, Rowe C, Philips CA. The survival of three strains of *Arcobacter butzleri* in the presence of lemon, orange and bergamot essential oils and their components in vitro and on food. *Lett Appl Microbiol* 2007;**44**:495–9.
23. Fisher K, Phillips C. The effect of lemon, orange and bergamot essential oils and their components on the survival of *Campylobacter jejuni*, *Escherichia coli* O157, *Listeria monocytogenes*, *Bacillus cereus* and *Staphylococcus aureus* in vitro and in food systems. *J Appl Microbiol* 2006;**101**:1232–40.
24. Cosentino S, Tuberoso CIG, Pisano B, Satta M, Mascia V, Arzedi E, et al. In vitro antimicrobial activity and chemical composition of *Sardinian thymus* essential oils. *Lett Appl Microbiol* 1999;**29**:130–5.
25. Pathan RK, Gali PR, Patham P, Gowtham T, Pasupuleti S. In vitro antimicrobial activity of *Citrus aurantifolia* and its phytochemical screening. *Asian Pac J Trop Dis* 2012;S328–31.
26. Espina L, Somolinos M, Lorán S, Conchello P, García D, Pagán R. Chemical composition of commercial citrus fruit essential oils and evaluation of their antimicrobial activity acting alone or in combined processes. *Food Control* 2011;**22**:896–902.
27. Subba MS, Southmithri TC, Suryanarayana R. Antimicrobial action of citrus oils. *J Food Sci* 1967; **32**:225–7.
28. Prabuseenivasan S, Jayakumar M, Ignaciumuth S. In vitro antibacterial activity of some plant essential oils. *BMC Complement Altern Med* 2006;**6**:39.
29. Hammer KA, Carson CF, Riley TV. Antimicrobial activity of essential oil and other plant extracts. *J Appl Microbiol* 1999;**86**:985–90.
30. Jiang Y, Wu N, Fu YJ, Wang W, Luo M, Zhao CJ, et al. Chemical composition and antimicrobial activity of the essential oil of Rosemary. *Environ Toxicol Pharmacol* 2011;**32**:63–8.
31. Ojeda-Sana AM, van Baren CM, Elechosa M, Juárez MA, Moreno S. New insights into antibacterial and antioxidant activities of Rosemary essential oils and their main components. *Food Control* 2013;**31**: 189–95.
32. Fumiere-Lemos M, Fumiere-Lemos M, Poltronieri-Pacheco H, Coutinho-Endringer D, Scherer R. Seasonality modifies rosemary's composition and biological activity. *Ind Crops Prod* 2015;**70**:41–7.

33. Bayramoglu B, Sahin S, Sumnu G. Solvent-free microwave extraction of essential oil from oregano. *J Food Eng* 2008;**88**:535–40.
34. Gündüz GT, Gönül SA, Karapinar M. Efficacy of oregano oil in the inactivation of *Salmonella* Typhimurium on lettuce. *Food Control* 2010;**21**:513–17.
35. Sarikurkcu C, Zengin G, Oskay M, Uysal S, Ceylan R, Aktumsek A. Composition, antioxidant, antimicrobial and enzyme inhibition activities of two *Origanum vulgare* subspecies (subsp. Vulgare and subsp. Hirtum) essential oils. *Ind Crops Prod* 2015;**70**:178–84.
36. Gutierrez J, Barry-Ryan C, Bourke P. The antimicrobial efficacy of plant essential oil combinations and interactions with food ingredients. *Int J Food Microbiol* 2008;**124**:91–7.
37. Lv F, Lian H, Yuan Q, Li Ch. In vitro antimicrobial effects and mechanism of action of selected plant essential oil combination against four food-related microorganisms. *Food Res Int* 2011;**44**:3057–64.
38. Firouzi R, Shekarforuush SS, Nazer AH, Bourumand Z, Jooyandeh AR. Effects of essential oils of oregano and nutmeg on growth and survival *Yersinia enterocolitica* and *Listeria monocytogenes* in barbecued chicken. *J Food Prot* 2007;**70**(11):2626–30.
39. Moore-Neibel K, Gerber C, Patel J, Friedman M, Jaroni D, Ravishankar S. Antimicrobial activity of oregano oil against antibiotic-resistant *Salmonella enterica* on organic leafy greens at varying exposure times and storage temperatures. *Food Microbiol* 2013;**34**:123–9.
40. Arranz E, Jaime L, López de las Hazas MC, Reglero G, Santoyo S. Supercritical fluid extraction as an alternative process to obtain essential oils with anti-inflammatory properties from marjoram and sweet basil. *Ind Crops Prod* 2015;**67**:121–9.
41. Hussain AI, Anwar F, Hussain-Sherazi ST, Przybylski R. Chemical composition, antioxidant and antimicrobial activities of basil (*Ocimum basilicum*) essential oils depends on seasonal variations. *Food Chem* 2008;**108**:986–95.
42. Opalchenova G, Obreshkova D. Comparative studies on the activity of basil—an essential oil from *Ocimum basilicum* L.—against multidrug resistant clinical isolates of the genera *Staphylococcus*, *Enterococcus* and *Pseudomonas* by using different test methods. *J Microbiol Methods* 2003;**54**:105–10.
43. Koga T, Hirota N, Takumi K. Bactericidal activities of essential oil of basil and sage against a range of bacteria and the effect of these essential oils on *Vibrio parahaemolyticus*. *Microbiol Res* 1999;**154**:267–73.
44. Singh R, Shushni MAM, Belkheir A. Antibacterial and antioxidant activities of *Mentha piperita* L. *Arab J Chem* 2015;**8**:322–8.
45. Tyagi AK, Malik A. Antimicrobial potential and chemical composition of *Mentha piperita* oil in liquid and vapour phase against food spoiling microorganisms. *Food Control* 2011;**22**:1707–14.
46. Mahboubi M, Haghi G. Antimicrobial activity and chemical composition of *Mentha pulegium*, L. essential oil. *J Ethnopharmacol* 2008;**119**:325–7.
47. Ahmad A, Khan A, Samber N, Manzoor N. Antimicrobial activity of *Mentha piperita* essential oil in combination with silver ions. *Synergy* 2014;**1**:92–8.
48. Furneri PM, Mondello L, Mandalar G, Paolino D, Dugo P, Garozzo A, et al. In vitro antimycoplasmal activity of *Citrus bergamia* essential oil and its major components. *Eur J Med Chem* 2012;**2**:66–9.

CHAPTER

12

SENSITIVITY OF ESBL-PRODUCING GRAM-NEGATIVE BACTERIA TO ESSENTIAL OILS, PLANT EXTRACTS, AND THEIR ISOLATED COMPOUNDS

Gy. Horváth[1], T. Bencsik[1], K. Ács[1] and B. Kocsis[2]

[1]*Institute of Pharmacognosy, Medical School, University of Pécs, Pécs, Hungary*

[2]*Institute of Medical Microbiology and Immunology, Medical School, University of Pécs, Pécs, Hungary*

INTRODUCTION

The discovery of antimicrobial drugs, chemotherapeutics (eg, sulfonamides) and antibiotics (eg, penicillins) was a revolutionary step in the struggle against infectious diseases.[1] These effective medicines can stop the multiplication of microbes or they can kill the sensitive strains and saved the lives of millions of people suffering from infections.

Science solved a problem, but another one was created: the resistance of microbes to antimicrobial drugs. The microbiologists encountered resistant strains early. There are two types of bacterial resistance: (1) natural resistance occurs when an antimicrobial drug cannot kill a bacterium: for example, *Pseudomonas aeruginosa* is naturally resistant to penicillin-G, as its cell envelope is impenetrable for this antibiotic and (2) acquired resistance develops when a bacterial strain becomes resistant to a drug to which it had been sensitive before. Penicillin-G has been used in therapy since 1940. Alexander Fleming found a penicillin-G resistant *Staphylococcus aureus* strain as early as 1944. Methicillin was approved in 1958, and the first methicillin-resistant strain was isolated in 1961.[1] The resistant strains have spread all over the world, mostly in hospitals where the so-called nosocomial polyresistant, or multidrug-resistant (MDR) strains are present. They cause a growing problem with the treatment of both Gram-positive and Gram-negative bacterial infections.

The genetic background for acquired resistance is a mutation in the bacterial genome[1]: (1) a point mutation in chromosome or (2) motile genetic elements, plasmids, and transposons, are built into the genome. In particular, the plasmids are dangerous. They are common in nosocomial strains and they can spread from bacterium to bacterium, possibly by horizontal and vertical transmission. A plasmid can jump from one species to the same one (eg, from *Escherichia coli* into *E. coli*) or from one species

Antibiotic Resistance. DOI: http://dx.doi.org/10.1016/B978-0-12-803642-6.00012-5

to another one (eg, from *E. coli* into *Klebsiella pneumoniae*). Plasmids can multiply in a bacterial cell, and after being released from the host, they can infect a great number of other new bacterial cells. They not only transport one resistance gene, but also two or three resistance genes, causing polyresistance. They can simultaneously transport genes of different virulent factors as well, increasing the virulence of the recipient nosocomial strains. This gene-complex is called pathogenicity island, which plays an important role in the quick development of difficult-to-treat, hospital-acquired infections. During the treatment, sensitive bacteria are killed, but resistant ones can survive. We select the resistant bacterial clones, which is a selective pressure for the spreading of resistant strains.

After a mutation, the bacterial strain will change, and we can follow the mechanism of resistance as follows:

1. One possibility is a modification in the target molecule; for example, mutation in a penicillin-binding protein is in the background of methicillin-resistant *S. aureus*, and mutation in ribosome can induce resistance to macrolides.
2. A by-pass mechanism can cause resistance to sulfonamides.
3. The membrane permeability of carbapenems may be diminished after a porin mutation in the outer membrane of *P. aeruginosa* strains.
4. In macrolide resistance, a speeded-up efflux mechanism can be found.
5. Another possibility is that the bacterium can start to produce the following:
   a. *Modifying enzymes:* for instance, *trans*-acetylase can cause resistance to aminoglycosides or macrolides.
   b. *Destroying enzymes:* for instance, β-lactamase produced by a bacterium can inactivate β-lactamring by splitting it. The first β-lactamases were discovered as penicillinases by E. P. Abraham and E. Chain from *Bacillus/E. coli* in 1940[2] and by A. Fleming from a *S. aureus* strain in 1944. These bacteria produced an enzyme named penicillinase that could destroy penicillin-G. Now it is known that this enzyme is a β-lactamase that attacks and hydrolyzes the amide bond in the β-lactam ring and inactivates it.

It has been proved that nearly 1000 β-lactamases[1] are produced by both Gram-positive and Gram-negative strains. Gram-negative bacteria, including members of Enterobacteriaceae like *K. pneumoniae*, *E. coli*, *Enterobacter* spp., and *Acinetobacter* spp., can cause serious, life-threatening infections both in the community, and in hospitals. The so-called hospital-acquired nosocomial infections include urinary tract, lower respiratory tract, ventilation-associated pneumonia, bloodstream infections associated with the intravascular catheters used in intensive care units, and intra-abdominal infections after perforation of the appendix or diverticulum in the gastrointestinal tract. These infections are treated with β-lactam antibiotics like penicillins, cephalosporins, and carbapenems. Resistance to these antibiotics, especially in third-generation cephalosporins, is based on the production of a serial type of so-called extended-spectrum β-lactamases (ESBLs). These resistant strains threaten the success of treatment and the life of the patient.

The proper classification of ESBLs synthetized by Gram-negative bacteria is debatable, and there are different points of view on the matter. There are at least three systems constructed by Amber, Bush, and Giske: R.P. Ambler[3] divided β-lactamases into four classes (A, B, C, and D) based on their amino acid sequences; K. Bush[4,5] separated β-lactamases into groups based on their substrate hydrolytic and inhibition properties; and C.G. Giske, balancing between science and clinical needs, set up a simplified system of β-lactamases.[6] Table 12.1 compares the basic characteristics of these systems.

Table 12.1 Types and Classifications of β-Lactamases

Functional Group	Molecular Class	Clinical Aspect	β-Lactamase Genotype	Substrate and Inh. for β-Lactamase
1	C	$ESBL_M$	AmpC	Cephalosporins Inh.clav.acid: *No*
2b	A	$ESBL_A$	SHV TEM	Penicillins Inh.clav.acid: *Yes*
2be	A	$ESBL_A$	CTX-M SHV TEM	Cephalosporins Inh.clav.acid: *Yes*
2d	D	$ESBL_M$	OXA-ESBL	Cephalosporins Inh.clav.acid: *Yes*
2df	D	$ESBL_{CARBA}$		Cephalosporins Inh.clav.acid: *Yes*
2f	A	$ESBL_{CARBA}$	Serine-type KPC	Cephalosporins + carbapenems Inh.clav.acid: *Yes*
3	B	$ESBL_{CARBA}$	Zinc type MBL	Cephalosporins + carbapenems Inh.clav.acid: *No*

Functional Groups data from: *Bush K. Bench-to-bedside review: the role of β-lactamases in antibiotic-resistant Gram-negative infections. Crit Care 2010;14:224–31.*
Molecular Class data from: *Ambler RP. The structure of β-lactamases. Philos Trans R Soc Lond B Biol Sci 1980;289:321–31.*
Clinical Aspect data from: *Giske CG, Sundsfjord AS, Kahlmeter G, Woodford N, Nordmann P, Paterson DL, et al. Redefining extended-spectrum β-lactamases: balancing science and clinical need. J Antimicrob Chemother 2009;63:1–4.*
Inh.clav.acid, inhibition of β-lactamase enzyme by clavulanic acid.

In the European Committee on Antimicrobial Susceptibility Testing (EUCAST) definition of ESBL,[7] ESBLs are enzymes that hydrolyze most penicillins and cephalosporins, including oxyimino-β-lactam compounds (cefuroxime, third- and fourth-generation cephalosporins, and aztreonam), but not cephamycins or carbapenems. Most ESBLs belong to the Ambler class A of β-lactamases and are inhibited by β-lactamase inhibitors (clavulanic acid, sulbactam, and tazobactam).[7]

The first ESBL-producing strain was isolated in 1983. This type of strain has been spread all over the world by clonal expansion of ESBL-producing bacteria or horizontal transfer of ESBL genes on plasmids. ESBL production has been observed mostly in Enterobacteriaceae. These strains were isolated from inpatients and outpatients, the hospital environment, nursing homes, healthy carriers, sick and healthy animals, and food products.[8,9] The screening breakpoint of >1 g/L is recommended for cefotaxime, ceftriaxone, ceftazidime, and cefpodoxime.[7] ESBL production may be combined with other resistance mechanisms (such as increased efflux or diminished outer membrane permeability). Consequently, these strains become MDR.

DIAGNOSTIC APPROACH

The EUCAST guidelines[7] give a detailed description of the diagnostic steps for a correct determination of ESBL-producing, Gram-negative bacteria. This section briefly summarizes the possibilities: the phenotyping screening and confirmation tests are proposed for the routine laboratories, and the genotyping confirmation tests are used for final diagnosis in specialized laboratories.

PHENOTYPING SCREENING TESTS

Group 1 of Enterobacteriaceae (*E. coli*, *K. pneumoniae*, *Proteus mirabilis*, *Salmonella* spp., and *Shigella* spp.). Cefotaxime, ceftriaxone, and ceftazidime are used as indicator cephalosporins.

Group 2 of Enterobacteriaceae (*Enterobacter* spp., *Serratia* spp., *Citrobacter freundii*, *Morganella morganii*, *Providencia* spp., and *Hafnia alvei*) Chromosomal AmpC β-lactamase is characteristic of this group of bacteria, and cefepime is stable for AmpC hydrolysis. It can be used in phenotypic testing with clavulanic acid inhibition.

PHENOTYPING CONFIRMATION TESTS

A combination of disk test, double-disk synergy test, gradient test method, and broth microdilution is used for the purpose of confirming phenotypes.

GENOTYPING CONFIRMATION TESTS

For the detection of the presence of ESBL genes PCR and ESBL gene sequencing are used,[7] or a DNA microarray-based methods are recommended.

EPIDEMIOLOGY

Third-generation cephalosporins are widely used to treat common infections, like pneumonia, meningitis, and bloodstream and urinary tract infections (UTIs). The bacteria reacts by producing ESBLs. ESBL-producing bacteria can degrade all types of cephalosporins, including the third generation of cephalosporins. In medical therapy, carbapenems (eg, imipenem, meropenem, ertapenem, and doripenem) can be used against these bacteria. Unfortunately, carbapenem-resistant strains appeared as a reaction. The mechanism of this type of resistance may be a porin mutation in the outer membrane of Gram-negative bacteria and diminished permeability for carbapenems. The other possible mechanism is carbapenemase production: serine-type carbapenemase (*K. pneumoniae* carbapenemase, KPC) may be detected in *P. aeruginosa* strains (eg, metallo-β-lactamase, MBL) or in *K. pneumoniae* strains. Colistin (polymyxin E) is one of the last options in the treatment of infections caused by these bacteria.

The prevalence of ESBL-producing strains in Northern European countries is low,[10] but the number of resistant strains is increasing. It is more common in southern and eastern European countries. CTX-M-15 is the most dominant genotype all over the world and in Scandinavian countries, followed by CTX-M-14. In Spain, CTX-M-14 is the most prevalent, but CTX-M-9 was also detected in Barcelona[11] and Poland. In India and Australia, CTX-M-15 and CTX-M-14 are the leaders. In the United States, the SHV phenotype starts to disappear and the number of CTX-M-15 increases. In South America, CTX-M 2 and CTX-M 8 are the characteristic phylo groups.

As a result, the number of ESBL-producing *E. coli* type CTX-M is increasing worldwide. It is important to at least slow down the rapid dissemination and persistence of these problematic strains.[12] The gut flora of healthy people and colonized patients is an ideal reservoir for the persistence of antibiotic-resistant genes, a source of clonal spreading of resistant strains, or horizontal gene transfer by plasmids. To minimize the development, persistence, and dissemination of resistant strains, we have to plan and follow an optimal antibiotic treatment. At the same time, we can diminish the selective pressure of the available antibiotics on bacteria and new treatment options are needed as well.

When we try to slow down and stop the spreading of resistant strains, a strict infection control is essential. Washing hands with disinfectant is one of the most effective methods in this aspect. The vaccination of risk groups may be an important step in the prevention of infections. In the treatment of diseases caused by resistant strains, the combination of available drugs or the combination of drugs with nondrugs (eg, essential oils (EOs)) may provide a solution. EOs present a number of new possibilities in the treatment of infections. In this chapter, we summarize the knowledge about this topic: the antibacterial and β-lactamase inhibitory activity of EOs, plant extracts, and their isolated compounds.

The next section summarizes the assays of antimicrobial activity of plant extracts frequently used in the cited articles.

FREQUENTLY USED METHODS FOR EVALUATING ANTIBACTERIAL ACTIVITY OF EOs, PLANT EXTRACTS, AND ISOLATED COMPOUNDS

There are several in vitro methods for detecting the antibacterial activity of EOs and plant extracts. Generally, disk diffusion, agar diffusion, and broth dilution methods (BDMs) are used, but in other cases time-killing and checkerboard assays are applied by researchers.

DISK DIFFUSION METHOD

The disk diffusion method (DDM) is classified as an agar diffusion method (ADM) because the plant extract to be tested diffuses from its reservoir through the agar medium seeded with the test microorganism. Generally, the reservoir is a filter paper disk, which is placed on top of an agar surface. If tested plant extracts or isolated compounds are microbiologically active, an inhibition zone develops around the filter paper disk after incubation. The diameter of the inhibition zone properly describes the antimicrobial potency of plant extracts or individual compounds. It should be mentioned that DDM is not an appropriate method for lipophilic extracts (eg, EOs) because the diffusion of the water-insoluble EO and its compounds from a filter paper disk into the agar medium is insufficient. Therefore, EOs give negative results or smaller inhibition zones with DDM than in other tests, such as the BDM.[13]

AGAR DIFFUSION METHOD

In the other type of ADM technique, plant extracts or solutions of individual compounds are passed through a hole in the agar plate. During the incubation period, the tested material diffuses from the hole into the agar medium seeded with the test microorganism. The active antimicrobial extracts or compounds result in zones of inhibition around the hole, which give information about the value of the minimum inhibitory concentration (MIC). Factors influencing the size of inhibition zones in DDM and AMD include the size of the filter paper disk or hole, the amount of compound placed onto the disk or into the hole, the type and concentration of the agar, the thickness and pH of the medium, the microbial strain tested, and the incubation temperature. Generally, DDM and ADM can be regarded as prescreening tests, which are appropriate for evaluating a high number of plant extracts or individual components during parallel examinations, but their results should not be overemphsized.[13]

BROTH DILUTION METHOD

The BDM is also suitable for the determination of MIC values. In this method, microorganisms are tested for the ability to produce visible growth on a series of agar plates (agar dilution, AD) or in broth (broth dilution, BD) containing dilutions of the antimicrobial agent. The MIC is the lowest concentration of an antimicrobial agent (in mg/mL), which prevents the appearance of visible growth of a microorganism. Each tube is inoculated with a known number of test microorganisms. In the classic tube or macrodilution method, 1 mL of the medium is diluted in test tubes. Nowadays, the compounds can be diluted in plates with a multichannel dilutor. Only 0.1 mL of the medium is needed for dilution, it is the so-called microdilution method. After one night of incubation, the result (turbidity) can determined by the naked eye or a plate reader. This is an economical and effective test. In the case of EOs examined with BDM, the main difficulty is their low water solubility. Therefore, different solvents (eg, dimethyl sulfoxide/DMSO or ethanol) or detergents (eg, Tween 20) are added to the tubes containing EO to be tested. However, these auxiliary materials may influence the multiplication of test bacterium and the MIC value, especially when their concentration is >5% (v/v).[13] We have to use careful control tests to avoid or minimize this effect.

CHECKERBOARD AND TIME-KILL ASSAYS

If two antimicrobial agents are combined, indifferent, additive, synergistic, and antagonistic effects may occur. The effect is indifferent when the combination of an antimicrobial constituent and an inactive substance has an identical effect to that of the effective compound. In the case of an additive effect, a mixture of antimicrobial compounds has an activity equal to the sum of the effects of each component. The synergistic effect occurs when a combination of antimicrobial constituents has a greater effect than the added activities of each compound. An antagonistic effect is observed when a reduced activity develops relative to the effect of the most efficient individual constituent.[14] These effects can be measured and quantified by the use of fractional inhibitory concentration (FIC) and fractional bactericidal concentration (FBC). If there are two antibacterial compounds, A and B, FIC can be calculated using the following expression: $FIC_A = MIC(_{A \text{ in the presence of } B})/MIC(_{A \text{ alone}})$; $FIC_B = MIC(_{B \text{ in the presence of } A})/MIC(_{B \text{ alone}})$. The FIC index is the sum of FIC_A and FIC_B. The FBC index can be similarly calculated using the FBC values of the individual bactericidal compounds. If the FIC index is <0.5, it means a synergistic effect, >0.5–1 indicates an additive effect, >1–2 means indifference, and ≥2 indicates an antagonistic effect.[14]

The checkerboard assay can be used to determine the FIC index experimentally. In this case, two antimicrobial agents are added to Mueller-Hinton broth in 96-well microtiter plates to give twofold dilutions in the horizontal and vertical directions, respectively. Then bacterial cells (approximately 8×10^5/mL) are added, and the plates are incubated.[15] The turbidity will develop when the antimicrobial drug concentration is not enough to stop the multiplication of bacteria or to kill microbes. The result can be detected by the naked eye or in a plate reader at 600 nm.

The time-kill assay is also an appropriate method for studying the effects after the combination of two antimicrobial agents. This method gives information about a direct correlation between exposure time to a test material and the extent of pathogenic death rate. Antagonism may be defined as at least a 100-fold increase in the number of colony-forming units (CFUs) per milliliter, whereas synergism occurs in the case of at least a 100-fold decrease in colony counts.[15]

MOLECULAR DOCKING

In some cases, the most developed and state-of-the-art techniques are used to predict the antimicrobial activity of a natural compound. The method of molecular docking predicts the best orientation of one molecule (eg, proteins) to another (eg, plant compound) when their bond to each other forms a stable complex. In a study, the in vitro microbiological assays and molecular docking experiments showed antibacterial activity and significant molecular interactions of eugenol and cinnamaldehyde with ESBL enzymes of pathogenic bacteria.[16] Therefore, the potential antimicrobial activity of a plant compound can be predicted without doing a lot of experiments, which saves both time and money. There are different software programs that can be used to perform molecular docking experiments.

The following sections introduce the application of various plant extracts involving EOs and their individual substances as potential antibacterial agents against ESBL-producing, Gram-negative bacteria including *K. pneumoniae*, *E. coli*, *Acinetobacter* spp., and *Enterobacter* spp. Each one will provide information on the plant sources of the antibacterial substances, their active compounds, and evidence for their antimicrobial activity based on in vitro assays. We grouped the previously published results according to the ESBL-producing, Gram-negative bacterial strains involved in the experiments.

EOs AND THEIR INDIVIDUAL COMPOUNDS AGAINST *K. PNEUMONIAE*

EOs contain hundreds of individual constituents (mainly monoterpenoids, sesquiterpenoids, and phenylpropanoids). Because of their multiple composition, they possess a complex mode of action, which makes them valuable agents in the treatment of MDR bacteria. It is well known that their composition is affected by several factors (eg, chemotype of plant species, environmental conditions, soil type, plant part, distillation process, etc.).[17] Gas chromatography–mass spectrometry (GC-MS) is a frequently used technique to analyze the chemical composition of EOs. In some articles, the antimicrobial activity of the EO or individual components were tested,[18] and in other cases, EO was combined with an antibiotic to test their synergistic effect.[19]

As previously mentioned, EOs are volatile, viscous, and water-insoluble substances. Therefore, the common microbiological methods (eg, disk diffusion, agar absorption, etc.) are not suitable for their antimicrobial testing without modifications. The most important limitation to these methods is that in the case of the DDM, only the more water-soluble components diffuse into the agar,[20] so it does not provide a true indication of the activity. In the agar absorption assay, EOs may be ineffectively absorbed by the agar; therefore, an oil layer remains on the agar surface. This process produces inconsistent and irreproducible results.[20] Among the frequently used microbiological assays, the broth microdilution method provides more reliable results, but the Tween concentration must be taken into consideration because it may influence the bacterial growth or alter cell membrane permeability. It may act antagonistically against the active oil components or may act synergistically with the oil constituents to produce higher activity.[20] Recent studies focusing on the antimicrobial activity of EOs in vitro described a wide range of assays with different parameters (agar recipes, incubation time, solvents, microorganisms, etc.),[21–25] so the results of the assays are very different and sometimes difficult to compare with each other, and their reliability is questionable. However, optimized and reproducible assays are needed for assessing the antibacterial effects of these oils.[26]

After 2008, the number of papers focusing on ESBL-producing *K. pneumoniae* is increasing. Nowadays, more and more research groups have started to examine the antimicrobial activity of plant extracts against ESBL-producing strains and their possible role in the treatment of troublesome infections. However, these results come from in vitro experiments. It should be mentioned that in some cases, the researchers did not indicate whether the *K. pneumoniae* strain was an ESBL producer.

In one study, the antibacterial activity of *Origanum vulgare* EO (OVEO) was demonstrated against multiresistant strains, including four *K. pneumoniae* (ESBL) isolated from hospitalized patients. The MIC values were determined by microdilution. Of the four *K. pneumoniae* strains, three were inhibited by OVEO at 0.125% (MIC) and one strain was inhibited at 0.25% (MIC). The chemical analysis of OVEO by GC-MS was not performed. The authors did not find differences in MIC values between Gram-positive and Gram-negative bacteria.[27]

The evaluation of antimicrobial activity of plants from the Lamiaceae family is very popular. In 2009, Jazani et al.[28] investigated the anti-*Klebsiella* activity of *Mentha pulegium* EO (MPEO) having 38% of piperitone content and 3.1% of menthone content. The 39 *Klebsiella* isolates were collected from urine samples of hospitalized patients in Iran. The antibiotic sensitivity of isolates was determined using disk diffusion assaying. Most of the *Klebsiella* isolates were resistant to ampicillin (79.5%), kanamycin (53.8%), ceftazidime (51.3%), and tobramycin (48.7%). The MIC and minimum bactericidal concentration (MBC) of MPEO were investigated by broth microdilution. The MPEO could inhibit the growth of every isolate. A 0.002% (v/v) of Tween 80 was used to increase the dispersion of the MPEO, but this chemical was not tested as a control. The authors concluded that MPEO may be applied to control infections caused by MDR *Klebsiella* spp., but further toxicological examinations are necessary to prove the safe application of MPEO in human therapy.

Orhan et al.[18] demonstrated the antibacterial activity of EOs obtained from other popular medicinal plants of the Lamiaceae family (eg, *Mentha* × *piperita*, *Mentha spicata*, *Ocimum basilicum*, *Origanum majorana*, *Origanum onites*, *O. vulgare*, and *Satureja cuneifolia*) and the Apiaceae family (eg, *Foeniculum vulgare*). Furthermore, several individual components (eg, linalool, *trans*-anethole, carvacrol, thymol, isomenthone, and carvone) of these EOs were tested for their antibacterial effects as well. Strains of *K. pneumoniae* producing ESBL enzymes were used in the study. The suspension of all bacteria was 10^5 CFU/mL. These microorganisms showed resistance to the following antibiotics: clavulanate-amoxicillin, ceftriaxone, cefepime, imipenem, tobramycin, ofloxacin, and ciprofloxacin. The MIC values of EOs and their components ranged between 32 and 64 μg/mL when tested by microdilution assays, and were similar to the MIC values of the reference antibiotics (ampicillin and ofloxacin) tested. The authors suggested the use of the abovementioned EOs as food preservatives or ingredients.

K. pneumoniae can be isolated from difficult-to-heal wounds. The topical application of EOs as antiseptic agents may be a way to treat hospital-acquired infections. Warnke et al.[29] examined the antibacterial effect of EOs of tea tree, eucalyptus, and lemongrass against hospital-acquired pathogens frequently found in surgical wards, including five ESBL-producing *K. pneumoniae* isolated from infected patients. The anti-*Klebsiella* activity of EOs was compared with an undiluted olive oil, as well as ethanol (70%) and chlorhexidine (0.1%) used as clinical antiseptics. The ADM method was used in the study. Lemongrass EO showed poor inhibition, with a 2–4 mm inhibition zone. In contrast, tea tree EO demonstrated the largest zone of inhibition (11–15 mm) against ESBL-producing *K. pneumoniae*. Eucalyptus EO was characterized by moderate inhibition (7–13 mm). Among the controls, chlorhexidine (2–4 mm) and ethanol (0–2 mm) showed weak anti-*Klebsiella* activity. GC-MS analysis of the EOs was not performed.

It has been demonstrated that clove (*Syzygium aromaticum*) EO (CEO) is an effective antimicrobial agent against human pathogens.[30] Sienkiewicz et al.[31] has proved the antibacterial activity of CEO against *K. pneumoniae* ESBL + isolated from hospitalized patients using the AD method. Therefore, a CEO may become an effective material to prevent the development and spreading of the antibiotic-resistant strains in a hospital environment.

In several cases, researchers investigated the anti-*Klebsiella* activity of an endemic plant species or a medicinal plant with the significant ethnomedicinal application. For instance, Köse et al.[32] examined the antibacterial activity of the EO of *Sideritis erythrantha* var. *erythrantha* (SE) and *S. erythrantha* var. *cedretorum* (SC), which are endemic in Turkey. However, they did not observe that EOs isolated from SE and SC were active against two *K. pneumoniae* (one of them was ESBL+) tested. In another study, the EOs of *Bidens tripartita* obtained from the fresh herb or flower heads were analyzed.[33] This plant is used in Polish folk medicine for the treatment of candidosis, skin diseases, and inflammations caused by bacteria or fungi. Unfortunately, these oils were not active against *K. pneumoniae* (MIC: >100 mg/mL).[33]

Currently, the body of knowledge about the mode of action of EOs against pathogens is increasing, but only a few papers focused on the *Klebsiella* ESBL + bacterium. In one study, *Plectranthus amboinicus* EO (PAEO) was tested in subinhibitory and inhibitory concentrations to reveal its effect on *K. pneumoniae* membrane permeability, capsule expression, urease activity, and cell morphology.[34] Carvacrol was the main constituent of PAEO. The results showed that PAEO in 2 × MIC (0.16%) could inhibit the growth of *Klebsiella* spp. after 2 h of incubation. When alkaline or neutral pH broth was applied, the authors could detect better MIC values of PAEO. Moreover, the PAEO inhibited the capsule expression and urease activity at subinhibitory concentrations.[34] In some cases, modern techniques (eg, molecular docking analysis) are also involved in the experiments. Using this method, the significant molecular interactions of eugenol (MIC: 63–999 μg/mL) and cinnamaldehyde (MIC: 245 μg/mL) with ESBL enzymes of pathogenic bacteria, including *Klebsiella*, could be determined.[16] Not only did these two EO constituents show strong anti-*Klebsiella* (ESBL+) effect, but clove and cinnamon EOs did as well, in which they are found in high concentrations. To date, studies have demonstrated that the bacterial cell targets of EOs include the cell wall and membrane, thereby disturbing adenosine triphosphate production and pH homeostasis. Furthermore, EOs can influence the cellular transcriptome, proteome, and quorum sensing systems.[35] Due to the large number of constituents, EOs seem to have no specific cellular targets.[36]

Many studies are concerned with the possible interactions between antibiotics and EOs or the individual components of EOs. However, some researchers studied this effect against

Klebsiella spp. Fadli et al.[19] tested EOs of two Moroccan thyme species, *Thymus maroccanus* and *Thymus broussonetii*, in combination with antibiotics by checkerboard testing. There was a synergistic effect between thyme EOs (MIC: 0.3 mg/mL of TMEO and 0.7 mg/mL of TBEO) and the tested antibiotics. Carvacrol was the main constituent of thyme oils, which showed synergistic activity in combination with ciprofloxacin as well. Radulović et al.[37] proved that *Filipendula ulmaria* EO, administered alone, showed lower antibacterial activity than the mixture of its individual compounds. They received the highest MIC value (<0.009 mg/mL) in the case of salicylaldehyde:linalool—namely, a 60/40 combination using the microdilution method. In contrast, antagonism was found between salicylaldehyde and methyl salicylate. They used the *K. pneumoniae* ATCC 10031 strain, but there was no information about the ESBL production of this bacterium. The therapeutic value of these experiments are highly important because the proved synergism between antibiotics and EOs or their components can reduce the minimum effective dose of the antibiotics. Therefore, their side effects and treatment cost can be reduced.

PLANT EXTRACTS AND THEIR INDIVIDUAL COMPOUNDS AGAINST *K. PNEUMONIAE*

Several research groups focus on the anti-*Klebsiella* activity of different plant extracts or tested isolated pure compounds. These plants play an important role in ethnomedicine and have been used in traditional medicine (for instance, *Terminalia alata*, an ethnomedicinal plant used for the treatment of gastrointestinal tract diseases).[38] The authors generally perform anti-*Klebsiella* examinations to find potential phytochemicals from natural sources, such as plants, that can be used against MDR microbes causing nosocomial and community-acquired infections. From the group of microbiological assays, DDM and ADM, including agar well diffusion methods, are used very frequently. The BDM is applied to determine the MIC and MBC values. Sometimes only plant extracts prepared by using different solvents (eg, ethanol, methanol, ethyl acetate, etc.) are tested, and there is no information about the microbiological active compounds.[38–40] In other cases, pure compounds (eg, flavonoids) are examined in combination with antibiotics.[41]

Table 12.2 summarizes the plant extracts that were found to be active against *K. pneumoniae.* It should be highlighted that we do not receive information about the ESBL-producing potency of *K. pneumoniae* in every case. In most cases, researchers indicated the MDR property of a Gram-negative bacterium.[39,40]

Table 12.3 summarizes the results of the studies performed by Rath and Padhy[46] and Mishra and Padhy[40]. They investigated 47 Indian ethnomedicinal plants against MDR bacteria (eg, *K. pneumoniae* and *Acinetobacter baumannii*), causing UTIs. *Murraya koenigii* (Rutaceae, MIC: 1.51 mg/mL), *Papaver somniferum* (Solanaceae, MIC: 1.51 mg/mL), *Cassia tora* (Leguminosae, MIC: 3.41 mg/mL), and *Eugenia jambolana* (Myrtaceae, MIC: 1.51 mg/mL) showed promising anti-*Klebsiella* activity, with lower MIC values compared to the others.

Research programs focusing on the combination of extracts or their isolated compound and antibiotics are more valuable. Dey et al.[47] examined the methanolic extract of *Punica granatum*

Table 12.2 Plant Extracts and Their Individual Compounds Effective Against ESBL and MDR *Klebsiella pneumoniae* Strains

Source Plant and Plant Part Tested	Extract	Method	Results	Type of *K. pneumoniae*	Reference
Acacia nilotica (L.) P.J.H. Hurter & Mabb., leaf	50% and 90% ethanol	Twofold serial dilution	MIC (μg/mL) = 156–313; MBC (μg/mL) = 313–1250	Clinical isolates (ESBL+)	39
Syzygium aromaticum (L.) Merr. & L.M. Perry, flower buds	50% and 90% ethanol	Twofold serial dilution	MIC (μg/mL) = 780–1560; MBC (μg/mL) = 1560–3130	Clinical isolates (ESBL+)	39
Cinnamomum zeylanicum J. Presl, bark	50% and 90% ethanol	Twofold serial dilution	MIC (μg/mL) = 390–780; MBC (μg/mL) = 1560–3130	Clinical isolates (ESBL+)	39
Pericopsis laxiflora (Benth.) Van Meeuwen, bark	Distilled water (Etaq), 70% ethanol ($Eeth_{70\%}$), methanol (Emet), ethyl acetate (Eace)	Agar well diffusion and tube dilution	Inhibition zones (mm): 0 (Etaq), 8 ($Eeth_{70\%}$), 9 (Emet), 10 (Eace) MIC, MBC (mg/mL) = 25 to >100	ESBL+	42
Terminalia alata Heyne ex. Roth, leaf and bark	Methanol (Em), ethanol (Ee), acetone (Ea)	Microbroth dilution	Leaf: MIC (mg/mL) = 1.56 (Em), 6.25 (Ee), 3.12 (Ea) MBC (mg/mL) = 12.5 (Em), 25.0 (Ee), 25.0 (Ea); Bark: MIC (mg/mL) = 1.56 (Em), 6.25 (Ee), 3.13 (Ea) MBC (mg/mL) = 12.5 (Em), 50.0 (Ee), 25.0 (Ea)	Clinical isolates (ESBL+)	38
Illicium griffithii Hook. f. & Thoms, fruit	Hexane (Eh), ethyl acetate (Ee), methanol (Em)	Disk diffusion, microbroth dilution	Inhibition zones (mm): 16–30 (Eh), 10–12 (Ee), 10 (Em); MIC (mg/mL): 0.15 (Eh), 0.62 (Ee), 2.5 (Em)	Clinical isolates (ESBL+)	43
Holarrhena antidysenterica Wall., leaf and bark	Ethyl acetate (Eea), ethanol (Ee), acetone (Ea)	Microbroth dilution	Leaf: MIC (mg/mL) = 6.25 (Eea), 1.56 (Ee), 3.12 (Ea) MBC (mg/mL) = 25.0, 12.5, 25.0; Bark: MIC (mg/mL) = 3.12 (Eea), 12.5 (Ee), 12.5 (Ea) MBC (mg/mL) = 12.5 (Eea), 25.0 (Ee), 25.0 (Ea)	Clinical isolates (ESBL+)	44

(Continued)

Table 12.2 Plant Extracts and Their Individual Compounds Effective Against ESBL and MDR *Klebsiella pneumoniae* Strains *Continued*

Source Plant and Plant Part Tested	Extract	Method	Results	Type of *K. pneumoniae*	Reference
Flindersine isolated from *Toddalia asiatica* (L.) Lam, leaf	Ethyl acetate	Broth dilution	MIC (μg/mL) = >250	*Klebsiella pneumoniae* ATCC 700603 (ESBL+)	45
Ulopterol isolated from *Toddalia asiatica* (L.) Lam, leaf	Ethyl acetate	Disk diffusion, broth microdilution	Inhibition zones (mm): 9–10; MIC (μg/mL) = 62.5 to > 250	Clinical isolates (ESBL+)	63
Bidens tripartita L., flower and herb	Water (Ew), methanol/ water (Emw), acetone/water (Eaw)	Broth microdilution, disk diffusion	Flower: inhibition zones (mm) = 0 (Ew), 8 (Emw), 8 (Eaw); MIC (mg/mL) = > 100 (Ew), 25.0 (Emw), 12.5 (Eaw); Herb: inhibition zones (mm) = 6 (Ew), 8 (Emw), 7 (Eaw); MIC (mg/mL) = 50.0 (Ew), 25.0 (Emw), 12.5 (Eaw)	*K. pneumoniae* ATCC 700603 (ESBL+)	33

Table 12.3 In Vitro Antibacterial Activity of Indian Medicinal Herbs Against *Acinetobacter baumannii*, MDR *Klebsiella pneumoniae,* and *Enterobacter aerogenes* Using Agar Well Diffusion and Dilution Method

Medicinal Plant and Plant Part Used	*Acinetobacter baumannii*		*Klebsiella pneumoniae*		*Enterobacter aerogenes*	
	Inhibition Zone (mm)	MIC (mg/mL)	Inhibition Zone (mm)	MIC (mg/mL)	Inhibition Zone (mm)	MIC (mg/mL)
Albizia lebbeck (L.) Benth., leaf	12	No activity	12	No activity	13	No activity
Allium sativum L., stem	17	9.63	17	9.63	17	9.63
Allium cepa L., stem	23	3.41	20	3.41	22	3.41
Alstonia scholaris L. R. Br., bark	11	No activity	12	No activity	No activity	No activity
Amomum aromaticum Roxb., seed	No activity	No activity	No activity	No activity	21	3.41
Anogeissus acuminata (Roxb. ex. DC.) Wall. ex. Guill. & Perr., leaf	23	4.27	25	4.27	27	1.51
Anthocephalus cadamba (Roxb.) Miq., leaf	24	4.27	20	9.63	25	4.27
Artocarpus heterophyllus Lam., leaf	22	9.63	21	9.63	19	9.63
Azadirachta indica L. Adelb., leaf	26	4.27	18	9.63	26	3.41

Table 12.3 In Vitro Antibacterial Activity of Indian Medicinal Herbs Against *Acinetobacter baumannii*, MDR *Klebsiella pneumoniae,* and *Enterobacter aerogenes* Using Agar Well Diffusion and Dilution Method *Continued*

	Acinetobacter baumannii		*Klebsiella pneumoniae*		*Enterobacter aerogenes*	
Medicinal Plant and Plant Part Used	**Inhibition Zone (mm)**	**MIC (mg/mL)**	**Inhibition Zone (mm)**	**MIC (mg/mL)**	**Inhibition Zone (mm)**	**MIC (mg/mL)**
Brassica juncea (L.) Vas. Matv. Czer. seed	24	1.51	22	3.41	16	4.27
Bridelia retusa (L.) Spreng., bark	26	3.41	22	9.63	28	1.51
Capsicum annuum L., fruit	22	3.41	No activity	No activity	No activity	No activity
Cassia tora L., leaf, seed	26	1.51	26	3.41	29	1.51
Cinnamomum tamala (Buch.-Ham.) T. Nees & C.H. Eberm., leaf	No activity	No activity	17	9.63	No activity	9.63
Cinnamomum zeylanicum J. Presl, bark	26	1.51	26	1.51	17	3.41
Coriandrum sativum L., seed	26	1.51	21	3.41	21	3.41
Cucumis melo L., seed	No activity	No activity	17	9.63	21	3.41
Cuminum cyminum L., seed	26	1.51	22	3.41	19	4.27
Curcuma amada Roxb., rhizome	23	3.41	17	9.63	No activity	No activity
Curcuma longa L., rhizome	22	3.41	22	3.41	21	3.41
Dalbergia latifolia Roxb., bark	18	9.63	No activity	No activity	19	9.63
Eucalyptus citriodora Hook., leaf	26	3.47	No activity	No activity	28	0.67
Elettaria cardamomum (L.) Maton, seed	No activity	No activity	12	No activity	12	No activity
Eugenia jambolana Lam., seed	23	4.27	26	1.51	26	1.51
Foeniculum vulgare L., seed	No activity	No activity	17	9.63	13	No activity
Ferula asafoetida L., seed	19	4.27	18	9.63	19	4.27
Gmelina arborea Roxb, root.	19	9.63	20	9.63	No activity	No activity
Holarrhena antidysenterica (L.) R.Br., leaf	Not tested	Not tested	21.6	6.25	23.5	3.25
Illicium verum Hook. f., flower	19	4.27	17	9.63	18	9.63
Melia azedarach L., leaf	19	9.63	No activity	No activity	No activity	No activity
Mentha spicata L., leaf	15	9.63	22	3.41	28	1.51
Mimusops elengi L., leaf	19	9.63	16	9.63	19	9.63

(*Continued*)

Table 12.3 In Vitro Antibacterial Activity of Indian Medicinal Herbs Against *Acinetobacter baumannii*, MDR *Klebsiella pneumoniae,* and *Enterobacter aerogenes* Using Agar Well Diffusion and Dilution Method *Continued*

	Acinetobacter baumannii		*Klebsiella pneumoniae*		*Enterobacter aerogenes*	
Medicinal Plant and Plant Part Used	**Inhibition Zone (mm)**	**MIC (mg/mL)**	**Inhibition Zone (mm)**	**MIC (mg/mL)**	**Inhibition Zone (mm)**	**MIC (mg/mL)**
Murraya koenigii L., leaf	26	1.51	26	1.51	17	3.41
Myristica fragrans L., seed	No activity	No activity	12	9.63	22	3.41
Nigella sativa L., seed	No activity	No activity	15	9.63	23	3.41
Papaver somniferum L., seed	19	4.27	28	1.51	No activity	No activity
Piper nigrum L., seed	No activity	No activity	22	3.41	25	1.51
Pongamia pinnata L., leaf	12	No activity	No activity	No activity	18	9.63
Pterocarpus marsupium Roxb., bark	22	9.63	13	No activity	19	9.63
Pterocarpus santalinus L., leaf	25	3.41	27	3.41	22	9.63
Schleichera oleosa (Lour.) Oken., seed	19	4.27	26	0.67	24	9.63
Shorea robusta Roth., leaf	16	9.63	18	9.63	No activity	No activity
Stereospermum kunthianum Cham., leaf	24	3.41	19	9.63	22	4.27
Syzygium aromaticum (L.) Merr. & L.M. Perry, flower bud	24	1.51	21	3.41	29	1.51
Tectona grandis L., bark	26	1.51	22	9.63	19	9.63
Terminalia alata Heyne ex. Roth, bark	Not tested	Not tested	26	1.56	25	3.13
Trachyspermum ammi Sprague, seed	No activity	No activity	21	3.41	21	3.41
Trigonella foenum L., seed	No activity	No activity	21	3.41	23	1.51
Zingiber officinale L., rhizome	18	9.63	20	3.41	23	3.41

Data from:
Mishra MP, Padhy RN. In vitro antibacterial efficacy of 21 Indian timber-yielding plants against multidrug-resistant bacteria causing urinary tract infection. Osong Public Health Res Perspect *2013;4:347–57.*
Rath S, Padhy RN. Monitoring in vitro antibacterial efficacy of 26 Indian species against multidrug resistant urinary tract infecting bacteria. Integrative Med Res *2014;3:133–41.*
Rath S, Padhy RN. Monitoring in vitro efficacy of Holarrhena antidysenterica against multidrug resistant enteropathogenic bacteria. Asian Pac J Trop Dis *2014;4(Suppl.1):S54–63.*

(pomegranate) fruit pericarp (PGME) in combination with ciprofloxacin against ESBL-producing bacteria, including *K. pneumoniae*. The checkerboard assay and determination of FIC were used. PGME showed the highest phenolic (229.23 ± 1.50 mg gallic acid equivalent/g) and flavonoid content (6.71 ± 0.01 mg quercetin equivalent/g) compared with fruit juice and water extract of pericarp. In some cases, PGME could enhance the antibacterial activity of ciprofloxacin due to the supposed bacterial efflux pump inhibitor activity of the polyphenolic constituents of PGME. The antibacterial activity of a natural compound can be increased in combination with an efflux pump inhibitor.[48] Kuete et al.[48] demonstrated the antibacterial activity of the methanol extract prepared from *Dioscorea bulbifera* and its isolated compounds; for example, bafoudiosbulbins A, B, C, F, G (diterpenes), and 2,7-dihydroxy-4-methoxyphenantrene on MDR bacteria. In the presence of an efflux pump inhibitor (phenylalanine arginine β-naphthylamide (PAβN)), the MIC values of the tested compounds could be increased.

In another study, the epigallocatechin gallate showed a synergistic effect in combination with imipenem against imipenem-resistant *K. pneumoniae* (IRKP).[49] A 4- to 64-fold reduction was observed in the MIC value of imipenem after coincubation with 0.25 × MIC of epigallocatechin gallate. This component affected the outer membrane protein composition of IRKP. Furthermore, the bacterial cells showed irregular forms and perforations on their surfaces detected by scanning electron microscope after the treatment with epigallocatechin gallate.

Among the natural compounds, flavonoids showed antimicrobial activity on ESBL-producing *K. pneumoniae* in combination with antibiotics.[41] Myricetin (a flavonol compound) showed synergistic activity in separate combination with amoxicillin/clavulanate, ampicillin/sulbactam and cefotoxin, but other flavonoids were not active.[41]

EOs, Plant Extracts, and Their Individual Compounds Against *E. coli*

The emergence and distribution of ESBL-producing *E. coli* is a major concern worldwide. These bacteria are widely isolated in hospitals, and they are also increasingly found in community-acquired infections[39,50–52] because antibiotics are frequently misused and the spreading of these mainly plasmid encoded enzymes among bacteria is very fast.[53,54] Some *E. coli* strains are harmless and colonize the healthy human intestine, but the pathogenic strains are among the most common causes of enteric and UTIs (cholecystitis, cholangitis, traveler's diarrhea, pyelonephritis, bacteremia, etc.),[55] and *E. coli* is also frequently isolated from burn wounds.[56]

There are several studies dealing with the antibacterial effects of EOs, plant extracts, and isolated compounds against MDR *E. coli*, but here, only the ones in which the authors stated that they used ESBL strains are included. Table 12.4 summarizes the EOs, Table 12.5 the plant extracts, and Table 12.6 isolated plant compounds that were found to be active against ESBL-producing *E. coli*.

The most frequently used methods are the simplest disk or agar diffusion and BDMs, but we could also find a modern technique, molecular docking, applied to investigate the interaction between the catalytic amino acids of bacterial proteins and active compounds aimed at identifying

Table 12.4 Essential Oils Effective Against ESBL and MDR *Escherichia coli* Strains

Source Plant	Method	Results	Type of *E. coli*	Resistance of *E. coli* strains	Reference
Allium sativum L.	Disk diffusion method	IZD = 6–30 mm	Clinical isolates from urine ($n = 49$), blood ($n = 1$), pus ($n = 6$), throat swab ($n = 1$); 57.8% (33/57) ESBL	Ca, Ca + Cl, Cfo, Cfo + Cl, Cfp	16
Bidens tripartita L., flower and herb	Broth microdilution and disk diffusion methods	Flower: IZD = 0 mm; MIC = 25 mg/mL; Herb: IZD = 0 mm; MIC = >100 mg/mL	ATCC 35218 (β-lactamase+, but not ESBL)	Not given	33
Cinnamomum cassia (Nees & T. Nees) J. Presl	Disk diffusion method	IZD = 7–30 mm	Clinical isolates from urine ($n = 49$), blood ($n = 1$), pus ($n = 6$), throat swab ($n = 1$); 57.8% (33/57) ESBL	Ca, Ca + Cl, Cfo, Cfo + Cl, Cfp	16
Eucalyptus oil (source plant not given)	Agar diffusion method	IZD = 7–13 mm	Clinical isolates ($n = 5$)	Not given	29
Lemongrass oil (source plant not given)	Agar diffusion method	IZD = 2–13 mm	Clinical isolates ($n = 5$)	Not given	29
Ocimum sanctum L.	Disk diffusion method	IZD = 6–20 mm	Clinical isolates from urine ($n = 49$), blood ($n = 1$), pus ($n = 6$), throat swab ($n = 1$); 57.8% (33/57) ESBL	Ca, Ca + Cl, Cfo, Cfo + Cl, Cfp	16
Origanum vulgare L.	Microdilution method	MIC = 0.125%, MBC = 0.5%	Clinical isolates ($n = 4$)	Ak, Cpm, Ca, Cf, Lev, P, Am, Am + Sb, Ac + Cl, G, Cfo, Cft, Azt, I, Mer, Pip, Pip + Tz, Tc + Cl, O, V	27
Origanum vulgare L.	Microdilution method	MIC = 0.5 μL/mL	Isolated from chicken livers from a chicken farm	D, L, Ma, Ff, Ac	57
Syzygium aromaticum (L.) Merrill & Perry	Disk diffusion method	IZD = 7–30 mm	Clinical isolates from urine ($n = 49$), blood ($n = 1$), pus ($n = 6$), throat swab ($n = 1$); 57.8% (33/57) ESBL	Ca, Ca + Cl, Cfo, Cfo + Cl, Cfp	16
Tea tree oil (source plant not given)	Agar diffusion method	IZD = 12–15 mm	Clinical isolates ($n = 5$)	Not given	29
Thymus vulgaris L. cv. Deutscher winter	Agar well diffusion method	1%: IZD = 2.5–3.1 mm; 10%: IZD = 10.3–10.6 mm	Clinical and food isolates	Fluoroquinolones	58

MIC, minimum inhibitory concentration; MBC, minimum bactericidal concentration; IZD, inhibition zone diameter; Ac, amoxicillin; Ak, amikacin; Am, ampicillin; Azt, aztreonam; C, chloramphenicol; Ca, ceftazidime; Cf, ciprofloxacin; Cfo, cefotaxime; Cph, cephalothin; Cfm, cefuroxime; Cfp, cefpodoxime; Cft, ceftriaxone; Cl, clavulanic acid; Cpm, cefepime; Cz, cefazolin; D, doxycycline; Dor, doripenem; Ff, florfenicol; Fox, cefoxitin; G, gentamycin; I, imipenem; L, lincomycin; Lev, levofloxacin; Ma, maquindox; Mer, meropenem; Net, netilmicin; O, oxacillin; P, penicillin; Pip, piperacillin; Sb, sulbactam; Sm, sulfamethoxazole; T, tetracycline; Tb, tobramycin; Tc, ticarcillin; Tg, tigecycline; Tr, trimethoprim; Tz, tazobactam; V, vancomycin.

Table 12.5 Plant Extracts Effective Against ESBL and MDR *Escherichia coli* Strains

Source Plant	Extract	Method	Results	Type of *E. coli*	Resistance of *E. coli* Strains	Reference
Acacia nilotica (L.) P.J.H. Hurter & Mabb., leaf	50% and 90% ethanol	Twofold serial dilution method	MIC (μg/mL) = 156 (3/10), 313 (7/10); MBC (μg/mL) = 313 (3/10)–625 (7/10)	Nosocomial ($n = 10$)	Ch, Ca, Cft, Cpm, Ac, Azt, Pip, G, Tb, Na, Cf, T, C	39
Acacia nilotica (L.) P.J.H. Hurter & Mabb., leaf	50% and 90% ethanol	Twofold serial dilution method	MIC (μg/mL) = 19.5 (3/16), 39 (13/16); MBC (μg/mL) = 39 (3/16)–156 (13/16)	Community acquired ($n = 16$)	Ch, Ca, Cft, Cpm, Ac, Azt, Pip, Ak, G, Tb, Na, Cf, Nf, T, C	39
Acorus calamus L., rhizome	Extracted with 70% ethanol, dried, and dissolved in DMSO	Agar well diffusion and broth dilution methods	IZD (mm) = 20; MIC (mg/mL) = 1.3–2.6	Clinical isolates ($n = 2$)	P, Am, Cx, M, Cfo, Cfx, Cfp, Cfc, Pip, Fox, Cph, Cp, Cb, Cfm, V, Azt, Co, Nv, Na, T, E	59
Bidens tripartita L., flower and herb	Water, methanol/water, acetone/water	Broth microdilution and DDM	Flower: IZD (mm) = 0, 9, 8, resp.; MIC (mg/mL) = > 100, 12.5, 6.2, resp.; Herb: IZD (mm) = 7, 9, 8, resp.; MIC (mg/mL) = 12.5, 25.0, 25.0, resp.	*E. coli* ATCC 35218 (β-lactamase +, but not ESBL)	Not given	33
Camellia sinensis (L.) Kuntze, leaf	Extracted with 70% ethanol, dried, and dissolved in DMSO	Agar well diffusion and broth dilution methods	IZD (mm) = 14–15; MIC (mg/mL) = 5.1	Clinical isolates ($n = 2$)	P, Am, Cx, M, Cfo, Cfx, Cfp, Cfc, Pip, Fox, Cph, Cp, Cb, Cfm, V, Azt, Co, Nv, Na, T, E	59
Cichorium intybus L., root	Extracted with 70% ethanol, dried, and dissolved in DMSO	Agar well diffusion and broth dilution methods	IZD (mm) = 13; MIC (mg/mL) = 5.1	Clinical isolates ($n = 2$)	P, Am, Cx, M, Cfo, Cfx, Cfp, Cfc, Pip, Fox, Cph, Cp, Cb, Cfm, V, Azt, Co, Nv, Na, T, E	59
Cinnamomum zeylanicum Blume, bark	50% and 90% ethanol	Twofold serial dilution method	MIC (μg/mL) = 3130 (7/10), 6250 (3/10); MBC (μg/mL) = 6250 (7/10)–12,500 (3/10)	Nosocomial ($n = 10$)	Ch, Ca, Cft, Cpm, Ac, Azt, Pip, G, Tb, Na, Cf, T, C	39
C. zeylanicum Blume, bark	50% and 90% ethanol	Twofold serial dilution method	MIC (μg/mL) = 195 (11/16), 780 (5/16); MBC (μg/mL) = 1560 (11/16)–3130 (5/16)	Community acquired ($n = 16$)	Ch, Ca, Cft, Cpm, Ac, Azt, Pip, Ak, G, Tb, Na, Cf, Nf, T, C	39

(*Continued*)

Table 12.5 Plant Extracts Effective Against ESBL and MDR *Escherichia coli* Strains *Continued*

Source Plant	Extract	Method	Results	Type of *E. coli*	Resistance of *E. coli* Strains	Reference
Curtisia dentata C.A. Sm., stem bark	Ethanol, dichloromethane, hexane, chloroform, acetone	Disk diffusion and broth dilution methods	Relative IZD = 4–28%; MIC (μg/mL) = 100–2500	Obtained from wastewater ($n = 18$), abattoir water ($n = 18$), Berg River ($n = 13$), Plankenberg River ($n = 13$)	Ca, Cfo	60
Delonix regia (Hook.) Raf., flower	Extracted with 70% ethanol, dried, and dissolved in DMSO	Agar well diffusion and broth dilution methods	IZD (mm) = 14–16; MIC (mg/mL) = 10.2	Clinical isolates ($n = 2$)	P, Am, Cx, M, Cfo, Cfx, Cfp, Cfc, Pip, Fox, Cph, Cp, Cb, Cfm, V, Azt, Co, Nv, Na, T, E	59
Hemidesmus indicus (L.) R.Br., stem	Extracted with 70% ethanol, dried, and dissolved in DMSO	Agar well diffusion and broth dilution methods	IZD (mm) = 24–25; MIC (mg/mL) = 1.3–2.6	Clinical isolates ($n = 2$)	P, Am, Cx, M, Cfo, Cfx, Cfp, Cfc, Pip, Fox, Cph, Cp, Cb, Cfm, V, Azt, Co, Nv, Na, T, E	59
Holarrhena antidysenterica Wall., bark	Extracted with 70% ethanol, dried, and dissolved in DMSO	Agar well diffusion and broth dilution methods	IZD (mm) = 19–22; MIC (mg/mL) = 2.6–5.1	Clinical isolates ($n = 2$)	P, Am, Cx, M, Cfo, Cfx, Cfp, Cfc, Pip, Fox, Cph, Cp, Cb, Cfm, V, Azt, Co, Nv, Na, T, E	59
Holarrhena antidysenterica Wall., leaf and bark	Ethyl acetate, ethanol, acetone	Microbroth dilution method	Leaf: MIC (mg/mL) = 3.1, 12.5, 3.1, resp. MBC (mg/mL) = 12.5, 50.0, 12.5, resp.; Bark: MIC (mg/mL) = 3.1, 12.5, 6.3, resp. MBC (mg/mL) = 12.5, 12.5, 25.0, resp.	Clinical isolates	Ak, G, Am, Ac + Cl, Pip + Tz, Cft, Cfo, Gfx, Na, Nor, Ox, Nf, T	44
Holarrhena antidysenterica Wall., leaf and bark	Petroleum ether, ethyl acetate, chloroform, hexane, acetone, ethanol, methanol, water	Agar well diffusion method	Leaf: IZD (mm) = not investigated, 23, 14, 6, 24, 21, 12, 13, resp.; Bark: IZD (mm) = 7, 15, not investigated, 17, 10, 21, 9, resp.	Clinical isolates	Ak, G, Am, Ac + Cl, Pip + Tz, Cft, Cfo, Gfx, Na, Nor, Ox, Nf, T	44

Lawsonia inermis L., leaf	Extracted with 70% ethanol, dried, and dissolved in DMSO	Agar well diffusion and broth dilution methods	IZD (mm) = 11–12; MIC (mg/mL) = 5.1	Clinical isolates ($n = 2$)	P, Am, Cx, M, Cfo, Cfx, Cfp, Cfc, Pip, Fox, Cph, Cp, Cb, Cfm, V, Azt, Co, Nv, Na, T, E	59
Mangifera indica L., leaf	Extracted with 70% ethanol, dried, and dissolved in DMSO	Agar well diffusion and broth dilution methods	IZD (mm) = 18–21; MIC (mg/mL) = 2.6–5.1	Clinical isolates ($n = 2$)	P, Am, Cx, M, Cfo, Cfx, Cfp, Cfc, Pip, Fox, Cph, Cp, Cb, Cfm, V, Azt, Co, Nv, Na, T, E	59
Pericopsis laxiflora (Benth.) Van Meeuwen, bark	Distilled water, 70% ethanol, methanol, ethyl acetate	Agar# well diffusion and tube dilution methods	IZD (mm) = 0–12; MIC, MBC (mg/mL) = 12.5– > 100	No 150C/12, collection of Department of Bacteriology and Virology, Institut Pasteur de Côte d'Ivoire	O, Fox	42
Phyllanthus amarus Schumacher, leaf and root	95% ethanol (5, 10, 20, 40, 80 mg/mL)	Disk and agar diffusion method	Root: IZD (mm) = 8–25; MIC (mg/mL) = 5–20, MBC (mg/mL) = 5–30; Leaf: IZD (mm) = 8–26; MIC (mg/mL) = 5–10, MBC (mg/mL) = 10–30	Isolated from HIV patients with or without diarrhea ($n = 8$)	Cph, Cfo, Ca, Cpm	61
Plumbago zeylanica L., root	Extracted with 70% ethanol, dried, and dissolved in DMSO	Agar well diffusion and broth dilution methods	IZD (mm) = 13–21; MIC (mg/mL) = 2.6	Clinical isolates ($n = 2$)	P, Am, Cx, M, Cfo, Cfx, Cfp, Cfc, Pip, Fox, Cph, Cp, Cb, Cfm, V, Azt, Co, Nv, Na, T, E	59
Punica granatum L., rind	Extracted with 70% ethanol, dried, and dissolved in DMSO	Agar well diffusion and broth dilution methods	IZD (mm) = 17–19; MIC (mg/mL) = 5.1	Clinical isolates ($n = 2$)	P, Am, Cx, M, Cfo, Cfx, Cfp, Cfc, Pip, Fox, Cph, Cp, Cb, Cfm, V, Azt, Co, Nv, Na, T, E	59
Punica granatum L., ripe fruit pericarp	Methanol and boiling water	Agar well diffusion, broth microdilution method	IZD (mm) = 8–20; MIC (μg/mL) = 640–2560	Clinical isolates ($n = 20$)	Ca, Cfo, Ac + Cl, Cf, Ox, Nor, Gfx, Lev	47
Syzygium aromaticum (L.) Merr. & L.M. Perry, bud	50% and 90% ethanol	Twofold serial dilution method	MIC (μg/mL) = 6250; MBC (μg/mL) = 12,500 (10/10)	Nosocomial ($n = 10$)	Ch, Ca, Cft, Cpm, Ac, Azt, Pip, G, Tb, Na, Cf, T, C	39

(*Continued*)

Table 12.5 Plant Extracts Effective Against ESBL and MDR *Escherichia coli* Strains *Continued*

Source Plant	Extract	Method	Results	Type of *E. coli*	Resistance of *E. coli* Strains	Reference
Syzygium aromaticum (L.) Merr. & L.M. Perry, bud	50% and 90% ethanol	Twofold serial dilution method	MIC (μg/mL) = 390 (2/16), 1560 (14/16); MBC (μg/mL) = 1560 (2/16)–3130 (14/16)	Community acquired ($n = 16$)	Ch, Ca, Cft, Cpm, Ac, Azt, Pip, Ak, G, Tb, Na, Cf, Nf, T, C	39
Terminalia alata Heyne ex. Roth, leaf and bark	Methanol, ethanol, acetone	Microbroth dilution method	Leaf: MIC (mg/mL) = 3.1, 3.1, 6.2, resp. MBC (mg/mL) = 12.5, 12.5, 25.0, resp.; Bark: MIC (mg/mL) = 1.6, 3.1, 6.3, resp. MBC (mg/mL) = 12.5, 12.5, 25.0, resp.	Clinical isolates	Ak, G, Am, Ac + Cl, Pip + Tz, Cft, Cfo, Gfx, Na, Nor, Ox, Nf, T	38
Terminalia alata Heyne ex. Roth, leaf and bark	Petroleum ether, ethyl acetate, chloroform, hexane, acetone, ethanol, methanol, water	Agar well diffusion method	Leaf: IZD (mm) = 0, 11, 16, 15, 18, 21, 21, 15, resp.; Bark: IZD (mm) = 7, 17, 18, 17, 20, 22, 26, 21, resp.	Clinical isolates	Ak, G, Am, Ac + Cl, Pip + Tz, Cft, Cfo, Gfx, Na, Nor, Ox, Nf, T	38
Terminalia bellirica (Gaertn.) Roxb., fruit	Extracted with 70% ethanol, dried, and dissolved in DMSO	Agar well diffusion and broth dilution methods	IZD (mm) = 16–23; MIC (mg/mL) = 2.6–5.1	Clinical isolates ($n = 2$)	P, Am, Cx, M, Cfo, Cfx, Cfp, Cfc, Pip, Fox, Cph, Cp, Cb, Cfm, V, Azt, Co, Nv, Na, T, E	59
Terminalia chebula Retz., fruit	Extracted with 70% ethanol, dried, and dissolved in DMSO	Agar well diffusion and broth dilution methods	IZD (mm) = 15–17; MIC (mg/mL) = 10.2	Clinical isolates ($n = 2$)	P, Am, Cx, M, Cfo, Cfx, Cfp, Cfc, Pip, Fox, Cph, Cp, Cb, Cfm, V, Azt, Co, Nv, Na, T, E	59

MIC, minimum inhibitory concentration; MBC, minimum bactericidal concentration; IZD, inhibition zone diameter; Ac, amoxicillin; Ak, amikacin; Am, ampicillin; Azt, aztreonam; C, chloramphenicol; Ca, ceftazidime; Cb, carbenicillin; Cf, ciprofloxacin; Cfc, cefaclor; Cfo, cefotaxime; Cfm, cefuroxime; Cfp, cefpodoxime; Cft, ceftriaxone; Cfx, cefixime; Cl, clavulanic acid; Co, co-trimoxazole; Cp, cefoperazone; Cph, cephalothin; Cpm, cefepime; Cx, cloxacillin; E, erythromycin; Fox, cefoxitin; G, gentamycin; Gfx, gatifloxacin; Lev, levofloxacin; M, methicillin; Na, Nalidixic acid; Nf, Nitrofurantoin; Nor, norfloxacin; Nv, novobiocin; O, oxacillin; Ox, ofloxacin; P, penicillin; Pip, piperacillin; T, tetracycline; Tb, tobramycin; Tz, tazobactam; V, vancomycin.

Table 12.6 Isolated Compounds Effective Against ESBL and MDR *Escherichia coli* strains

Compound	Method	Results	Type of *E. coli*	Resistance of *E. coli* strains	Reference
Cinnamaldehyde	Macrobroth-dilution method	MIC = 122 μg/mL	Clinical isolates from urine (*n* = 49), blood (*n* = 1), pus (*n* = 6), throat swab (*n* = 1); 57.8% (33/57) ESBL	Ca, Ca + Cl, Cfo, Cfo + Cl, Cfp	16
Eugenol	Macrobroth-dilution method	MIC = 499 μg/mL	Clinical isolates from urine (*n* = 49), blood (*n* = 1), pus (*n* = 6), throat swab (*n* = 1); 57.8% (33/57) ESBL	Ca, Ca + Cl, Cfo, Cfo + Cl, Cfp	16
Luteolin	Broth macrodilution method	MIC = >200 μg/mL	Amoxicillin-resistant *E. coli* (AREC)	Ac	62
Ulopterol	Disk diffusion and broth dilution methods	IZD = 11 mm MIC = 62.5–250 μg/mL	Clinical isolate ESBL-3984	Not given	63

MIC, minimum inhibitory concentration; MBC, minimum bactericidal concentration; IZD, inhibition zone diameter; Ac, amoxicillin; Ca, ceftazidime; Cfo, cefotaxime; Cfp, cefpodoxime; Cl, clavulanic acid.

their mode of action.[16] The majority of the investigated pathogens are clinical isolates, but unfortunately the ESBL+ strain can also be isolated from food[57] or water.[60] The ESBL + strains are resistant to several antibiotics, but sometimes ESBL-bacteria show similar antibiotic polyresistance,[31] and many times, the laboratory strains and ESBL-producing/MDR bacteria isolated from clinical samples show similar sensitivity to the tested EOs or extracts.[39,31,47]

There are a few reports stating that the antibacterial effects of isolated compounds against ESBL+ *E. coli* leads to the suggestion that this effect may result from the synergism of several compounds. We still have only limited knowledge on how the EOs, plant extracts, and their isolated compounds exert their antibacterial effect. For example, Eumkeb et al.[62] have found that luteolin can reverse the bacterial resistance to amoxicillin via three possible mechanisms: inhibition of protein and peptidoglycan synthesis, inhibition of the activity of certain ESBLs, and alteration of outer and inner membrane permeability. Due to their resistance-modifying effects, EOs, plant extracts, and compounds may also be used in combination with antibiotics. For instance, oregano EO can enhance the activity of sarafloxacin, levofloxacin, maquindox, florfenicol, doxycycline (synergism), lincomycin, amoxicillin, ceftiofur, ceftriaxone (additive effects) against ESBL-producing *E. coli*[57]; and extracts of pomegranate pericarp may increase the susceptibility of ESBL+ *E. coli* to ciprofloxacin (8 of 20 strains responded positively).[47] And the combination of plant extracts may also lead to increased antimicrobial activity.[59]

ANTIBACTERIAL EFFECT OF EOs AND PLANT EXTRACTS AGAINST *ACINETOBACTER* SPP.

Acinetobacter spp. are opportunistic pathogens, which can cause severe nosocomial infections, especially in the case of burn patients.[64] Nowadays, it produces serious problems in medicinal health care due to its ESBL and MBL enzymes, which are responsible for the high resistance to a wide spectrum of antimicrobial agents. As a result, in the last few years, several studies have appeared, which examined medicinal plant extracts or EOs as a possible solution for the problem of *Acinetobacter* infection.

In several cases, the application of EOs and antibiotics in combination would be considerable because it can produce stronger inhibition in comparison with their single administration. For instance, the combination of *Myrtus communis* EO with ciprofloxacin and polymyxin B has been shown to have a synergistic activity tested by checkerboard assaying.[65] As a result, fractional inhibitory concentration indexes (FICIs) were determined against all MDR clinical wound isolates of *A. baumannii*. Researchers took into consideration the fact that synergy was strain- and antibiotic-dependent, which was probably due to the different mode of action of the antimicrobial agents. Enhanced permeability of bacterial cell wall and membrane structures that are related to the possible mechanism of myrtle oil and terpenoid components has been described earlier.[66,67] Therefore, the authors suggest that this mechanism could improve the entrance of antibiotics through the bacterial cell wall, which finally leads to their increased sensitivity. The synergistic potential of myrtle EO in combination with polymyxin B was also confirmed with time-kill assay.[65] In another study *Coriandrum sativum* EO in combination with chloramphenicol, ciprofloxacin, gentamicin, or tetracycline resulted in synergistic activity against two *A. baumannii* strains.[68] Among the tested agents, the lowest FIC index (FICI: 0.047) was detected in the case of the combination of coriander oil and chloramphenicol. Nevertheless, coriander oil combined with piperacillin or cefoperazone could produce an additive effect, which was also confirmed by the obtained FICI values.[68]

In an in vitro study, the EOs of *F. vulgare*, *Nigella sativa*, and *Juniperus osteosperma* did not show any activity against *Acinetobacter* sp.[69] In contrast, Guerra et al.[70] found that the application of EO of *Citrus limon* (156.25 μg/mL) and *Cinnamomum zeylanicum* (78.125 μg/mL) at subinhibitory concentrations in combination with amikacin had synergistic activity. Moreover, an additive effect was observed with the combinations of both oils and gentamicin.[70] *Helichrysum italicum* EO manifested the ability to increase the susceptibility of *Acinetobacter* strains to chloramphenicol. This presumption was confirmed by the decreased MIC values of antibiotics (MIC_{CA}: 32 mg/L, MIC_{CA+EO}: 4 mg/L) in the presence of 2.5% EO.[71,72] Consequently, these results offer great opportunity to develop new medical strategies in potentially lethal infections caused by MDR strains of *A. baumannii*.

Against *Acinetobacter* spp. further in vitro results of EOs were summarized in Table 12.7.

In 2012, Doughari et al.[60] investigated the antimicrobial potential of *Curtisia dentata*, one of the well-known traditional medicinal plants in South Africa. In the in vitro antimicrobial tests, 14 strains of *Acinetobacter haemolyticus* and 29 strains of *Acinetobacter lwoffii* were involved. As a result, the ethanol extract of the stem bark showed antibacterial activity against *A. haemolyticus* (MIC: 100–850 μg/mL) and *A. lwoffii* (MIC: 150–2500 μg/mL), respectively. The water extract of

Table 12.7 Activity of Some Essential Oils and Their Components Against *Acinetobacter* spp. Using In Vitro Techniques

Essential Oil	Main Components	Methods	Antibacterial Activity	Reference
Three ecotypes of *Myrtus communis* L. MyHN: Herceg Novi, MyK: Kotor MyB: Bar	α-Pinene, α-terpineol, 1,8-cineole, linalool, myrtenyl acetate, geranyl acetate	Broth microdilution	MIC (MyHN): 1.41–4 μL/mL MIC (MyB): 0.25–4 μL/mL MIC (MyK): 0.71–4 μL/mL	65
Coriandrum sativum L. (from Sigma–Aldrich)	Not determined	Broth microdilution	MIC: 1–4 μL/mL	68
Eucalyptus globulus L. fruits (EGF) *Eucalyptus globulus* L. leaves (EGL); *Eucalyptus citriodora* Hook leaves (ECL) *Eucalyptus radiata* Sieber ex. DC leaves (ERL)	EGF: aromadendrene 1,8-cineole EGL: 1,8-cineole, α-pinene ECL: citronellal, citronellol ERL: 1,8-cineole, α-terpineol	Broth microdilution	MIC (EGF): 1 mg/mL MIC (EGL): 2 mg/mL MIC (ECL): 2 mg/mL MIC (ERL): 1 mg/mL MIC (aromadendrene): 2 mg/mL MIC (1,8-cineole): 8 mg/mL MIC (citronellal): 2–4 mg/mL MIC (citronellol): 0.125–0.25 mg/mL	23
Aniba rosaeodora (AR), *Melaleuca alternifolia (MA)*, *Origanum vulgare (OV)*, *Pelargonium graveolens (PG)* (from Erbamea)	AR: limonene, linalool, geraniol MA: γ-terpinen, terpinen-4-ol, caryophyllene OV: α-pinene, cymene, cymenol PG: linalool, citronellol, geraniol	Broth microdilution	MIC (AR): 0.25 mg/mL MIC (MA): 3.4 mg/mL MIC (OV): 0.6 mg/mL MIC (PG): 0.5 mg/mL	73
Iranian Fennel oil (from Barije Essence Pharmaceutical company)	Anethole, fenchone	Broth microdilution	Most strains sensitive to all test dilutions in the range of 0.03–25 v/v%	74
Origanum vulgare L. (from Ferquima-Industry and Commerce Ltd)	Not determined	Broth microdilution	MIC: 0.125–0.5% solution	27

this herb did not show any activity against the investigated strains. In addition, after 1 h of exposure to 30 mg/mL of ethanol extract, an increased cation (Na^+ and K^+) presence were detected in the medium, which is probably due to the bacterial cell wall damage in the case of both strains. This could be one of the possible mechanisms of action of *Curtisia dentata*.[60]

The leaf extract of *Lantana camara* L., which is native of South Africa, could be one of the next promising effective agents against ESBL-producing *A. baumannii*. According to Dubey and Padhy,[75] methanol and dichloromethane leaf extracts were the most potent antibacterial agents against

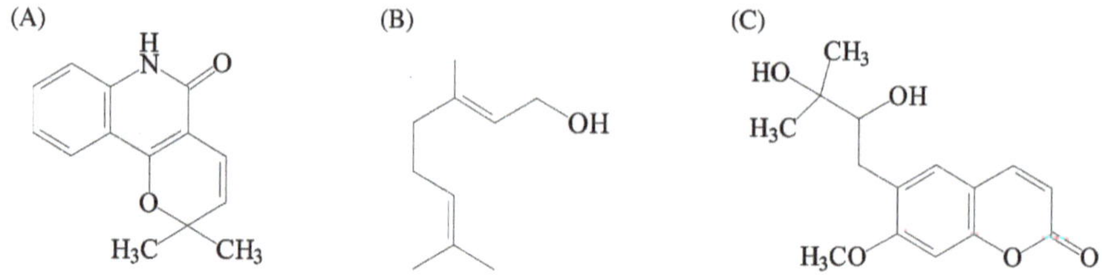

FIGURE 12.1

Chemical structures of some effective isolated plant compounds against ESBL-producing, Gram-negative bacteria. (A) Flindersine, (B) geraniol, and (C) ulopterol.

A. baumannii using the agar well diffusion model. As a result, MIC and MBC values (mg/mL) were calculated in the case of methanol (MIC: 1.562; MBC: 12.5) and dichloromethane extracts (MIC: 3.125; MBC: 25). In addition, the extracts presented activity against other ESBL-positive strains (eg, *Citrobacter freundii*, *P. mirabilis*, *Proteus vulgaris*, and *P. aeruginosa*).[75]

Considering the research of Hasson et al.[76] the ethanol extract of *Saussurea lappa* could be a new antibacterial agent for several human pathogens, including *A. baumannii*. Using the AD method, 6 mg/mL concentration was detected as a bactericide to *A. baumannii*. The water extract did not show any activity in the previously mentioned studies. During the examination of antiresistant activity with well diffusion assays, the test bacteria developed resistance against all concentrations of the extracts except for 300 μL of 40 mg/mL stock solution.[76]

In Ayurvedic medicine, leaves of *Toddalia asiatica* are generally used to treat lung diseases and rheumatism. The flindersine (2,6-dihydro-2,2-dimethyl-5H-pyrano [3,2-c] quinoline-5-one-9cl, Fig. 12.1) component, which was isolated from the ethyl acetate extract of this herb, also possessed great anti-*Acinetobacter* activity (MIC: 125 μg/mL) using the broth microdilution method.[45]

In 2013 and 2014, two studies were published,[40,46] which presented a huge biological activity screening of 47 Indian traditional herbs against MDR urinary tract pathogens, including *A. baumannii*. The main purpose of this study was the evaluation of methanol extracts of Indian spices and herbs using in vitro agar well diffusion and dilution methods. As a result, they determined the visible inhibition zones and MIC values of each plant extract against several resistant Gram-positive and Gram-negative pathogens. The most effective herbs, which presented the lowest MIC values (1.51 mg/mL), were *Brassica juncea*, *Cassia tora*, *Cinnamomum zeylanicum*, *C. sativum*, *Cuminum cyminum*, *M. koenigii*, *Syzygium aromaticum*, and *Tectona grandis*. Summarized anti-*Acinetobacter* results of these two studies were presented in Table 12.3.[40,46]

ANTIBACTERIAL EFFECT OF EOs AND PLANT EXTRACTS AGAINST *ENTEROBACTER* SPP.

Similarly to the previously mentioned bacteria, *Enterobacter* spp. are also becoming increasingly antibiotic resistant related to their ESBL and tripartite drug efflux pump production. Therefore, the discovery of alternative antimicrobial agents that could directly attack these defensive mechanisms would be desirable.

Köse et al.[32] studied the EOs of *S. erythrantha* var. *erythrantha* and *S. erythrantha* var. *cedretorum* and found that it did not show any activity against *Enterobacter cloacae* using the DDM. Fadli et al.[19] detected the anti-*Enterobacter* activity of EO from the aerial parts of *T. broussonetii*, *T. maroccanus*, *Thymus pallidus*, and *Rosmarinus officinalis* against several MDR strains. The EO of *T. broussonetii* (MIC: 0.234–0.468 mL/L), and *T. maroccanus* (MIC: 0.234–1.875 mL/L) presented the best activity. In the presence of phenylalanine arginine β-naphthylamide (PAβN, 20 mg/L), which is a well-known efflux pump inhibitor, decreased MIC values were observed. Furthermore, 0.31 mL/L of *T. broussonetii* and *T. maroccanus* oil strongly enhanced the activity of chloramphenicol. A year later, studies also found that EO of *T. maroccanus* can alter the bacterial inner and outer membrane permeability similar to polymyxin B. For the detection of intracellular protein release and membrane permeability increasing the properties of EO, sodium dodecyl sulfate-polyacrylamide gel electrophoresis, an outer membrane permeability test, was performed in combination with Western-blot analysis. As a result, alteration of the inner and outer membranes was detected after EO administration without degradation of the cellular constituents, β-galactosidase, β-lactamase, and intracellular protein release.[19,77]

Lorenzi et al.[71] observed that EO of *H. italicum* could successfully decrease the chloramphenicol resistance in the *Enterobacter aerogenes* strains, which are overexpressed in the bacterial efflux pumps. In the presence of the EO, the MIC of chloramphenicol has been reduced to 128 mg/L from 1024 mg/L. The same effect was observed in the case of PaβN supplement. *Helichrysum italicum* could reduce intrinsic resistant activity to chloramphenicol in wild-type control stains of *E. aerogenes*. Among the tested EO components, geraniol presented the strongest efflux pump inhibitor activity. Moreover, geraniol showed an ability to enhance and restore the activity of not only chloramphenicol, but also β-lactams and fluoroquinolones equally, which are very important antimicrobial agents in clinical therapy.[71]

The potential efflux pump inhibitor activity of the acyclic monoterpene geraniol (Fig. 12.1) and its saturated or unsaturated monoterpene derivatives (eg, nerol, geranyl acetate, neryl acetate, neral, geranial, citronellol, and linalool) was also investigated. This could be a promising step for the further investigation and human application of geraniol and its derivatives.[72]

In a recently published study by Kuete et al.[48] the antimicrobial activity of isolated natural compounds (terpenoids, alkaloids, coumarins, naphthoquinones, anthraquinones, flavonoids, and xanthones) was investigated on resistant strains of *E. aerogenes*. The naphthoquinone plumbagin produced the highest activity (MIC: 16–64 mg/L) in comparison with 4-hydroxylonchocarpin (flavonoids, MIC: 128–256 mg/L), 7-methoxycoumarin (coumarin, MIC: 128–256 mg/L), norcassaide (alkaloid, MIC: 64–256 mg/L), and laurentixanthone B (xanthone, MIC: 128–256 mg/L). In the presence of an efflux pump inhibitor (PAβN), increased activity was observed for each component. However, terpenoids did not show activity for every test strain. Researchers have suggested that the presence of nitrogen in the molecule may enhance the efficacy of terpenoids.[48]

The synergistic effect of isolated naturally occurring flavonoids (eg, apigenin, quercetin, and naringenin) was determined in combination with antibiotics against ceftazidime-resistant *E. cloacae*.[78] Results indicated that apigenin (FICI: <0.01) and naringenin (FICI: <0.14) showed a synergistic effect in combination with ceftazidime. In the case of quercetin, synergy was not detected (FICI: 1.00). Nevertheless, the antibacterial activity of quercetin (MIC: 256–512 μg/mL) has been shown for singular applications, as well as in the case of apigenin and naringenin (equal MIC: >526 μg/mL). Electron microscopy with inner- and outer-membrane permeability tests clearly showed that application of 3 μg/mL ceftazidime with equal amounts of apigenin significantly increased membrane penetration and disrupted cytoplasm membranes, which was followed by leakage of intracellular contents and cell death. The result of enzyme assays indicated that in addition to direct effect on the membrane structure, apigenin inhibits β-lactamase activity, which is possibly associated with the mode of action of this component. Research also suggests that the chemical structure (eg, 5,7-OH group of A ring and one 4′-OH group of the B ring) of apigenin and naringenin plays an important role in synergism.[78]

In the following section, the anti-*Enterobacter* activity of some plant extracts will be summarized. A new coumarin derivative, ulopterol (6-(2,3-dihydroxy-3-methylbutyl)-7-methoxy-2H-chromen-2-one, Fig. 12.1) was isolated from *T. asiatica*, and it was tested against *E. aerogenes*.[63] Ulopterol resulted in a significant inhibition zone (250 μg/disk: 12 mm) and MIC value (62.5 μg/mL). Research has revealed that ulopterol has a potential antibacterial effect in addition to the previously mentioned flindersine against not only *E. aerogenes*, but also against several Gram-negative and Gram-positive strains and fungi.[63]

As was mentioned previously, the antibacterial activity of Indian medicinal plants has been published in two extensive studies[40,46] involving Gram-negative and Gram-positive pathogens (eg, *K. pneumoniae*, *A. baumannii*, and *E. aerogenes*). The most effective anti-*Enterobacter* plant was *Eucalyptus citriodora*, which had the lowest MIC value (0.67 mg/mL). Further results of this research are presented in Table 12.3. In another study by Rath and Padhy, the antibacterial activity of *Holarrhena antidysenterica* (Roxb. ex. Fleming) Wall[44] and *T. alata* Heyne ex. Roth[38] was successfully confirmed by in vitro methods against ESBL-producing *E. aerogenes*. The inhibition zones and MIC values of the most active extracts from both herbs were also summarized in Table 12.3. The ethyl acetate extract of leaves from *H. antidysenterica* and its methanol bark sample were most effective against the selected pathogens. However, in the case of *T. alata*, no difference was observed in MIC values of methanol leaf and bark extracts.[38,44]

Dioscorea bulbifera and its isolated diterpene compounds (bafoudiosbulbins A–G) were effective antibacterial agents against MDR bacteria, including, *K. pneumoniae*.[48] Bafoudiosbulbins B and C (MIC: 64 μg/mL) presented the best anti-*Enterobacter* action. Nevertheless, besides the crude extract, bafoudiosbulbins F and G also performed moderate activity (MIC: 64–256 μg/mL). It was also remarkable that every component proved to have a stronger effect than the reference antimicrobial agent, chloramphenicol. However, the exact mechanism of action is not clear yet; studies have suggested that the efflux pump inhibitor potential and membrane disrupting ability of isolated components may play an important role in their antibacterial activity.[48]

CONCLUSIONS AND FUTURE OUTLOOK

There is no doubt that multiple antibiotic resistance of bacteria is a significant public health issue. ESBL-producing, Gram-negative bacteria present a high risk among infections in both hospital and community environments. Plant materials are popular remedies for many infectious diseases; therefore, researchers have started to focus on these materials. Their EOs, extracts, or isolated components are promising tools for the development of new drugs against MDR microbes, either solely or in combination with already existing antibiotics. In addition, natural antimicrobial agents are preferred in the food industry[58] and veterinary fields.[57]

Studies demonstrated that medicinal plants can produce constituents with different chemical structures, which are active against ESBL-producing, Gram-negative strains. Despite the large number of studies focusing on the investigation of plants and their isolated compounds on different microbes, including *K. pneumoniae*, *E. coli*, *A. baumannii*, and *Enterobacter* spp., there is a huge gap between the present results and their usefulness in clinical practice. Most of the results are only based on in vitro studies; however, in vivo and clinical studies are also necessary. The combination of plant secondary metabolites and antibiotics seems to be a promising solution to minimize and control the growing problem of antibiotic resistance.

ACKNOWLEDGMENT

This work was supported by a grant from the Hungarian Scientific Research Fund (OTKA PD 104660).

ABBREVIATIONS

ESBL	Extended-spectrum β-lactamase
EUCAST	European Committee on Antimicrobial Susceptibility Testing
KPC	*Klebsiella pneumoniae* carbapenemase
MBL	Metallo-β-lactamase
MDR	Multidrug-resistant
MRSA	Methicillin-resistant *Staphylococcus aureus*

REFERENCES

1. Opal SM, Pop-Vicas A. *Molecular mechanisms of antibiotic resistance in bacteria Chapter 18*. 8th ed. *Mandell, Douglas, and Bennett's principles and practice of infectious diseases*, vol. 1. New York, NY: Elsevier; 2015. p. 235–51.
2. Abraham EP, Chain E. An enzyme from bacteria able to destroy penicillin. *Nature* 1940;**146**:837.
3. Ambler RP. The structure of β-lactamases. *Philos Trans R Soc Lond B Biol Sci* 1980;**289**:321–31.
4. Bush K, Fish JF. Epidemiological expansion, structural studies, and clinical challenges of new β-lactamases from Gram-negative bacteria. *Annu Rev Microbiol* 2011;**65**:455–78.
5. Bush K. Bench-to-bedside review: the role of β-lactamases in antibiotic-resistant Gram-negative infections. *Crit Care* 2010;**14**:224–31.
6. Giske CG, Sundsfjord AS, Kahlmeter G, Woodford N, Nordmann P, Paterson DL, et al. Redefining extended-spectrum β-lactamases: balancing science and clinical need. *J Antimicrob Chemother* 2009;**63**:1–4.
7. Giske CG, Martinez-Martinez L, Cantón R, Stefani S, Skov R, Glupczynski Y, et al. *EUCAST guidelines for detection of resistance mechanisms and specific resistances of clinical and/or epidemiological importance* Version 1.0; December 2013. p. 1–40.
8. Petternel C, Galler H, Zarfel G, Luxner J, Haas D, Grisold AJ, et al. Isolation and characterization of multidrug-resistant bacteria from minced meat in Austria. *Food Microbiol* 2014;**44**:41–6.

9. Zurfluh K, Nüesch-Inderbinen M, Morach M, Berner AZ, Hächler H, Stephan R. Extended-spectrum-β-lactamase-producing Enterobacteriaceae isolated from vegetables imported from the Dominican Republic, India, Thailand and Vietnam. *Appl Environ Microbiol* 2015;**81**:3115–20.
10. Brolund A. Overview of ESBL-producing Enterobacteriaceae from a Nordic perspective infection. *Ecol Epidemiol* 2014;**4**:1–9.
11. Coelho A, Phedra-Carrasco N, Bartolomé R, Quintero-Zarate JN, Larrosa N, Cornejo-Sánchez T. Role of IncH12 plasmids harbouring bla_{VIM-1}, $bla_{CTX-M-9}$, *aac*(6′)-*Ib* and *qnrA* genes in the spread of multi-resistant *Enterobacter* and *Klebsiella pneumoniae* strains in different units at Hospital Vall d'Hebron, Barcelona, Spain. *Int J Antimicrob Agents* 2012;**39**:514–17.
12. Bradford PA. Extended-spectrum β-lactamases in the 21st century: characterization, epidemiology and detection of their important resistance threat. *Clin Microbiol Rev* 2001;**14**:933–51.
13. Pauli A, Schilcher H. In vitro antimicrobial activities of essential oils monographed in the European Pharmacopoeia. In: Can Baser KH, Buchbauer G, editors. *Handbook of essential oils. Science, technology, and application.* 6th ed. New York, NY: CRC Press, Taylor & Francis Group; 2010. p. 353–548.
14. European Committee for Antimicrobial Susceptibility Testing (EUCAST) of the European Society of Clinical Microbiology and Infectious Diseases (ESCMID). Terminology relating to methods for the determination of susceptibility of bacteria to antimicrobial agents. *Clin Microbiol Infect* 2000;**6**:503–8.
15. Osburne MS, Rothstein DM, Farquhar R, Murphy CK. In vitro time-kill activities of rifalazil, alone and in combination with vancomycin, against logarithmic and stationary cultures of *Staphylococcus aureus. J Antibiot (Tokyo)* 2006;**59**:80–5.
16. Dhara L, Tripathi A. Antimicrobial activity of eugenol and cinnamaldehyde against extended spectrum β-lactamase producing Enterobacteriaceae by in vitro and molecular docking analysis. *Eur J Integr Med* 2013;**5**:527–36.
17. Stefanello MEA, Pascoal ACRF, Salvador MJ. Essential oils from neotropical Myrtaceae: chemical diversity and biological properties. *Chem Biodivers* 2011;**8**:73–94.
18. Orhan IE, Ozcelik B, Kan Y, Kartal M. Inhibitory effects of various essential oils and individual components against Extended-Spectrum-Beta-Lactamase (ESBL) produced by *Klebsiella pneumoniae* and their chemical compositions. *J Food Sci* 2011;**76**:M538–46.
19. Fadli M, Saad A, Sayadi S, Chevalier J, Mezrioui N-E, Pages J-M, et al. Antibacterial activity of *Thymus maroccanus* and *Thymus broussonetii* essential oils against nosocomial infection—bacteria and their synergistic potential with antibiotics. *Phytomedicine* 2012;**19**:464–71.
20. Hood JR, Wilkinson JM, Cavanagh HMA. Evaluation of common antibacterial screening methods utilized in essential oil research. *J Essent Oil Res* 2003;**15**:428–33.
21. Stamenic M, Vulic J, Djilas S, Misic D, Tadic V, Petrovic S, et al. Free-radical scavenging activity and antibacterial impact of Greek oregano isolates obtained by SFE. *Food Chem* 2014;**165**:307–15.
22. Celik A, Aydinlik N, Arslan I. Phytochemical constituents and inhibitory activity towards methicillin-resistant *Staphylococcus aureus* strains of *Eryngium* species (Apiaceae). *Chem Biodivers* 2011;**8**:454–9.
23. Mulyaningsih S, Sporer F, Reichling J, Wink M. Antibacterial activity of essential oils from *Eucalyptus* and of selected components against multidrug-resistant bacterial pathogens. *Pharm Biol* 2011;**49**:893–9.
24. Zomorodian K, Saharkhiz MJ, Rahimi MJ, Bandegi A, Shekarkhar G, Bandegani A, et al. Chemical composition and antimicrobial activities of the essential oils from three ecotypes of *Zataria multiflora. Pharmacogn Mag* 2011;**7**:53–9.
25. Buru AS, Pichika MR, Neela V, Mohandas K. In vitro antibacterial effects of *Cinnamomum* extracts on common bacteria found in wound infections with emphasis on methicillin-resistant *Staphylococcus aureus. J Ethnopharmacol* 2014;**153**:587–95.
26. Horváth G, Ács K, Kocsis B. TLC-direct bioautography for determination of antibacterial activity of *Artemisia adamsii* essential oil. *JAOAC Int* 2013;**96**:1209–13.

27. da Costa AC, Cavalcanti dos Santos BH, Filho LS, de Oliveira EL. Antibacterial activity of the essential oil of *Origanum vulgare* L. (Lamiaceae) against bacterial multiresistant strains isolated from nosocomial patients. *Rev Bras Farmacogn* 2009;**19**:236–41.
28. Jazani NH, Ghasemnejad H, Sagedpoor S. Antibacterial effects of Iranian *Mentha pulegium* essential oil in isolates of *Klebsiella* sp. *Pak J Biol Sci* 2009;**12**:183–5.
29. Warnke PH, Lott AJS, Sherry E, Wiltfang J, Podschun R. The ongoing battle against multi-resistant strains: In-vitro inhibition of hospital-acquired MRSA, VRE, *Pseudomonas*, ESBL *E. coli* and *Klebsiella* species in the presence of plant-derived antiseptic oils. *J Craniomaxillofac Surg* 2013;**41**:321–6.
30. Horváth G, Jámbor N, Kocsis E, Böszörményi A, Lemberkovics É, Héthelyi É, et al. Role of direct bioautographic method for detection of antistaphylococcal activity of essential oils. *Nat Prod Commun* 2011;**6**:1379–84.
31. Sienkiewicz M, Danuta K, Malgorzata W. The susceptibility of *Klebsiella pneumoniae* ESBL + strains to clove (*Syzygium aromaticum*) essential oil. *Anti-Infect Agents* 2013;**11**:159–67.
32. Köse EO, Deniz IG, Sarikürkcü C, Aktas Ö, Yavuz M. Chemical composition, antimicrobial and antioxidant activities of the essential oils of *Sideritis erythrantha* Boiss. and Heldr. (var. *erythrantha* and var. *cedretorum* P. H. Davis) endemic in Turkey. *Food Chem Toxicol* 2010;**48**:2960–5.
33. Tomczykowa M, Tomczyk M, Jakoniuk P, Tryniszewska E. Antimicrobial and antifungal activities of the extracts and essential oils of *Bidens tripartita*. *Folia Histochem Cytobiol* 2008;**46**:389–93.
34. Goncalves TB, Braga MA, de Oliveira FFM, Santiago GMP, Carvalho CBM, Cabral PB, et al. Effect of subinhibitory and inhibitory concentrations of *Plectranthus amboinicus* (Lour.) Spreng essential oil on *Klebsiella pneumoniae*. *Phytomedicine* 2012;**19**:962–8.
35. Nazzaro F, Fratianni F, De Martino L, Coppola R, De Feo V. Effect of essential oils on pathogenic bacteria. *Pharmaceuticals* 2013;**6**:1451–74.
36. Faleiro ML, Miguel MG. Use of essential oils and their components against multidrug-resistant bacteria. In: Rai MK, Kon KV, editors. *Fighting multidrug resistance with herbal extracts, essential oils and their components*. USA: Elsevier; 2013. p. 65–94.
37. Radulović N, Mišić M, Aleksić J, Doković D, Palić R, Stojanović G. Antimicrobial synergism and antagonism of salicylaldehyde in *Filipendula vulgaris* essential oil. *Fitoterapia* 2007;**78**:565–70.
38. Rath S, Padhy RN. Monitoring in vitro antibacterial efficacy of *Terminalia alata* Heyne ex. Roth, against MDR enteropathogenic bacteria isolated from clinical samples. *JACME* 2013;**3**:93–102.
39. Khan R, Islam B, Akram M, Shakil S, Ahmad A, Ali SM, et al. Antimicrobial activity of five herbal extracts against multidrug resistant (MDR) strains of bacteria and fungus of clinical origin. *Molecules* 2009;**14**:586–97.
40. Mishra MP, Padhy RN. In vitro antibacterial efficacy of 21 Indian timber-yielding plants against multidrug-resistant bacteria causing urinary tract infection. *Osong Public Health Res Perspect* 2013;**4**:347–57.
41. Lin R-D, Chin Y-P, Lee M-H. Antimicrobial activity of antibiotics in combination with natural flavonoids against clinical extended-spectrum β-lactamase (ESBL)-producing *Klebsiella pneumoniae*. *Phytother Res* 2005;**19**:612–17.
42. Abou Q, Karamoko Q, Adama C, Augustin AA. Phytochemical screening and evaluation of the antibacterial activity of bark extracts of *Pericopsis* (Afrormosia) *laxiflora* (Benth.) of *Escherichia coli* and *Klebsiella pneumoniae* ESBL. *J Chem Pharm Res* 2013;**5**(1):86–90.
43. Vijayakumar A, Duraipandiyan V, Jeyaraj B, Agastian P, Raj MK, Ignacimuthu S. Phytochemical analysis and in vitro antimicrobial activity of *Illicium griffithii* Hook. f. & Thoms extract. *Asian Pac J Trop Dis* 2012;**2**(3):190–9.
44. Rath S, Padhy RN. Monitoring in vitro efficacy of *Holarrhena antidysenterica* against multidrug resistant enteropathogenic bacteria. *Asian Pac J Trop Dis* 2014;**4**(Suppl. 1):S54–63.

45. Duraipandiyan V, Ignacimuthu S. Antibacterial and antifungal activity of Flindersine isolated from the traditional medicinal plant, *Toddalia asiatica* (L.) Lam. *J Ethnopharmacol* 2009;**123**:494–8.
46. Rath S, Padhy RN. Monitoring in vitro antibacterial efficacy of 26 Indian species against multidrug resistant urinary tract infecting bacteria. *Integrative Med Res* 2014;**3**:133–41.
47. Dey D, Debnath S, Hazra S, Ghosh S, Ray R, Hazra B. Pomegranate pericarp extract enhances the antibacterial activity of ciprofloxacin against extended-spectrum β-lactamase (ESBL) and metallo-β-lactamase (MBL) producing Gram-negative bacilli. *Food Chem Toxicol* 2012;**50**:4302–9.
48. Kuete V, Teponno RB, Mbaveng AT, Tapondjou LA, Meyer JJM, Barboni L, et al. Antibacterial activities of the extracts, fractions and compounds from *Dioscorea bulbifera*. *BMC Complement Altern Med* 2012;**12**:228–35.
49. Cho Y-S, Oh J-J, K-H. OH. Synergistic anti-bacterial and proteomic effects of epigallocatechin gallate on clinical isolates of imipenem-resistant *Klebsiella pneumoniae*. *Phytomedicine* 2011;**18**:941–6.
50. Khan AU, Musharraf A. Plasmid mediated multiple antibiotic resistances in *Proteus mirabilis* isolated from patients with urinary tract infection. *Med Sci Mont* 2004;**10**:598–602.
51. Akram M, Shahid M, Khan AU. Etiology and antibiotics resistance pattern of community acquired urinary infections in J N M C Hospital Aligarh India. *Ann Clin Microbiol Antimicrob* 2007;**6**:4.
52. Livermore DM, Canton R, Gniadkowski M, Nordmann P, Rossolini GM, Arlet G, et al. CTX-M: changing the face of ESBLs in Europe. *J Antimicrob Chemother* 2007;**59**:165–74.
53. Livermore DM. Beta-Lactamases in laboratory and clinical resistance. *Clin Microbiol Rev* 1995; **8**:557–584.s.
54. Harbottle H, Thakur S, Zhao S, White DG. Genetics of antimicrobial resistance. *Anim Biotechnol* 2006;**17**(2):111–24.
55. Borellio SP, Murray RP, Ferike G. Topley and Wilson's microbiology and microbial infections. Part V, chapter 52. In: Ajello L, Hay RJ, editors. *Medical mycology*. New York, NY: Wiley; 2010. p. 1360–85.
56. Guggenheim M, Zbinden R, Handschin AE, Gohritz A, Altintas MA, Giovanoli P. Changes in bacterial isolates from burn wounds and their antibiograms: a 20-year study (1986–2005). *Burns* 2009;**35**(4):553–60.
57. Si H, Hu J, Liu Z, Zeng ZL. Antibacterial effect of oregano essential oil alone and in combination with antibiotics against extended-spectrum beta-lactamase-producing *Escherichia coli*. *FEMS Immunol Med Microbiol* 2008;**53**:190–4.
58. Anžlovar S, Baričevič D, Avguštin JA, Koce JD. Essential oil of common thyme as a natural antimicrobial food additive. *Food Technol Biotechnol* 2014;**52**(2):263–8.
59. Ahmad I, Aqil F. In vitro efficacy of bioactive extracts of 15 medicinal plants against ESBL-producing multidrug-resistant enteric bacteria. *Microbiol Res* 2007;**162**:264–75.
60. Doughari JH, Ndakidemi PA, Human IS, Benade S. Antioxidant, antimicrobial and antiverotoxic potentials of extracts of *Curtisia dentata*. *J Ethnopharmacol* 2012;**141**(3):1041–50.
61. Akinjogunla OJ, Eghafona NO, Enabulele IO, Mboto CI, Ogbemudia FO. Antibacterial activity of ethanolic extracts of *Phyllanthus amarus* against extended spectrum β-lactamase producing *Escherichia coli* isolated from stool samples of HIV sero-positive patients with or without diarrhoea. *Afr J Pharm Pharmacol* 2010;**4**(6):402–7.
62. Eumkeb G, Siriwong S, Thumanu K. Synergistic activity of luteolin and amoxicillin combination against amoxicillin-resistant *Escherichia coli* and mode of action. *J Photochem Photobiol B Biol* 2012;**117**:247–53.
63. Raj MK, Balachandran C, Duraipandiyan V, Agastian P, Ignacimuthu S. Antimicrobial activity of Ulopterol isolated from *Toddalia asiatica* (L.) Lam.: a traditional medicinal plant. *J Ethnopharmacol* 2012;**140**:161–5.

64. Owlia P, Azimi L, Gholami A, Asghari B, Lari AR. ESBL- and MBL-mediated resistance in *Acinetobacter baumannii*: a global threat to burn patients. *Infez Med* 2012;**3**:182–7.
65. Aleksic V, Mimica-Dukic N, Simin N, Stankovic NN, Knezevic P. Synergistic effect of *Myrtus communis* L. essential oils and conventional antibiotics against multi-drug resistant *Acinetobacter baumannii* wound isolates. *Phytomedicine* 2014;**21**:1666–74.
66. Oyedemi SO, Okoh AI, Mabinya LV, Pirochenva G, Afolayan AJ. The proposed mechanism of bactericidal action of eugenol, α-terpineol and γ-terpinene against *Listeria monocytogenes, Streptococcus pyogenes, Proteus vulgaris* and *Escherichia coli. Afr J Biotechnol* 2009;**8**:1280–90.
67. Amensour M, Bouhdid S, Fernandez-Lopez J, Idaomar M, Senhaji NS, Abrini J. Antibacterial activity of extracts of *Myrtus communis* against food-borne pathogenic and spoilage bacteria. *Int J Food Prop* 2010;**13**:1215–24.
68. Duarte A, Ferreira S, Silva F, Domingues FC. Synergistic activity of coriander oil and conventional antibiotics against *Acinetobacter baumannii. Phytomedicine* 2012;**19**:236–8.
69. Purkayastha S, Narain R, Dahiya P. Evaluation of antimicrobial and phytochemical screening of Fennel, Juniper and Kalonji essential oils against multidrug-resistant clinical isolates. *Asian Pac J Trop Biomed* 2012;**2**:S1625–9.
70. Guerra FQS, Mendes JM, de Sousa JP, Morais-Braga MFB, Santos BHC, Coutinho HDM, et al. Increasing antibiotic activity against multi-drug *Acinetobacter* spp. by essential oils of *Citrus limon* and *Cinnamomum zeylanicum. Nat Prod Res* 2011;**26**:2235–8.
71. Lorenzi V, Muselli A, Bernardini AF, Berti L, Pagès JM, Amaral L, et al. Geraniol restores antibiotic activities against multidrug-resistant isolates from Gram-negative species. *Antimicrob Agents Chemother* 2009;**53**:2209–11.
72. Bertli L, Lorenzi V, Casanova J, Muselli A, Pagès JM, Bolla JM. *European Patent application. Geraniol as bacterial efflux pump inhibitor*. EP 2184061 A1; May 12, 2010.
73. Rosato A, Piarulli M, Corbo F, Muraglia M, Carone A, Vitali ME, et al. In vitro synergistic action of certain combinations of gentamicin and essential oils. *Curr Med Chem* 2010;**17**:3289–95.
74. Jazani NH, Zartoshti M, Babazadeh H, Ali-daiee N, Zarrin S, Hosseini S. Antibacterial effects of Iranian fennel essential oil on isolates of *Acinetobacter baumannii. Pak J Biol Sci* 2009;**12**:738–41.
75. Dubey D, Padhy RN. Antibacterial activity of *Lantana camara* L. against multidrug resistant pathogens from ICU patients of a teaching hospital. *J Herb Med* 2013;**3**:65–75.
76. Hasson SS, Al-Balushi MS, Alharthy K, Al-Busaidi JZ, Aldaihani MS, Othman MS, et al. Evaluation of anti-resistant activity of Auklandia (*Saussurea lappa*) root against some human pathogens. *Asian Pac J Trop Biomed* 2013;**3**:557–62.
77. Fadli M, Chevalier J, Bolla JM, Mezrioui NE, Hassani L, Pages JM. *Thymus maroccanus* essential oil, a membranotropic compound active on Gram-negative bacteria and resistant isolates. *J Appl Microbiol* 2012;**113**:1120–9.
78. Eumkeb G, Chukrathok S. Synergistic activity and mechanism of action of ceftazidime and apigenin combination against ceftazidime-resistant *Enterobacter cloacae. Phytomedicine* 2013;**20**:262–9.

CHAPTER

ANTIMICROBIAL AND ANTIBIOTIC POTENTIATING ACTIVITY OF ESSENTIAL OILS FROM TROPICAL MEDICINAL HERBS AND SPICES

13

Z. Aumeeruddy-Elalfi[1], A. Gurib-Fakim[2] and M.F. Mahomoodally[1]

[1]*Department of Health Sciences, Faculty of Science, University of Mauritius, Réduit, Mauritius*

[2]*CIDP Research & Innovation, BioPark, Socota Phoenicia, Mauritius*

INTRODUCTION

Antimicrobial resistance to antibiotics is gaining much momentum and has been reported in the literature as a major problem affecting many populations worldwide.[1] Scientists have been dedicating much efforts geared towards the quest for new phytochemicals obtained naturally from plant sources as alternative antimicrobial treatments. Plants used traditionally in the past are nowadays more commonly used in the treatment of numerous human infectious diseases, based on ethnobothanical sources.[2,3] According to recent literature, herbs are still found in 40% of prescription drugs.[4] For instance, plants from the Lamiaceae family are cultivated worldwide, mainly for culinary purposes but also as medicinal herbs since they are enriched in polyphenols.[5] Their potent bioactivity and moderately low toxicity have rendered them beneficial ingredients in complementary/alternative medicine and as dietary supplements.[5]

Essential oils (EOs) are natural volatile complex mixtures of biologically active substances, used since time immemorial as flavoring agents and as the active ingredients in a plethora of commercial products.[6] These plant metabolites are also classified as natural products having pharmacological potential that can be of therapeutic benefit in the management of human diseases.[2,7] Recently, natural products such as EOs have attracted much attention and are increasingly falling within the scope of phytomedicines as complementary and alternative therapy.[8] The Lamiaceae family includes several aromatic plants from which EOs of medicinal quality are extracted.[6,9] On the whole, EOs extracted from plants of different families have been reported as potential candidates with major antimicrobial properties.[1,9]

Moreover, the biological properties of EOs derived from natural products have been recognized and used for centuries, although scientific confirmation has only been reviewed recently.[10] More than 300 of the approximately 3000 known EOs worldwide have gained importance for their pluripotential and wide range of biological activities.[11] In addition, EOs have been widely appraised as potential antimicrobial agents, having the ability to overcome the resistant phenotype of multidrug-resistant (MDR) bacteria and to act against food-borne pathogens.[7]

Antibiotic Resistance. DOI: http://dx.doi.org/10.1016/B978-0-12-803642-6.00013-7

The aim of this chapter is to provide an overview of the methods of extractions of EOs, their proficiency as antimicrobials, their mode of action, and their efficacy in potentiating conventional antibiotics to counteract the action of resistant pathogens.

EOs: A PLURIPOTENTIAL NATURAL PRODUCT

EOs, which are also known as *ethereal oils*,[12,13] are obtained mostly from vegetable organs (namely, flowers, leaves, barks, woods, roots, rhizomes, fruits, and seeds) through processes such as expression, fermentation, enfleurage, and extraction.[1,14] The source of the term EO is thought to exist since the 16th century, named as Quinta Essentia, by Paracelsus von Hohenheim, a physician and alchemist who founded the role of chemistry in medicine.[13]

EOs found their place since ancient times in medicine and as part of rituals.[15] Egyptians were using aromatic plant materials for preserving mummies, while the Chinese valued aromatic plants for their vast curative purposes and also used them as perfume.[15] In India, the Ayurveda philosophy was based upon a variety of herbs, including a number of scented substances mentioned in the research literature. Humans have been using aromatic plants since ancient times; however, the distillation of EOs to extract their medicinal and fragrant components were not widely practiced until the 18th century.[15] In fact, distillation as a method of extracting EOs was first used in Egypt and Persia over 2000 years ago.[7,12] The first authentic written proof of distillation of EO was illustrated by Arnold of Villanova (c. 1235–1311), a Catalan physician.[7,12] Even though they were not well known in Europe until the 16th century, EOs obtained by the distillation process made their first appearance in pharmacies since the 13th century, with their pharmacological properties available as part of the pharmacopoeias.[7,12]

Nowadays, EOs have been used for a variety of applications, such as their key role in the cosmetic industry (due to their rich and diverse fragrances), in the food industry (as preservatives and flavorings), and in the pharmaceutical industry for their pharmacological properties.[16,17] EOs have found importance in the agro-industry as biofertilizers, and have also become popular due to their eco-friendly attributes, rapid volatility, and minimal residual activity, which are less detrimental to human health.[18]

Moreover, there are a vast amount of in vitro data indicating that plant-derived natural products such as EOs are valuable sources of novel treatments for MDR organisms.[9,19–21]

EXTRACTION OF EOs

EOs are naturally volatile and complex blends of biologically active compounds obtained through a form of distillation from aromatic plant materials of a single botanical species and form.[6,13] The constituents of EOs are produced in the plant through the mevalonic and shikimic acid pathways as secondary metabolites, and they are stocked in the plants in different organs depending on the plant: glandular trichomes, oil cells, or ducts in plant tissue.[13,22]

The well-known EOs are isolated from various aromatic plants generally distributed in the Mediterranean and tropical countries from around the world. This ability of a plant to accumulate

EOs is important in both gymnosperms and angiosperms and is sometimes a form of defense for the plant against fungal, bacterial, and viral phytopathogens.[13,23] EOs obtained from aromatic plants of Mauritius come from a large variety of families, of which the most common are the Lamiaceae, Zingiberaceae, Myrtaceae, Apiaceae, Rutaceae, Lauraceae, Piperaceae, and Verbenaceae.[24]

The method of recovering EOs is crucial for the proper characterization and evaluation of EOs. In 1928, Clevenger introduced a circulatory distillation apparatus that was designed for rapid quantitative determination of volatile EOs. This apparatus, now known as the Clevenger apparatus, is still the most efficient device used to obtain EOs for study.[6,25] Nowadays, the method of choice for the isolation of EOs remains the hydrodistillation process in the Clevenger apparatus.[26,27] In addition, though, there are several new revolutionary methods used for the extraction of EO in developed countries. However, these methods are expensive and not affordable on a large scale.

In the process of hydrodistillation, the plant material is totally immersed in water, which is brought to the boiling point to break the plant cytoplasm and liberate the EO components. The extraction process lasts about 90 min (depending on the source of the plant material). An azeotropic mixture is obtained, and the volatile EO escapes the surface of the water and is carried along with the steam from the boiling water. Upon condensation, oil- and water-rich layers are formed and separated. The EOs are dried and protected from sunlight to avoid oxidation. The temperature used for hydrodistillation is monitored, as elevated temperatures lead to the degradation of natural constituents (especially monoterpenes), which may undergo structural changes under these conditions. EO is collected, dried using anhydrous magnesium sulfate, and stored at 4°C until further use.[6,27,28]

The term EO is often used to refer to different natural volatile aromatic materials, although the technical terms depend on the process by which the material was obtained. For instance, solvent-extracting plant materials result in a natural volatile aromatic product called *concrete.* Likewise, cold expression of citrus fruit results in expressed oil, which is often termed EO.[13]

CHARACTERIZATION AND COMPOSITION OF EOs

Characterization of natural materials started only since the last five decades by using chemical reactions to identify new compounds. It was not until the introduction of spectroscopic methods that many more EO constituents (including those present in traces) were identified. Nowadays, different types of apparatus are used to identify and quantify the metabolites present in EOs: high-performance liquid chromatography (HPLC), gas chromatography (GC), gas chromatography—mass spectrometry (GC-MS), GC with a flame ionization detector (FID), liquid chromatography—massspectrometry (LC-MS), fast GC, GC-MS coupled with MS, GC coupled with Fourier transform infrared spectroscopy (FTIR), and head space (HS)-GC.[29]

The chemical composition of EOs are characterized and analyzed nowadays mostly by GC-MS.[30] GC is certainly a very rapid method of separation since no preliminary procedures are required. It is also a method of choice when only a very small quantity of oil is available.[29]

The major component found in EOs is in the class of organic compound known as terpenes, associating together to form macromolecules of terpenes (namely, monoterpenes (C10) and sesquiterpenes (C15)). For instance, diterpenes (C20) are two monomers of monoterpenes bonded together, and triterpenes (C30) are two monomers of associated sesquiterpenes. Monoterpenes represent 80–90% of the molecules found in EOs. These oils also contain many other components with different functional groups (eg, aldehyde, ketone, esters, alcohol, ethers, and carbures).[31]

EOs can be extracted from different families, among which the most popular are Apiaceae, Asteraceae, Cupressaceae, Lamiaceae, Lauraceae, Myrtaceae, Pinaceae, Piperaceae, Rutaceae, Santalaceae, and Zingiberaceae. The composition of these EOs can be classified in two distinctive groups: terpene hydrocarbons (monoterpenes and sesquiterpenes) and aromatic compounds, as shown in Table 13.1.[13,24] The monoterpenes consists mainly of monocyclic carbure (cymene, α-pinene, β-pinene), bicyclic carbure (sabinene), alcohol (citronellol, geraniol), and phenol (carvacrol, thymol). The sesquiterpenes consists of carbure (caryophyllene) and alcohol (farnesol). The aromatic compounds consist of aldehyde (cinnamaldehyde), alcohol (cinnamyl alcohol), phenol (chavicol, eugenol), methoxy derivative (anethole, estragole), methylene dioxy compound (safrole), and terpenoides (ascaridol, menthol).

The composition of EOs from different parts of the same plant can differ widely. For instance, EOs obtained from the seeds of coriander (*Coriandrum sativum* L.) is reportedly quite different in composition to the EO of cilantro, which is obtained from the immature leaves of the same plant.[32] Previous studies have reported that the above mentioned major components are the ones that determine the biological properties of the EOs.[33]

Table 13.1 Examples of Some Key Families of Plant Producing EOs and Their Major Components

Family	Plant Species	Identification	Chemical Constituents
Apiaceae (Umbelliferae)	*Coriandrum sativum* (coriander), *Cuminum cyminum* (cumin), *Centella asiatica* (centella), *Myrrhis odorata* (fennel), *Petroselinum crispum* (parsley)	Group of aromatic plants, often with hollow stems, distributed all around the world; consist of more than 4000 species; annuals, biennials, or perennials herbs, shrubs, or trees	Monoterpenes, sesquiterpenes, and phenylpropanoid compounds, coumarins, benzofluran derivatives, nonterpene, and alkaloid compounds
Asteraceae (Compostiae)	*Chrysothamnus nauseosus* (rabbit brush), *Ageratum conyzoides* (goat weed), *Artmisia* spp. (armoise)	The largest plant family with over 30,000 species consisting of evergreen shrubs, herbs, tuberous perennials, and tree herbs	Monoterpenes, coumarins
Cupressaceae	*Thujopsis dolabrata* (hiba), *Cryptomeria japonica* (Japanese cedar), *Cupressus macrocarpa* (Monterey cypress)	Family of conifers dispensed worldwide. Trees or shrubs, resinous and aromatic	Terpenes, phenols
Lamiaceae (Labiates)	*Hyssopus officinalis* (hyssop), *Lavandula angustifolia* (lavender), *Ocimum basilicum* (basil), *Mentha* spp. (mint), *Rosemary officinalis* (rosemary), *Salvia officinalis* (sage), *Thymus vulgaris* (thyme), *Tectona grandis* (teak)	Diverse family of aromatic herbs (many widely used culinary herbs) and shrubs	Terpenes, coumarins

Table 13.1 Examples of Some Key Families of Plant Producing EOs and Their Major Components *Continued*

Family	Plant Species	Identification	Chemical Constituents
Lauraceae (Laurel family)	*Laurus nobilis* (bay leaf), *Sassafras albidum* (sassafras), *Cinnamomum verum* (cinnamon)	Flowering plants, found in tropical and warm areas	Terpenes, aldehydes, esters and alcohols
Myrtaceae	*Backhousia citriodora* (lemon myrtle), *Eucalyptus* spp., *Leptospermum* spp. (lemon tea tree and manuka), *Pimenta dioica* (allspice), *Melaleuca alternifolia* (tea tree)	Consist of 3000 species of trees and plants, aromatic, flowering and also evergreens	Terpenoids, phenols, alcohols, aldehydes
Pinaceae	*Abies* spp. (fir) *Cedrus* spp. (cedar), *Pinus* spp. (pine oil)	Consist of 300 species Mostly evergreen, resinous plant	Turpentine and terpenoids
Pipperaceae	*Piper nigrum* (pepper), *Peperomia* spp. (radiator plant)	Family of fleshy, soft shrubs and occasionally small trees	Terpenoids
Rutaceae	*Fortunella* spp. (kumquat), *Citrus* spp. (citrus)	Plants are mostly shrubs and trees; frequently aromatic with glandular punctate, responsible for the aromatic smell of the family's members	Coumarins, esters, terpenes, aldehydes, alcohols
Santalaceae	*Santalum* spp. (sandalwood)	Family of woody flowering plants and trees	Esters, alcohols, aldehydes
Zingiberaceae	*Zingiber* spp. (ginger)	Family consisting of aromatic rhizomes	Terpenes and sesquiterpenes in rhizomes and leaves

BIOLOGICAL PROPERTIES OF EOs

EOs and extracts of aromatic plants have been utilized and recognized for many years as useful sources of pharmaceutical agents and food additives.[34] Much more importance has been given to EOs from different plant species in the last few decades due to the rising interest in their antibacterial, antifungal, insecticidal, and antioxidant properties.[7] These volatile oils also function in chemical defense, act as insecticides and acaricides, avoid fungi phytopathogen colonization, attract the natural enemies of herbivores[35,36], and have numerous pharmacological properties.

Biologically active compounds of EOs have been shown to possess numerous medicinal properties that are useful in the pharmaceutical and nutraceutical industries. The major composition of EOs and their biological properties are summarized in Table 13.2.

Table 13.2 Overview of Major Components Present in EOs and Their Biological Properties

Properties	Molecules	Major Composition of EO	Classification	Other Uses	Reference
Antioxidant	Linalool	*Ocimum basilicum* (sweet basil), *Cinnamomum altissimum* (mountain cinnamon)	Terpene alcohol	Fixative agent, floral and nonfloral fragrance, used in the synthesis of Vitamin E	27
	Eugenol	*Syzygium aromaticum* (clove), *O. basilicum* (sweet basil)	Phenol	In fragrance and flavors, as anesthetic for tropical fish, in dentistry	37
	Eucalyptol	*Melaleuca armillaris* (bracelet honey myrtle), *Piper betel* (betel), *M. alternifolia* (tea tree)	Alcohol	To make medicated balm, fragrances, dental preparation	38
	Thymol	*Thymus vulgaris* (thyme), *Salvia lavandulifolia* (sage)	Phenol	Fungicide; fragrance in soap ad household products	39,40
	Rosmanol	*Rosmarinus officinalis* (romarin)	Diterpenes	–	41
Antibacterial	Carvacrol and thymol	*Oreganum vulgare* (oregano), *T. vulgaris* (thyme)	Carvacrol: Phenol Thymol: Phenol	Carvacrol: in dental preparations, Thymol: in cosmetics, in traces for herbaceous fragrances, as fungicide	7,42,43
	Eugenol	*Syzygium aromaticum* (clove), *L. nobilis* (bay leaves), *Cananga odorata* (ylang ylang), *Pogostemon cablin* (patchouli), *Jasminum grandiflorum* (jasmine), *Myristica fragrans* (nutmeg)	Phenol	In fragrance and flavors, as anesthetic for tropical fish, in dentistry	7,29,44
	p-cymene and carvacrol	*O. basilicum* (sweet basil), *T. vulgaris* (thyme)	Carvacrol: Phenol *p*-cymene: alkylbenzene	*p*-cymene: analgesic	45–47
	Carvone	*Enterolobium contortisiliquum* (pacara earpode), *Mentha spicata* L (spearmint)	Ketone	Food flavors and fragrances	48,49
	Estragole	*Enterolobium contortisiliquum* (pacara earpode), *Artemisia dracunculus* (tarragon), *O. basilicum* (sweet basil), *Illicium verum* (star anise)	Aromatic ether	Perfumes and food additives	50
	Cinnamaldehyde	*Cinnamomum verum* (cinnamon)	Aldehyde	Agrichemical (fungicide), food flavors, perfumes	51
	Terpinene	*Melaleuca alternifolia* (narrow leaved tea tree), *Cuminum cyminum* (cumin)	Monoterpene hydrocarbon	Fragrance	52
Anticancer	β-elemene	*Boswellia* (frankincense)	Sesquiterpenes	Insect pheromones	53
Antiacetyl-cholinesterase	Isoeugenol, eugenol	*Eucalyptus* spp.	Phenylpropene	Fragrance	54

EMERGENCE OF DRUG-RESISTANT PATHOGENS: ANTIMICROBIAL RESISTANCE

Antibiotic resistance is the cause of common infections. The emergence of bacterial infections (eg, urinary tract infections, pneumonia, and bloodstream infections) is increasing throughout the world. MDR bacteria cause a high percentage of hospital-acquired infections, making them very difficult to treat.

Microorganisms are categorized as resistant when their minimal inhibitory concentrations (MICs) are high above a predefined threshold. Antimicrobial resistance is different from antibiotic resistance, which is defined as the resistance to drugs to treat infections caused by other microbes, such as parasites (eg, malaria), viruses (eg, HIV), and fungi (eg, *Candida*). That is, the potential of microorganisms to withstand attack by antimicrobial agents such as antibiotics, antivirals, antifungals, and antimalarials.[55] The breakthrough of new resistance mechanisms are spreading globally, putting at risk the treatment of common infectious diseases. The outcomes of these new resistance mechanisms are the death and disability of many individuals. Potential anti-infective treatments are becoming more within the scope of research, as standard medical treatments are failing to give significant results.

The leading cause of antimicrobial resistance is the excessive use of antimicrobial agents combined with the lack of new products, particularly for antibiotics, especially in developing countries. The outcome of this uncontrolled diagnostic tends lead to the emergence of multiple antibiotic-resistant strains. In health care, this results in a problematic to treat nosocomial infections.[56]

The resistance of infectious diseases to common drugs increases health-care costs, as alternative more expensive therapies are the only alternatives. In some developing countries, these therapies are often not available. The fight for antimicrobial resistance has nowadays gained in importance. Many important interventions, such as organ transplantation, cancer chemotherapy, major surgery, and the simple treatment of infections, are jeopardized. A return to the preantibiotic era must be considered since many infectious diseases risk becoming uncontrollable and could reverse the progress that has been made to date toward managing their spread.[55]

Studies have shown that bacteria developed resistance by the acquisition of new mechanisms or genetic material from other resistant organisms. This may progressively lead to a change in the nature of proteins expressed by the organism. In some cases, it has been reported that alteration in the structural and functional features of the bacteria followed the acquisition of genetic material. As described in the literature, microorganisms have undergone a change in several of their mechanisms, which have modified their capability and given rise to antimicrobial resistance.[57,58]

The principal mechanism of resistance as described in the literature,[59] is the acquisition of resistance genes, which is one of the most documented mechanisms. There are numerous amounts of genetic elements, which functions by counteracting the toxic effects of antibiotic drugs.

Moreover, there are different types of genes that have been reported as having the potential to develop into the resistant genes. These genes have been termed *protoresistance genes.* The resistant genes are present in pathogenic clinical isolates, as well as in the nonpathogenic bacteria that constitute the environmental pool. These simultaneously acquired resistances are explained by the modification of the targets, antibiotic binding proteins, which produces altered bacterial cell walls that no longer contain the binding site of the antimicrobial agent. It also includes the acquisition of gene-encoding enzymes, such as β-lactamases, that inactivate the antibacterial agent before it can produce an effect.

Also, there is downregulation of the genes responsible for the encoding of the protein on the outer membrane, which prevents the drug from entering the cells and to meet its target. Another result of genetic modification contributing to antimicrobial resistance is mutations that limit the access of antimicrobial agents to the intracellular target site via downregulation of porin genes. The organism also may acquire efflux pumps for this purpose that pump out the antibiotics that have been able to enter the cells, thereby preventing their intracellular accumulation and exerting its effect at the target site.[59]

It has been theorized that some organisms acquire resistance via horizontal evolution (ie, the acquisition of new genetic material from other resistant intraspecific or interspecific organisms, which is made possible by the transposons that contain the resistance genes). These transfers of genetic material are achieved via the process of conjugation, transduction, and transformation.[59]

In addition to the conventional resistance mechanisms mentioned previously (such as chromosomal β-lactamase, acquisition of efflux pumps, and the mutations in antibiotic target molecules), the formation of biofilms is another strategy adopted by microorganisms to defend against drugs. A biofilm, which is a structured consortium of bacteria niched in a self-produced polymer matrix consisting of polysaccharide, protein, and DNA, is the cause of multiple chronic infections (cystic fibrosis, endocarditis, infections caused by indwelling medical devices, and cystitis, among others) due to the increased tolerance to antibiotics and disinfectant chemicals that it induces. The formation and maintaining of a biofilm is often linked with a rise in the level of mutations, as well as with quorum sensing–regulated mechanisms.[60,61]

However, the mechanisms of resistance in biofilms are different compared to that of plasmids, transposons, and mutations, which confer distinctive resistance to the individual bacterial cells. In biofilms, resistance has been reported to depend on multicellular strategies due to the heterogenous nature of the biofilms. The biofilm matrix has been documented to act as a diffusion barrier, preventing the antibiotics from reaching their targets.

Furthermore, the heterogeneity of biofilms, which has been shown in vitro by high oxygen concentration at the surface and low concentrations in the center of the biofilm, can explain the different types of activity in these regions. It has also been shown that growth, protein synthesis, and metabolic activity are stratified in biofilms. These established microenvironments inside the biofilms have been proposed to explain the antagonizing action of antibiotic.

Also, the small subpopulations of bacteria within the layers of biofilm have been thought to differentiate into resistant organisms, and this innate resistance prevents the complete destruction of the biofilms to an extent that high-level antibiotic treatment fails to provide effective results.[62–64] Moreover, it is documented that oxidative stress lead to an increase in mutation of the bacteria within the biofilms. The microcolonies undergo distinctive change, which promotes antibiotic resistance.

ANTIBACTERIAL SPECTRUM OF EOs

EOs have been reported to be slightly more active against Gram-positive than Gram-negative bacteria.[32,65] Gram-negative organisms possess an outer membrane surrounding the cell wall, which restricts the diffusion of hydrophobic compounds through its lipopolysaccharide layer[66] which can explain why Gram-negative organisms are less susceptible to the action of EOs/antibiotics. However, it is to be taken into consideration that the individual components of EOs can have different action toward Gram-positive and Gram-negative organisms.[67] In addition, the chemical composition of EOs from different plant species tend to vary with their geographical positions and harvesting period. This variation in composition can possibly cause variability in the degree of susceptibility of Gram-negative and Gram-positive bacteria.[68]

Depending on the range of bacterial species susceptible to them, EOs analog to antibacterial can be classified as broad-spectrum, intermediate-spectrum, or narrow-spectrum. However, it is to be noted that that the spectra of activity of EOs may change with the acquisition of resistance genes by the bacteria as follows:

1. Broad-spectrum EOs are those EOs or components of EOs that are active against several phyla (which can cover from the firmicutes to the actinobacteria and proteobacteria). Examples of corresponding antibiotic with broad-spectrum activities are tetracyclines, phenicols, and fluoroquinolones.
2. Narrow-spectrum EOs are those EOs that have restricted activity and are solely only useful against specific species of microorganisms. For example, some EOs are only effective against Gram-positive bacteria, while others can only be effective against Gram-negative bacteria.[13]

ANTIMICROBIAL POTENTIAL OF EOs: A TARGET FOR MDR MICROORGANISMS

The emergence of MDR pathogens and their impact on health care, along with the inappropriate use of antibiotics, has become an urgent issue for patient safety.[28–69] Indeed, infectious diseases including MDR-related diseases are considered as one of the leading causes of global morbidity and mortality, especially in developing countries.[55] In addition, antibiotic resistance has become a major public health problem of increasing magnitude, and the discovery and development of novel antimicrobial agents from natural products to address this problem is of uttermost importance.[69,70]

The quest for novel effective antimicrobial agents from natural products has attracted much attention particularly in the health-care sector, where microbial resistance is increasing at an alarming rate and offering new challenges. Antimicrobial phytochemicals isolated from medicinal plants are thus being explored and their components probed in view of medical applications to fight fatal opportunistic infections.[6] Moreover, the problem of MDR has been observed from a wide range of pathogens and the most common example is methicillin-resistant *Staphylococcus aureus* (MRSA). The prevalence of MRSA infection has experienced a breakthrough in acute care and chronic care, thereby increasing the number of patients at risk of potential complications from contact precautions.[71] Also, the appearance of MDR among the Gram-negative bacteria, which have been correlated to the production of extended spectrum β-lactamase, is of growing concern worldwide; hence, such infectious diseases are becoming more challenging and difficult to manage and treat.[69] The establishment of new antibiotics is too expensive and even when the cost is affordable, the time for its development and implementation are very slow compared to the rate of increase of MDR pathologies.[69–71]

EOs have been used mostly in its liquid form, which penetrates the skin easily and binds to other biomolecules when integrated in lotions, powders, or pills. Recently, the vapors of some EOs have been studied for their antiseptic potential.[69–73] In health care, for instance, vapors of EOs can be exploited for decontaminating the environment and reducing the propagation of nosocomial bacteria.[73]

The considerable number of reports on the use of EOs and their components against multidrug-resistant bacteria, such as MRSA, *Mycobacterium tuberculosis*, *Acinetobacter baumannii*, and *Pseudomonas aeruginosa* indicate the potential of these natural products to overcome the development of antibacterial resistance.

S. aureus is a Gram-positive bacterium that is also part of the human microbial skin flora, but when present in consequent amounts, it results in minor infections. However, in some cases, *S. aureus* also causes severe diseases such as pneumonia, sepsis, endocarditis, or meningitis, particularly in hospitalized patients. The resistance of *S. aureus* against common drugs is a serious problem, particularly in the hospital parameters and EOs have been reported as good alternative agents to manage *S. aureus* infections. Many EOs have been reported to be efficient against MRSA.[13] For instance, *Eucalyptus globulus, Salvia rosifolia, Thymus vulgaris*, and *Lavandula stoechas* have been proved to inhibit the growth of MRSA, with the MIC values 85.6, 125, 18.5, and 31.2 μg/mL, respectively; their main constituents are detailed in Table 13.3.

Likewise, plants belonging to different families have been reported to have antifungal activity. In the literature, EOs have been well documented as effective antifungal agents.[74–76] Common conventional antifungal agents, such as chlorhexidine and imidazole derivatives, have restricted uses for children and infants, as well as in pregnant women, due to their multiple known side effects such as hepatotoxicity, nausea, diarrhea, and impotency.[77] Also, synthetic fungicides are not eco-friendly, and if their use can be limited, it would benefit the environment, as they cause environmental pollution owing to their slow biodegradation.[78] The realm of plant-derived materials such as EOs thus offers a wide spectrum of plausible effective alternatives to the available synthetic antifungals.

However, only limited data are available on the potential of EOs against human and plant fungal pathogens. Fungal species of the genera *Botrytis, Aspergillus, Fusarium*, and *Alternaria* have been considered to be major plant pathogens worldwide.[79,80] Meanwhile, among the common

Table 13.3 EO Constituents and Multidrug-Resistant Bacteria

EO	Family	Main Constituents
Cleistocalyx operculatus	Myrtaceae	γ-Terpinene (5.8%); globulol (5.6%); *cis*-linalool oxide (5.2%)
Eucalyptus globulus	Myrtaceae	1,8-Cineole (47.2%)
Kadsura longipedunculata	Schisandraceae	δ-Cadinene (21.8%)
Lavandula stoechas	Lamiaceae	α-Fenchone (39.2%); myrtenyl acetate (9.5%); α-pinene (6.1%); camphor (5.9%)
Salvia rosifolia	Lamiaceae	α-Pinene; 1,8-cineole
Thymus vulgaris	Lamiaceae	Thymol (48.1%)

human pathogens are the genus *Candida, Malassezia, Trichophyton, Epidermophyton,* and *Microsporum.*[81] Millions of people throughout the world, particularly in tropical countries where the climate is hot and humid, are affected by superficial fungal infections, which are the most common skin diseases. These diseases occur in both healthy and immunocompromised persons. Many skin diseases, which result from fungal infections caused by dermatophytes, are present in tropical and semitropical areas.

Several researchers have reported on mono- and sesquiterpenes hydrocarbons as the major components of plant EOs with enormous potential in the inhibition of microbial pathogens.[80] Among the reported active antimicrobial compounds, the phenolic terpenes present in EOs have been documented to be effective by interacting directly with the cell wall and cell membrane of the microbes. Different phytomolecules isolated from EOs of different families have been reported previously as antifungal agents; carvacrol, the most documented one, is present in a large group of families such as the Asteraceae, Lamiaceae, Lauraceae, and Verbenaceae.[82] There are also terpineol, *m*-cymene, myrcene, sabinene, linalool, thymol, menthol, eugenol, camphor, 1,8-cineole, geranial, limonene, and β-caryophyllene.[83–87]

In addition, the use of EOs offer potential novel template molecules and mixtures of bioactive compounds that can be exploited industrially as bio-products for the wellness, pharmaceutical, and food industries. The rich blend of different groups of chemical compounds present in EOs, which act as antimicrobial agents provide several targets within the microbial cell. The antimicrobial activity is thus attributed to several mechanisms of action, as detailed next.[88]

MODE OF ACTION OF EOs AS ANTIMICROBIAL AGENTS

Different EOs possess different modes of action, owing to their chemical composition, dose of antimicrobial major components, and degree of affinity to certain target sites within bacterial cells. The advantage of EOs in their fight against bacteria is that they are hydrophobic, which enables them to partition in the lipids of the cell membrane and mitochondria.[88] The major modes of actions of EOs toward bacterial cell are as follows:

1. The rupture of the cell wall. Once the EO achieves partition in the cell membrane, it will cause the leakage of the cell contents and ions. The cell wall is a critical structure for the life and survival of bacterial species. Lysis of the cell wall, if followed by the massive leakage of important solutes and molecules, will eventually result in cell death. Carvacrol, citral, *p*-cymene, and thymol have been reported to increase membrane permeability and the swelling of cellular membranes.[89] Also, eugenol has been studied[90] and indicated as a component of EO that induces damage in the cell wall of the bacteria.
2. The cytoplasmic membrane is also the target of EOs. More specifically, the proteins present in the cytoplasmic membrane are the targets of some EO components. It has been reported that the compounds present in the EO of cinnamon has the potential of inhibiting amino acid decarboxylase in *Enterobacter aerogenes.*

3. Cell membranes are vital barriers that delimit and regulate the intracellular and extracellular flow of substances. The destruction of one of its structure, the electrons transport system, by the action of EO and components has been reported[91] to inhibit bacteria.
4. Phenolic compounds present in the EOs have been reported to cause the disruption of the proton motive force, electron flow, and active transport, and to induce the coagulation of the cell contents.[92]

These are the major reported mode of actions of EOs on bacterial cells. However, the steps that follow one inactivation or lysis result in a series of consequences that are often fatal to the bacteria.

ANTIBIOTIC-POTENTIATING ACTIVITY OF EOs

The resistance and virulence of pathogenic bacteria are on the increase, particularly for the pathogenic Gram-negative bacteria *Enterobacter*, *Pseudomonas,* and *Acinetobacter*. In addition to this, the developments of new antibiotics that are effective are becoming rare. Much effort is being made to find new treatments for bacterial resistance. EOs have been reported in several scientific papers, through strong in vitro studies, to be potential antibacterial agents against a wide spectrum of bacteria.[93–95]

New techniques such as the enhancement of antibiotics using phyto-compounds are being studied. In this process, EOs have been recognized as antibiotic activity–modifying agents (ie, compounds with the potential to enhance the capacity of antibiotics).[95] The interactions between the EOs and antibiotics have shown to be effective against both the Gram-positive and Gram-negative bacteria, including MDR pathogens.[9,94,96–98]

The interactions between EO components and the antibiotics are assessed through different in vitro assays, among which the checkerboard, graphical, and time-kill methods are the most commonly reported. The combinations of EOs and antibiotics or EO components and antibiotics give different outcomes according to their degree of effectiveness. The results are evaluated by calculating the fractional inhibitory concentration (FIC) index, in order to determine cut offs for detecting the type of interaction resulting in the association of the different components.

The interactions can be of additive, synergistic, antagonistic, or indifferent type. The term *synergism* is applied when the addition of the EO to the antibiotic result in a marked increase in the rate of bactericidal effect; that is, the bactericidal rate is more important in the case of the combination (EO + antibiotic) in comparison to the effect observed when twice the concentration of EO alone is used. The term *additive* is used to qualify the interaction when the mixture of antibacterial agents exhibits an effect, which is equal to the sum of the effects of each component. The antagonist effect is used in the situation where the combination result a decrease in bactericidal effect; that is, the bactericidal rate is less in the case of the combination (EO + antibiotic) in comparison to the effect observed when twice the concentration of EO alone is used. Indifferent effect is applied to the situation where the combination of antibacterial agents (antibiotic) and that of the inactive substance. A FIC index <0.5 indicates synergism, $0.5<\text{FIC}<1$ indicates an additive effect, $1<\text{FIC}<4$ indicates indifference, and $\text{FIC}\geq 4$ indicates an antagonist effect between the components.[97,98]

INTERACTION BETWEEN THE DIFFERENT COMPONENTS OF EOs

The bioactivity of EOs is a result of complex interactions that are involved between the different classes of compounds constituting EOs. Among these compounds are phenols, aldehydes, ketones, alcohols, esters, ethers, or hydrocarbons. However, in some cases, the bioactivity tends to depend on the major components present in the EO. For instance, EOs containing cinnamaldehyde, citral, carvacrol, or eugenol as major components showed the highest antibacterial activity.[27,98] The potentiating effect of antibiotics has been investigated using EOs or components of EOs to fight MDR in pathogenic bacteria. Table 13.4 lists several EOs extracted from tropical herbs and spices and studied for their potentiating effects on antibiotics.

A few studies have been dedicated to the cause of the synergistic and antagonistic effects of EOs and its components. Theoretical mechanisms of antimicrobial interactions exhibiting a synergistic effect are thought to involve the inhibition of enzymes acting on the degradation or excretion of antimicrobials. Another possibility suggested in the literature is that antimicrobials with different sites and modes of action may be acting simultaneously on two sites both inside and on the cell to exhibit a synergistic effect. These simultaneous actions are thought to be dependent on each other, thus producing a significant effect. In addition, the interaction of one component with the cell wall or cell membrane is thought to increase the uptake of other antimicrobials.[103]

Table 13.4 EOs and Their Potentiating Activity

EO	Plant Family	Antibiotics	Bacteria Screened	Effect	Reference
Salvia fruticosa	Lamiaceae	Oxacillin	Methicillin-resistant *Staphylococcus epidermidis*	Synergistic	95
Salvia officinalis	Lamiaceae		Methicillin-resistant *S. epidermidis*	Synergistic	
Salvia sclarea	Lamiaceae		Methicillin-resistant *S. epidermidis*	Synergistic	
Piper betle	Piperaceae	Gentamicin	*Escherichia coli* *S. epidermidis*	Synergistic	9
Psiadia arguta	Asteraceae		*E. coli* *S. epidermidis*	Synergistic	
Pimenta dioica	Myrtaceae		*E. coli* *S. epidermidis*	Synergistic	
Rosmarinus officinalis	Lamiaceae	Ciprofloxacin	*Klebsiella pneumoniae* *Staphylococcus aureus*	Synergistic Antagonistic	98
Mentha piperita	Lamiaceae		*K. pneumoniae* *S. aureus*	Synergistic Synergistic	
Thymus vulgaris	Lamiaceae		*K. pneumoniae* *S. aureus*	Synergistic Synergistic	
Melaleuca alternifolia	Myrtaceae		*K. pneumoniae* *S. aureus*	Synergistic Antagonistic	

(*Continued*)

Table 13.4 EOs and Their Potentiating Activity *Continued*

EO	Plant Family	Antibiotics	Bacteria Screened	Effect	Reference
Citrus limon	Rutaceae	Amikacin	*Acinetobacter* spp.	Synergistic	99
Cinnamomum zeylanicum	Lauraceae			Synergistic	
Eucalyptus spp.		Chlorhexidine Digluconate	*S. epidermidis*	Synergistic	100
Coriandrum sativum	Apiaceae	Chloramphenicol Ciprofloxacin Gentamicin Tetracycline	*Acinetobacter baumannii*	Synergistic	101
Origanum vulgare	Lamiaceae	Fluoroquinolones Doxycycline Lincomycin Maquindox	*E. coli*	Synergistic	102

Moreover, the synergistic effect that exists between different compounds acting together may sometimes be due to reduced metabolic activity in the bacteria. It is thought that the metabolic activity decreases with time when the bacteria has been resisting antibiotics for a long time. In that state, they are more vulnerable to an antibacterial acting on several sites (eg, on the cell membrane and in the cell) simultaneously, as is expected during synergistic interactions.[104]

On the other hand, the antagonistic effects induced during the interaction of two different components are less well known. It is thought to occur when a bacteriostatic compound interacts with a bactericidal compound, but also when the two compounds in the interaction share the same site of action.[103]

Several detailed research papers have suggested that the activities of EOs are predominantly influenced by whole EOs, rather than a mixture of the major components isolated from the EOs. This concept thus implies that the minor components interacting with the major elements in the EOs play a useful role in inducing the synergistic activity.[98,103]

CONCLUSION

A wide variety of reports have investigated into the emergence of microorganisms that can overcome the effect of antibiotic targets, as well as their frequencies of adaptation to these antibiotics. The new insight of antibacterial therapy, which involve combination therapy instead of the monotherapy of EO in view of potentiating antibiotics, have been studied recently to improve the efficacy of modulating drug resistance. The introduction of natural alternatives like EOs seems a potential avenue that can reduce the treatment intensities and doses of synthetic antimicrobials. Consequently, the potential adverse effects, as well as the incidence of resistance, are alleviated.

REFERENCES

1. Aleksic V, Knezevic P. Antimicrobial and antioxidative activity of extracts and essential oils of *Myrtus communis* L. *Microbiol Res* 2014;**169**(4):240–54.
2. Palombo EA. Traditional medicinal plant extracts and natural products with activity against oral bacteria: potential application in the prevention and treatment of oral diseases. *Evid Based Complement Alternat Med* 2011;**2011**:1–15.
3. Lai P, Roy J. Antimicrobial and chemopreventive properties of herbs and spices. *Curr Med Chem* 2004;**11**(11):1451–60.
4. Newman DJ, Cragg GM. Natural products as sources of new drugs over the last 25 years. *J Nat Prod* 2007;**70**(3):461–77.
5. Delamare APL, Moschen-Pistorello IT, Artico L, Atti-Serafini L, Echeverrigaray S. Antibacterial activity of the essential oils of *Salvia officinalis* L. and *Salvia triloba* L. cultivated in South Brazil. *Food Chem* 2007;**100**:603–8.
6. Bakkali F, Averbeck S, Averbeck D, Idaomar M. Biological effects of essential oils-a review. *Food Chem Toxicol* 2008;**46**:446–75.
7. Burt SA, Reinders RD. Antibacterial activity of selected plant essential oils against *Escherichia coli* O157:H7. *Lett Appl Microbiol* 2004;**36**:162–7.
8. Zu Y, Yu H, Liang L, Fu Y, Efferth T, Liu X, et al. Activities of ten essential oils towards *Propionibacterium acnes* and PC-3, A-549 and MCF-7 cancer cells. *Molecules* 2010;**15**:3200–10.
9. Aumeeruddy-Elalfi Z, Gurib-Fakim A, Mahomoodally F. Antimicrobial, antibiotic potentiating activity and phytochemical profile of essential oils from exotic and endemic medicinal plants of Mauritius. *Ind Crops Prod* 2015;**71**:197–204.
10. Neveen H, El Soud A, Deabes MM, Abou El Kassem LT, Khalil MY. Antifungal activity of family *Apiaceae* essential oils. *Res J Appl Sci* 2012;**8**:4964–73.
11. Raut JS, Karuppayil SM. A status review on the medicinal properties of essential oils. *Ind Crops Prod* 2014;**62**:250–64.
12. Guenther E. *The essential oils, vol. 1*. New York, NY: D. Van Nostrand Company Inc.; 1948.
13. Hunter VM. *Essential oils: art, agriculture, science, industry and entrepreneurship: A focus on the Asia-Pacific region*. New York, NY: Nova Science Publishers and University of Malaysia Perli; 2009.
14. Celiktas OY, Hames Kocabas EE, Bedir E, Vardar Sukan F, Ozek T, Baser KHC. Antimicrobial activities of methanol extracts and essential oils of *Rosmarinus officinalis*, depending on location and seasonal variations. *Food Chem* 2007;**100**(2):553–9.
15. Lawless J. *The illustrated encyclopedia of essential oils: the complete guide to the use of oils in aromatherapy and herbalism*. London: Harper Collins Publishers; 1995.
16. Schmidt B, Ribnicky DM, Poulev A, Logendra S, Cefalu WT, Raskin L. A natural history of botanical therapeutics. *Metab Clin Exp* 2008;**57**:53–9.
17. Nychas GJE. Natural antimicrobials from plants. In: Gould GW, editor. *New methods of food preservation*. 1st ed. London: Blackie Academic & Professional; 1995. p. 58–89.
18. Lubbe A, Verpoorte R. Cultivation of medicinal and aromatic plants for specialty industrial materials—a review. *Ind Crops Prod* 2011;**34**:785–801.
19. Kon KV, Rai MK. Plant essential oils and their constituents in coping with multidrug-resistant bacteria. *Expert Rev Anti Infect Ther* 2010;**10**:775–90.
20. Sienkiewicz M, Poznańska-Kurowska K, Kaszuba A, Kowalczyk E. The antibacterial activity of geranium oil against Gram-negative bacteria isolated from difficult-to-heal wounds. *Burns* 2014;**40**(5):1046–51.
21. Zenati F, Benbelaid F, Khadir A, Bellahsene C, Bendahou M. Antimicrobial effects of three essential oils on multidrug resistant bacteria responsible for urinary infections. *J Appl Pharm Sci* 2014;**4**(11):15–18.

22. Chaiyana W, Okonogi S. Inhibition of cholinesterase by essential oil from food plant. *Phytomedicine* 2012;**19**:836–9.
23. Fisher MC, Henk DA, Briggs CJ, Brownstein JS, Madoff LC, Mccraw SL. Emerging fungal threats to animal, plant and ecosystem health. *Nature* 2012;**484**:186–94.
24. Gurib-Fakim A. *Medicinal plants of Mauritius and of the world.* Baie du Tombeau, Mauritius: Caractère LTD; 2007.
25. Sandra P, Bicchi C. *Capillary gas chromatography in essential oil analysis.* Heidelberg: Huethig; 1987.
26. Kulisic T, Radonic A, Katalinic V, Milos M. Use of different methods for the testing activity of oregano essential oil. *Food Chem* 2004;**85**:633–40.
27. Hussain IA, Anwar F, Iqbal STH, Przybylski R. Chemical composition; antioxidant and antimicrobial activities of basil (*Ocimum basilicum*) essential oils depends on seasonal variations. *Food Chem* 2008;**108**:986–95.
28. Bedi S, Tanuja S, Vyas P. *A handbook of aromatic and essential oil plants: cultivation, chemistry, processing and uses.* Jodhpur: Agrobios; 2010.
29. Maree J, Kamatou G, Gibbons S, Viljoen A, Van Vuuren S. The application of GC–MS combined with chemometrics for the identification of antimicrobial compounds from selected commercial essential oils. *Chemometr Intell Lab Syst* 2014;**130**:172–81.
30. Vankar PS. Essential oils and fragrances from natural sources. *Resonance* 2004;**9**(4):35.
31. Bozin B, Mimica-Dukic N, Siminn N, Anackov G. Characterization of the volatile composition of essential oils of some lamiaceae species and the antimicrobial and antioxidant activities of the entire oils. *J Agric Food Chem* 2006;**54**:1822–8.
32. Delaquis PJ, Stanich K, Girard B, Mazza G. Antimicrobial activity of individual and mixed fractions of dill, cilantro, coriander and eucalyptus essential oils. *Int J Food Microbiol* 2002;**74**:101–9.
33. Celikel N, Kavas G. Antimicrobial properties of some essential oils against some pathogenic microorganisms. *Czech J Food Sci* 2008;**26**:174–81.
34. Juliano C, Mattana A, Usai M. Composition and in vitro antimicrobial activity of the essential oil of *Thymus herba-barona* Loisel growing wild in Sardinia. *J Essent Oil Res* 2000;**12**:516–22.
35. Yadav RKP, Papatheodorou EM, Karamanoli K, Constantinidou H-IA, Vokou D. Abundance and diversity of the phyllosphere bacterial communities of Mediterranean perennial plants that differ in leaf chemistry. *Chemoecology* 2008;**18**:217–26.
36. Karamanoli K, Menkissoglu-Spiroudi U, Bosabalidis AM, Vokou D, Constantinidou H-IA. Bacterial colonization of the phyllosphere of nineteen plant species and antimicrobial activity of their leaf secondary metabolites against leaf associated bacteria. *Chemoecology* 2005;**15**:59–67.
37. Porres-Martínez M, González-Burgos E, Carretero Accame ME, Gómez-Serranillos P. Phytochemical composition, antioxidant and cytoprotective activities of essential oil of *Salvia lavandulifolia* Vahl. *Food Res Int* 2013;**54**:523–31.
38. Chabir N, Romdhane M, Valentin A, Moukarzel B, Ben Marzoug HN, Ben Brahim N, et al. Chemical study and antimalarial, antioxidant, and anticancer activities of *Melaleuca armillaris* (Sol Ex Gateau) Sm essential oil. *J Med Food* 2011;**14**(11):1383–8.
39. Shan B, Cai YZ, Sun M, Corke H. Antioxidant capacity of 26 spice extracts and characterization of their phenolic constituents. *J Agric Food Chem* 2005;**53**:7749–59.
40. Wojdylo A, Oszmianski J, Czemerys R. Antioxidant activity and phenolic compounds in 32 selected herbs. *Food Chem* 2007;**105**:940–9.
41. Wang SY. Antioxidant capacity of berry crops, culinary herbs and medicinal herbs. *Acta Hortic* 2003;**620**:461–73.
42. Ultee A, Bennik MH, Moezelaar R. The phenolic hydroxyl group of carvacrol is essential for action against the food-borne pathogen *Bacillus cereus*. *Appl Environ Microbiol* 2002;**68**:1561–8.
43. Lambert RJW, Skandamis PN, Coote P, Nychas G-JE. A study of the minimum inhibitory concentration and mode of action of oregano essential oil, thymol and carvacrol. *J Appl Microbiol* 2001;**91**:453–62.

44. Wendakoon CN, Sakaguchi M. Inhibition of amino acid decarboxylase activity of *Enterobacter aerogenes* by active components in spices. *J Food Prot* 1995;**58**(3):280–3.
45. Solórzano-Santos F, Miranda-Novales V. Essential oils from aromatic herbs as antimicrobial agents. *Curr Opin Biotechnol* 2012;**23**(2):136–41.
46. Veldhuizen EJA, Tjeerdsma-Van Bokhoven JLM, Zweijtzer C, Burt SA, Haagsman HP. Structural requirements for the antimicrobial activity of carvacrol. *J Agric Food Chem* 2006;**54**:1874–9.
47. Kiskó G, Roller S. Carvacrol and *p*-cymene inactivate *Escherichia coli* O157:H7 in apple juice. *BMC Microbiol* 2005;**5**:36.
48. Shahat AA, El-Barouty G, Hassan RA, Hammouda FM, Abdel-Rahman FH, Saleh MA. Chemical composition and antimicrobial activities of the essential oil from the seeds of *Enterolobium contortisiliquum* (leguminosae). *J Environ Sci Health B* 2008;**43**(6):519–25.
49. Aggarwal KK, Khanuja SPS, Ahmad A, Santha Kumar TRK, Gupta V, Kumar S. Antimicrobial activity profiles of the two enantiomers of limonene and carvone isolated from the oils of *Mentha spicata* and *Anethum sowa*. *Flavour Fragr J* 2002;**17**(1):59–63.
50. Ariamuthu S, Balakrishnan V, Lavanya Srinivasan M. Chemical composition and antibacterial activity of essential oil from fruits of *Illicium verum* Hook. *Int J Res Phytochem Pharmacol* 2013;**3**(2):85–9.
51. Yossa N, Patel J, Macarisin D, Millner P, Murph C, Bauchan G, et al. Antibacterial activity of cinnamaldehyde and sporan *against Escherichia coli* O157:H7 and *Salmonella*. Publications from USDA-ARS/UNL Faculty Paper 2012; 1136.
52. Oyedemi SO, Bradley G, Afolayan AJ. Ethnobotanical survey of medicinal plants used for the management of diabetes mellitus in the Nkonkobe municipality of South Africa. *J Med Plant Res* 2009;**3**:1040–4.
53. Chen Y, Zhou C, Ge CZ, Liu Y, Feng W, Li S, et al. Composition and potential anticancer activities of essential oils obtained from myrrh and frankincense. *Oncol Lett* 2013;**6**(4):1140–6.
54. Yeom HJ, Kang J, Kim SW, Park IL-K. Fumigant and contact toxicity of Myrtaceae plant essential oils and blends of their constituents against adults of German cockroach (*Blattella germanica*) and their acetylcholinesterase inhibitory activity. *Pestic Biochem Physiol* 2013;**107**:200–6.
55. World Health Organization. Antimicrobial resistance. Fact sheet N 194; 2015.
56. Fleischer T, Mensah MLK, Mensah AY, Komlaga G, Gbedema SY, Skaltsa H. Antimicrobial activity of essential oils of *Xylopia Aethiopica*. *Afr J Tradit Complement Altern Med* 2008;**5**(4):391–3.
57. Tenover FC. Mechanisms of antimicrobial resistance in Bacteria. *Am J Med* 2006;**119**(6A):S3–10.
58. Jayaraman R. Antibiotic resistance: an overview of mechanisms and a paradigm shift. *Curr Sci* 2009;**96**(11):1475–84.
59. Claudio O, Gualerzi Letizia B, Attilio F, Cynthia Pon L. *Antibiotics targets mechanisms and resistance*. Weinheim: John Wiley & Sons, Medical; 2013.
60. Wingender J, Strathmann M, Rode A, Leis A, Flemming HC. Isolation and biochemical characterization of extracellular polymeric substances from *Pseudomonas aeruginosa*. *Methods Enzymol* 2001;**336**:302–14.
61. Costerton W, Veeh R, Shirtliff M, Pasmore M, Post C, Ehrlich G. The application of biofilm science to the study and control of chronic bacterial infections. *J Clin Invest* 2003;**112**(2003):1466–77.
62. Stewart PS, Costerton JW. Antibiotic resistance of bacteria in biofilms. *Lancet.* 2001;**358**:135–8.
63. Patel R. Biofilms and antimicrobial resistance. *Clin Orthop Relat Res* 2005;**437**:41–7.
64. Anderson GG, O'Toole GA. Innate and induced resistance mechanisms of bacterial biofilms. In: Romeo T, editor. *Bacterial biofilms. Current topics in microbiology and immunology 322*. Berlin and Heidelberg: Springer-Verlag; 2008. p. 85–105.
65. Harpaz S, Glatman L, Drabkin V, Gelman A. Effects of herbal essential oils used to extend the shelf life of fresh water reared Asian sea bass fish (*Lates calcarifer*). *J Food Prot* 2003;**66**:410–17.
66. Baur AK. The lung tumor promoter, butylated hydroxytoluene (BHT), causes chronic inflammation in promotion-sensitive BALB/cByJ mice but not in promotion-resistant CXB4 mice. *Toxicology* 2001;**169**(1):1–15.

67. Dorman HJD, Deans SG. Antimicrobial agents from plants: antibacterial activity of plant volatile oils. *J Appl Microbiol* 2000;**88**:308–16.
68. Pintore G, Usai M, Bradesi P, Juliano C, Boatto G, Tomi Chessa M, et al. Chemical composition and antimicrobial activity of *Rosmarinus officinalis* L. oils from Sardinia and Corsica. *Flavour Fragr J* 2001;**17**:15–19.
69. Edward-Jones V. Alternative antimicrobial approaches to fighting multidrug-resistant infections. In: Rai M, Kon K, editors. *Fighting multidrug resistance with herbal extracts, essential oils and their components*. Amsterdam: Academic Press; 2013. p. 1–8.
70. Yala D, Merad AS, Mohamedi D, Ouar Korich MN. Classification et mode d'action des antibiotiques. *Méd Maghreb* 2001;**91**:5–12.
71. Ahmad A, Van Vuuren S, Viljoen A. Unravelling the complex antimicrobial interactions of essential oils—the case of *Thymus vulgaris* (Thyme). *Molecules* 2014;**19**(3):2896–910.
72. Doran AL, Morden WE, Dunn K, Edwards-Jones V. Vapour–phase activities of essential oils against antibiotic sensitive and resistant bacteria including MRSA. *Lett Appl Microbiol* 2009;**48**:387–92.
73. Laird K, Phillips C. Vapour phase: a potential future use for essential oils as antimicrobials? *Lett Appl Microbiol* 2012;**54**:169–74.
74. Adam K, Sivropoulu A, Kokkini S, Lanaras T, Arsenakis M. Antifungal activities of *Origanum vulgare* subsp. hirtum, *Mentha spicata*, *Lavandula angustifolia* and *Salvia fruticosa* essential oils against human pathogenic fungi. *J Agric Food Chem* 1998;**46**:1739–45.
75. Rasooli I, Rezaei MB, Allameh A. Growth inhibition and morphological alterations of *Aspergillus niger* by essential oils from *Thymus eriocalyx* and *Thymus x-porlock*. *Food Control* 2006;**17**:359–64.
76. De Lira Mota K, de Oliveira Pereira F, de Oliveira W, Lima I, de Oliveira Lima E. Antifungal activity of *Thymus vulgaris* L. essential oil and its constituent phytochemicals against *Rhizopus oryzae*: interaction with ergosterol. *Molecules* 2012;**17**(12):14418–33.
77. Curtis C. Use and abuse of topical dermatological therapy in dogs and cats. Part 1. shampoo therapy. *In Pract* 1998;**20**:244–51.
78. Hawksworth DL, Kirsop BE, Jong SC, Pitt JI, Samson RA, Kirsop BE. *Living resources for biotechnology. Filamentous fungi*. Cambridge: Cambridge University Press; 2008.
79. Rosslenbroich H, Stuebler D. *Botrytis cinerea*—history of chemical control and novel fungicides for its management. *Crop Prot* 2000;**19**:557–61.
80. Prabhu RM, Patel R. Mucormycosis and entomophthoramycosis: a review of the clinical manifestations, diagnosis and treatment. *Clin Microbiol Infect* 2004;**10**:31–47.
81. Omoruyi B, Afolayan A, Bradley G. The inhibitory effect of *Mesembryanthemum edule* (L.) bolus essential oil on some pathogenic fungal isolates. *BMC Complement Altern Med* 2014;**14**(1):168.
82. Abbaszadeh S, Sharifzadeh A, Shokri H, Khosravi AR, Abbaszadeh A. Antifungal efficacy of thymol, carvacrol, eugenol and menthol as alternative agents to control the growth of food-relevant fungi. *Med Mycol J* 2014;**24**(2):51–6.
83. Bullerman LB, Lieu FY, Seire AS. Inhibition of growth and aflatoxin production by cinnamon and clove oils, cinnamic aldehyde and eugenol. *J Food Sci* 1977;**42**:1107–16.
84. Xu J, Zhou F, Ji BP, Pei RS, Xu N. The antibacterial mechanism of carvacrol and thymol against *Escherichia coli*. *Lett Appl Microbiol* 2008;**47**(3):174–9.
85. Cabral C, Gonçalves MJ, Cavaleiro C, Sales F, Boyom F, Salgueiro L. Composition and anti-fungal activity of the essential oil from Cameroonian *Vitex rivularis* Gürke. *Nat Prod Res* 2010;**23**:1478–84.
86. Tzasna H, Margarita C, Ana maria G, Angel D, Sameul M, Patricia DJ. Antifungal activity of the essential oils of two verbenaceae: *Lantana achyranthifolia* and *Lippia graveolens* of Zapotitlan de las Salinas, Puebla (Mexico). *B Latinoam Caribe PL* 2008;**7**(4):202–6.
87. Sumonrat C, Suphitchaya C, Tipparat H. Antimicrobial activities of essential oils and crude extracts from tropical *Citrus* spp. against food-related microorganisms. Songklanakarin. *J Mat Sci Technol* 2008;**30**(1):125–31.

88. Derwich E, Benziane Z, Boukir A. GC/MS analysis and antibacterial activity of the essential oil of *Mentha pulegium* grown in Morocco. *Res J Agric Biol Sci* 2010;**6**:191–8.
89. Burt SA. *Antibacterial activity of essential oils: potential application in food*. Utrecht: Utrecht University; 2007.
90. Bennis S, Chamin F, Chami N, Bouchikhi T, Remmal A. Surface alteration of *Saccharomyces cerevisiae* induced by thymol and eugenol. *Lett Appl Microbiol* 2004;**38**(6):454–8.
91. Tassou CC, Koutsoumanis K, Nychas GJE. Inhibition of *Salmonella enteritidis* and *Staphylococcus aureus* on nutrient both by mint essential oil. *Food Res Int* 2000;**48**:273–80.
92. Denyer SP, Hugo WB. Mechanisms of antibacterial action—a summary. In: Denyer SP, Hugo WB, editors. *Mechanisms of action of chemical biocides*. Oxford: Blackwell Scientific Publications; 1991. p. 331–4.
93. Rapper DS, Kamatou G, Viljoen A, Van Vuuren S. The in vitro antimicrobial activity of *Lavandula angustifolia* essential oil in combination with other aroma-therapeutic oils. *Evid Based Complement Alternat Med* 2013;**2013**:1–10.
94. Bassolé I, Juliani H. Essential oils in combination and their antimicrobial properties. *Molecules* 2012; **17**(12):3989–4006.
95. Chovanová R, Mikulášová M, Vaverková S. In vitro antibacterial and antibiotic resistance modifying effect of bioactive plant extracts on Methicillin-Resistant *Staphylococcus epidermidis*. *Int J Microbiol* 2013;**2013**:1–7.
96. Coutinho HDM, Costa JG, Falcão-Silva VS, Siqueira JP, Lima Jr EO. Effect of *Momordica charantia* L. in the resistance to aminoglycosides in methicillin-resistant *Staphylococcus aureus*. *Comp Immunol Microbiol Infect Dis* 2010;**33**(2010):467–71.
97. Rosato A, Piarulli M, Corbo F. In vitro synergistic antibacterial action of certain combinations of gentamicin and essential oils. *Curr Med Chem* 2010;**17**:3289–95.
98. Van Vuuren SF, Suliman S, Viljoen AM. The antimicrobial activity of four commercial essential oils in combination with conventional antimicrobials. *Lett Appl Microbiol* 2009;**48**(4):440–6.
99. Guerra F, Mendes J, Sousa J, Morais-Braga M, Santos B, Melo Coutinho H, et al. Increasing antibiotic activity against a multidrug-resistant *Acinetobacter* spp by essential oils of *Citrus limon* and *Cinnamomum zeylanicum*. *Nat Prod Res* 2012;**26**(23):2235–8.
100. Karpanen TJ, Worthington T, Hendry ER, Conway BR, Lambert PA. Antimicrobial efficacy of chlorhexidine digluconate alone and in combination with eucalyptus oil, tea tree oil and thymol against planktonic and biofilm cultures of *Staphylococcus epidermidis*. *J Antimicrob Chemother* 2008; **62**(5):1031–6.
101. Duarte A, Ferreira S, Silva F, Domingues FC. Synergistic activity of coriander oil and conventional antibiotics against *Acinetobacter baumannii*. *Phytomedicine* 2012;**19**(3–4):236–8.
102. Si H, Hu J, Liu Z, Zeng Z. Antibacterial effect of oregano essential oil alone and in combination with antibiotics against extended-spectrum β-lactamase-producing *Escherichia coli*. *FEMS Immunol Med Microbiol* 2008;**53**:190–4.
103. Eliopoulos G, Gilbert D, Sanford J. Antimicrobial therapy. In: Gilbert DN, Moellering Jr RC, Eliopoulos GM, Chambers HF, Saag MS, editors. *Sanford guide to antimicrobial therapy*. Sperryville, VA: Antimicrobial Therapy; 2010.
104. Gunnison J, Kunishige E, Coleman V, Jawetz E. The mode of action of antibiotic synergism and antagonism: the effect in vitro on bacteria not actively multiplying. *J Gen Microbiol* 1995; **13**(3):509–18.

CHAPTER

ANTIMICROBIAL NATURAL PRODUCTS AGAINST BACTERIAL BIOFILMS

14

V.G. Beoletto[1], M. de las Mercedes Oliva[1], J.M. Marioli[2], M.E. Carezzano[1] and M.S. Demo[1]

[1]*Department of Microbiology and Immunology, Universidad Nacional de Río Cuarto, Río Cuarto, Argentina*

[2]*Department of Chemistry, Universidad Nacional de Río Cuarto, Río Cuarto, Argentina*

INTRODUCTION

Nowadays, the continual appearance of bacterial resistance to antibiotics has become the major problem that scientists have to deal with. Pathogenic microorganisms responsible for human, animal, and plant diseases are acquiring resistance due to the inadequate use of antimicrobial drugs. The indiscriminate prescription of antibiotics creates selective pressure for the emergence of resistant strains, resulting in a growing problem. There is an urgent need to discover novel classes of antimicrobial substances unable to generate resistance that can effectively work on a target site or in different target sites of the cell. In this way, plants constitute natural sources with potential therapeutic or medicinal values that have been successfully used as alternative treatments to solve health ailments.[1]

For a long period of time, plants have been a valuable source of natural products for maintaining human and animal health.[2] Ancient texts of India and China contain exhaustive depictions of the use of a variety of plant-derived medications.[3] Moreover, there is evidence that Neanderthals living 60,000 years ago in present-day Iraq used plants such as hollyhock, which are still widely used in ethnomedicine around the world.[4] Approximately 300,000 species are known in the plant kingdom, although the total number of species is much higher. Considering that only 1% of these species has been studied extensively in terms of chemical composition and therapeutic use, there is still enormous potential for discovering bioactive molecules.

Plants have an almost limitless ability to synthesize substances. Some of them may serve as plant defense mechanisms against predation by microorganisms, insects, and herbivores. Some substances give plants their odors, while others are responsible for plant pigment. Many compounds give plants their flavor, and some of the same herbs and spices used by humans to season food yield useful medicinal compounds. Many plant extracts, such as essential oils (EOs), have been shown to exert biological activity in vitro and in vivo, mainly antimicrobial activity justifying the research on traditional medicine.[1] EOs are aromatic oily liquids obtained from plant material like flowers, buds, seeds, leaves, twigs, bark, herbs, wood, fruits, and roots. They are also known as *volatile* or *ethereal oils*.

Antibiotic Resistance. DOI: http://dx.doi.org/10.1016/B978-0-12-803642-6.00014-9

Phytochemicals with biological activities can be divided into the following categories depending on the chemical structure and the mechanism of action:

- *Phenols and phenolic acids*: They inhibit microbial enzymes interacting with sulfhydryl groups or through nonspecific interactions with the proteins. These are effective against viruses, bacteria, and fungi.
- *Quinones*: They affect cell surface adhesins, cell wall polypeptides, and membrane-bound enzymes.
- *Flavones and flavonoids*: Their biological activity is attributed to the ability to form a complex with bacterial cell walls and to disrupt microbial membranes.
- *Tannins*: These compounds inactivate microbial adhesins, enzymes, cell envelope, transport proteins, polysaccharides, etc.
- *Coumarins*: They were found to inhibit microorganisms in vitro.
- *Terpenoids and EOs*: Their mechanism of action involves membrane disruption by lipophilic compounds.
- *Alkaloids*: They have the ability to intercalate into DNA, resulting in an antimicrobial effect.
- *Lectins and polypeptides*: They formed ion channels in the microbial membrane and foster competitive inhibition of adhesion of microbial proteins to host polysaccharide receptors.[4]

This chapter is organized into two sections; the first focuses on the description of bacterial biofilms as a resistance structure to antimicrobial substances, presenting plant-derived products such as EOs, terpenes and extracts as new strategies to avoid biofilm formation or to eliminate attached biofilm. In the second section, different methods are described for obtaining the plant-derived products, with emphasis on EOs and aqueous and organic extracts.

PLANT-DERIVED PRODUCTS WITH ANTIBIOFILM ACTIVITY

Several mechanisms are described in this section, explaining the way that antibiotics exert their microbial action, including the inhibition of cell wall synthesis (eg, β-lactams), the inhibition of protein synthesis (eg, macrolides and tetracyclines), the interference of nucleic acid synthesis (eg, quinolones), the inhibition of metabolic pathways (eg, sulfamides), and the disruption of bacterial membrane structure (eg, polymyxins).

Bacteria can be naturally resistant to antibiotics, or they can acquire this resistance. Naturally occurring resistance to antibiotics is attributed to the lack of a drug target (eg, *Mycoplasma* cells that do not have cell walls) or being impermeable to the drug (eg, penicillin in Gram-negative bacteria). In the other way, they can acquire resistance by mutations or incorporating multiple resistance genes into the host's genome or plasmids by the transference from other organisms through conjugation, transformation, or transduction mechanisms. Consequently, cells became capable of producing enzymes that destroy antibiotics (eg, β-lactamases), expressing efflux systems that prevent drugs from reaching its intracellular target (eg, multidrug efflux systems of *Pseudomonas aeruginosa*), modifying the drug's target site (eg, modification of penicillin-binding proteins in *Staphylococcus aureus*), or producing an alternative metabolic pathway that avoids the action of the drug.[5]

The latest research about microorganisms and resistance structures has focused on the ability of microbial communities to attach to alive or inert surfaces. These communities of microorganisms, known as *biofilms*, develop when unicellular organisms adhere to a solid surface and reproduce themselves to form a community encased in an exopolysaccharide (EPS) matrix. These sessile cells acquire properties distinct from planktonic ones that give them some advantages, like an increased resistance to antimicrobial agents, causing an enormous impact on medicine.[6]

CHARACTERISTICS OF BACTERIAL BIOFILMS

Evidence of biofilm's existence was described in the 1970s, when many strains of coagulase-negative staphylococci were reported to form mucoid deposits on cerebrospinal fluid shunts.[7] The use of scanning electron microscopy (SEM) helped in the observation of rounded cells located in intravascular catheters, peritoneal dialysis catheters, and pacemaker leads, lately identified as *Staphylococcus epidermidis*. These bacteria were forming agglomerations of cells embedded into an amorphous material first referred to as "slime," a structure that was consequently described for other microorganisms such as *S. aureus*. Further studies using molecular tools helped in the detection of a locus responsible for the production of this biofilm in Gram-positive cocci (namely, the *ica* locus).[8,9]

This slimy material structure consists of an EPS matrix containing microcolonies of bacteria that develop in organized communities with functional heterogeneity. Biofilms can be constituted by beneficial or opportunistic pathogenic bacteria and depending on the species, strain, and environmental conditions, the constituents could vary. It is an important factor in infectious diseases caused by bacterial pathogens of animals or plants. Life within biofilms confers advantages to cells living inside it, including acquiring resistance to adverse environmental conditions and increasing protection against predation, desiccation, exposure to antibacterial substances (antibiotics/disinfectants), phagocytosis, and other components of the innate/adaptive immune and inflammatory defense system of the host.[8,10,11]

The content of biofilm structure consists of substances of various chemical natures such as exopolysaccharides, proteins, teichoic acids, and extracellular DNA, with channels that allow nutrients circulation. These describe biofilms as organized structures, where differentiated cells could be recognized within the matrix with their own patterns of gene expression and complex metabolic activity, the reasons for which are compared with tissues of higher organisms.[8,10,11]

The properties of biofilm will depend on the tissue involved and the surrounding conditions. They can form on living tissues of animals, such as mouth, heart, and plant surfaces, but also in dead tissues, such as dead bone, medical devices, and inert surfaces. The presence of biofilm in animal tissues stimulates the host humoral response, producing antibodies and defense molecules that are unable to pass through the exopolysaccharide structure. Similar situations occur with antibiotic and other chemotherapeutics, where biofilm and cells inside it are protected, but free cells are sensitive to them.[11] In plant biofilm, surface and extracellular bacterial components have been extensively studied, and different molecules involved in the process have been found that play crucial roles, independent of the development of a beneficial or pathogenic relationship. These bacterial components are also key molecules in the establishment, maturation, and dispersal of biofilms.[10]

Biofilms can be established in different environments in which bacterial cells have the ability to attach to biological or inert surfaces, responding to environmental signals that trigger the process.[10] One of these signals is an increase in the number of cells in the habitat involved that is sensed by bacteria. This high density of cellular population allows the performance of certain processes that individual cells cannot perform efficiently, like metabolite or exoenzyme excretion, which are only effective above a certain concentration threshold. This is a strategy in which bacteria maintain a critical mass of cells in a specific location for long enough periods to initiate beneficial or antagonistic interactions with the host.[12]

The development of a biofilm is often described in four steps that involve many specific functional requirements during the process, including the expression of highly specialized factors characterized by different profiles of biological functions. The four steps are (1) initial attachment of bacterial cells, (2) aggregation and accumulation of cells in multiple layers, (3) biofilm maturation, and (4) detachment of cells from the biofilm into a planktonic state to initiate a new cycle of biofilm formation elsewhere.[8,9]

In the first step, an initial attachment of bacteria to the surface occurs, which is highly dependent on the hydrophobicity of cells and biomaterials surface; cells adhere reversibly to natural or inert surfaces to be colonized using such physicochemical forces as hydrophobic, electrostatic, and Lifshitze van der Waals.[8] A number of studies of the infection process of medical devices found that bacteria expressed cell surface proteins that interacted with the host extracellular matrix components, constituting a tight association. Specific autolysin molecules involved in this step, AtlA and AtlE, were recognized in staphylococcal Gram-positive bacteria performing enzymatic and adhesive functions that degraded the bacterial cell wall and were bound to unmodified polystyrene.[8,9] This initial step in plant tissues have particular characteristics depending on the surface, developing high specificity for their host in some cases, whereas others function on a variety of surfaces. The interaction with plant tissues is a recognition process between the surface of bacteria and the plant that involves adhesions attributed to polysaccharides and surface proteins mediated by motility structures. These cellular structures and molecules, like lipopolysaccharides, pili, and membrane proteins, are necessary in this initial union representing a highly specific reaction essential in cell–cell communication for the colonization of the leaf surface.[10,12,13]

The second step is characterized by the aggregation and accumulation of cells in multiple layers. This is an active process where biofilm is established progressively as molecules of the cells surface recognize adhesive molecules in the matrix, encouraging the union between them and the development of a matrix.[8,11] There are factors with adhesive function that serve as glue, like polysaccharide intercellular adhesin, whose synthesis is mediated by an intercellular adhesion (icaADBC) locus, which was initially discovered and studied in *Staphylococcus epidermidis*.[9] There are also additional components and specific molecules required for aggregation of cells into undifferentiated microcolonies, such as flagella and pili, which are necessary for the movement of cells inside the matrix.[10,11]

Biofilm maturation is the third step, and it is specific to each microorganism species. In this process, microcolonies inside the matrix developed in a true biofilm structure. This biofilm is mainly constituted of cells inserted in exopolysaccharides that acquires high resistance to antimicrobial substances. In this step, signal molecules needed for maturation are produced by sessile cells playing a crucial role for the expression of specific genes required for the biofilm development. These molecules are produced individually by cells and when the concentration is enough to be sensed, the expression of specific genes is triggered. This phenomenon is known as *quorum sensing (QS)* and involves several different signal molecules, such as acylatedhomoserine lactones (AHLs) among proteobacteria; γ-butyrolactones in *Streptomyces* spp., and oligopeptides among Gram-positive microbes.[11,12]

The last step is the detachment of cells from the surface to a planktonic state. In this process, sessile cells within biofilm are set free to begin a new colonization process with the development of biofilm on a new surface. For this stage, enzymes that digest EPS (such as alginate, a matrix component of *P. aeruginosa* biofilm) has been described.[8,11]

ANTIMICROBIAL RESISTANCE OF BIOFILMS

Life within biofilms represents a great advantage for bacteria, becoming more persistent and very difficult to treat with antimicrobial agents due to the resistance that they encounter in this community. In fact, when cells exist in a biofilm, they can become 10–1000 times more resistant to the effects of antimicrobial agents.[6]

Several mechanisms have been proposed to explain this reduced susceptibility of biofilms to antibiotics, attributing this resistance to many factors that are beginning to be elucidated. Three main resistance mechanisms are proposed:

1. *Exopolysaccharide matrix*: These compounds contribute to delay the entry of antibiotics or allow an incomplete penetration into the biofilm because they adsorb into the exopolysaccharide matrix and having a retarded entry; for example, positively charged aminoglycosides bind to negatively charged polymers in the biofilm resulting in a slower penetration.
2. *Chemical microenvironment*: As biofilms mature, the amount of cells grows, inducing the synthesis of the σ factor, RpoS, which is expressed only in the stationary phase. The increase in cell density also leads to the activation of QS systems and the environmental conditions in the biofilm induce or select cells that are phenotypically resistant to high concentrations of antimicrobials. This high density of the cell population produces gradients of nutrient and oxygen availability. The oxygen is consumed by cells located in the surface layers of the biofilm, creating anaerobic zones in the deeper layers. This situation is unfavorable for aminoglycosides because they are less effective in anaerobic conditions than aerobic conditions. In addition, pH gradients are generated by the local accumulation of acidic waste products interfering directly with the action of antibiotics. As a result of the lack of oxygen and the alteration of pH, some bacteria stop reproducing and enter a nongrowing state, in which they are protected from being killed. A clear example is penicillin, which becomes inefficient because it kills only growing bacteria as the target site is the cell wall synthesis. Another consequence of the slow growth rate is a general response to this environmental stress situation producing changes to avoid heat and cold shocks and changes in pH and many chemical agents. In addition, there is an osmotic stress response that induces changes in porins that reduce the permeability to antibiotics of the cell envelope.
3. *Phenotypic/persister variants*: When cells attach to a surface in biofilm development, many changes in gene expression are induced, originating the biofilm-specific phenotype of resistance. This phenotypic state is originated by a subpopulation of cells resistant to high concentrations of antimicrobials induced or selected by the environmental conditions present in this matrix.[6,14,15]

These mechanisms of resistance appear only if a critical size of a group of cells inside biofilms is reached. Consequently, metabolic changes due to variations in substrates trigger these resistance strategies and the aggregation of cells that will finally alter the passage of antimicrobial substances across the barrier. The fact that these antibiotic resistance mechanisms are inherently multicellular explains the lack of resistance in a planktonic state.[6]

INHIBITION OF BIOFILM BY NATURAL PRODUCTS

Elimination of biofilms from their habitat is not an easy task. This constitutes a real problem in clinics when the antimicrobial therapy being used is unable to remove biofilm from the site of infection. For this reason, the research on alternative compounds with the ability to inhibit the formation or to eradicate the biofilm is needed. Research studies are looking for substances that can effectively suppress the formation of biofilm but cannot generate resistance based on the interference or blocking of regulatory mechanisms or molecules involved in biofilm development. The process of biofilm formation requires many regulatory elements, implying that multiple antibiofilm tools would have to be used to suppress it.[16] Natural compounds derived from plants are being considered as antimicrobial and antibiofilm drugs that constitute a potential alternative to antibiotic therapy.[17] The following section describes some of the most recent studies of antibiofilm activity performed with EOs, terpenes, and extracts from plants.

Essential oils

EOs and their constituents are attributed with many biological properties, including antimicrobial, anticancer, and antimutagenic ones. They are generally regarded as safe because they have low toxicity, reduced genotoxicity, the ability to act on multiple cellular targets, and low production costs.[18] Nowadays, research work is focusing on the mechanisms of action of EOs, but few studies have examined the effects of EOs on bacterial biofilms.

Streptococcus pyogenes (Group A) is an important human pathogen that causes several diseases and has the ability to form biofilms, which is a key virulence factor of this pathogen. Studies with EOs of *Pogostemon cablin* and *Pogostemon heyneanus* exhibited antibiofilm activity against strains of *S. pyogenes* that are responsible for pharyngitis. It was observed that both EOs inhibited the biofilms considerably, depending on this activity of the *S. pyogenes* serotype.[19]

Gram-negative bacteria are considered more resistant to antimicrobial agents due to the presence of the outer membrane of the cell wall, which is impermeable to the diffusion of molecules with a molecular mass greater than 600–1000 Da. However, many EOs have antimicrobial activity on Gram-negative bacteria.[20] *P. aeruginosa* is Gram-negative and naturally resistant to many antibiotics. It is an opportunistic pathogen, ubiquitous, and responsible for intrahospitalary infections and burned patients. It has the ability to form biofilms in natural tissues and medical devices.[21] Nowadays, scientists are focusing on the developing of novel methods that control biofilms. Current researches describe the ability of EOs to pass through the exopolysaccharide matrix of biofilm; a study on cinnamon bark oil showed that it exerted strong inhibition activity on biofilm formation of *P. aeruginosa* and *Escherichia coli* O157:H7 (EHEC).[22]

The EO of leaves and cones from *Cupressus sempervirens*, a medicinal tree in the Northern Hemisphere, are used externally for headaches, colds, coughs, and bronchitis. This EO was tested for its antimicrobial and antibiofilm activity against *Klebsiella pneumoniae*, a Gram-negative bacteria member of the coliform group considered as an opportunistic pathogen responsible for urinary infections, pneumonia, and sepsis. The results showed moderate antibacterial activity, but it presented effective biofilm eradication.[23]

There are common seafood bacterial pathogens that form biofilms like *Vibrio* sp., *Aeromonas hydrophila*, *Salmonella* sp., and *Listeria monocytogenes*, among others. Mizan et al.[24] cited several experiences with EOs of *Cymbopogon* sp. over listerial biofilms formed on stainless steel surfaces, reporting a total biofilm reduction when the EOs were combined with a disinfectant solution (ethanol and saline solution with 0.5% (v/v) of Tween 80). Other experiences with this bacterium described treatment with cinnamon EO on a stainless steel surface that reduced the number of sessile listerial cells.[24]

One alternative approach to defeat biofilms is targeting the bacterial communication system via QS. This is a process in which bacteria produce and detect signal molecules and coordinates their behavior in a cell-density-dependent manner. QS plays an important role in the virulence and survival of *P. aeruginosa* and other pathogenic bacteria, constituting a novel and potential target for anti-infective agents. EOs with anti-QS activity might be important in reducing the virulence and pathogenicity of drug-resistant bacteria. The first report on anti-QS activity of EOs was performed evaluating the inhibitory activity of plant EOs using strains of *Chromobacterium violaceum* (CV12472 and CVO26) and *P. aeruginosa* (PAO1) with 21 EOs. The results indicated that *Lavandula angustifolia* (lavender), *Mentha piperita* (peppermint), *Syzygium aromaticum* (clove), and *Cinnamomum verum* (cinnamon) oils showed anti-QS activity on *C. violaceum* pigment production and reduction in the swarming motility in PAO1.[25] In other investigations with clove oil, the effects of the subinhibitory concentrations on QS regulation, virulence factors, and biofilm formation against *P. aeruginosa* PAO1 and *A. hydrophila* WAF-38 were evaluated. The biofilm-forming capability of both species was reduced in a concentration-dependent manner at all tested subminimum inhibitory concentration values and significantly reduced the AHL levels.[26] As described previously, cinnamon bark oil reduced biofilm formation and also had the ability to lower QS regulation and swarming in *P. aeruginosa*.[22] The studies reviewed in this chapter clearly reveal the anti-QS and biofilm inhibitory activity of EOs against Gram-negative pathogens, which are known to be more resistant to antimicrobial substances.

Terpenes

In the search for natural products, terpenes constitute one of the most promising antimicrobial and antibiofilm substances. Staphylococci are considered major pathogens of humans and animals, developing resistance to antibiotics as they are indiscriminately used. In addition, some members of this genus have the ability to form biofilm, becoming less susceptible to chemotherapeutics. The effect of carvacrol and thymol on *S. aureus* and *Staphylococcus epidermidis*, including antimicrobial activity of both compounds in planktonic cells and biofilm cells, were studied. These terpenes inhibited the growth of preformed biofilm and interfered with biofilm formation during planktonic growth.[27] Investigations on methicillin-resistant *S. aureus* (MRSA) and methicillin-susceptible *S. aureus* (MSSA) strains with farnesol showed an inhibition effect on biofilm formation and alteration of cell membrane integrity. In addition, the synergistic effects between this terpene and the antibiotic gentamicin were tested, exposing the biofilms to different concentrations of both agents and demonstrating a reduction in bacterial populations and biofilm formation.[28]

The maintenance of a healthy mouth microenvironment is one of the major interests of dentists, so natural products are being considered as an attractive alternative to provide sanitation due to their antimicrobial properties and the ability to inhibit virulence factors. *Streptococcus mutans* is the causal agent of dental caries in animals and humans and is responsible for the formation and accumulation of plaques. Studies on this pathogen exposed to farnesol showed that this sesquiterpene affected glucan synthesis, which is a component of the biofilm matrix. However, the inhibition of this polysaccharide synthesis was attributed to the disruption of the cell membrane caused by farnesol more than to the direct inhibition of the glucan synthesis because this terpene is a poor enzyme inhibitor. These promissory results show that natural compounds have the ability to affect the biofilm formation in infective processes.[29] Other investigations on *S. mutans* attempted to determine the penetration into biofilms of different commercial antiseptic mouth rinses containing eucalyptol, menthol, methyl salicylate, and thymol. The results demonstrated that this mouth rinses had the best penetration velocity, and after 5 min of exposure, they removed the biofilm from the surface.[30]

The terpenes carvacrol, thymol, geraniol, and eugenol were analyzed with the aim of testing their ability to cross through biofilm and perform antimicrobial activity on MRSA and MSSA strains. These terpenes were able to inhibit MRSA when slime was produced, what means that these terpenes could pass through the exopolysaccharide layer and maintain the inhibitory and bactericidal effects. The antimicrobial activity of some monoterpenes showed that the structure and chemical substituent are important factors for this biological activity. In this experience, aromatic monoterpenes (carvacrol, eugenol, and thymol) showed the best inhibitory activity.[31] Carvacrol and thymol are hydrophobic, so they interact with the lipid bilayer of cytoplasmic membranes, causing loss of integrity and leakage of cellular material.[32]

In experiences with the terpenes carveol and carvone over *Rhodococcus erythropolis*, a reduction in the volume of biofilm and cell aggregation was observed, demonstrating the ability of these terpenes to prevent biofilm formation.[17] In a study involving eugenol, carveol, and carvone, they showed antibacterial activity against *P. aeruginosa* and antibiofilm activity at subinhibitory concentrations.[21] In other research, cinnamaldehyde markedly inhibited *P. aeruginosa* and *E. coli* O157:H7 (EHEC) biofilm formation, and eugenol significantly decreased the biofilm formation of EHEC.[22] These results clearly showed the potential application of terpenes to prevent biofilm production.

Aqueous and organic extracts

The elucidation of the antimicrobial activity of plant-derived compounds and their mechanisms of action is a complex problem. The possibility that plant extracts bring about reduction of bacterial growth does not necessarily mean that they will do so in other cellular systems. One of the suggested hypotheses that explains these situations is that plants contain multiple chemicals that may be causing distinct effects on different aspects of the cell or the QS system. A study of the anti-QS activities against *P. aeruginosa* PAO1 involving the aqueous extracts of six medicinal plants (*Conocarpus erectus*, *Chamaesyce hypericifolia*, *Callistemon viminalis*, *Bucida buceras*, *Tetrazygia bicolor*, and *Quercus virginiana*) focused on the potential ability of these extracts to interfere with the QS-dependent production of virulence factors, biofilm formation, and pyoverdin production. The results showed a reduction of the properties analyzed, but with a differentiated effect on virulence factors, biofilms, and QS gene expression. These findings mean that some extracts were able to inhibit biofilm production, but not any other QS-dependent process.[33] Similarly, the influence of aqueous extracts of raspberry, blueberry, blackberry, cranberry, grape, oregano, thyme, basil, kale, ginger, and turmeric on QS, inhibiting the swarming motility of *P. aeruginosa*, was demonstrated.[34] Curcumin is one of the major constituents of turmeric (*Curcuma longa* L.) roots/rhizomes and has been used as a repellent. The extracts of this plant demonstrated antimicrobial activity, and in recent studies, the ability to affect QS-dependent systems on *P. aeruginosa* PAO I was evaluated. The results showed an effective inhibition of the production of biofilm and other virulence factors such as pigments, enzymes, and signal molecules (homoserine lactone or HSL).[35]

The water extracts of the flowers of *Calendula officinalis* were assayed for their antibacterial and antibiofilm activity against *Salmonella* sp., *Shigella dysenteriae*, *Shigella flexneri*, *Shigella sonnei*, and *E. coli*, producing a decrease in the adherence of bacteria to the surface and in the absorbance values. The chemical composition was also analyzed, finding that alkaloids, saponins, flavonoids, terpenoids, glycosides, and phenols are potentially responsible for these inhibitory activities.[36]

The antimicrobial activity of the methanolic extracts of the medicinal tree *C. sempervirens* was analyzed, showing a strong inhibition of the growth of *K. pneumoniae*. In the same study, the antibiofilm potential was evaluated, demonstrating the biofilm elimination from the surface tested.[23] In the process of finding out the biological properties of plant extracts, the identification of the bioactive compounds is necessary. In a study on the methanolic extract of *Andrographis paniculata* compounds such as diterpenoid lactone and rographolide were identified, elucidating their mechanisms of action on the formation of *P. aeruginosa* biofilm. It was found that the substances of this methanolic extract inhibit the biofilm production, mimicking QS molecules.[37] *Cuminum cyminum*, often used for food flavoring, has recognized antimicrobial properties. Methyl eugenol contained in the methanolic extract obtained from this plant could damage the biofilm structure and formation of *P. aeruginosa* PAO1 biofilm, interfering with AHL.[38]

Different water–ethanolic fractions obtained from the leaves of *Arctium lappa* (burdock) were found to significantly inhibit the formation of biofilm by *E. coli*. The 70% ethanolic amount showed complete inhibitory activity against biofilm formation. This study confirmed the absence of biofilm formation in cultures treated with burdock leaf fractions with SEM. In addition, high correlation between the antibiofilm capacity and the phenolic contents of the different fractions was observed.[39]

Investigations on flavonoids obtained from the *Moringa oleifera* seed coat were carried on to evaluate the antibiofilm potential of *S. aureus* and *P. aeruginosa*. The Gram-negative cell was more resistant to the action of the flavonoid, and with lower inhibitory values, than the Gram-positive one. However, there was no difference in the ability of this natural compound to avoid the initial cell attachment of both species. Furthermore, flavonoids were shown to disrupt the preformed biofilms and reduce the metabolic activity of biofilm.[40] These results demonstrate the importance of isolating the compounds present in plant extracts in order to know the particular structure and biological activity of each one, identifying their target site.

ISOLATION AND CHEMICAL ANALYSIS OF PLANT-DERIVED COMPOUNDS

The vast array of natural organic compounds in plants can be arranged in five large groups: carbohydrates, lipids, alkaloids, phenolics, and terpenoids.[41] However, the task of isolating and identifying a plant compound may be overwhelming. Yet the biological activity of a medicinal plant or plant extract can hardly be attributed to a single plant product; rather, it is generally due to a combination of plant products or a family of plant products. Thus, the approach to isolate a single biologically active plant product involves several steps, starting from the selection of the plant of interest and ending with the identification of the molecules responsible for the bioactivity.

There are at least two approaches to begin drug discovery using higher plants: random selection of plants and selection of plants based on their use as traditional medicines. The latter approach includes plants used in herbalism, folklore, and shamanism.[42] Once the plant (or group of plants) has been selected, it is then necessary to choose an extraction method to isolate the bioactive compound (or compounds) from the plant. Traditional methods used for the separation of metabolites from plant materials include the use of hot water extracts to make teas or ethyl alcohol extracts to make tinctures. However, more sophisticated and reliable extraction methods must be used to facilitate the complete identification of the bioactive compounds in the plant. This section briefly describes some protocols generally used in the area of phytoanalysis, as well as some state-of-the-art extraction methods recently described in the literature.

COLLECTION OF PLANTS

The collection of plant specimens is the first step once the plant of interest has been selected, and the preservation of the nature of bioactive compounds in plant specimens is very important to facilitating their isolation and identification. The best option is to freeze the collected plant specimens in liquid nitrogen. However, it is rare to go on a field trip with a convenient amount of liquid nitrogen, so alternative ways to preserve the plant specimens should be used. One such option is to use plastic bags with hermetic seals to put the samples after collecting plants on site. These bags keep the plants alive until they can be frozen. Collecting representative parts of the plant is also important to obtain different extracts from the plant. Thus, roots, vegetative shoots, flowers, fruits, and seeds can be collected and stored in separate plastic bags. Detailed records of the plants collected in the field should be kept. It is also advisable that a botanist identifies the plant and to deposit a voucher specimen in the local herbarium. Finally, photographs taken on the field where the specimens were collected are of great help for plant identification and recollection.

GRINDING OF PLANTS

The best way to rupture plant tissues is probably to freeze them in liquid nitrogen and pulverize them with a pestle in a mortar. A frozen powder with very small particles can be obtained in this way, facilitating the extraction of the compounds of interest. The extraction of bioactive compounds can be done by simply adding the frozen powder to an appropriate extraction mixture or by using different extraction protocols. Alternatively, the powder can be stored in a freezer for later use. However, the extraction of bioactive compounds should be made as soon as possible because many compounds can be degraded, even at −80°C.

The grinding of the plant or plant part should be made in case liquid nitrogen is not available. A manual or mechanical mill can be used to grind hard material, like dried bark and stems, and a manual or mechanical chopper may be used to grind the soft parts of the plant, like leaves and flowers. Sieving of the ground-up plant is always advisable to keep gross material from being involved in the extraction steps.

EXTRACTION PROTOCOLS

There are many extraction protocols, and the selection of the best one depends on the chemical characteristics of the bioactive compounds of interest. Many compounds are water-soluble, and they may be extracted easily using hot water. However, there also are many water-insoluble compounds, and they have to be extracted using a different solvent or solvent mixtures.

At first glance, two types of extracts may be obtained from a plant: EOs and the so-called organic extracts. EOs are usually obtained by steam distillation, although they can also be obtained by expression, fermentation, enfleurage, or extraction. EOs are complex mixtures, constituted by terpenoid hydrocarbons, oxygenated terpenes, and sesquiterpenes. They originate from the plant's secondary metabolism and are responsible for its characteristic aroma. An organic extract of a plant is a concentrated preparation of the plant obtained by removing the active constituents with a suitable solvent, evaporating all (or nearly all) of it and adjusting the residual mass or powder to a prescribed concentration. This kind of extracts can contain a vast number of constituents, mainly defined by the chemical characteristics of the solvent used for extraction. Organic extracts from a plant are usually obtained by solid–liquid extraction methods, although liquid–liquid extractions of a plant macerate are also used to narrow the type of chemicals to be extracted. Brief descriptions of the methods usually employed for obtaining EOs and organic extracts of plants are given in the following sections.

METHODS FOR OBTAINING EOs

Steam distillation

The selective evaporation of the volatile part of a plant can be performed by the steam distillation method. In this method, the plant material is placed upon a grid inside the still, and water steam is put in contact with the finely divided plant material and the volatile constituents are removed from the plant. Then they are collected in a special part of the apparatus. The rationale behind steam distillation is that water condenses on the plant parts and forms an immiscible phase, where they are heated, allowing the volatile areas of the plant to evaporate with the water. Steam distillation enables a compound or mixture of compounds to be distilled at a temperature substantially below that of the boiling points of the individual components. The vapor mixture of water and oil is then condensed by indirect cooling with water. From the condenser, distillate flows into a separator (eg, a Clevenger-type trap) where oil separates from water by the difference in their densities.

Supercritical fluid extraction

Another way to obtain the volatile fraction of a plant is to extract it with supercritical fluids.[43] A supercritical fluid is characterized by physical and thermal properties that are among those of the pure liquid and gas phases. These physical and thermal properties can be adjusted by controlling the temperature and the pressure of the system. Thus, fluid density, diffusivity, and other physical parameters may be adjusted to completely penetrate the solid and to solubilize the material of interest and selectively extract it. Moreover, a cosolvent can also be used to expand the ability to extract the supercritical fluid.

In its supercritical state, carbon dioxide is one of the most used extractants in supercritical fluid extraction (SFE) because it is safe, readily available with good purity, and inexpensive. Carbon dioxide has a critical temperature of 304.25K and a critical pressure of 7.39 MPa. An advantageous physical characteristic of carbon dioxide for its use in SFE is that it can be removed easily from the extracted material just by decompressing the system. Carbon dioxide in its supercritical state is used frequently in the tea and coffee decaffeination industries, in the extraction of EOs, in the extraction of flavors and fragrances from natural sources, and in the extraction of ingredients from spices.

Microwave-assisted hydrodistillation

Microwave-assisted hydrodistillation (MAHD) works essentially the same as traditional hydrodistillation, but in MAHD, the heating of the solvent is attained by using microwaves. The vessel containing the solvent (generally water) and the parts of the plant is located inside a microwave oven (usually operating at 2.45 GHz), and various output powers and reaction times may be applied to optimize the extraction process. Again, the use of microwaves for the heating process accelerates the extraction of compounds, and shorter times are needed to obtain similar quantities of extracts. Moreover, the chemical composition of the extracts obtained by traditional hydrodistillation and by MAHD does not differ substantially.

Solvent-free microwave extraction

Solvent-free microwave-assisted extraction (SFME) is a new technique proposed for plants that has been developed in recent years as a way to further shorten extraction times, lower extraction costs, and simplify the process. The fundamental principle of SFME is that there is sufficient water in the fresh plant material to be heated by microwaves and to evaporate the volatile compounds contained in the vegetal. However, the fresh material is not easy to preserve, and herbs are frequently dried before they are conserved and used. Thus, EO in these dried plant materials cannot be evaporated simply by in situ water by conventional SFME. An alternative may be to moisten the dried plant materials before EOs extraction, but this pretreatment makes the whole process complex and time consuming. This problem has been elegantly solved by the simple addition of microwave-absorbing material, such as carbonyl iron powders, which absorb the electromagnetic radiation and provide a calefaction speed much greater than that of water.[44]

METHODS FOR OBTAINING AQUEOUS AND ORGANIC EXTRACTS OF PLANTS

Aqueous extractions

Aqueous extractions are often called *traditional extractions* considering that these extraction methods were used by human beings since ancient times in traditional medicine. Infusions are made either with hot or cold water. In hot water infusions, a small amount of ground-up and dried plant material is placed in almost boiling water. A milder treatment, called *cold water infusion*, is performed using cold or room temperature water. More severe treatment with boiling water may be necessary to extract some chemicals from hard parts of the plant. In decoction, a small amount of plant material is placed in boiling water for a period of time. Polar molecules and small peptides or proteins can be extracted using water.

Organic solvent extractions

More sophisticated apparatuses are often used for organic solvent extractions. For example, the Soxhlet extractor is generally used for liquid–solid extractions. The ground-up solid is put in the extractor thimble and an organic solvent with a low boiling point, like hexane, ethyl ether, acetone, or chloroform, is poured into the distillation flask. Then, the flask is heated, and the solvent evaporates, condenses in the condenser attached to the top of the Soxhlet extractor, and falls over the thimble containing the sample. The level of the solvent in the extractor body rises until it reaches a given height, and then it returns to the flask by siphoning. The cycle is repeated over time, and finally, the solvent is evaporated from the extract. Thus, various components can be extracted based on their solubility in the solvent used for the extraction. The Soxhlet extractor has been extensively used to extract different compounds from dried samples.

Liquid–liquid extraction is another way to obtain extracts from plants. The first step is to macerate the ground-up plant in alcohol–water mixtures. A 1:1 volume-by-volume alcohol–water mixture is often used for this purpose, and ethanol is frequently the alcohol chosen. Many plant components are extracted in this way. Then, the alcohol is evaporated under low pressure and a water concentrate with some solids in suspension or precipitated is obtained. Solids are removed by vacuum filtration, and the concentrate is submitted to liquid–liquid extraction with organic solvents.

The organic solvent used for the liquid–liquid extraction may be lighter or heavier than water, and different extraction apparatuses are used in each case. The precise design of the apparatus depends on the manufacturer. Figs. 14.1 and 14.2 show apparatus designs commonly used for liquid–liquid extractions. Fig. 14.1 shows a design used for liquid–liquid extractions using organic solvents that are lighter than water, such as ethyl ether, benzene, or hexane. The solvent is distilled in the heated flask attached to the lower end (B) and vapors are condensed at the reflux condenser (A) located at the top of the apparatus. The condensed solvent passes through a funnel down a tube with a sintered glass disk at the end (C). There, the organic solvent is delivered as small droplets into water concentrate contained in a glass thimble, and the solvent rises by upward displacement. In this way, the components of the plant are partitioned between the small droplets of organic solvent and the aqueous phase and then carried with the organic solvent into the round flask, where the organic solvent evaporates to continue the process. It is important to be aware that pure organic solvent condenses every time at the reflux condenser at the top of the apparatus. Therefore, it is the pure solvent that contacts the water concentrate, and even compounds with small partition constants can be extracted in this way by using a relatively low volume of solvent.

Fig. 14.2 shows a common design of apparatus for liquid–liquid extractions using organic solvents that are heavier than water. The apparatus works essentially in the same way as previously described, but the tube with the sintered glass disk ending is located just above the water concentrate, so the organic solvent is delivered into this concentrate as a fine rain of small droplets. The organic solvent goes to the bottom of the glass thimble by downward displacement, and then it returns to the round flask through the small return tube (D) located at the bottom of the glass thimble. The organic solvent evaporates in the round flask, and the process continues.

The organic solvent used for the extraction process (in either of the abovementioned types of apparatus) transfers the soluble components of the plants into the round flask located at the side part of the apparatus. A constant concentration of extracted components is reached after several

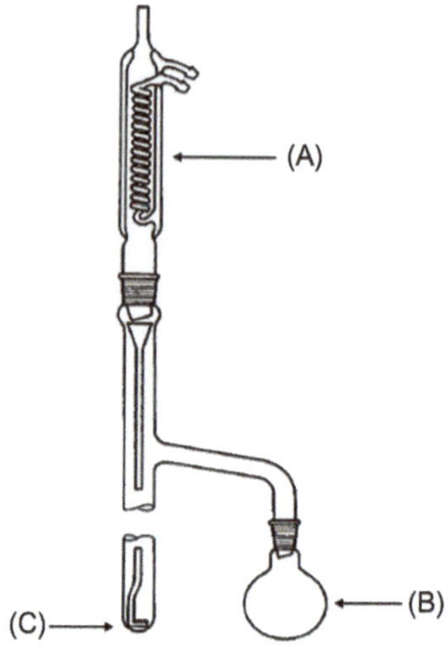

FIGURE 14.1

Apparatus used for liquid–liquid extractions using lighter-than-water organic solvents; (A) condenser; (B) heated flask for solvent evaporation; (C) condensed solvent delivery tube with sintered glass disc ending.

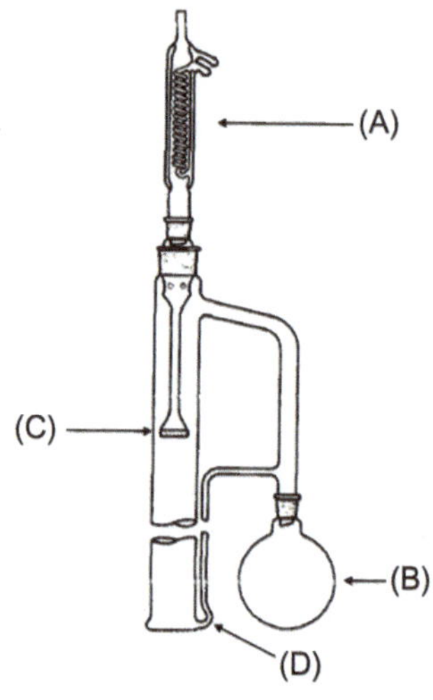

FIGURE 14.2

Apparatus used for liquid–liquid extractions using heavier-than-water organic solvents; (A) condenser; (B) heated flask for solvent evaporation; (C) condensed solvent delivery tube with sintered glass disc ending; (D) organic solvent return tube.

hours of continuous extraction. Then the organic solvent may be evaporated by vacuum distillation to obtain a plant extract consisting of the plant components that are soluble or partially soluble in the solvent used. Thus, organic solvents with different polarities are used to extract different components of the plant in the water concentrate.

Microwave-assisted extraction

Microwave-assisted extraction (MAE) is a relatively new extraction technique that combines microwave and traditional solvent extraction. In MAE, the extraction occurs as a result of changes in the cell structure caused by electromagnetic waves. It has been proposed that the extraction acceleration observed in MAE may be due to the heat and mass transfer gradients working in the same direction.[45] Using microwaves for heating the solvents and plant tissues increases the kinetics of extraction, and various advantages are thus obtained over traditional solvent extraction, including shorter extraction times, higher extraction rates, lower costs, and less solvent use. MAE has been employed in the extraction of various compounds from natural sources, including EOs from leaves, terpenes from must, and gossypol from cottonseeds.

CONCLUSIONS

The emergence of multiple drug-resistant strains in the last decade is due to the excessive and indiscriminate use of antibiotics. Furthermore, some of these bacteria have the ability to produce biofilm, an exopolysaccharide structure that confers protection against environmental adversities helping to make antimicrobial substances less efficient. Consequently, there is an urgent need to discover new substances capable of defeating these resistant structures. Having more effective antimicrobial agents that are also active against biofilm and can prevent (or at least interfere with) biofilm formation would be a considerable achievement.[27]

Efforts to disrupt biofilms have enabled the identification of bioactive molecules, representing an enormous challenge for scientists, who are considering plants as a vast source of antimicrobial substances.[46] Natural products obtained from plants are being studied for their capacity to affect biofilm formation or destruction without creating a selective pressure that would lead to the appearance of resistant mutants.[17] The biofilm inhibitory activity of plant-derived products throw light on the search of new substances with antimicrobial properties that do not exert strong selective pressure for developing resistance.

Many extraction techniques are available to obtain plant extracts, and chemists must choose the appropriate one to attain the goal in question. Traditional methods of extraction have proved to be effective and have been extensively tested. However, many times they use considerable amounts of organic solvents, which are not ecofriendly. On the other hand, there is a growing number of new methods which are ecofriendly, but they need to be further tested. Anyway, phytochemists have at hand a considerable number of extraction methods, and the purification of biologically active plant compounds is becoming less complicated.

REFERENCES

1. Demo MS, Oliva MM. Antimicrobial activity of medicinal plants from South America. In: Watson RR, Preedy VR, editors. *Botanical medicine in clinical practice*. Wallingford: CAB International; 2008. p. 152–63.
2. Nascimento GGF, Locatelli J, Freitas PC, Silva GL. Antibacterial activity of plant extracts and phytochemicals on antibiotic resistant bacteria. *Braz J Microbiol* 2000;**31**:247–56.

3. Ahmad I, Aqil F, Owais M, editors. *Modern phytomedicine. Turning medicinal plants into drugs.* Weinheim: Wiley-VCH Verlag GmbH & Co. KGaA; 2006.
4. Cowan MM. Plant products as antimicrobial agents. *Clin Microbiol Rev* 1999;**12**:564–82.
5. Tenover FC. Mechanisms of antimicrobial resistance in bacteria. *Am J Med* 2006;**119**(6A):S3–10.
6. Mah T-FC, O'Toole GA. Mechanisms of biofilm resistance to antimicrobial agents. *TIM* 2001;**9**(1):34–9.
7. Christensen GD, et al. Adherence of coagulase-negative *Staphylococci* to plastic tissue culture plates: a quantitative model for the adherence of *Staphylococci* to medical devices. *J Clin Microbiol* 1985; **22**(6):996–1006.
8. Arciola CR, Campoccia D, Speziale P, Montanaro L, Costerton JW. Biofilm formation in staphylococcus implant infections. A review of molecular mechanisms and implications for biofilm-resistant materials. *Biomaterials* 2012;**33**:5967–82.
9. Büttener H, Mack D, Rohde H. Structural basis of *Staphylococcus epidermidis* biofilm formation: mechanism and molecular interactions. *Front Cell Infect Microbiol* 2015;**14**:1–15.
10. Bogino PC, Oliva MM, Sorroche FG, Giordano W. The role of bacterial biofilms and surface components in plant-bacterial associations. *Int J Mol* 2013;**14**:15838–59.
11. Costerton JW, Stewart PS, Greenberg EP. Bacterial biofilms: a common cause of persistent infections. *Science* 1999;**284**:1318–22.
12. Danhorn T, Fuqua C. Biofilm formation by plant-associated bacteria. *Annu Rev Microbiol* 2007;**61**:401–22.
13. Morris CE, Monier JM. The ecological significance of biofilm formation by plant-associated bacteria. *Annu Rev Phytopathol* 2003;**41**:429–53.
14. Drenkard E. Antimicrobial resistance of *Pseudomonas aeruginosa* biofilms. *Microb Infect* 2003;**5**:1213–19.
15. Stewart PS, Costerton JW. Antibiotic resistance of bacteria in biofilms. *Lancet* 2001;**358**:135–8.
16. Masák J, Cejková A, Schreiberova O, Rezanka T. *Pseudomonas* biofilms: possibilities of their control. *FEMS Microbiol Ecol* 2014;**89**:1–14.
17. Carvalho CCCR, Da Fonseca MMR. Preventing biofilm formation: promoting cell separation with terpenes. *FEMS Microbiol Ecol* 2007;**61**:406–13.
18. Raut JS, Karuppayil SM. A status review on the medicinal properties of essential oils. *Ind Crops Prod* 2014;**62**:250–64.
19. Nithyananda P, Shafreenc RMB, Muthamilc S, Murugand R, Pandian SK. Essential oils from commercial and wild Patchouli modulate Group A *Streptococcal* biofilms. *Ind Crops Prod* 2015;**69**:180–6.
20. Demo M, Oliva MM, López ML, Zunino MP, Zygadlo JA. Antimicrobial activity of essential oils obtained from aromatic plants of Argentina. *Pharm Biol* 2005;**43**(2):129–34.
21. El Abed S, Houari A, Latrache H, Remmal A, Koraichi SI. In vitro activity of four common essential oil components against biofilm-producing *Pseudomonas aeruginosa*. *Res J Microbiol* 2011;**6**(4):394–401.
22. Kim YG, Lee JH, Kim SI, Baek KH, Lee J. Cinnamon bark oil and its components inhibit biofilm formation and toxin production. *Int J Food Microbiol* 2015;**195**:30–9.
23. Selim SA, Adam ME, Hassan SM, Albalawi AR. Chemical composition, antimicrobial and antibiofilm activity of the essential oil and methanol extract of the Mediterranean cypress (*Cupressus sempervirens* L.). *BMC Complement Altern Med* 2014;**14**:179.
24. Mizan MdFR, Jahid IK, Ha S-D. Microbial biofilms in seafood: a food-hygiene challenge. *Food Microbiol* 2015;**49**:41–55.
25. Khan MSA, Zahin M, Hasan S, Husain FM, Ahmad I. Inhibition of quorum sensing regulated bacterial functions by plant essential oils with special reference to clove oil. *Lett Appl Microbiol* 2009;**49**:354–60.
26. Husain FM, Ahmad I, Asif M, Tahseen Q. Influence of clove oil on certain quorum-sensing-regulated functions and biofilm of *Pseudomonas aeruginosa* and *Aeromonas hydrophila*. *J Biosci* 2013;**38**:835–44.

27. Nostro A, et al. Effects of oregano, carvacrol and thymol on *Staphylococcus aureus* and *Staphylococcus epidermidis* biofilms. *J Med Microbiol* 2007;**56**:519–23.
28. Jabra-Rizk MA, Meiller TF, James CE, Shirtliff ME. Effect of farnesol on *Staphylococcus aureus* biofilm formation and antimicrobial susceptibility. *Antimicrob Agents Chemother* 2006;**50**(4):1463–9.
29. Koo H, et al. Inhibition of *Streptococcus mutans* biofilm accumulation and polysaccharide production by apigenin and *tt*-farnesol. *JAC* 2003;**52**:782–9.
30. Wakamatsu R, Takenaka S, Ohsumi T, Terao Y, Ohshima H, Okiji T. Penetration kinetics of four mouthrinses into *Streptococcus mutans* biofilms analyzed by direct time-lapse visualization. *Clin Oral Invest* 2014;**18**:625–34.
31. Gallucci N, Oliva M, Carezzano E, Zygadlo J, Demo M. Terpenes antimicrobial activity against slime producing and non-producing staphylococci. *MMC* 2010;**21**:132–6.
32. Ultee A, Kets EPW, Smid EJ. Mechanisms of action of carvacrol on the food-borne pathogen *Bacillus cereus*. *Appl Environ Microbiol* 1999;**65**(10):4606–10.
33. Adonizio A, Kong K-F, Mathee K. Inhibition of quorum sensing-controlled virulence factor production in *Pseudomonas aeruginosa* by south Florida plant extracts. *Antimicrob Agents Chemother* 2008;**52**(1):198–203.
34. Vattem DA, Mihalik K, Crixell SH, McLean RJC. Dietary phytochemicals as quorum sensing inhibitors. *Fitoterapia* 2007;**78**:302–10.
35. Rudrappa T, Bais H. Curcumin, a known phenolic from *Curcuma longa*, attenuates the virulence of *Pseudomonas aeruginosa* PAO1 in whole plant and animal pathogenicity models. *J Agric Food Chem* 2008;**56**(6):1955–62.
36. Ghaima KK, Rasheed SF, Ahmed EF. Antibiofilm, antibacterial and antioxidant activities of water extract of *Calendula officinalis* flowers. *IJBPR* 2013;**4**(7):465–70.
37. Murugan K, Sangeetha S, Kalyanasundaram VB, Al-sohaibani S. In vitro and in silico screening for *Andrographis paniculata* quorum sensing mimics: new therapeutic leads for cystic fibrosis *Pseudomonas aeruginosa* biofilms. *Plant Omics* 2013;**6**:340–6.
38. Packiavathy SV, Abraham I, Palani A, Ramaswamym BR, Shunmugiah KP, Arumugam VR. Antiquorum sensing and antibiofilm potential of *Capparis spinosa*. *ARCMED* 2012;**42**(8):658–68.
39. Lou Z, Song X, Hong Y, Wang H, Lin Y. Separation and enrichment of burdock leaf components and their inhibition activity on biofilm formation of *E. coli*. *Food Control* 2013;**32**:270–4.
40. Onsare JG, Arora DS. Antibiofilm potential of flavonoids extracted from *Moringa oleifera* seed coat against *Staphylococcus aureus*, *Pseudomonas aeruginosa* and *Candida albicans*. *J Appl Microbiol* 2014;**118**:313–25.
41. Stecher G, Huck CW, Stöggl WM, Bonn GK. Phytoanalysis: a challenge in phytomics. *TrAC* 2013;**22**:1–14.
42. Ramos Mendonça-Filho R. Bioactive phytocompounds: new approaches in the phytosciences. In: Ahmad I, Aqil F, Owais M, editors. *Modern phytomedicine. Turning medicinal plants into drugs. Chapter 2.* Weinheim: Wiley-VCH Verlag GmbH & Co. KGaA; 2006.
43. Reverchon E, De Marco I. Supercritical fluid extraction and fractionation of natural matter. *J Supercrit Fluid* 2006;**38**:146.
44. Wang Z, et al. Improved solvent-free microwave extraction of essential oil from dried *Cuminum cyminum* and *Zanthoxylum bungeanum* Maxim. *J Chromatogr A* 2006;**1102**:11–17.
45. Veggi PC, Martinez J, Meireles AA. Fundamentals of microwave extraction. In: Chemat F, Crovotto G, editors. *Microwave-assisted extraction for bioactive compounds: theory and practice. Food engineering series*, vol. 4. New York, NY: Springer Science; 2013.
46. Kalia VC. Quorum sensing inhibitors: an overview. *Biotech Adv* 2013;**31**:224–45.

CHAPTER

15

THE POTENTIAL USE OF BACTERIOPHAGE THERAPY AS A TREATMENT OPTION IN A POST-ANTIBIOTIC ERA

R.R. Bragg, C.E. Boucher, W.A. van der Westhuizen, J.-Y. Lee, E. Coetsee, C. Theron and L. Meyburgh

Department of Microbial, Biochemical and Food Biotechnology, University of the Free State, Bloemfontein, Republic of South Africa

INTRODUCTION

Bacteriophages are viruses that infect and kill bacterial cells with great efficacy and are present in all ecosystems that support the growth of bacteria.[1] The use of bacteriophages and their products to eradicate pathogens has attracted increasingly more interest over the past three decades, on account of the global emergence of antibiotic-resistant bacterial strains.

The International Committee for Taxonomy of Viruses classifies phages in the order *Caudovirales*.[2] Phages have capsid protein heads, which carry and protect the genomic material of the viruses, as shown in Fig. 15.1. The genomic material can vary in size, arrangement (circular, linear, or segmented), and structure (ie, single-stranded DNA (ssDNA), double-stranded DNA (dsDNA), single-stranded RNA (ssRNA), and double-stranded RNA (dsRNA)).[3]

The use of bacteriophages to treat bacterial infections is by no means a new concept. The earliest reported discovery of bacteriophages was by English bacteriologist Ernest Hankin, who reported antibacterial activity against *Vibrio cholerae* in the Ganges and Jumna rivers in India in 1896; unfortunately, he could not identify the agent responsible for the phenomenon.[4,5] Although the true discoverer of bacteriophages is a contested issue (between Felix d'Herelle and Frederick Twort in the early 20th century),[5,6] d'Herelle is presently considered the father of applied bacteriophage science, as Twort did not pursue his discoveries. At that time, d'Herelle proposed that the bactericidal agent was possibly similar to viruses already known at that time, such as the tobacco mosaic virus, and that they were able to multiply in bacterial cells. He was also responsible for proposing the name *bacteriophage*, a combination of *bacteria* and *phagein*, which means "to eat" in Greek. Initial research into bacteriophage therapy during pre-ethics times saw d'Herelle administering phage lysate, first to himself and then to those willing among his staff, to investigate its efficacy. He later used the bacteriophages to treat dysentery, which was the first reported therapeutic use of bacteriophages.[7]

Antibiotic Resistance. DOI: http://dx.doi.org/10.1016/B978-0-12-803642-6.00015-0

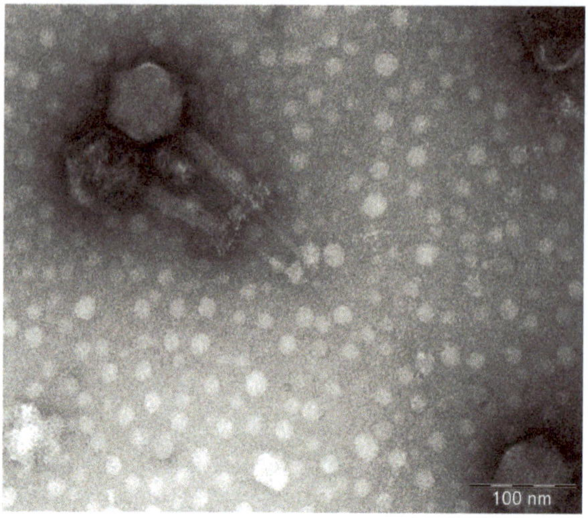

FIGURE 15.1

Two *Listeria* phages P100 (Listex P100) showing contracted tails; the left capsid (darker) is empty and the right capsid (lighter) is packed with DNA.

Soon after research on bacteriophages (hereafter referred to as *phages*) began, penicillin was discovered by Alexander Fleming in 1928.[8] Although antibiotics had been used before Fleming's discovery, he was first to identify their source and effect. At that time, antibiotics were easily produced with more reliable, stable, and reproducible effects than phage therapy.[9] Furthermore, the mode of action of phages was still not completely understood at those times, which lead to unreliable results.[10] For these reasons, penicillin was mass produced to prevent countless deaths of wounded soldiers during World War II. Due to the success of antibiotic treatment, research in phage therapy waned in the West as more effort was invested in the development of antibiotics. Despite this, the Society of Friends of Felix d'Herelle was founded to continue phage research,[11] and countries in Eastern Europe continued their investigation into phages, successfully creating medical applications to treat and prevent bacterial infections.

Collaborations between d'Herelle and a Georgian microbiologist named George Eliava led to the founding of a bacteriology institute later to be named after Eliava, which became the cornerstone of phage therapy that is integral to medicine in Georgia to this day. Progress made in phage therapy during this time, however, was largely lost because it was not properly documented and due to publications being written in Russian and Polish, which inhibited the transfer of knowledge to the English-speaking world.[5] The aim of this chapter is the summarization of the most outstanding characteristics of phages and the benefits of phage therapy over antibiotic administration. This chapter discusses the potential disadvantages of phage therapy and possible solutions, as well as regulatory issues that have impeded the successful production of phage products in the West thus far.

BACTERIOPHAGE LIFE CYCLES

Apart from morphological characteristics, phages are also classified according to the mode of replication. Replication of phages occurs via one of two cycles, the lytic or lysogenic cycle.[5] The main focus in phage therapy research is on lytic phages, which redirect host cell metabolism to produce phage progeny and ultimately cause cell death by lysis. Attachment of phage via tail proteins to host cell receptors initiates infection. The tail proteins of a commercial *Listeria* phage P100 (Listex P100) are shown in Fig. 15.2. This is followed by the injection of phage nucleic acid into the host cytoplasm, replication of phage nucleic acid by host cell machinery, expression of phage genes, assembly of viral particles, and release of mature phages by host cell lysis.[5] The alternative life cycle, termed *lysogeny*, involves the replication of dormant phage nucleic acids, along with host genes for several generations, until induction of the lytic mode of replication occurs in response to host cell damage or environmental stress.[5]

Temperate phages favor the lysogenic state. The term *lysogeny* describes the integration of phage nucleic acid into the host's genome, with the resulting integrated phage referred to as a *prophage*. Alternatively, the phage genome in some cases may be maintained as circular DNA or RNA in the host cytoplasm.[12] The lysogeny-lysis bi-switch is encoded by the *cI/Cro* genes, which are mainly influenced by the state of the host cell upon infection by phage.[13] Under favorable physiological conditions, the temperate phage will revert to lysogeny and integrate its genome into that of the host cell via chromosomal integration or episomal maintenance. Gene regulation in the temperate phage λ was extensively reviewed by Dodd and colleagues.[13] In short, the lysogenic

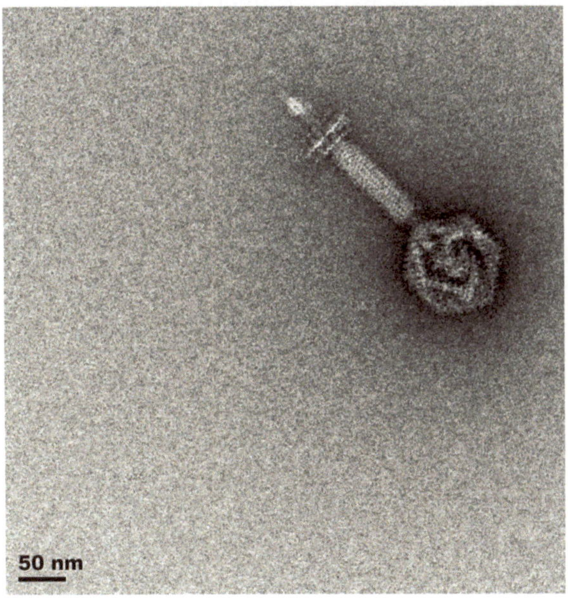

FIGURE 15.2

Listeria phage P100 (Listex P100) with contracted tail and visible capsid and tail proteins.

state is maintained by the repression of promoters P_R and P_L by the repression of *cI*, preventing transcription of early lytic genes. Lysogeny is indefinitely maintained in this way until conditions become unfavorable to the host cell.

The SOS repair system, which is activated once DNA is damaged, ultimately results in the derepression of the phage genome.[14] This mechanism involves autoproteolysis of the λ *cI* repressor, which is stimulated by the host cell DNA-repair protein RecA. Following sufficient inactivation of *cI*, prophage induction occurs and lytic genes are transcribed.[13] Once this process begins, the phage genome is replicated and essential proteins are expressed in order for the virions to be packaged before lysing the host cell and being released into the environment.

PHAGE THERAPY AS AN ALTERNATIVE TO ANTIBIOTICS

Penicillin resistance in *Staphylococcus aureus* was discovered in the middle of the 20th century. By the 1960s, antibiotic resistance had become widespread and clinically significant.[15] In 1996, bacteria were reported to show resistance against third-generation antibiotics such as cephalosporin and vancomycin.[16] Growing global awareness of antibiotic resistance in the latter half of the 20th century increased interest in phages and their products in the West. In the 1980s, Ślopek and his colleagues in Wrocław, Poland, started clinical trials to treat patients infected with antibiotic-resistant bacteria with phages.[15] Phages were able to protect mice from a lethal level of infection of *Escherichia coli* better than various antibiotics,[17] restoring interest in this field. The present applications of phage therapy range from surface and food sterilization to medical and veterinary treatment by preventing infection. In the food industry, the application of phages as pesticides against plant pathogens was approved in 2005 by the US Environmental Protection Agency (EPA), with the first phage-containing pesticide released under the trade name AgriPhage.[18]

Intricacies of antibiotic resistance are further compounded by the fact that once a bacterium is resistant to an antibiotic, the probability of it resisting all antibiotics in the same class is very high. More important, it has been found that once bacteria acquire resistance to multiple antibiotic classes (ie, become multidrug-resistant), their resistance is greater against the individual antibiotics than against their parental strains.[19] The repercussions of antibiotic misuse and resultant resistance development have resulted in bans against the use of antibiotics as general prophylactic and growth factors in food animals (ie, animals raised for human consumption) in the European Union and United States.[20] Since the end of 2013, all antibiotic use in animal feed for growth factors has been banned in the European Union, and these substances reserved for use to treat illness.[21] These bans were put into place to reduce the possibility of antibiotic-resistant bacteria being formed from the improper use of antibiotics and residual sublethal concentrations, which make medically important antibiotics null and void.

However, several discussions have begun because there are differing opinions on whether antibiotic misuse in food animals has any influence on medically important resistant bacteria.[20,22] Thus, there are opposing arguments against the total ban of antibiotic use in food animals, as some argue that this does not affect the severity of the resistance found in hospitals and nosocomial cases of infection, as there is no direct line of infection from food animals to hospitals. Furthermore, reports have been written to monitor whether the bans have had an effect on the levels of antibiotic resistance and to manage and assess the usage of antibiotics.[23] The bans appear to have alleviated the presence of resistant bacteria affecting food animals, but this has not reverberated in the medical sector.

From the report of the joint scientific assessment by the Food and Agriculture Organization (FAO), World Organization for Animal Health (OIE), and the World Health Organization (WHO) it was concluded that nonhuman antimicrobial use should be assessed in conjunction with human health to determine how veterinary antimicrobials will be used. In addition, efforts are being made to standardize the use of antibiotics internationally, which should further alleviate the problem of escalating antibiotic resistance.

The multiple-antibiotic-resistant bacteria that emerged in the 1990s foreshadowed an impending post-antibiotic era, endangering human and animal health. An alternative to antibiotics is evidently required, a recourse that is highly effective in killing harmful bacteria while having no toxic effects on the host. Phage therapy appears to meet these criteria, as discussed in the following sections of this chapter. Phages, to our knowledge, are a much safer treatment than antibiotics and are possibly more effective due to the lower occurrences of resistant strains. In addition, bacteria have a 10-fold less probability of becoming resistant to phages than to antibiotics.[24]

ADVANTAGES OF PHAGE THERAPY OVER ANTIBIOTIC TREATMENT

One of the key features of phages to be used in therapy is the specificity with which they infect bacterial cells, to such an extent that certain phages only attack specific strains within a species.[25] This contrasts the mode of operation observed in antibiotics, which work broadly against either Gram-negative or Gram-positive bacteria, regardless of pathogenicity. This can often lead to gaps in the host microecology, which promotes colonization by other potentially more pathogenic bacteria, causing secondary infections. The narrow host range displayed by phages can be advantageous, as beneficial bacteria are not disturbed during administration. Specificity is mainly determined by the ability of phage receptors to recognize host target proteins involved in adsorption during initial stages of infection.

In addition, the specificity of phages means that they show negligible toxicity toward host cells, other than some rare, reversible allergic reactions. In 1987, researchers reported a less than 3% occurrence of allergic reaction to phage administration in a group of 138 human patients[26]; however, these allergic reactions were fully reversible. By contrast, several antibiotics can cause multiple side effects, including fatal allergies and permanent tissue damage, particularly when applied in high doses.[27] Phages also have the added advantage of remaining in the environment in a dormant state until their hosts are available for replication.

The vascular dissemination of intravenously administered phages has been demonstrated,[28] which is ideal when treating localized infections in different parts of the body. This is an obvious advantage over certain antibiotics, which only reach specific organs during treatment. Phages have also been effective in penetrating biofilms layer by layer as they infect, lyse, and proliferate, as opposed to antibiotics that are not able to infiltrate biofilms effectively.[29] Phage therapy has proved effective in the elimination of membrane biofouling caused by antibiotic-resistant *Delftia tsuruhatensis* biofilms in a prototype lab-scale membrane bioreactor, demonstrating the potential to solve the problem of membrane biofouling in full-scale wastewater treatment plants, as well as pathogenic biofilms on medical devices.[30]

Phage therapy has more longevity potential than antibiotics, due to lower occurrences of resistant strains, since bacteria have 10-fold less probability of becoming resistant to phages than to antibiotics.[24] It has been hypothesized that phages naturally evolved alongside their host bacteria.[31] Therefore, if a bacterium develops resistance to a phage, the phage could adapt through mutation and evolve to once again infect the resistant bacterium, as opposed to antibiotics that obviously cannot adapt. In addition, there is likely to be a number of other phages capable of infecting a bacterium that developed resistance to one phage, since phages are ubiquitous in the biosphere—estimated at about 10^{31} phages[31]—making them readily available for discovery in common and accessible environments.[32] For example, sewerage is an abundant source of phages specific to enteric bacteria, like *E. coli* phages shown in Fig. 15.3. In contrast, the available number of approved antibiotic compounds available is limited, and the development and especially the approval of new antibiotics are very long processes.

Phage therapy is potentially a more sustainable treatment option since phages are self-managing in the sense that they will proliferate as long as there are host cells in an environment. This is in contrast to the need to maintain antibiotic pressure long enough to prevent bacteria from becoming resistant. In addition, phages will remain in the environment in a state of dormancy until their hosts are available for replication.

The safety of phage therapy in immunocompromised individuals has been demonstrated.[33] It has been proposed that phage therapy might be less efficient in immunocompromised patients

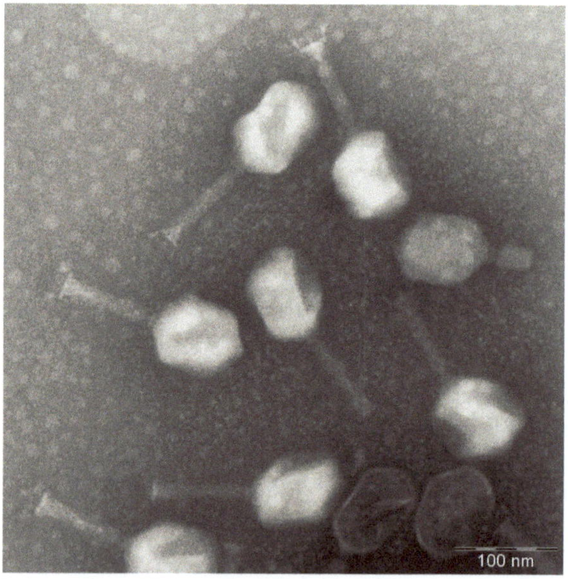

FIGURE 15.3

E. coli phages BSP (isolated from sewage water in Bloemfontein, Free State, RSA).

due to the development of resistance by bacteria, based on the observation of resistant mutants emerging after exposure to phages in certain cases.[17] In immunocompetent individuals, phage-resistant bacteria can be cleared by the immune system following the eradication of phage-sensitive bacteria, whereas these remaining bacteria will be less efficiently cleared in hosts with impaired immunity.[34] Despite these arguments, it has been showed that immunodeficiency could be advantageous to phage therapy owing to the delayed clearance of virions from the body. The half-life of phage T7 in B-cell-deficient mice was shown to be significantly prolonged in comparison to immunocompetent mice.[35]

LYTIC PHAGES AND THEIR ENZYMES AS POTENTIAL TREATMENT OPTIONS

Just as phages have been researched as an alternative to antibiotic therapy, their lytic enzymes (endolysins) have been identified and researched to serve as potential treatment options. The idea is to use the isolated enzymes, which accomplish bacterial lysis, rather than the entire phage.[36] The ultimate goal is to express and engineer these enzymes for use in almost all environments that bacterial cells inhabit.[37]

The majority of phages (including the most studied phage, λ) produce a soluble, muralytic endolysin that is responsible for degradation of the host peptidoglycan layer, allowing mature phages to exit the host cells to search for new ones.[38,39] Although there are some exceptions, phage enzymes generally have a cell wall-binding domain at the C-terminal and an enzymatically active domain at the N-terminal.[40] The mode of activity can be either through endo-β-*N*-acetylglucosaminidase or *N*-acetylmuramidase (acting on the sugar moiety), an endopeptidase (acting on the peptide moiety), or an *N*-acetylmuramoyl-L-alanine amidase, which hydrolyzes the glycan–peptide bonds.[41]

In cases involving muralytic enzymes, secondary proteins (known as *holins*) are produced to form a pore in the internal membrane of the host cell, allowing the endolysin access to the outer peptidoglycan layer. Holins are also the determining factors for the onset of host cell lysis and represent one of the most diverse groups of proteins with shared function.[42] Simpler, single-stranded nucleic acid phages produce single-protein lysis systems consisting of nonmuralytic, holin-independent enzymes. One such example is protein E of phage φX174, which achieves cell lysis by inhibiting peptidoglycan biosynthesis by the host cell.[43]

It has been found that the phage endolysins have a much broader range of infectivity than complete phages,[44] which relieves the issue of phage host specificity. As biotherapeutic agents, phages have been found to be effective but unreliable for traceability, as the possible emergence of endemic phages after phage treatment can interfere with accounting for phage numbers.[45]

Lytic enzymes work quickly on the bacterial cells with great affinity for their substrates, as illustrated by one study that demonstrated complete reduction in viability of a culture of 10^7 group A streptococci only seconds after the addition of nanogram quantities of enzymes.[46] Isolated and recombinant lysins were found to be successful against Gram-positive bacteria when applied extracellularly, providing an effective solution to antibiotic-resistant, Gram-positive bacteria.[47] Furthermore, it has been shown that these enzymes can be used on mucosal surfaces in animal control tests against *Streptococcus pneumoniae*, *S. pyogenes*, and *S. agalactiae* in order to prevent or eliminate infection.[47] Gram-negative bacteria, however, are protected from these types of lysins by their outer membrane layers that prevent lysin attachment and thus inhibit lytic activity.

Naturally occurring lysins exhibiting antibacterial activity against Gram-negative bacteria have also been described,[48] as well as lysins engineered to access Gram-negative substrates by fusion of catalytic domains to outer membrane-permeabilizing agents.[49,50] The first therapeutic lysin specific for Gram-negative organisms in a collection of native lysins found in *Acinetobacter* phages was shown to efficiently kill all tested clinical isolates of *Acinetobacter baumannii* and reduce planktonic and biofilm *A. baumannii* in vitro and in vivo.[51]

Phage lysins have species- and subspecies-level specificity, which is beneficial because there will be minimal damage to the natural microbiota colonizing an organism while pathogenic cells are destroyed.[46] There have been reports on chimeric rearrangements of the N-terminal and C-terminal domains of different phages to increase the range of endolysins available as additional antibacterial agents.[52] To date, there have been no reports of resistance to lytic enzymes, as observed in *S. pneumoniae* continuing to yield lysin-sensitive bacteria after continual exposure to low concentrations of the enzyme.[47]

LYSOGENIC PHAGES: SPECIFICITY AND POSSIBLE APPLICATIONS

In contrast to virulent phages, which operate as obligate killers in an antagonistic relationship with their host, the interaction between the lysogenic phage and its host is synergistic in nature. Procurement of a lysogenic phage is potentially beneficial to the fitness of the host,[53] as discussed in further detail later in this chapter. Unlike virulent phages, temperate phages are wholly dependent on the host, as they are integrated into the host nucleic acid replication mechanisms, resulting in dependence on host genomic replication.[53] Accordingly, temperate phages have an even narrower host specificity range than their virulent counterparts. Specificity is regulated by multiple factors, including host membrane receptors for phage attachment, host nucleic acid replication machinery, and phage mutation or variance. If all factors coincide and are favorable to phage integration and replication, a phage will be able to proliferate in the host. These factors are slightly more lax for virulent phages than for temperate ones, as they do not have to integrate into the host genome for survival.

Temperate phages are not ideal for use as therapeutic agents, as they are not capable of maintaining a lytic cycle, unlike virulent phages. Their ultraspecific nature means that if modified to become virulent, they would be as close as possible to a "magic bullet" for a pathogenic bacterial strain. The implication is that other beneficial strains would be allowed to thrive and not be destroyed in the process of eliminating pathogens. On the other hand, the specificity can also be severely limiting, as several temperate phages would have to be identified and genetically manipulated to target a broader host range. Even more important, their integration into the host genome allows gene transfer to and from the host, potentially making them more pathogenic by transferring genes, including virulence factors, to their genomes.[54,12] This topic is discussed in further detail later in this chapter.

While both temperate and virulent phages have their individual advantages, temperate phages require more investment of time, funds, and research when compared to using virulent ones. Furthermore, the host range may be too narrow and defined for temperate phages to be used to treat bacterial infections.

POSSIBLE ADVERSE EFFECTS OF PHAGE THERAPY AND POTENTIAL SOLUTIONS

LYTIC/VIRULENT PHAGES

The narrow host specificity of phages, although an attractive feature, may become problematic in treatment when several bacterial strains are simultaneously involved in a particular infection. Phage therapy, therefore, needs to be accompanied by highly accurate microbial diagnostics, which might explain the unreliable results obtained in the early years of phage therapy research.[55] It should be acknowledged that diagnostics has come a long way since the days of d'Herelle, with techniques such as 16S ribosomal RNA (rRNA) sequencing[56] and real-time polymerase chain reaction[57] making accurate and rapid diagnoses of pathogens a possibility. To overcome obstacles relating to specificity, phages can be administered in a cocktail mixture of different phages that broaden the host range to be treated.

Phages are not consistent in their replication numbers, as this depends on the amount of host cells present. This is one of the reasons why phage therapy in the early years did not compare well to antibiotics. However, several companies have supplied a solution to this problem via thorough research in either the use of cocktails of phages to overcome diverse bacterial receptors,[58] or finding optimal concentrations of phages that deliver results. Administration of such a mixture of phages also further diminishes the probability of resistance occurring. Phage cocktails were found to be more effective than separate phages, with less resistant cells evolving after the longest incubation period.[59]

The destruction of harmless bacterial strains could allow the growth of pathogenic strains due to reduced competition for adhesion areas and nutrients. This effect could worsen clinical symptoms in target patients. However, the high level of specificity of phages would make this an unlikely scenario. Unlike phage therapy, treatment with antibiotics generally destroys all bacteria nonspecifically, which can lead to secondary infections because there is less competition for space and nutrients with other pathogens entering the host. If highly specific phages cannot be obtained, a method to decrease the chances of secondary infections during phage therapy, as with antibiotic therapy, is to administer probiotics during the therapy. This method was used during a Russian trial on human infants suffering from dysbacteriosis to improve the efficacy of phage therapy.[60]

A natural consequence of both antibiotic and phage therapy is the potential release of cell wall and intracellular bacterial components into the environment after lysis of bacterial cells. Peptidoglycan and lipopolysaccharides contained in bacterial cell walls are capable of causing local and general inflammatory responses.[9] Lysis-deficient phages, which kill bacteria without endotoxin release, have been created to overcome this problem.[61] In one such example, a lysis-deficient *S. aureus* phage P954 was created by insertional inactivation of the endolysin gene, resulting in a phage incapable of lysis.[62] In this case, the phage holin protein is responsible for cell death by perforating the inner membrane, thus ceasing respiration.[63]

LYSOGENIC/TEMPERATE PHAGES

Transfer of virulence factors by phages

It has been established that the presence of prophages has an effect on the virulence and pathogenicity of bacterial species by their involvement in gene transfer between strains.[64] Prophage genes integrated into the host genome can code for various virulence factors, including toxins,[65] regulatory factors that enhance the production of virulence genes,[66] and enzymes with the ability to alter properties related to virulence.[67] Therefore, phages have the potential to convert bacterial hosts from nonpathogenic strains to virulent strains or strains with increased virulence through lysogenic conversion.[68]

Bacterial virulence properties involved in any of the different stages of infection can be altered by means of prophages.[69] Prophages contribute greatly to exotoxin production, and exotoxins are considered the most widely recognized example of phage-encoded virulence factors.[65] One such example includes the case of avirulent strains of *Corynebacterium diphtheriae* infected with a phage yielding virulent lysogens that produced the diphtheria toxin.[70] It was determined that these virulence genes are encoded on phage β.[71] Another example of this principle is the production of the scarlatinal exotoxin by a temperate phage located within the genome of nontoxigenic *Streptococcus pyogenes*.[72] The presence of prophage genes in the host genome can also cause strains of the same species to be associated with different diseases, as typified by two *S. pyogenes* strains that belong to different M serotypes, where the differences in the genomes are a result of prophage sequences.[54] In certain cases, prophage induction is required for the release of toxins.[73] The production of *Shiga* toxin (Stx2) carried by *E. coli* phages H-19B and 933W depend upon lysis of the host for release into the external environment.[64]

Other steps influenced by prophages during the stages of bacterial infection are bacterial adhesion and colonization. Transposon mutagenesis was used to identify a locus encoding two surface proteins possibly involved in platelet adherence, PblA and PblB, in *Streptococcus mitis*.[74] Sequence analysis revealed that the genes resemble phage capsid and tail fiber genes and reside on an inducible prophage SM1.[75] Bacterial enzymes that play a role during the invasion of host tissues are commonly phage encoded.[64] For example, the hyaluronidase of *S. pyogenes* is phage encoded[76] and hyaluronidase activity is also associated with phage particles themselves.[77] Hyaluronidase benefits phage infection by means of penetrating the hyaluronic acid capsule of *S. pyogenes*, as well as phage particle release by cell lysis. During bacterial infections, attachment and adhesion is often aided by biofilm formation.[78] The most strongly induced genes during *Pseudomonas aeruginosa* biofilm formation are within a filamentous prophage, Pf4,[79] and the role of the prophage in biofilm structural integrity and virulence was confirmed by the study of a Pf4 deletion mutant.[80]

Prophages also play a role in antibiotic resistance, although mobile antibiotic resistance genes are typically encoded by plasmids and transposons.[69] Resistance to several antibiotic classes is transferred by streptococcal phages via the generalized transduction of nonphage-encoded resistance genes.[81] While not much is known about the direct contribution of prophages to antibiotic resistance, two β-lactamase genes (*blaTEM* and *blaCTX-M9*) and one encoding a penicillin-binding protein (mecA) have been identified in bacteriophage DNA isolated from sewerage and river water samples, suggesting that phages are resistance gene reservoirs in aquatic environments.[82] The presence of prophage genes in a bacterial genome may also enhance the transmission of pathogens, such as the prophage-encoded cholera toxin of *V. cholerae*,[83] which upregulates enterocyte adenylate cyclase activity during infection, leading to the development of profuse, watery diarrhea[84] that aids the spread of cholera.[85]

IMMUNOGENICITY OF PHAGES

If the immune system inactivates the phages before enough time has passed for pathogens to be eradicated, phage therapy will probably fail. Therefore, studying interactions between phages and the immune system is crucial to the successful development of specific phage therapies. One such study concerned the formation of phage antibodies produced in patients undergoing antistaphylococcal phage therapy.[86] Antibody titer analyzes of 44 of the 57 patients showed no measurable antiphage antibodies for up to 21 days during treatment. In 8 of the remaining patients, antiphage antibody titers never exceeded 1:40, and in the remaining 2 patients, titers of between 1:320 and 1:1280 were found during the therapy. It was concluded from the study that phage immunogenicity would not hinder phage therapy.

Conversely, a study of immune interference by poultry during phage therapy against *E. coli* concluded that the poultry host would have gained some immunity to the phage after the first treatment, which decreased efficacy in secondary treatments.[87] To overcome this problem, using different phages for a follow-up treatment would be advisable. Immunity to phages in poultry is unlikely to be a major problem in the broiler industry, as the birds do not live longer than 42 days in most countries. Thus, treating an initial *E. coli* infection should be enough to increase yields in poultry production by decreasing the incidence of disease with only a single treatment. There is also the possibility that the phage used in this particular study is highly immunogenic; thus, other phages might react differently to host immune responses.

BACTERIAL RESISTANCE TO PHAGE INFECTION

One of the main advantages of the use of phages for the control of bacterial diseases is that resistance to phages is far less of a concern than resistance to antibiotics.[88] However, bacteria are not passive targets for viral predation. Bacteria are able to become resistant against phages by inhibiting steps in almost all stages of infection,[1] exhibiting several analogies to both the innate and adaptive immune responses in vertebrates.[89] The main modes of resistance to phage infection observed in bacteria are outlined in Table 15.1.

ANALOGIES TO INNATE IMMUNITY

Host nucleic acid modification

Due to its numerous applications in molecular biology, the restriction/modification system is the most extensively studied bacterial defense mechanism. It involves the modification of host DNA.[90,91] Two enzymes, a methyltransferase (MTase) and a restriction endonuclease (REase), are responsible for the host bacterium's ability to recognize and destroy foreign nucleic acid. The MTase methylates the host's genomic DNA to prevent degradation by REase, which recognizes only nonmethylated DNA. Restriction/modification systems are classified into four major types (I–IV) based on the subunit composition, ATP/GTP requirement, and cleavage mechanism of the endonucleases. The most simple and common systems, type II, cleave target DNA at highly specific sites. Type II systems are also classified into subtypes based on cleavage specificity.

Table 15.1 Bacterial Mechanisms of Resistance to Phage Infection

Resistance Mechanism	Description
Adsorption resistance	The absence of receptor molecules on bacterial surfaces prevents binding of phage to host cell.
Immunity to superinfection	Inhibition of phage replication as a result of recognition of phage-associated motifs.
Abortive infection	Inhibition of phage replication by suicide of infected bacteria.
Restriction/modification enzymes	Restriction enzymes cleave incoming foreign DNA at specific sequences. The modification enzyme protects host DNA from cleavage by biochemical modification.
CRISPR/Cas proteins	Phage DNA is integrated into CRISPR loci and transcribed into short, noncoding RNAs. These RNAs and Cas proteins recognize and silence foreign genetic elements.

Data from: Abedon ST. Bacterial "immunity" against bacteriophages. Bacteriophage *2012;2:50–4. Labrie SJ, Samson JE, Moineau S. Bacteriophage resistance mechanisms.* Nat Rev Microbiol *2010;8:317–27.*

The Type II systems consist solely of the MTase-REase pair that is characteristically encoded within the same operon; however, there have been reports of apparent cases of disjointed localization of the two genes. Type I systems are the most complex, depend on ATP, and consist of three genes encoding the R (restriction), M (modification), and S (specificity) subunits of the R-MA complex; and the R subunit also comprises a distinct ATPase domain belonging to the helicase Superfamily II. Type III systems are similar to type II systems, in that they consist of only R and M subunits; however, like type I systems, they are ATP-dependent and the R subunit also includes the helicase domain. Type IV systems are a discrete, two-subunit complex consisting of an AAA + family GTPase and an endonuclease cleaving the target DNA nonspecifically.[92] The DNA phosphorothioation (DND) system labels host DNA by phosphorothioation and destroys unmodified DNA. The modification genes, *dndABCDE*, encode proteins involved in site-specific S-modification of the host DNA backbone.[92]

ANALOGIES TO ADAPTIVE IMMUNITY

CRISPR/Cas systems

The clustered regularly interspaced short palindromic repeats (CRISPR)/CRISPR-associated gene (Cas) systems utilize the self-/nonself-discrimination principle to mediate immunity against invading pathogens and mobile genetic elements.[93] These systems are encoded in the genomes of most archaea and many bacteria and function as true adaptive immunity mechanisms that target specific sequences in plasmid or viral genomes. Implementing the RNA interference (RNAi) principle,[94] the CRISPR–Cas system obtains short DNA sequences or fragments of nonself genetic material (spacers) from mobile genetic elements and integrates them at specific locations in the CRISPR locus in the host genome.

Essential constituents of the CRISPR–Cas system are the CRISPR loci with spacers and diverse *Cas* genes nearby. The CRISPR memory expresses small CRISPR RNAs when the cell is infected by a previously encountered mobile genetic element. These RNAs are integrated into the Cascade complex, are comprised of Cas protein subunits, and are used to recognize and cleave similar foreign DNA. These RNAs guide effector proteins to complementary invading nucleic acids, which eventually neutralizes infection.[95] The CRISPR–Cas systems are very diverse; some systems target the invading DNA, while other systems target the invading RNA. The Cas proteins have DNase activity, RNase activity, or both and are involved in one or more of the stages. There are three types of CRISPR–Cas systems, each of which contains a different effector complex that binds to the crRNA and activates cleavage.

WILL BACTERIAL RESISTANCE TO PHAGES AFFECT PHAGE THERAPY?

Despite the proposed advantages of using phage cocktails to overcome bacterial resistance to phage infection, the emergence of bacteria resistance to such phage cocktails remains a concern.[96] Mechanisms of antibiotic resistance and phage resistance in bacteria may be radically different, but both result in failure to inhibit bacterial proliferation. Phages and antibiotics are both driving forces of bacterial evolution.[97] In the antagonistic, coevolutionary weapons race between bacteria and phages, it has been shown that phages evolve to reinfect previously resistant hosts,[98] suggesting that infectious phages are continuously available, even in the event of total resistance to a certain phage occurring in a given bacterial population. After all, maintenance of a phage-resistant phenotype comes at a cost to a bacterium.

Phage resistance in bacteria is in some cases associated with attenuated virulence due to the absence of important surface features associated with virulence.[99] Several examples exist that demonstrate the ability of a phage to infect a bacterium regardless of their lack of recent contact or geographical origin.[100] Not surprisingly, phages are also capable of launching counterattacks on both innate and adaptive bacterial immunity. A phage-encoded CRISPR/Cas system capable of counteracting the phage inhibitory chromosomal island of its host, *V. cholerae*, has recently been described,[101] as well as anti-CRISPR/Cas genes in *P. aeruginosa*.[102] These discoveries offer insight into the survival of phages despite widespread and effective immune mechanisms in bacteria.

Interestingly, the CRISPR/Cas system has been used to sensitize and selectively kill antibiotic-resistant bacteria.[103] A temperate lambda phage was used to deliver a CRISPR/Cas system into the genome of antibiotic-resistant *E. coli*, resulting in the destruction of β-lactam resistance-conferring genes, as well as lytic phages encoding CRISPR/Cas systems targeting β-lactamase genes. This approach renders antibiotic-sensitized bacteria resistant to infection by genetically modified lytic phages and allows the selective eradication of antibiotic-resistant strains.[103]

CHALLENGES CONCERNING BACTERIOPHAGE THERAPY

Phage therapy implementation in the United States and western Europe has met numerous obstacles.[104] More than two decades following its reappraisal by the West, no phase III disease-treatment clinical trials have been conducted.[104,105] Apart from previously discussed intrinsic challenges of phage therapy, several issues arise regarding its authorization for routine application in the West. As yet, commercial phage preparations are readily available only in two former Soviet republics, Russia and Georgia.[106,107] Considered an experimental treatment in Poland, phage therapy is covered by the Physician Practice Act (Polish Law Gazette No. 28 of 1997) and the declaration of Helsinki, where alternative treatment options are not available.[108] However, phage therapies implemented by early Soviet clinical trials did not obtain regulatory approval because trials were not conducted within the framework of regulatory authorities.[109]

The absence of distinct guidelines for regulatory approval of phage therapies is perhaps the primary difficulty encountered when integrating phages into conventional medicine. Currently, phage therapy is regulated according to guidelines developed for antibacterials, which require clinical testing of every individual component of the phage preparation and preclude modification of the therapy without reapproval.[109] Long-term phage research has indicated that phage preparations need to be updated regularly because continued exposure to a specific phage preparation results in bacterial resistance.[107,110] Preparations are updated by "training" ineffective phages by selection of phage mutants more effective against resistant bacteria, or by acquiring new phages from the environment or clinical samples. This rapid and inexpensive *sur-mesure* (meaning "made-to-measure" in French) approach is unfortunately incompatible with licensing procedures due to the extent of clinical testing and research required by regulators. In contrast, a *prêt-à-porter* (ready-to-wear) approach might be more amenable to approval by rigorous regulatory pathways, although such uniform, mass-produced preparations would be useful for only limited periods due to development of resistance.[107,110] A regulatory framework applicable to phage therapy may exist; the US Food and Drug Administration (FDA) has approved the influenza vaccine to be reformulated yearly in concurrence with the circulating flu strain.[111]

The restrictions of drug licensing encourage the development of phage products for application in the agricultural sector, with the aim of acquainting the populace and authorities with the concept of phage therapy and generating revenue to fund research and development for medical applications.[112] A US-based company, Intralytix, Inc., developed anti-*E. coli* and anti-*Listeria monocytogenes* phage cocktails for the food industry and obtained FDA approval and patents for the anti-*L. monocytogenes* phage cocktail as a food additive.[113] The food safety product ListShield is approved for direct application to ready-to-eat meat and has recently received generally recognized as safe status for application to fish, processed and fresh fruit and vegetables, and dairy products.[114] Another commercial *Listeria* phage (Listex P100) is shown in Fig. 15.4.

Troublesome concerns regarding the protection of intellectual property by patent law further discourages investment into the development of human phage therapeutics.[107,112,115] A phage or phage cocktail is considered novel by patent law only if it has not been isolated or produced before.[116,113] The concept of phage therapy is more than a century old, and literature pertaining to the isolation of and application of phages as natural entities to treat bacterial infections is extensive, invalidating the novelty of phages and such applications.[113,116] Naturally, private companies will be hesitant to invest in development of phage therapies, as competitors can readily isolate similar phages from environmental sources. Additionally, US and European patenting pathways for biological materials are disparate. The requirements of novelty, industrial applicability, and inventive steps are uniform in the Europe and the United States, but European patent law also requires a technical intervention to isolate a phage from its environment and proper characterization of the phage. Phages claimed in US patents are required to be considerably different from their counterparts in nature, and it appears that only genetically modified phages meet this specification.[116] Phage lysins

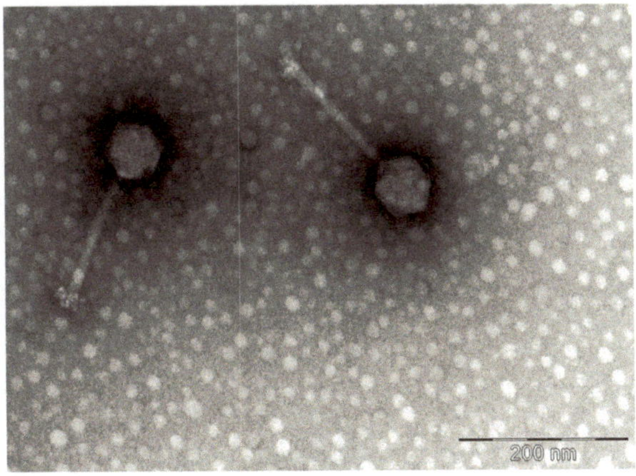

FIGURE 15.4

Two *Listeria* phages P100 (Listex P100) with noncontracted tails.

present an intriguing case, however, because even though they have proven effective in eradicating several antibiotic-resistant bacteria and are covered by numerous patents, no clinical trials have been initiated.[113]

CONCLUSION

The potential of phage therapy as antimicrobial treatment was not recognized for many years by the Western world. Numerous accounts of the capacity of phages to control bacterial proliferation in clinical, veterinary, and industrial settings have accumulated over the years. The present research mentions no serious adverse effects on phage therapy to control bacterial infection, even demonstrating its virtue in individuals with compromised immune systems. The occurrence of various mechanisms of bacterial resistance to phage infection is widely acknowledged; still, it is maintained that phage resistance will not reach the drastic proportions of antibiotic resistance. The use of engineered phage enzymes for the destruction of bacteria is also under investigation. Despite promising results, phage therapy presents distinctive challenges concerning patentability and regulation, discouraging investment and development. Favorable outcomes of phage therapy and the ever-increasing occurrence of antibiotic resistance urge the creation of innovative solutions to these unique challenges.

ACKNOWLEDGMENTS

The authors would like to acknowledge Amy Strydom and Thembakazi Noguda for all images in this chapter.

REFERENCES

1. Elbreki M, Ross RP, Hill C, O'Mahony J, McAuliffe O, Coffey A. Bacteriophages and their derivatives as biotherapeutic agents in disease prevention and treatment. *J Virol [Internet]* 2014. [cited June 3, 2015]; 2014:382539 [20 p.]. Available from: http://www.hindawi.com/journals/jvi/2014/382539/
2. Lavigne R, Molineux IJ, Kropinski AM. Order Caudovirales. In: King AMQ, Adams MJ, Lefkowitz EJ, Carstens EB, editors. *Virus taxonomy: ninth report of the international committee on taxonomy of viruses.* Amsterdam: Elsevier; 2011. p. 39–45.
3. Ackermann H. 5500 Phages examined in the electron microscope. *Arch Virol* 2007;**152**:227–43.
4. Hankin EH. L'action bactericide des eaux de la Jumna et du Ganges sur le vibrion du cholera. *Ann Inst Pasteur* 1896;**10**:511–23.
5. Sulakvelidze A, Alavidze Z, Morris JG. Bacteriophage therapy. *Antimicrob Agents Chemother* 2001;**45**:649–59.
6. Duckworth D. Who discovered bacteriophage? *Bacteriol Rev* 1976;**40**:793–802.
7. Chanishvili N. Phage therapy—history from Twort and d'Herelle through Soviet experience to current approaches. *Adv Virus Res* 2012;**83**:400–602.
8. Fleming A. On antibacterial action of culture of *Penicillium*, with special reference to their use in isolation of *B. influenzae*. *Br J Exp Pathol* 1929;**10**:226–36.
9. Summers WC. Bacteriophage therapy. *Annu Rev Microbiol* 2001;**55**:437–51.
10. Radetsky P. The good virus. *Discover* 1996;**17**:50–8.
11. Raiga-Clemenceau A. d'Herelle's bacteriophage and its therapeutic property. *Sem Hop Ther* 1974;**50**:229–31.
12. Gill JJ, Hyman P. Phage choice, isolation and preparation for phage therapy. *Curr Pharm Biotechnol* 2010;**11**:2–14.
13. Dodd IB, Shearwin KE, Egan JB. Revisited gene regulation in bacteriophage. *Curr Opin Gen Dev* 2005;**15**:145–52.
14. Ptashne M. *A genetic switch.* Cambridge, MA: Cell Press & Blackwell Scientific Publications; 1992.
15. Ślopek S, Durlakowa I, Kucharewicz-Krakowska B, Dąbrowski M, Biskiewicz R. Results of bacteriophage treatment of suppurative bacterial infection. *Arch Immunol Ther Exp* 1983;**31**:293–327.
16. Tenover FC, Hughes JM. The challenges of emerging infectious diseases. Development and spread of multiply-resistant bacterial pathogens. *JAMA* 1996;**275**:300–4.
17. Smith HW, Huggins MB. Successful treatment of experimental *Escherichia coli* infections in mice using phage: its general superiority over antibiotics. *J Gen Microbiol* 1982;**128**:307–18.
18. Nagy JK, Király L, Schwarczinger I. Phage therapy for plant disease control with a focus on fire blight. *Cent Eur J Biol* 2011;**7**:1–12.
19. Cohen SP, McMurry LM, Hooper DC, Wolfson JS, Levy SB. Cross-resistance to fluoroquinolones in multiple-antibiotic-resistant (Mar) *Escherichia coli* selected by tetracycline or chloramphenicol: decreased drug accumulation associated with membrane changes in addition to OmpF reduction. *Antimicrob Agents Chemother* 1989;**33**:1318–25.
20. Phillips I. Withdrawal of growth-promoting antibiotics in Europe and its effects in relation to human health. *Int J Antimicrob Agents* 2007;**30**:101–7.
21. Jarlier V, Carlet J, McGowan J, Goossens H, Voss A, Harbarth S, et al. Priority actions to fight antibiotic resistance: results of an international meeting. *Antimicrob Resist Infect Control* 2012;**1**:17.
22. Casewell M, Friis C, Marco E, McMullin P, Phillips I. The European ban on growth-promoting antibiotics and emerging consequences for human and animal health. *J Antimicrob Chemother* 2003;**52**:159–61.
23. Boucher HW, Talbot GH, Bradley JS, Edwards JE, Gilbert LD, Rice LB, et al. Bad bugs, no drugs: no ESKAPE! An update from the Infectious Diseases Society of America. *Clin Infect Dis* 2009;**48**:1–12.

24. Carlton RM. Phage therapy: past history and future prospects. *Arch Immunol Ther Exp (Warsz)* 1999;**47**:267–74.
25. Joerger RD. Alternatives to antibiotics: bacteriocins, antimicrobial peptides and bacteriophages. *Poult Sci* 2003;**82**:640–7.
26. Ślopek S, Weber-Dabrowska B, Dąbrowski M, Kucharewicz-Krukowska A. Results of bacteriophage treatment of suppurative bacterial infections in the years 1981–1986. *Arch Immunol Ther Exp* 1987;**35**:569–83.
27. Kalghatki S, Spina SC, Costello JC, Liesa M, Morones-Ramirez JR, Slomovic S, et al. Bactericidal antibiotics induce mitochondrial dysfunction and oxidative damage in mammalian cells. *Sci Transl Med* 2013;**5** 192ra85
28. Dąbrowska K, Świtała-Jeleń K, Opolski A, Weber-Dąbrowska B, Górski A. Bacteriophage penetration in vertebrates. *J Appl Microbiol* 2005;**98**:7–13.
29. Kutter E, De Vos D, Gvasalia G, Alavidze Z, Gogokhia L, Kuhl S, et al. Phage therapy in clinical practice: treatment of human infections. *Curr Pharm Biotechnol* 2010;**11**:69–86.
30. Bhattacharjee AS, Choi J, Motlagh AM, Mukherji ST, Goel R. Bacteriophage therapy for membrane biofouling in membrane bioreactors and antibiotic-resistant bacterial biofilms. *Biotechnol Bioeng* 2015;**112**:1644–54.
31. Hendrix RW, Smith MC, Burns RN, Ford ME, Hatfull GF. Evolutionary relationships among diverse bacteriophages and prophages: all the world's a phage. *Proc Natl Acad Sci* 1999;**96**:2192–7.
32. Ackermann HW. Bacteriophage taxonomy. *Microbiol Aust* 2011;**32**:90–4.
33. Weber-Dąbrowska B, Mulczyk M, Górski A. Bacteriophage therapy for infections in cancer patients. *Clin Appl Immunol Rev* 2001;**1**:31.
34. Borysowski J, Górski A. Is phage therapy acceptable in the immunocompromised host? *Int J Infect Dis* 2008;**12**:466–71.
35. Srivastava AS, Kaido T, Carrier E. Immunological factors that affect the in vivo fate of T7 phage in the mouse. *J Virol Methods* 2004;**115**:99–104.
36. Fenton M, Ross P, McAuliffe O, O'Mahoney J, Coffey A. Recombinant bacteriophage lysins as antibacterials. *Bioeng Bugs* 2010;**1**:9–16.
37. Yang H, Yu J, Wei H. Engineered bacteriophage lysins as novel anti-infectives. *Front Microbiol [Internet]* 2014;**5** [cited May 18, 2015]; Article 542 [6 p.]. Available from: http://www.ncbi.nlm.nih.gov/pmc/articles/PMC4199284/pdf/fmicb-05-00542.pdf. Free full text article
38. Young R, Wang I, Roof WD. Bacteriophage holins: deadly diversity. *Trends Microbiol* 2000;**8**:120–8.
39. Catalão MJ, Gil F, Moniz-Pereira J, São-José C, Pimentel M. Diversity in bacterial lysis systems: bacteriophages show the way. *FEMS Microbiol Rev* 2013;**37**:554–71.
40. Tišáková L, Godány A. Bacteriophage endolysins and their use in biotechnological processes. *J Microbiol Biotech Food Sci* 2014;**3**:164–70.
41. Young RY. Bacteriophage lysis: mechanism and regulation. *Microbiol Rev* 1992;**56**:430–81.
42. Wang I, Smith DL, Young R. Holins: the protein clocks of bacteriophage infections. *Annu Rev Microbiol* 2000;**54**:799–825.
43. Tanaka S, Clemons WM. Minimal requirements for inhibition of MraY by lysis protein E from bacteriophage ΦX174. *Mol Microbiol* 2012;**85**:975–98.
44. Sable S, Lortal S. The lysins of bacteriophages infecting lactic acid bacteria. *Appl Microbiol Biotechnol* 1995;**43**:1–6.
45. Kropinski AM, Lingohr EJ, Moyles DM, Ojha S, Mazzocco A, She Y-M, et al. Endemic bacteriophages: a cautionary tale for evaluation of bacteriophage therapy and other interventions for infection control in animals. *Virol J* 2012;**9**:207–15.
46. Nelson D, Loomis L, Fischetti VA. Prevention and elimination of upper respiratory colonization of mice by group A streptococci by using a bacteriophage lytic enzyme. *Proc Natl Acad Sci USA* 2001;**98**:4107–12.
47. Fischetti VA. Bacteriophage lytic enzymes: novel anti-infectives. *Trends Microbiol* 2005;**13**:491–6.

48. Lai MJ, Lin NT, Hu A, Soo PC, Chen LK, Chen LH, et al. Antibacterial activity of *Acinetobacter baumannii* phage ϕAB2 endolysin (LysAB2) against both Gram-positive and Gram-negative bacteria. *Appl Microbiol Biotechnol* 2011;**90**:529–39.
49. Lukacik P, Barnard TJ, Keller PW, Chaturvedi KS, Seddiki M, Fairman JW. Structural engineering of a phage lysin that targets Gram-negative pathogens. *Proc Natl Acad Sci USA* 2012;**109**:9857–62.
50. Briers Y, Walmagh M, Van Puyenbroeck V, Cornelissen A, Cenens W, Aertsen A, et al. Engineered endolysin-based "Artilysins" to combat multidrug-resistant Gram-negative pathogens. *mBio* 2014;**5**:0e01379-14
51. Lood R, Winer BY, Pelzek AJ, Diez-Martinez R, Thandar M, Euler CW, et al. Novel phage lysin capable of killing the multidrug-resistant Gram-negative bacterium *Acinetobacter baumannii* in a mouse bacteremia model. *Antimicrob Agents Chemother* 2015;**59**:1983–91.
52. García P, García J, García E, Sánchez- Puelles JM, López R. Modular organization of the lytic enzymes of *Streptococcus pneumoniae* and its bacteriophages. *Gene* 1990;**86**:81–8.
53. Koskella B, Meaden S. Understanding bacteriophage specificity in natural microbial communities. *Viruses* 2013;**5**:806–23.
54. Brüssow H, Canchava C, Hardt WD. Phages and the evolution of bacterial pathogens: from genomic rearrangements to lysogenic conversion. *Microbiol Mol Biol Res* 2004;**68**:560–602.
55. Keen E. Phage therapy: concept to cure. *Front Microbiol [Internet]* 2012;**3**:238 [cited May 18, 2015]; [3 p.]. Available from: http://www.ncbi.nlm.nih.gov/pmc/articles/PMC3400130/pdf/fmicb-03-00238.pdf. Free full text article
56. Rhoads D, Wolcott R, Sun Y, Dowd S. Comparison of culture and molecular identification of bacteria in chronic wounds. *Int J Mol Sci* 2012;**13**:2535–50.
57. Espy M, Uhl J, Sloan L, Buckwalter S, Jones M, Vetter E. Real-time PCR in clinical microbiology: applications for routine laboratory testing. *Clin Microbiol Rev* 2006;**19**:165–256.
58. Goodridge LD. Designing phage therapeutics. *Curr Pharm Biotechnol* 2010;**11**:15–27.
59. Tanji Y, Shimada T, Yoichi M, Miyanaga K, Hori K, Unno H. Toward rational control of *Escherichia coli* O157:H7 by a phage cocktail. *Appl Microbiol Biotechnol* 2004;**64**:270–4.
60. Litvinova AM, Chtetsova VM, Kavtreva IG. Evaluation of efficacy of the use of *E. coli-Proteus* bacteriophage in intestinal dysbacteriosis in premature infants. *Vopr Okhr Materin Det* 1979;**23**:42–4.
61. Matsuda T, Freeman TA, Hilbert DW, Duff M, Fuortes M, Stapleton PP, et al. Lysis-deficient bacteriophage therapy decreases endotoxin and inflammatory mediator release and improves survival in a murine peritonitis model. *Surgery* 2005;**137**:639–46.
62. Paul VD, Sundarrajan S, Rajagopalan S, Hariharan S, Kempashanaiah N, Padmanabhan S, et al. Lysis-deficient phages as novel therapeutic agents for controlling bacterial infection. *BMC Microbiol [Internet]* 2011;**11**:195 [cited June 3, 2015]; [9 p]. Available from: http://www.biomedcentral.com/1471-2180/11/195. Free full text article
63. Young RJ. Bacteriophage holins: deadly diversity. *Mol Microbiol Biotechnol* 2002;**4**:21–36.
64. Wagner PL, Livny J, Neely MN, Acheson DWK, Friedman DI, Waldor MK. Bacteriophage control of Shiga toxin 1 production and release by *Escherichia coli*. *Mol Microbiol* 2002;**44**:957–70.
65. Bishai WR, Murphy JR. Bacteriophage gene products that cause human disease. In: Calendar R, editor. *The bacteriophages*. New York, NY: Plenum Press; 1988. p. 683–724.
66. Spanier JG, Cleary PP. Bacteriophage control of antiphagocytic determinants in group A streptococci. *J Exp Med* 1980;**152**:1393–406.
67. Guan S, Bastin DA, Vrma NK. Functional analysis of the O-antigen glucosylation gene cluster of *Shigella flexneri* bacteriophage SfX. *Microbiology* 1999;**145**:1263–73.
68. Boyd EF, Brüssow H. Common themes among bacteriophage-encoded virulence factors and diversity among the bacteriophages involved. *Trends Microbiol* 2002;**10**:521–9.
69. Wagner PL, Waldor MK. Bacteriophage control of bacterial virulence. *Infect Immun* 2002;**70**:3985–93.
70. Tinsley CR, Bille E, Nassif X. Bacteriophages and pathogenicity: more than providing a toxin. *Microbes Infect* 2006;**8**:1365–71.

71. Freeman VJ. Studies on the virulence of bacteriophage-infected strains of *Corynebacterium diphtheriae*. *J Bacteriol* 1951;**61**:675–88.
72. Zabriskie JB. The role of temperate bacteriophage in the production of erythrogenic toxin by group A streptococci. *J Exp Med* 1964;**119**:761–79.
73. Fortier LC, Sekulović O. Importance of prophages to evolution and virulence of bacterial pathogens. *Virulence* 2013;**4**:354–65.
74. Bensing BA, Rubens CE, Sullam PM. Genetic loci of *Streptococcus mitis* that mediate binding to human platelets. *Infect Immun* 2001;**69**:1373–80.
75. Bensing BA, Siboo RM, Sullam PM. Proteins PblA and PblB of *Streptococcus mitis*, which promote binding to human platelets, are encoded within a lysogenic bacteriophage. *Infect Immun* 2001;**69**:6186–92.
76. Hynes WL, Ferretti JJ. Sequence analysis and expression in *Escherichia coli* of the hyaluronidase gene of *Streptococcus pyogenes* bacteriophage H4489A. *Infect Immun* 1989;**57**:533–9.
77. Benchetrit LC, Gray ED, Wannamaker LW. Hyaluronidase activity of bacteriophages of group A streptococci. *Infect Immun* 1977;**15**:527–32.
78. Lewis K. Persister cells, dormancy and infectious disease. *Nat Rev Microbiol* 2007;**5**:48–56.
79. Whiteley M, Bangera MG, Bumgarner RE, Parsek MR, Teitzel GM, Lory S, et al. Gene expression in *Pseudomonas aeruginosa* biofilms. *Nature* 2001;**413**:860–4.
80. Rice SA, Tan CH, Mikkelsen P, Kung V, Woo J, Tay M, et al. The biofilm life-cycle and virulence of *Pseudomonas aeruginosa* are dependent on a filamentous prophage. *ISME J* 2009;**3**:271–82.
81. Ubukata K, Konno M, Fujii R. Transduction of drug resistance to tetracycline, chloramphenicol, macrolides, lincomycin and clindamycin with phages induced from *Streptococcus pyogenes*. *J Antibiot (Tokyo)* 1975;**28**:681–8.
82. Colomer-Lluch M, Jofre J, Muniesa M. Antibiotic resistance genes in the bacteriophage DNA fraction of environmental samples. *PLoS ONE [Internet]* 2011;**6**:e17549 [cited August 21, 2015]; [about 10 p.]. Available from: http://journals.plos.org/plosone/article?id=10.1371/journal.pone.0017549.
83. Waldor MK, Mekalanos JJ. Lysogenic conversion by a filamentous phage encoding cholera toxin. *Science* 1994;**272**:1910–14.
84. Finkelstein RA. Cholera enterotoxin (choleragen): a historical perspective. In: Barua D, Greenborough WB, editors. *Cholera*. New York, NY: Plenum Press; 1992. p. 155–87.
85. Mintz ED, Popovic T, Blake PA. Transmission of *Vibrio cholerae* O1. In: Wachsmuth I, Blake PA, Olsvik O, editors. *Vibrio cholerae and cholera: molecular to global perspectives*. Washington, DC: ASM Press; 1994. p. 345–56.
86. Kucharewicz-Krukowska A, Ślopek S. Immunogenic effect of bacteriophage in patients subjected to phage therapy. *Arch Immunol Ther Exp* 1987;**35**:553–61.
87. Huff WE. Immune interference of bacteriophage efficacy when treating colibacillosis in poultry. *Poult Sci* 2010;**89**:895–900.
88. Golkar Z, Bagasra O, Pace DG. Bacteriophage therapy: a potential solution for the antibiotic resistance. *J Infect Dev Ctries* 2014;**8**:129–36.
89. Abedon ST. Bacterial "immunity" against bacteriophages. *Bacteriophage* 2012;**2**:50–4.
90. Labrie SJ, Samson JE, Moineau S. Bacteriophage resistance mechanisms. *Nat Rev Microbiol* 2010;**8**:317–27.
91. Makarova KS, Wolf YI, Snir S, Koonin EV. Defense islands in bacterial and archaeal genomes and prediction of novel defense systems. *J Bacteriol* 2011;**193**:6039–56.
92. Makarova KS, Wolf YI, Koonin EV. Comparative genomics of defense systems in archaea and bacteria. *Nucleic Acids Res* 2013;**41**:4360–77.
93. Makarova KS, Grishin NV, Shabalina SA, Woolf YI, Koonin EV. A putative RNA-interference-based immune system in prokaryotes: computational analysis of the predicted enzymatic machinery, functional analogies with eukaryotic RNAi, and hypothetical mechanisms of action. *Biol Direct [Internet]* 2006;**1** [cited June 5, 2015]; [26 p.]. Available from: http://www.biologydirect.com/content/pdf/1745-6150-1-7.pdf. Free full text article

94. Carthew RW, Sontheimer EJ. Origins and mechanisms of miRNAs and siRNAs. *Cell* 2009;**36**:642–55.
95. Makarova KS, Haft DH, Barrangou R, Brouns SJJ, Charpentier E, Horvath P, et al. Evolution and classification of the CRISPR-Cas systems. *Nat Rev Microbiol* 2011;**9**:467–77.
96. Örmälä A, Jalasvuori M. Should bacterial resistance to phages be a concern, even in the long run? *Bacteriophage* 2013;**3**:e24219.
97. Clokie MR, Millard AD, Letarov AV, Heaphy S. Phages in nature. *Bacteriophage* 2011;**1**:31–45.
98. Buckling A, Rainey PB. Antagonistic coevolution between a bacterium and a bacteriophage. *Proc Biol Sci* 2002;**296**:931–6.
99. Meyer JR, Agrawal AA, Quick RT, Dobias DT, Schneider D, Lenski RE. Parallel changes in host resistance to viral infection during 45,000 generations of relaxed selection. *Evolution* 2010;**64**:3024–34.
100. Wolf A, Wiese J, Jost G, Witzel KP. Wide geographic distribution of bacteriophages that lyse the same indigenous freshwater isolate (*Sphingomonas* sp. strain B18). *Appl Environ Microbiol* 2003;**69**:2395–8.
101. Seed KD, Lazinski DW, Calderwood B, Camilli A. A bacteriophage encodes its own CRISPR/Cas adaptive response to evade host immunity. *Nature* 2013;**494**:489–91.
102. Bondy-Denomy J, Pawluk A, Maxwell KL, Davidson AR. Bacteriophage genes that inactivate the CRISPR/Cas bacterial immune system. *Nature* 2013;**493**:429–32.
103. Yosef I, Manor M, Kiro R, Qimron U. Temperate and lytic bacteriophages programmed to sensitize and kill antibiotic-resistant bacteria. *Proc Natl Acad Sci USA* 2015;**112**:7267–72.
104. Kutter EM, Kuhl SJ, Abedon S. Re-establishing a place for phage therapy in western medicine. *Future Microbiol* 2015;**10**:685–8.
105. Abedon ST, Kuhl SJ, Blasdel BG, Kutter EM. Phage treatment of human infections. *Bacteriophage* 2011;**1**:66–85.
106. Chanishvili N. *A literature review of the practical application of bacteriophage research*. 1st ed. New York, NY: Nova Science Publishers, Inc.; 2012.
107. Pirnay J, De Vos D, Verbeken G, Merabishvili M, Chanishvili N, Vaneechoutte M, et al. The phage therapy paradigm: *Prêt-à-Porter* or *Sur-mesure*? *Pharm Res* 2011;**28**:934–7.
108. Górski A, Miedzybrodzki R, Borysowski J, Weber-Dabrowska B, Lobocka M, Fortuna W, et al. Bacteriophage therapy for the treatment of infections. *Curr Opin Investig Drugs* 2009;**10**:766–74.
109. Sarhan WA, Azzazy HME. Phage approved in food, why not as a therapeutic? *Expert Rev Anti Infect Ther* 2015;**13**:91–101.
110. Pirnay J, Blasdel BG, Bretaudeau L, Buckling A, Chanishvili N, Clark JR, et al. Quality and safety requirements for sustainable phage therapy products. *Pharm Res* 2015;**32**:2173–9.
111. Sulakvelidze A. The challenges of bacteriophage therapy. *Eur Ind Pharm* 2011;**10**:14–18.
112. Thiel K. Old dogma, new tricks—21st century phage therapy. *Nat Biotechnol* 2004;**22**:31–6.
113. Brüssow H. What is needed for phage therapy to become a reality in Western medicine? *Virology* 2012;**434**:138–42.
114. Intralytix, Inc. *Intralytix receives additional regulatory approval for ListShield™—phage-based food safety product effective against* Listeria monocytogenes *[Internet]*. Baltimore, MD: Intralytix, Inc.; 2015 [cited August 21, 2015] Available from: http://www.intralytix.com/Intral_News_PR010515.htm
115. Henein A. What are the limitations on the wider therapeutic use of phage? *Bacteriophage [Internet]* 2013;e24872 [cited August 21, 2015]; [about 6 p.]. Available from: http://www.ncbi.nlm.nih.gov/pmc/articles/PMC3821673/.
116. Pirnay J, Verbeken G, Rose T, Jennes S, Zizi M, Huys I, et al. Introducing yesterday's phage therapy in today's medicine. *Future Virol* 2012;**7**:379–90.

CHAPTER

BACTERIOCINS AND ITS USE FOR MULTIDRUG-RESISTANT BACTERIA CONTROL

16

G.M. Preciado, M.M. Michel, S.L. Villarreal-Morales, A.C. Flores-Gallegos, J. Aguirre-Joya, J. Morlett-Chávez, C.N. Aguilar and R. Rodríguez-Herrera

Food Research Department, School of Chemistry, Universidad Autónoma de Coahuila, Saltillo, Coahuila, Mexico

INTRODUCTION

Bacteriocins are bioactive antimicrobial peptides (AMPs), also known as antimicrobial proteins, that are synthesized in the ribosome and extracellularly released. These compounds are produced by many bacteria and can kill or inhibit the growth of other bacteria. They show great potential against antibiotic-resistant strains.[1] Bacteriocins are produced by both Gram-positive and Gram-negative bacteria. These compounds confer to bacteria a competitive advantage in their environmental niche to reducing competition for resources and form an important component of its chemical defense system.[2] Bacteriocins were first identified in 1925. Production of these proteins is widespread among bacterial species, and it has been suggested that virtually all bacterial species could produce them.[3] A great diversity of bacteriocins has been reported in most bacterial species, and there are cases where one bacterial species can produce different types of bacteriocins.[4] The most studied bacteriocins are produced by lactic acid bacteria (LAB)[5] because the use of these bacteria and their metabolic products is generally considered as safe (GRAS, Grade One).[6,7] LAB are involved in fermentation and food preservation, and improvement of their hygienic quality by inhibiting the competitive flora, including pathogens,[4] makes them ideal for use as bio-preservatives or microbial biocontrol.[8] One concern about the use of bacteriocins as food preservatives is the possible appearance of natural or acquired bacteriocin resistance (BACR) by bacterial pathogens. Natural resistance to class IIa bacteriocins has been reported in 1−8% of the wild-type strains tested.[9]

Although bacteriocins could be categorized as antibiotics, they actually are not. The principal difference between bacteriocins and antibiotics is that bacteriocins straiten their activity against the bacterial strains of related species, and especially to strains of the same species.[7] Antibiotics, on the other hand, have a broader spectrum of activity, and even if its activity is limited, no preferential effect is seen on closely related strains.[7] The synthesis of bacteriocins is in the ribosome, while antibiotics are synthesized by multistep enzyme pathways.[9] Moreover, bacteriocins usually have a low molecular weight; they undergo posttranslational modification and can be easily degraded by proteolytic enzymes, especially by proteases of mammalian gastrointestinal tract, making them safe for human consumption.[7] Application of biological products as bacteriocins to inhibit or eliminate pathogenic microorganisms is a method of great interest in the food industry whose final objective is to obtain safer food for consumers.[10] Nisin is a unique bacteriocin that has been approved

Antibiotic Resistance. DOI: http://dx.doi.org/10.1016/B978-0-12-803642-6.00016-2

as a safe food preservative by the European Union (E234), as well as by the World Health Organization (WHO) and US Food and Drug Administration (FDA), and is being used as a practical food preservative in more than 50 countries. Nisin is a 34-amino acid antibiotic and one of the most studied bacteriocins. Up to now, five variants are already known: nisin A, Z, Q, U, and F; this compound is produced for the bacteria *Lactococcus lactis*.[11] Most LAB bacteriocins are small, cationic, and heat-stable, and can be divided into three principal classes.[7] However, other authors mentioned the class IV composed by complex bacteriocins, which contain carbohydrates or lipid moieties.[10]

Bacteriocins can be produced in the early or late steps of the growth cycle of the producer microorganism. They can release or hold them together extracellularly. These compounds have different temperatures and stabilities and are sensitive to pH and proteases.[9,12] Traditionally, new bacteriocins have been identified by screening bacterial isolates for antimicrobial activity, followed by purification and identifying the bacteriocin and estimating genetic determinants. This strategy is still essential for identification of powerful bacteriocins of various subclasses. Recent examples include (1) a class IIa bacteriocin named avicin A, which was identified from *Enterococcus avium* strains isolated from fecal samples of healthy human babies from Ethiopia and Norway; (2) a bacteriocin called *garvicin ML* produced by a *Lactococcus garvieae* strain isolated from mallard duck; (3) a class IIb *bacteriocin enterocin* called X isolated from a *Enterococcus faecium* voltage sugar apples; and (4) a glycosylated bacteriocin (F glycocin) that was identified from *Lactobacillus plantarum* isolated from fermented corn.[13] The aims of this chapter are to review and discuss the current information about bacteriocin classification, their mode of action and structure, diversity, biotechnological applications, molecular evolution, and methods of purification and their importance on control of MDR bacteria.

CLASSIFICATION OF BACTERIOCINS

The bacteriocins are classified according to Zacharof and Lovitt[7] as follows:

- Class I: Lantibiotics. The lantibiotics are small (5 kDa) and heat-stable peptides, which contain the following characteristic polycyclic thioether amino acids: lanthionine or methyllanthionine. Nisin and lactisin are the principal representatives of this group.
- Class II: Nonlantibiotics. These bacteriocins are also small (10 kDa), relatively heat stable, and with membrane active peptides. They are divided into two subclasses according to the mode of action. Subclass IIa is integrated by pediocinlike or listeria-active bacteriocins that possess an N-terminal. Subclass IIb includes two components: lactacin F and lactococcin G.
- Class III: Bacteriocins. This group includes the complex bacteriocins and consists of heat-labile proteins that have high molecular weight (30 kDa). Bacteriocins of this group include helveticin I, by *Lactobacillus helveticus,* and enterolysin, produced by *E. faecium.*

STRUCTURE AND MECHANISM OF ACTION

The general mechanism for bacteriocin inhibitory activity is that a bacteriocin recognizes either a general or specific receptor molecule on a sensitive cell, which allows it to get close to the cell membrane to mitigate membrane interaction and pore formation.[12] Bacteriocins are usually synthesized as inactive peptides with an N-terminal sequence guide. These precursors are transported to the cell surface during the exponential growth phase and become their active forms enzymatically. Carriers contain an N-terminal peptide part, responsible for the guide peptide cleavage, as well as a C-terminal part, responsible for adenosine triphosphate hydrolysis and energy supply. For Class II, accessory proteins are used to facilitate membrane translocation, cleave the peptide tab, or both. The control system is the production of bacteriocins composed of three components: a peptide that induces (or a pheromone-activating factor), transmembrane histidine kinase (a pheromone receptor), and a response regulator. Class III, the N-terminal portion of these bacteriocins, is homologous to an endopeptidase involved in cell wall synthesis, whereas the C-terminal portion is responsible for recognition of the target cell.[13]

DIVERSITY OF BACTERIOCINS IN GRAM-NEGATIVE BACTERIA

The first bacteriocin was isolated from Gram-negative bacteria in 1952, when Gratia identified a protein with antimicrobial characteristics from *Escherichia coli*. Later, this protein was classified as a colicin.[14] Actually, bacteriocins from Gram-negative bacteria are classified into four main categories (Table 16.1): colicins, colicinlike bacteriocins, phage-taillike bacteriocins, and microcins.[17] The colicins (produced by *E. coli*) have been used as the model for bacteriocin structure, function, and evolution. Cascales et al.,[18] Bakkal et al.,[16] and Gillor et al.[15] reported the existence of two types of colicins based on their mechanism of killing: (1) pore former colicins (A, B, E1, Ia, Ib, K, and 5), which kill antagonists by forming pores in the cell membranes and (2) nuclease colicins (E2, E3, E4, E5, E6, E7, E8, E9), which kill by such actins as DNases, RNases, and tRNAses.

Colicinelike proteinase bacteriocins are classified based on such structural and functional characteristics as nucleases (pyocins S1, S2) and pore formers (pyocin S5).[19] The most studied colicinlike bacteriocins are klebicins from *Klebsiella* spp., S-pyocins from *Pseudomonas aeruginosa*, and alveicins from *Hafnia alvei*.

Table 16.1 Classes of Gram-Negative Bacteriocins and Some Examples

Bacteriocin	Class	Size (KDa)	Example	Bacteria Producer
Colicins	Pore formers/ nucleases	20–80	Colicins A, B	*Escherichia coli*
Colicinlike	–	20–80	Klebcins	*Klebsiella pneumoneae*
Phage-taillike	–	>80	R and F pyocins	*Pseudomonas aeruginosa*
Micronics	Posttranslationally modified/unmodified	<10	Colicin V Microcin C7	

Data from:
Bakkal S, Robinson SM, Ordonez CL, Waltz DA, Riley MA. Role of bacteriocins in mediating interactions of bacterial isolates taken from cystic fibrosis patients. Microbiology *2010;156:2058–67.*
Gillor O, Kirkup BC, Riley MA. Colicins and microcins: the next generation antimicrobials. Adv Appl Microbiol *2004;54(1):129–46.*
Güllüce M, Karadayı M, Barış O. Bacteriocins: promising natural antimicrobials. In: Méndez-Vilas A, editor. Microbial pathogens and strategies for combating them: science, technology and education. *Extremadura: Formatex; 2013. p. 1016–27.*

Two of the most studied phage-taillike bacteriocins are the R- and F-pyocins from *P. aeruginosa*. They are encoded in a gene cluster, which comprise a DNA region greater than 40 kb.[20] There are 44 open reading frames with R2/F2 phenotypes, including regulatory, lysis, and toxin genes.

Microcins are small peptides (<10 kDa), that can be divided into two classes: posttranslationally modified (B17, C7, J25, and D93) and unmodified (E492, V, L, H47, and 24) microcins. Gillor et al.[15] reported that microcins are chromosomally encoded. Bakkal et al.[16] reported that the full function of Gram-negative bacteriocins is encoded via three (toxin, immunity, and lysis) linked genes. In fact, there are differences in the genetics of the different bacteriocins produced by Gram-negative bacteria. For example, Nakayama et al.[20] reported that the colicin gene cluster consists of three bacteriocin-related genes in close proximity, while colicinelike pyocin S3 does not possess a lysis gene. The close antimicrobial activity spectrum limits the use of Gram-negative bacteria bacteriocins in professional industries. Most studies have focused on more suitable types of bacteriocins from Gram-positive bacteria.[13]

DIVERSITY OF BACTERIOCINS IN GRAM-POSITIVE BACTERIA

Bacteriocins produced by Gram-positive bacteria are formed by fewer than 60 amino acids. In addition, they have a broad spectrum of action.[21–23] Production of bacteriocins in Gram-positive bacteria has been extensively studied; this group includes LAB, and bacteriocins from these bacteria have potential application in food.[21] The main genus producers of bacteriocins are *Carnobacterium, Enterococcus, Lactobacillus, Lactococcus, Leptosphaeria, Leuconostoc, Melissococcus, Oenococcus, Pediococcus, Streptococcus, Tetragenococcus, Vagococcus,* and *Weissella*.[24] Gram-positive bacteria also include other bacteriocin producer genera with biotechnological importance, such as *Aerococcus, Microbacterium, Propionibacterium,* and *Bifidobacterium*.[25] Bacteriocins are produced in vitro under controlled conditions (pH, time, and temperature), which ensure the safety of the final product.[26] Pure bacteriocins also can be added to nonfermented food to promote the quality and safety of food products and inhibit pathogenic microorganisms.[27]

The mechanisms of action of these bacteriocins are broadly known; it is known that these peptides and proteins bind to specific receptors on cells and form pores in the membrane, causing cell permeability and therefore bacteria death. Only a few receptors have been identified, including lipid II, which is a lipid-binding lantibiotics.[28] However, the effectiveness of bacteriocins have encouraged the use of these molecules in inhibiting pathogens during biotechnological processes. Some examples are shown in Table 16.2.

LAB AND BACTERIOCINS

FAO/WHO define a probiotic as "living microorganism, which when administered in adequate amounts confer a benefit to the host health." LAB are considered probiotics because they can withstand the acidity of the stomach longer in contrast to other bacteria.[29] There is evidence that LAB are able to inhibit the growth of pathogens by production of antimicrobial substances such as short-chain fatty acids, diacetyl, hydrogen peroxide, and several bacteriocins.[30]

Table 16.2 Bacteriocins Produced by Gram-Positive Bacteria

Class	Subclass	Bacteriocin	Producer Microorganism	Applications	References
Class I	IA	Nisin	*Lactococcus lactis* subsp. *Cremoris*	Inhibition of *Listeria monocytogenes* and *Staphylococcus aureus* in cheddar cheese	Dimitrieva-Moats et al. (2012)
	IA	Nukacin	*Staphylococcus simulans*	Prevention of bovine mastitis	Ceotto et al. (2010)
Class II	IIa	Carnobacteriocin X	*Carnobacterium maltaromaticum* C2	Antimicrobial activity against *L. monocytogenes*	Tulini et al. (2014)
	IIb	Enterocin X	*Enterococcus faecium*	Antimicrobial activity against *L. monocytogenes*	Hu et al. (2010)
	IIc	Carnocyclin A	*Carnobacterium maltaromaticum* UAL307	Antimicrobial activity against Gram-positive *Listeria* spp.	Martin-Visscher et al. (2008)
	IId	Aureocin	*Staphylococcus aureus*	Prevention and control of bovine mastitis	Coelho et al. (2007)
Class III	IIIa	Lysostaphin	*Staphylococcus simulans* subsp. *staphylolyticus*	Prevention and control of human and animal infections caused by *S. aureus*	Bastos et al. (2010)
Class IV	–	Enterocin	*Enterococcus faecalis*	Food biopreservative	Gálvez et al. (2007)
	–	Enterocina	*E. faecium*	Inhibition of *Micrococcus luteus, L. monocytogenes, Listeria innocua, Streptococcus agalactiae, S. aureus, Bacillus cereus, Bacillus subtilis, E. faecalis, Klebsiella pneumonia, Escherichia coli*	De la Fuente-Salcido et al. (2015)

Dimitrieva-Moats GY, Ünlü G. Development of freeze-dried bacteriocin-containing preparations from lactic acid bacteria to inhibit Listeria monocytogenes *and* Staphylococcus aureus. Probiotics Antimicrob Proteins *2012;**4**:27–38.*

Coelho MLV, Nascimento JS, Fagundes PC, Madureira DJ, Oliveira SS, Brito MAP, Bastos MCF. Activity of staphylococcal bacteriocins against Staphylococcus aureus *and* Streptococcus agalactiae *involved in bovine mastitis.* Res Microbiol *2007;**158**:625–30.*

Ceotto H, Holo H, Silva da Costa KF, Nascimento JS, Salehian Z, Nes IF, Bastos MCF. Nukacin 3299, a lantibiotic produced by Staphylococcus simulans *3299 identical to nukacin ISK-1.* Vet Microbiol *2010;**146**:124–31.*

Tulini FL, Lohans CT, Bordon KC, Zheng J, Arantes EC, Vederas JC. Purification and characterization of antimicrobial peptides from fish isolate Carnobacterium maltaromaticum *C2: carnobacteriocin X and carnolysins A1 and A2.* Int J Food Microbiol *2014;**173**:81–8.*

Hu CB, Malaphan W, Zendo T, Nakayama J, Sonomoto K. Enterocin X, a novel two-peptide bacteriocin from Enterococcus faecium *KU-B5, has an antibacterial spectrum entirely different from those of its component peptides.* Appl Environ Microbiol *2010;**76**:4542–5.*

Martin-Visscher LA, van Belkum MJ, Garneau-Tsodikova S, Whittal RM, Zheng J, McMullen LM. Isolation and characterization of carnocyclina, an oval circular bacteriocin produced by Carnobacterium maltaromaticum *UAL307.* Appl Environ Microbiol *2008;**74**:4756–63.*

Bastos MCF, Coutinho BG, Coelho MLV. Lysostaphin: a staphylococcal bacteriolysin with potential clinical applications. Pharmaceuticals (Basel) *2010;3:1139–61.*

Gálvez A, Abriouel H, López RL, Ben Omar N. Bacteriocin-based strategies for food biopreservation. Int J Food Microbiol *2007;**120**:51–70.*

De la Fuente-Salcido NM, Catañeda RJC, García ABE. Isolation and characterization of bacteriocinogenic lactic bacteria from Tuba and Tepache, two traditional fermented beverages in México. Food Sci Nutr *2015;**3**(2):1–9.*

Some of the main microorganisms used as probiotics are *Lactobacillus*, *Bifidobacterium*, *Enterococcus*, *Lactococcus*, *Pediococcus*, *Streptococcus,* and *Leuconostoc*.[31] Probiotic microorganisms are potential therapeutic agents for intestinal infections and antibiotic treatment complements, reducing side effects.[32] Some of the benefits of using LAB bacteriocins for food preservation are the following[6]:

- LAB have a long history of safe use in food preservation.
- They are listed as GRAS.
- Bacteriocins are inactive and nontoxic for eukaryotic cells.
- They do not cause changes to the intestinal flora because they are inactivated by digestive proteases (because of their protein nature).
- They have a broad antimicrobial spectrum.
- They can act synergistically with other bacteriocins, and even with antibiotics.
- They are stable at different pHs and temperatures used in food processing.

Nowadays, there is lot of work to do on the implementation of bacteriocins in different types of food in order to improve food safety (Table 16.3), to inactivate bacterial endospores, and to broaden the spectrum of antimicrobial treatments.[6] The antimicrobial action or bacteriocin spectrum of LAB includes spoilage microorganisms and foodborne pathogens, such as *Staphylococcus aureus* and *Listeria monocytogenes*, even such Gram-negative bacteria as *E. coli* and *Salmonella*. There are treatments where bacteriocins are combined with physical or chemical methods, such as osmotic shock, electrical pulse, high pressure, and being in the presence of a chelating agent, in order to disrupt the integrity of the cell membrane of Gram-negative bacteria and thus facilitate or ensure that the bacteriocin is able to inhibit these bacteria.[9]

Helicobacter pylori is a pathogen that is able to colonize in the acidic environment of the stomach because of its urease, causing chronic gastritis, peptic ulcers, and gastric cancer.[30] There is a treatment for this bacteria based on antibiotics (clarithromycin and amoxicillin) and an inhibitor of the proton pump. This treatment has proved to be efficient; however, large doses of antibiotics cause pathogen resistance.[33] An alternative or complement to this treatment is the use of LAB, as demonstrated by Lim[29] who evaluated in vitro fermentation broths of *L. plantarum* BK10, *Lactobacillus brevis* BK11, *Lactobacillus acidophilus* BK13, *Lactobacillus paracasei* BK57, and *Enterococcus faecalis* BK61, all of which are isolated from baik kimchi (fermented vegetable food) against *H. pylori* and prove to have bactericidal activity against this pathogenic species.

Lü et al.[34] reported antimicrobial activity by the bacteriocin lactocin MXJ 32A from *Lactobacillus coryniformis* against both Gram-positive and Gram-negative pathogenic bacteria transmitted by food, such as *Listeria monocytogenes*, *Salmonella*, *S. aureus,* and *E. coli.* This lactocin even inhibited the growth of *Salmonella* spp. and *S. aureus* that are resistant to ampicillin, amoxicillin, amikacin, chloramphenicol, ciprofloxacin, gentamicin, and levofloxacin, making it possible to use this bacteriocin as an alternative to antibiotics or as a food preservative.

An interesting proposal was offered by Acuña et al.[35,36] with the construction of a recombinant hybrid bacteriocin enterocin CRL35 produced by *Enterococcus mundtii* and *E. coli* microcin V, in order to broaden the range of action, since this bacteriocin has antimicrobial activity against Gram-positive bacteria. Meanwhile, microcin, which is an AMPs but is also produced by Gram-negative bacteria, is active against Gram-negative bacteria. This combination, called Ent35-MccV, resulted

Table 16.3 Antimicrobial Treatments with LAB Bacteriocins

Antimicrobial Treatment	Effect	References
Escherichia coli O157:H7, *Listeria monocytogenes,* and *Brochothrix thermosphacta* strains were incubated for 6 h in a preparation 0.2 mg/mL of lauric arginate and 320 AU/mL nisin Z.	By scanning electron microscopy and transmission electron microscopy, the formation of membrane channels and leak potassium and phosphate ions at fairly high levels was determined, which resulted in a decrease in the viable cells.	Pattanayaiying et al. (2014)
Synergy of three bacteriocins, nisin and two enterocins produced by *Enterococcus faecium* (resistant to irradiation with 100% residual activity at doses of 4 kGy and 82.35% to 40 kGy dose), combined with irradiation at 3.0 kGy	Synergistic antimicrobial effect in controlling the growth of *L. monocytogenes* in the meat when γ-irradiation and these two bacteriocins (nisin and enterocins) are combined.	Turgis et al. (2012)
Combination of bacteriocins produced by *Lactobacillus sakei* sakacin C2 and C2 at a concentration of 640 AU/mL in ham inoculated with *L. monocytogenes.*	Effective control of the growth of *L. monocytogenes,* without any adverse effect on the quality of vacuum-packed ham	Gao et al. (2015)
Cyclic bacteriocin combination of AS-48 (25 mg/mL); further heating at 60°C for 5 min. Combination of AS-48 (25 mg/mL) and 2-nitro-1-propanol (2NPOH) at a concentration of 25 mM.	Combination of enterocin AS-48 and 2NPOH or heat showed reduced counts of *Staphylococcus aureus* in sauces.	Grande et al. (2012)

Pattanayaiying R, H-Kittikun A, Cutter CN. Effect of lauric arginate, nisin Z, and a combination against several food-related bacteria. Int J Food Microb *2014;***188***:135–46.*
Turgis M, Stotz V, Dupont C, Salmieri S, Khan RA, Lacroix M. Elimination of Listeria monocytogenes *in sausage meat by combination treatment: radiation and radiation-resistant bacteriocins.* Rad Phys Chem *2012;***81***:1185–8.*
Gao Y, Li D, Liu X. Effects of Lactobacillus sakei *C2 and sakacin C2 individually or in combination on the growth of* Listeria monocytogenes, *chemical and odor changes of vacuum-packed sliced cooked ham.* Food Control *2015;***47***:27–31.*
Grande M, Abriouel H, Lucas R, Gálvez A. Increasing the microbial inactivation of Staphylococcus aureus *in sauces by a combination of enterocin AS-48 and 2-nitropropanol, and mild heat treatments.* Food Control *2012;***25***:740–4.*

in the inhibition of enterohemorrhagic *E. coli* and *Listeria monocytogenes*. Another proposal is the combination of bacteriocins and inactivation methods. Zhao et al.[37] demonstrated the synergy between nisin and high hydrostatic pressure (HHP) inactivation treatment in cucumber juice, and compared the results with the juice without nisin and juice with nisin both subject to HHP, revealing the effectiveness of the nisin combination with HHP to reduce the mesophilic aerobic microbiota in cucumber juice.

SIGNIFICANCE OF BACTERIOCINS IN NATURE

Bacteriocins are secondary metabolites produced by allelopathy of microorganisms, which means that they are capable to produce chemical substances that can suppress the growth and development of relative organisms (microorganisms in this case), which contributes to the invasiveness of the allelopathic organism.[38] Some applications of bacteriocins are listed in Table 16.4.

Table 16.4 Bacteriocins Applications According to Their Nature

Group of Bacteriocins	Examples	Producer Organisms	Applications
Lantibiotics	Nisin, subtilin, epidermin, mersacidin	*Lactococcus lactis, Leuconostoc mesenteroides, Staphylococcus epidermidis*	Blood pressure treatment, dental caries, herpes, mastitis infection, inflammation, allergies, and skin infection treatment
Colicins	Colicins E2, E7	*Escherichia coli*	Hemolytic uremic syndrome treatment, urogenital infection treatment, and hemorrhagic colitis
Microcins	Microcin V, Subtilosin A	*E. coli, B. subtilis*	Antimicrobial agent, salmonellosis treatment

Data from:
Balciunas EM, Martinez FAC, Todorov SD, de Melo Franco BDG, Converti A, de Souza Oliveira RP. Novel biotechnological applications of bacteriocins: a review. Food Control *2013;32(1):134–42.*
Güllüce M, Karadayı M, Barış O. Bacteriocins: promising natural antimicrobials. In: Méndez-Vilas A, editor. Microbial pathogens and strategies for combating them: science, technology and education. *Extremadura: Formatex; 2013. p. 1016–27.*

The importance of bacteriocins nowadays is that they represent a suitable substitution of traditional drugs used to control most infections worldwide. This substitution is needed because of the rising occurrence of MDR illnesses, so multiple studies have focused on exploring the promissory role of naturally or genetically modifying bacteriocins to replace traditional antibiotics and to use as food preservatives.[15]

MOLECULAR EVOLUTION OF BACTERIOCINS

Colicins of *E. coli* and other enteric bacteriocins, such as klebicins, remain the only bacteriocins for which detailed evolutionary investigations have been undertaken.[39] Protein sequence comparisons among them reveal two families, the pore former and nuclease colicins, with no sequence similarity. Studies that include DNA and protein sequence comparison, DNA polymorphism in natural isolates, experimental evolution, and mathematical modeling have revealed two ways for colicin evolution to occur.[40]

Colicin gene clusters are encoded on plasmids and include three (toxin, immunity, and lysis) genes.[40] Colicins are large proteins. Pore-forming colicins range in size from 449 to 629 amino acids. Nuclease bacteriocins have an even broader range, from 178 to 777 amino acids.[40] Pore-forming colicins are the most abundant (constituting approximately 80% of colicin-producing strains).[41] If the C-terminal residues are compared within the pore-forming family, three subfamilies emerge: Ia, E1, and A. Otherwise, two subfamilies of nuclease colicins can be distinguished based on slightly reduced levels of protein sequence similarity, which reflects the RNAse versus DNAse functions of the proteins.[42] For explaining an unusual pattern of divergence between two pairs of closely related nuclease colicins, a positive selection was proposed, which encompasses two steps: (1) the occurrence of coupled mutations in the immunity region that provided the host cell with a broadened immunity function and a novel colicin and (2) the rapid fixation of this evolved strain in the ancestral colicinogenic population.[42]

Pore-forming colicins differ significantly from nuclease colicins with respect to their killing action, translocation mechanism, and action of its immunity proteins. Immunity proteins of pore-forming colicins are integral inner-membrane proteins, while those from nuclease colicins bind to colicins in cytoplasm.[43] In pore-forming colicins, regions of colicin sequence that encode precise functional domains have recombined among the three groups to give rise to new colicins.[44] Recombination also seems to have occurred between chromosomally encoded bacteriocins and colicins. *Serratia marcescens* produces a chromosomally encoded bacteriocin (N28b) that shares a high degree of similarity in the first 45 residues with colicin, with 35 identical residues.[44] Sequence of a pore-forming colicin, colicin Y,[45] was closely related to colicin U, another pore-forming colicin. However, this pair of colicins has a pattern of DNA substitution identical to that observed among nuclease colicins, with an elevated level of substitution in the immunity region. This suggests that diversifying selection is not restricted to nuclease colicins.[41] This is also supported by several E2 colicins.[46]

Other well-studied bacteriocins are pyocins from *P. aeruginosa*.[47] They have been classified into three groups based in their structure: R, F, and S, of more than 1400 *P. aeruginosa* isolates. R- and F-pyocins are produced by over 90% of clinical isolates, while S-pyocins are produced by over 70% of these same strains. Four pyocin and immunity genes for S-pyocins have been characterized. They share functional similarity, nuclease activity, and some DNA sequence with a subclass of the nuclease colicins of *E. coli*. However, S-pyocins are chromosomally encoded, whereas nuclease colicins are always plasmid encoded. In contrast with nuclease colicins, substitutions between pyocins are clustered within the translocation and receptor-binding domains.[42]

The sequence obtained for other bacteriocins from *Klebsiella pneumoniae*, klebicin B, showed extensive similarity to both pyocins and nuclease colicins. However, as is the case for many pore-forming colicins, the klebicin B gene cluster appears to be chimeric, with different portions of the gene cluster arising from pyocinlike and colicinlike sources.[42]

Bacteriocin production in halophilic *Archaebacteria* is ubiquitous.[48] Two halocins from this group of bacteria, which inhabit hypersaline environments, have been characterized (H4 and H6), and two putative microcins (Hal R1 and S8) have been identified. Halocin H4 is encoded on a megaplasmid that acts at the membrane level. It has a signal peptide cleaved from the preprotein upon secretion. On the other hand, colicinlike proteins do not have leader sequences, and Gram-positive proteins have cleavable leader peptides that do not conform to typical signal sequences.[49] Halocin H6 reaches its maximal activity as the culture enters the stationary phase; as H4, killing is by single-hit kinetics, and the protein is about 32 kDa.[50] The evolutionary relationship between halocins and other bacteriocins could be clarified as new bacteriocin sequences are available, but regardless of whether halocins and other bacteriocins share a common ancestor, halocins apparently play a similar role in populations as that envisioned for traditional bacteriocins.

Bacteriocins-encoding plasmids demonstrate another aspect of bacteriocin evolution. They are chimeras with a plasmid "backbone" comprising replication and maintenance sequences typical of plasmids found in the bacteriocins' host species. The plasmid pKlebB isolated from *K. pneumoniae* contains similar sequences to pNBL63 (*Klebsiella oxytoca)* and pJHCMW1 (*K. pneumoniae*), encoding plasmid maintenance functions.[42] As more bacteriocins are sequenced and characterized, evolution models will become more complex, and perhaps other evolution models have to be proposed.

GENETICS AND PROTEIN ENGINEERING OF BACTERIOCINS

LAB are well known to produce bacteriocins active against foodborne pathogens and food spoilage bacteria. Thus, their possible use as natural preservatives have been the focus of research. The development of heterologous expression systems for bacteriocins in different hosts could lead to advantages over native systems, such as (1) facilitated control of bacteriocin gene expression or increased bacteriocin production; (2) production of bacteriocins in safer hosts, especially when bacteriocins will be applied in food; (3) construction of multibacteriocinogenic strains with a wider antagonistic spectrum or acquisition of antimicrobial properties; (4) better adaptation of the selected hosts to food environments; and (5) providing antagonistic properties for LAB already used as starter, protective, or probiotic cultures.[51]

The development of new biotechnological tools and recent advances in LAB genetics have sparked research on heterologous production of bacteriocins in the last decade. Among heterologous systems, *E. coli* has been extensively characterized. Many of its biological processes are well understood, and a wide range of genetic tools are available for its manipulation. As a consequence, these bacteria have invariably been selected as the first host for cloning a variety of genes, including those involved in the biosynthesis of many LAB bacteriocins.

Eukaryotic hosts have been also used for heterologous expression. Van Reenen et al.[52] reported the production of a bacteriocin called *plantaricin* by *L. plantarum* 423, which was active against *Listeria* spp., *Staphylococcus* spp., *Pediococcus* spp., and *Lactobacillus* spp. Gene-encoding plantaricin was cloned and heterologous was produced in *Saccharomyces cerevisiae,* resulting in a heat-resistant and pH-stable protein. The DNA sequence of this bacteriocin revealed an operon structure similar to pediocin from *Pediococcus acidilactici* and coagulin from *Bacillus coagulans*, a non-LAB,[53] but plantaricin was more active against foodborne pathogens, which could make it suitable for biological control, especially during alcohol fermentation.[49]

Another LAB bacteriocin producer is *Leuconostoc,* which is widely used in the food industry as a starter and protective culture. Many *Leuconostoc* strains produce bacteriocins, among which the class IIa bacteriocins leucocin A (LcnA) and leucocin C (LecC) are some of the most frequently studied.[53,54] In addition, other *Leuconostoc* bacteriocins have been found, such as the class IIb two-peptide bacteriocin leucocin H from *Leuconostoc* MF215B[55] and the class IIc cyclic bacteriocin leucocyclicin Q from *Leuconostoc mesenteroides* TK41401.[56] Within *Leuconostoc*, *Leuconostoc carnosum* 4010 is an antimicrobial strain used as a protective culture in vacuum-packed meats. Wan et al.[57] reported the heterologous expression of two novel genes encoding leucocin A and B from *Leuconostoc carnosum* in another LAB, *L. lactis*. Gene transcription of leucocin B was demonstrated by reverse transcription polymerase chain reaction (RT-PCR) analysis, and its secretion and activity was confirmed by sodium dodecyl sulfate (SDS) gel using a leucocin B-sensitive *Leuconostoc* strain.

Leuconostoc mesenteroides also produces a bacteriocin called mesentericin Y105 (MesY105), a class IIa anti-*Listeria* bacteriocin with potential food grade applications,[58] but Hechard et al.[59] reported the production of another class bacteriocin, mesentericin B105. Secretion of this bacteriocin was studied using the MesY105 dedicated transport system. *Leuconostoc mesenteroides* DSM20484 was used as a heterologous expression system and was transformed independently using two plasmids containing operons *mesYI* and *mesBHF*, which are required for MesY105 secretion and

MesB105 production, respectively. It is known that in most cases, heterologous bacteriocin producers lead to lower peptide yields than natural producers.[60] Exceptions in these low yields are described when high copy number plasmids are used[61] or when genes are under strong promoter control.[62] However, using strategies proposed by Morisset and Frére[63] high secretion levels of bacteriocins (or levels at least similar to those of natural producers) were reached. Thus, MesY105 secretion machinery appears to be a useful tool for the secretion of class II bacteriocins by *Leuconostoc*.

Enterococcus is another LAB genus. Gutiérrez et al.[64] reported on enterolysin A (EnlA) production by *E. faecalis* II/1. This is a heat-labile class III bacteriocin form with antilisterial activity, and its heterologous expression was studied in *E. coli*. The PCR-amplified products of enterolysin A structural gene, the N-terminal part of EnlA with endopeptidaselike activity, and the C-terminal part of EnlA (similar to a lysis gene of bacteriophage) were cloned in an expression vector, leading to the synthesis and secretion of a functional-active His-tagged enterolysin A protein, which was purified and shown to be fully active against the indicator strain. The expression of N-terminal or C-terminal part of EnlA and deletion of the last 58 amino acids from the C-terminal domain of EnlA led to the synthesis of biologically nonactive proteins. Thus, the development of heterologous expression systems has proved to be successful in many cases, and other cloning and expression systems could be helpful in bacteriocin production.

PURIFICATION AND CHARACTERIZATION OF BACTERIOCINS

The bacteriocin producers are LAB that have nutritional complexes needs to grow. This not only increases the cost of production, but also makes the bacteriocins purification process more difficult. Bacteriocins form an extremely heterogeneous group of substances, and specific purification protocols need to be designed for each of them, which may explain why only some bacteriocins were purified to homogeneity as nisin.[13] The most common purification techniques are salinization, solvent extraction, ultrafiltration, adsorption-desorption, liquid chromatography, ion exchange, and high-performance liquid chromatography (HPLC).[10]

Three main methods for purifying LAB bacteriocins can be distinguished.[13] First, purification can be performed by a conventional method, which is based on a laborious series of steps of ammonium sulfate precipitation, ion exchange, hydrophobic interaction, gel filtration, HPLC, and reverse phase chromatography. Second, a simple three-step protocol including precipitation with ammonium sulfate, extraction with chloroform/methanol/precipitation, and reverse-phase HPLC is performed. Third, bacteriocins can be isolated by single-unit operation—that is, expanded bed adsorption using a hydrophobic interaction gel after maximizes bio-available title bacteriocins by adjusting the pH crude fermentation medium.[13]

Authors such as Kittikun et al.[65] followed this method for a bacteriocin produced by *E. faecalis*. The first step was precipitation with 70% ammonium sulfate. In the second step, the active fraction was separated into a reverse-phase cartridge (Sep-Pak Vac 12cc cartridge C8). Three fractions were collected and checked by the agar well diffusion essay. The fraction was eluted with 100% acetonitrile. This active fraction was then applied to a cation exchange column. Finally, this active fraction was applied to a column of RP-HPLC. However, Sahoo et al.[1] used a three-step protocol involving positive salt precipitation, cation exchange chromatography, and reverse-phase chromatography. An et al.[66] performed purification of the bacteriocin CAMT2 using a two-step method: ammonium sulfate precipitation followed by column chromatography. The first step was similar to that of the previously mentioned method; the second step involved sodium dodecyl sulfate-polyacrylamide gel electrophoresis (SDS-PAGE), where the purity of the samples was determined.

For bacteriocins characterization, Ghanbari et al.[67] determined the biological nature of the antimicrobial activity of bacteria, using isolates incubated in MRS broth at 30°C for 24 h, which then were tested for their sensitivity to proteolytic enzymes. To clarify whether the detected antimicrobial activity stems from the production of hydrogen peroxide, 2600 IU/mL of catalase (C-100, Sigma, London) were added. To determine the temperature sensitivity of possible activities, bacteriocin cell-free supernatant samples were incubated under defined conditions.

APPLICATIONS OF BACTERIOCINS

These antimicrobial compounds have many applications. Some of the most important are biopreservation, shelf-life extension, clinical antimicrobials, and microbial control during fermentation.[13] Bacteriocins can be used in different food processes, especially those from LAB, because they are considered GRAS and can be used as bio-conservers.[68] Bacteriocins have been utilized during food fermentation[69]; in processes such as cheese elaboration[70] and meat products, which have the bacteriocin integrated into the package[71]; as antibotulinic agents in cheese and liquid eggs, sauces, and canned foods,[13] among other processes. Chen and Hoover[72] indicated three different methods for the incorporation of bacteriocins into food: (1) direct inoculation of LAB into the food, and these bacteria produce the bacteriocins; (2) incorporation into the food of purified or semipurified bacteriocins as preserving components; and (3) use of a previously fermented product that produces bacteriocin as an ingredient of the processed food. Bacteriocin obtained ex situ can be applied to a food substrate or in immobilized preparations. In this case, bacteriocins are attached to the support, which acts as a reservoir and diffuser of the AMP into the food. In addition, support can protect the bacteriocin from inactivation in an enzymatic way or for interaction of food components.[73]

Bacteriocins produced ex situ and obtained by fermentation have been used to supplement food-grade substrates such as milk or whey,[13] kimchi, mashed potatoes, fresh-cut products, cider, fruit and vegetable juices, canned vegetables, soy milk, and zucchini purée.[74] On the other hand, bacteriocin-producing bacteria have been used as coculture, protective, or starter cultures in fermented and nonfermented vegetables such as olives, sourdough, miso, sauerkrauts, refrigerated pickles, mung bean sprouts,[73] cheese, yogurt, and Portuguese fermented meat.

A method to prepare food packing films with bacteriocins added is to incorporate them directly into the films made with biodegradable proteins (eg, zeins of maize) or soybean. In addition, bacteriocin can be adsorbed or used to recover the surface polymer (eg, nisin adsorption on polyethylene), or on other polymers such as ethylene, polypropylene, vinyl acetate, polyamide, polyvinyl chloride, and polyester acrylic.[67] It has been demonstrated that combinations of storage temperature near the freezing point, in addition to antimicrobial packages, are very effective for maintaining the high microbiological quality of meat, inhibiting bacteria species such as

Brochothrix thermosphacta.[75] Moreover, nisin showed effective action against *Listeria monocytogenes*, *S. aureus*, *Bacillus cereus,* and other pathogens.[13]

Presently, both nisin and pedicin PA-1/AcH have been authorized for use as bio-preservatives.[76] Under the name Nisaplin,[77] this commercial product includes 2.5% of nisin, 7.5% of NaCl, and fat-free dry milk (12% protein and 6% carbohydrates). Results of toxicological studies demonstrated that nisin intake causes no toxic effects to human beings, and 6950 mg/kg was estimated as the lethal dose.[13] On the other hand, pediocin PA-17Arch has been employed as ALTA 2431 (Quest), which is obtained by *Pediococcus acidilactici* during fermentation.[60] Natamycin (under the commercial name of Natamax) inhibits fungi and yeast, and it is used in the commercial production of grated cheese.[78] Other commercial product based on bacteriocins is AvGard, which can inhibit Gram-negative pathogens and is used in meat products. These products are sold with different formulations, and they are used to produce foods such as meat, seafood, cheese, bread, and beverages.[77]

Pediocin AcH is used to control *Listeria monocytogenes* growth in the vacuum packing of Vienna sausages, where nitrites are used to stabilize the meat's red color and inhibit the development of spore microorganisms such as *Clostridium botulinum.* However, nitrites can react with meat amine compounds and produce nitrosamines, which are carcinogenic compounds. In this case, bacteriocins provide an alternative to nitrite use. In addition, bacteriocins have stability at cooking temperatures and are well digested (because of its protein nature) by the gastrointestinal tract proteases.[72] Actually, different types of bacteriocins have been used successfully in the food industry, but they are only a small fraction of the diversity available with bacteriocins. Continuing research will lead to obtaining more diverse and efficient antimicrobial bacteriocins.[79]

MULTIDRUG-RESISTANT BACTERIA

Strictly, the term multidrug-resistant bacteria (MDRB) refers to resistance to more than one antimicrobial agent.[80] But there are other definitions used for MDR in both Gram-positive and Gram-negative bacteria.[81] One is extensively drug-resistant bacteria (XDR), a term that first was used to describe extensively drug-resistant *Mycobacterium tuberculosis* (XDR MTB), and was defined as follows: resistance to the first-line drugs isoniazid and rifampicin, to a fluoroquinolone, and to at least one of the three second-line parenteral agents.[81,82] Pandemic drug resistance (PDR) is another definition frequently used for MDR organisms. It comes from the Greek prefix *pan,* meaning "all," so PDR means resistant to all antimicrobial agents. Because it is virtually impossible to be resistant to all antimicrobial agents, there are other examples of common definitions for PDR: resistant to almost all commercially available antimicrobials, and resistant to all antimicrobials that have been routinely tested.[83,84]

The clinical importance of MDR bacteria is that antibiotic-resistant organisms are a major public health problem in both developing and developed countries[85]. For example, in 2008, 28% of the population infected with tuberculosis (TB) in Russia presented with the MDR form of the illness (MDR TB), representing the highest level reported for the WHO.[86] It is estimated that almost 45% of the intrahospital infections are caused by *E. coli*, *P. aeruginosa*, *Acinetobacter* spp., *S. aureus,* and other bacteria previously reported with PDR, like *M. tuberculosis, Acinetobacter baumannii*, and *Klebsiella* spp.[87]

A strategy to control MDR organisms and delay the illnesses caused by them is the development of new drugs like a novel β-lactamase inhibitor such as avibactam.[88] A new cephalosporin combined with a β-lactamase inhibitor was recently approved by the FDA for anfractuous urinary and intraabdominal infections.[89] Other alternatives of treatments are the use of traditional medicine with nonedible plants in the form of extracts or infusions, because as reported by Rath and Padhy,[87] this type of medicine is often the most accessible means of health care in the developing countries of Asia, Africa, and Latin America. Nevertheless, a few drugs with hopeful prospects recently have been approved or are in development to combat of MDR infections. This bacterial control is a challenge because of the absence of trustworthy comparators.[90]

POTENTIAL OF BACTERIOCINS FOR DEVELOPMENT OF NOVEL ANTIBIOTICS

Bacteriocins have the potential for biotechnological applications in food preservation and infectious disease treatment.[27,91] There are a large number of infectious diseases that are difficult to treat with traditional antibiotics, mainly because of resistance of the strains to these compounds and horizontal gene transfer among bacterial strains. Bacteriocins are a good option against drug-resistant pathogens such as *P. aeruginosa, E. coli*, methicillin-resistant *Streptococcus aureus*, penicillin-resistant *Streptococcus pneumoniae,* and *M. tuberculosis*, because these AMPs have high potency (acting at picomolar and nanomolar concentrations) and specificity.[28,91]

Bacteriocins as a source of new antibiotics have the advantages of diversity, specific targets, being considered safe (food grade), the ability to act synergistically with antibiotics, and reduced cell toxicity. However, it has been pointed out that depending on their size, they may be susceptible to proteolytic enzymes and some may become toxic to mammalian cells, which are drawbacks.[92] Several studies indicate that bacteriocins present great potential in the clinical area and as therapeutic agents because they can be used as substitutes for antimicrobial agents or in combination therapy with traditional antibiotics; this is due to their high activity against pathogenic microorganisms and their stability.[28,93] Some in vivo studies in animal models such as mice have shown that bacteriocins can be a potential therapeutic agent against recurrent infections by *M. tuberculosis, Streptococcus pyogenes, S. aureus,* and *S. pneumoniae*.[28]

The direct use of bacteriocin-producing microorganisms such as *Lactobacillus* spp., *Bifidobacterium* spp., and nonpathogenic *E. coli* may be more effective for pathogen bacterial control than the use of pure bacteriocins. However, special measures should be taken to control the applications in order to prevent the development of resistance (due to horizontal gene transfer) and promote the use of bacteriocin cocktails.[23,28,92] A solution to prevent the degradation or inhibition of bacteriocins by the action of gastric enzymes can be their encapsulation with various gums or safe polymers such as alginate.

There are other important applications of bacteriocins as well, such as the treatment of cancer.[94] It has been shown that colicin A and E1 bacteriocins inhibit the growth of fibroblasts in tumor cell lines; also, the effectiveness of nisin to increase apoptosis in carcinoma cells has been verified. In addition, producers of bacteriocin probiotic supplements can prevent the onset of cancer.[23] However, the use of unpurified protein cell extract from bacteriocins can promote the resistance of microorganisms to these molecules. Therefore, during the development of new antibiotics, it is appropriate to administer purified bacteriocins in a somewhere safe carrier or excipient, which protects them against enzymes.

EVALUATION OF BACR AND SUSCEPTIBILITY OF MICROORGANISMS

Most bacteriocins exhibit a spectrum of inhibitory activity on different genera and species of pathogens in humans, animals, and plants. Bacteriocins include a variety of peptides and proteins with different microbial targets, mechanisms of action, and immunity levels.[26] The sensitivity of a microorganism against a bacteriocin depends on ecological conditions such as pH, salt concentration, the presence of molecules in the membrane, and induction culture.[95] However, bacteriocins can act at picomolar and nanomolar concentrations against microorganisms resistant to antibiotics,[26,91] and being cationic, they can interact with anionic components of the bacterial surface, promoting pore formation, permeability of membrane, and eventually bacteria death; in addition, they may have functions similar to DNase, RNase, and peptidoglycanase.[23]

Bastos et al.[26] stated that the mechanisms of resistance to bacteriocins found in Gram-positive bacteria involve changes in the bacterial cell envelope such as (1) reduction or loss of bacteriocin binding, (2) bacteriocin sequestering, (3) bacteriocin efflux pumping (export), and (4) bacteriocin degradation. On the other hand, bacteriocins can be susceptible to different enzymes, and even bacteriocin-producing microorganisms may have mechanisms that can protect their own bacteriocins,[23] such as gene-encoding proteins that confer immunity.[23,28] However, these immunity mechanisms are poorly studied. In class I bacteriocins such as nisin, immunity has been discussed because there are two mechanisms that work together to protect the bacteriocin-producing cell: one ABC system transports the bacteriocin outside the membrane, and the other mechanism includes an immunity protein that interacts specifically with extracellularly analog bacteriocin, preventing receptor-binding membranes.[23]

Purified bacteriocins when ingested along the gastrointestinal tract can be susceptible to the action of different proteolytic enzymes. However, the consumption of bacteriocin-producing probiotic strains allows direct interaction of these peptides in the intestine with pathogenic microorganisms.[28] Some examples of bacteriocin sensitivity to proteolytic enzymes have been cited by Beshkova and Frengova,[27] such as sensitivity of enterocin, pediocin, acidocin, and plantaricin, while other bacteriocins such as nisin Z, thermophilin, and plantaricin are sensitive to pepsin, trypsin, and papain. However, there cannot be generalized sensitivity of bacteriocins against specific enzymes because it can be specific, as in the case of inactivation of nisin by proteinase K and actinase E, while nisin A is stable to proteinase K and actinase E.

Resistance of microorganisms to AMPs may develop.[26] The resistance mechanisms involve changes in the surface of cells that affect the susceptibility of the bacteriocins, as in the case of nisin and in some cases of mutations that reduce the affinity of the interaction between the bacteriocin and the receptor in the membrane, and reduction of the expression of genes for receptor-binding bacteriocins, such as the Man-PTS receptor in *Listeria monocytogenes*.[28]

There have been reports of in vivo inhibition and reduction of microorganisms by bacteriocins, such as reduction of vancomycin-resistant *Enterococcus* in mouse intestine by bacteriocins such as pediocin PA-1 and la nisin Z produced by *Pediococcus acidilactici* and *L. lactis*, respectively.[96] In addition, interesting results have been observed in vitro as 58% reduction of *Giardia lamblia* trophozoites for bacteriocin produced by *L. acidophilus* (50 μg) and the 81.63% reduction in vivo in mice after 5 doses, promoting an alternative therapy against giardiasis.[93] Moreover, Ruiz et al.[97] proposes the use of a bacteriocin-like inhibitory substance (BLIS) L23 (from *Lactobacillus fermentum*) and L60 (from *Lactobacillus rhamnosus*) for designing potentially new bioproducts against *Neisseria gonorrhoeae*. These BLIS inhibited the growth of *N. gonorrhoeae* about 80% under in vitro conditions.

The degree of pathogen sensibility to different bacteriocins differs. It was shown that the antibacterial efficiency of the bacteriocins nisin, pediocin 34, and enterocin F99 against Gram-positive and pathogenic bacteria was different. Pediocin 34 was more effective than nisin and enterocin F99 against *Listeria monocytogenes*.[98] The control effectiveness of pathogen microorganisms by bacteriocins may vary because the time of contact, exposure of the pathogen to the bacteriocin, and nutritional conditions during bacteriocin-producing growth. Other assessments of inhibition of bacteria associated with mastitis in cows and goats by the action of bacteriocins of *Bacillus thuringiensis* proposed a curative formulation for application in cattle based on bacteriocins such as morricin 269, kurstacin 287, kenyacin 404, entomocin 420, and tolworthcin 524.[99]

CONCLUSIONS AND FUTURE PERSPECTIVES

There is a large number of infectious diseases that are difficult to treat with traditional antibiotics, mainly because of resistance of the pathogenic strains to these compounds and horizontal gene transfer among bacterial strains. Bacteriocins could constitute efficient new antibiotics with the advantages of diversity, specific targets, safety (food grade), ability to act synergistically with traditional antibiotics, and reduced cell toxicity. However, depending on their size, they may be susceptible to proteolytic enzymes and some of them may become toxic to mammalian cells. Nowadays, bacteriocins have been widely used, especially in the field of food preservation. These bacteriocins have been produced by LAB and have the potential to cover a very broad range of applications, including the food industry and the medical sector. But some cautions should be taken to avoid the development of resistance to these compounds in pathogens by horizontal transfer. Crude or purified proteins should be used to control MDRB instead of bacteriocin-produced microorganisms. Furthermore, depending on only one or few bacteriocins may increase the risk of developing resistance in the pathogens to those bacteriocins, and so diversity of bacteriocins should be investigated, especially in LAB microorganisms from traditional fermented foods and beverages. In addition, a disease should be treated with a cocktail of bacteriocins instead a single, purified bacteriocin. Fusion proteins may be an excellent tool to use against MDR pathogens. Our research found that still more effective methods are needed for the separation and purification of bacteriocins. Some of the methods and results described in this chapter may lead to finding other GRAS bacteriocins.

REFERENCES

1. Sahoo TK, Jena PK, Patel AK, Seshadri S. Purification and molecular characterization of the novel highly potent bacteriocin TSU4 produced by *Lactobacillus animalis* TSU4. *Appl Biochem Biotechnol* 2015;**177**:90–4.
2. Singh NP, Tiwari A, Bansal A, Thakur S, Sharma G, Gabrani R. Genome level analysis of bacteriocins of lactic acid bacteria. *Comput Biol Chem* 2015;**56**:1–6.

3. Cavera VL, Arthur TD, Kashtanov D, Chikindas ML. Bacteriocins and their position in the next wave of conventional antibiotics. *Int J Antimicrob Agents* 2015;1–8. In Press. http://dx.doi.org/10.1016/j.ijantimicag.2015.07.011
4. Cintas LM, Casaus MP, Herranz C, Nes IF, Hernández PE. Review: bacteriocins of lactic acid bacteria. *Food Sci Technol Int* 2001;**7**(4):281–305.
5. Rojas C, Vargas P. Bacteriocinas: sustituto de preservantes tradicionales en la industria alimentaria. *Tecnol Marcha* 2008;**21**(2):17.
6. Gálvez A, López R, Pulido L, Pérez R, Burgos M, Grande J. Application of lactic acid bacteria and their bacteriocins for food biopreservation. In: Hartel RW, Rodriguez D, Clark P, Topping D, Finley J, Roos Y, editors. *Food biopreservation*. New York, NY, Heidelberg, Dordrecht, and London: Springer; 2014. p. 15–22.
7. Zacharof MP, Lovitt RW. Bacteriocins produced by lactic acid bacteria: a review. *APCBEE Proc* 2012;**2**:50–6.
8. Todorov SD, Prévost H, Lebois M, Dousset X, LeBlanc JG, Franco BD. Bacteriocinogenic *Lactobacillus plantarum* ST16Pa isolated from papaya (*Carica papaya*)—from isolation to application: characterization of a bacteriocin. *Food Res Int* 2011;**44**(5):1351–63.
9. Duhan JS, Nehra K, Gahlawat SK, Saharan P, Duhan S. Bacteriocins from lactic acid bacteria. In: Salar RK, Gahlawat SK, Swach P, Duhan JS, editors. *Biotechnology. Prospects and applications*. New Delhi: Springer; 2013. p. 127–41.
10. Da Silva Sabo S, Vitolo M, González JMD, de Souza Oliveira RP. Overview of *Lactobacillus plantarum* as a promising bacteriocins producer among lactic acid bacteria. *Food Res Int* 2014;**64**:527–36.
11. Pisano MB, Fadda ME, Melis R, Ciusa ML, Viale S, Deplano M, et al. Molecular identification of bacteriocins produced by *Lactococcus lactis* dairy strains and their technological and genotypic characterization. *Food Control* 2015;**51**:511–18.
12. Macwana S, Muriana PM. Spontaneous bacteriocin resistance in *Listeria monocytogenes* as a susceptibility screen for identifying different mechanisms of resistance and modes of action by bacteriocins of lactic acid bacteria. *J Microbiol Methods* 2012;**88**(1):7–13.
13. Balciunas EM, Martinez FAC, Todorov SD, de Melo Franco BDG, Converti A, de Souza Oliveira RP. Novel biotechnological applications of bacteriocins: a review. *Food Control* 2013;**32**(1):134–42.
14. Güllüce M, Karadayı M, Barış O. Bacteriocins: promising natural antimicrobials. In: Méndez-Vilas A, editor. *Microbial pathogens and strategies for combating them: science, technology and education*. Extremadura: Formatex; 2013. p. 1016–27.
15. Gillor O, Kirkup BC, Riley MA. Colicins and microcins: the next generation antimicrobials. *Adv Appl Microbiol* 2004;**54**(1):129–46.
16. Bakkal S, Robinson SM, Ordonez CL, Waltz DA, Riley MA. Role of bacteriocins in mediating interactions of bacterial isolates taken from cystic fibrosis patients. *Microbiology* 2010;**156**:2058–67.
17. Chavan MA, Riley MA. Molecular evolution of bacteriocins in Gram-negative bacteria. In: Riley MA, Chavan MA, editors. *Bacteriocins: ecology and evolution*. Heidelberg: Springer; 2007. p. 5–18.
18. Cascales E, Buchanan SK, Duche D, Kleanthous C, Lloubes R, Postle K, et al. Colicin biology. *Microbiol Mol Biol Rev* 2007;**71**:158–229.
19. Michel-Briand Y, Baysse C. The pyocins of *Pseudomonas aeruginosa*. *Biochimie* 2002;**84**:499–510.
20. Nakayama K, Takashima K, Ishihara H, Shinomiya T, Kageyama M, Kanaya S, et al. The R-type pyocin of *Pseudomonas aeruginosa* is related to P2 phage, and the F-type is related to lambda phage. *Mol Microbiol* 2000;**38**:213–31.
21. Heng NCK, Wescombe PA, Burton JP, Jack RW, Tagg JR. The diversity of bacteriocins in Gram-positive bacteria. In: Riley MA, Chavan MA, editors. *Bacteriocins: ecology and evolution*. Berlin and Heidelberg: Springer-Verlag; 2007. p. 45–83.
22. Perez RH, Zendo T, Sonomoto K. Novel bacteriocins from lactic acid bacteria (LAB): various structures and applications. *Microb Cell Fact* 2014;**13**(1):1–13.

23. Yang S-C, Lin C-H, Sung CT, Fang JY. Antibacterial activities of bacteriocins: application in foods and pharmaceuticals. *Front Microbiol* 2014;**5**(241):1–10.
24. Thokchom S, Joshi SR. Probiotic and bacteriocin efficacy of lactic acid bacteria from traditional fermented foods: a review. *Assam Univ J Sci Technol* 2012;**10**(1):142–55.
25. Swetwiwathana A, Visessanguan W. Potential of bacteriocin-producing lactic acid bacteria for safety improvements of traditional Thai fermented meat and human health. *Meat Sci* 2015;**109**:101–5.
26. Bastos MCF, Coelho MLV, Santos OCS. Resistance to bacteriocins produced by Gram-positive bacteria. *Microbiology* 2015;**161**:683–700.
27. Beshkova D, Frengova G. Bacteriocins from lactic acid bacteria: microorganisms of potential biotechnological importance for the dairy industry. *Eng Life Sci* 2012;**12**(4):419–32.
28. Hassan M, Kjos M, Nes IF, Diep DB, Lotfipour F. Natural antimicrobial peptides from bacteria: characteristics and potential applications to fight against antibiotic resistance. *J Appl Microbiol* 2012;**113**:723–36.
29. Lim SM. Anti-*Helicobacter pylori* activity of antimicrobial substances produced by lactic acid bacteria isolated from Baikkimchi. *J Korean Soc Appl Biol Chem* 2014;**57**:621–30.
30. Hamilton-Miller JM. The role of probiotics in the treatment and prevention of *Helicobacter pylori* infection. *Int J Antimicrob Agents* 2003;**22**:360–6.
31. Ukeyima MT, Enujiugha VN, Sanni TA. Current applications of probiotic foods in Africa. *Afr J Biotechnol* 2010;**9**:394–401.
32. De Bortoli N, Leonard G, Ciancia E, Merlo A, Bellini M, Costa F, et al. *Helicobacter pylori* eradication: a randomized, prospective study of triple therapy versus triple therapy plus lactoferrin and probiotics. *Am J Gastroenterol* 2007;**102**:951–66.
33. Malfertheiner P, Mgraud F, O'Morain C, Bazzoli F, El-Omar E, Graham D, et al. Current concepts in the management of *Helicobacter pylori* infection. *Aliment Pharmacol Ther* 2002;**16**:167–80.
34. Lü X, Yi L, Dang J, Dang Y, Liu B. Purification of novel bacteriocin produced by *Lactobacillus coryniformis* MXJ 32 for inhibiting bacterial foodborne pathogens including antibiotic-resistant microorganisms. *Food Control* 2014;**46**:264–71.
35. Acuña L, Picariello G, Sesma F, Morero R, Augusto A. A new hybrid bacteriocin, Ent35–MccV, displays antimicrobial activity against pathogenic Gram-positive and Gram-negative bacteria. *FEBS Open Bio* 2012;**2**:12–19.
36. Acuña L, Corbalan NS, Fernandez-No IC, Morero RD, Barros-Velazquez J, Bellomio A. Inhibitory effect of the hybrid bacteriocin Ent35-MccV on the growth of *Escherichia coli* and *Listeria monocytogenes* in model and food systems. *Food Bioprocess Technol* 2015;**8**:1063–75.
37. Zhao L, Wang Y, Wang S, Li H, Huang W, Liao X. Inactivation of naturally occurring microbiota in cucumber juice by pressure treatment. *Int J Food Microbiol* 2014;**174**:12–18.
38. Belgeri A, Adkins SW. Allelopathic potential of invasive parthenium weed (*Parthenium hysterophorus* L.) seedlings on grassland species in Australia. *Allelopathy J* 2015;**36**(1):1–14.
39. Braun V, Pilsl H, Groß P. Colicins: structures, modes of actions, transfer through membranes, and evolution. *Arch Microbiol* 1994;**161**:199–206.
40. Riley MA, Wertz JE. Bacteriocins: evolution, ecology and application. *Annu Rev Microbiol* 2002;**56**:117–37.
41. Riley MA, Gordon DM. A survey of col plasmids in natural isolates of *Escherichia coli* and an investigation into the stability of col-plasmid lineages. *J Gen Microbiol* 1992;**138**:1345–52.
42. Riley MA. Bacteriocins, biology, ecology, and evolution. In: Schaechter M, editor. *Encyclopedia of microbiology*. Oxford: Elsevier; 2009. p. 32–44.
43. Osborne M, Lian L, Wallis R, Reilly A, James R, Kleanthous C, et al. Sequential assignments and identication of secondary structure elements of the colicin E9 immunity protein in solution by homonuclear and heteronuclear NMR. *Biochemistry* 1994;**33**:12347–55.

44. Viejo M, Gargallo D, Ferrer S, Enfedaque J, Regue M. Cloning and DNA sequence analysis of a bacteriocin gene of *Serratia marcescens*. *J Gen Microbiol* 1992;**138**:1737–43.
45. Riley MA, Cadavid M, Collett M, Neely M, Adams M, Philips C, et al. The newly characterized colicin Y provides evidence of positive selection in pore-former colicin diversification. *Microbiology* 2000;**146**:1671–7.
46. Tan Y, Riley MA. Nucleotide polymorphism in colicin E2 gene clusters: evidence for non-neutral evolution. *Mol Biol Evol* 1997;**14**:666–73.
47. Kuroda K, Kageyama M. Comparative study of F-type pyocins of *Pseudomonas aeruginosa*. *J Biochem* 1981;**89**:1721–36.
48. Torreblanca M, Meseguer I, Ventosa A. Production of halocin is a practically universal feature of archaeal halophilic rods. *Lett Appl Microbiol* 1994;**19**:201–5.
49. Jack RW, Tagg JR, Ray B. Bacteriocins of Gram-positive bacteria. *Microbiol Rev* 1995;**59**:171–200.
50. Shand R, Price L, O'Connor. Halocins: protein antibiotics from hypersaline environments. In: Oren A, Raton FLB, editors. *Microbiology and biogeochemistry of hypersaline environments*. Boca Raton, FL: CRC Press; 1998. p. 297–304.
51. Gutiérrez J, Larsen R, Cintas LM, Kok J, Hernández PE. High-level heterologous production and functional expression of the sec-dependent enterocin P from *Enterococcus faecium* P13 in *Lactococcus lactis*. *Appl Microbiol Biotechnol* 2006;**17**:1–11.
52. Van Reenen CA, Chikindas ML, Van Zyl WH, Dicks LMT. Characterization and heterologous expression of a class IIa bacteriocin, plantaricin 423 from *Lactobacillus plantarum* 423 in *Saccharomyces cerevisiae*. *Int J Food Microbiol* 2003;**81**:29–40.
53. Hastings JW, Sailer M, Johnson K, Roy KL, Vederas JC, Stiles ME. Characterization of leucocin A-UAL 187 and cloning of the bacteriocin gene from *Leuconostoc gelidum*. *J Bacteriol* 1991;**73**:7491–500.
54. Fimland G, Sletten K, Nissen-Meyer J. The complete amino acid sequence of the pediocin-like antimicrobial peptide leucocin C. *Biochem Biophys Res Commun* 2002;**295**:826–7.
55. Blom H, Katla T, Holck A, Sletten K, Axelsson L, Holo H. Characterization, production and purification of leucocin H, a two-peptide bacteriocin from *Leuconostoc* MF215B. *Curr Microbiol* 1999;**39**:43–8.
56. Masuda Y, Ono H, Kitagawa H, Ito H, Mu F, Sawa N, et al. Identification and characterization of leucocyclicin Q, a novel cyclic bacteriocin produced by *Leuconostoc mesenteroides* TK41401. *Appl Environ Microbiol* 2011;**77**:8164–70.
57. Wan X, Saris PEJ, Takala TM. Genetic characterization and expression of leucocin B, a class IId bacteriocin from *Leuconostoc carnosum* 4010. *Res Microbiol* 2015;**166**:494–503.
58. Hechard Y, Derijard B, Letellier F, Cenatiempo Y. Characterization and purification of mesentericin Y105, an anti-Listeria bacteriocin from *Leuconostoc mesenteroides*. *J Gen Microbiol* 1992;**138**:2725–31.
59. Hechard Y, Berjeaud JM, Cenatiempo Y. Characterization of the mesB gene and expression of bacteriocins by *Leuconostoc mesenteroides* Y105. *Curr Microbiol* 1999;**39**:265–9.
60. Ennahar S, Sashihara T, Sonomoto K, Ishizaki A. Class IIa bacteriocins: biosynthesis, structure and activity. *FEMS Microbiol Rev* 2000;**24**:85–106.
61 Biet F, Berjeaud JM, Worobo RW, Cenatiempo Y, Fremaux C. Heterologous expression of the bacteriocin mesentericin Y105 using the dedicated transport system and the general secretion pathway. *Microbiology* 1998;**144**:2845–54.
62 Chikindas ML, Venema K, Ledeboer AM, Venema G, Kok J. Expression of lactococcin A and pediocin PA-1 in heterologous hosts. *Lett Appl Microbiol* 1995;**21**:183–9.
63 Morisset D, Frére J. Heterologous expression of bacteriocins using the mesentericin Y105 dedicated transport system by *Leuconostoc mesenteroides*. *Biochimie* 2002;**84**:569–76.
64. Gutiérrez J, Criado R, Citti R, Martín M, Herranz C, Nes IF. Cloning, production and functional expression of enterocin P, a sec-dependent bacteriocin produced by *Enterococcus faecium* P13, in *Escherichia coli*. *Int J Food Microbiol* 2005;**103**:239–50.

65. Kittikun H, Biscola A, El-Ghaish V, Jaffrès S, Dousset E, Pillot G, et al. Bacteriocin-producing *Enterococcus faecalis* KT2W2G isolated from mangrove forests in southern Thailand: purification, characterization and safety evaluation. *Food Control* 2015;**54**:126–34.
66. An J, Zhu W, Liu Y, Zhang X, Sun L, Hong P, et al. Purification and characterization of a novel bacteriocin CAMT2 produced by *Bacillus amyloliquefaciens* isolated from marine fish *Epinephelus areolatus*. *Food Control* 2015;**51**:278–82.
67. Ghanbari M, Jami M, Kneifel W, Domig KJ. Antimicrobial activity and partial characterization of bacteriocins produced by lactobacilli isolated from Sturgeon fish. *Food Control* 2013;**32**(2):379–85.
68. Deegan LH, Cotter PD, Hill C, Ross P. Bacteriocins: biological tools for biopreservation and shelf-life extension. *Int Dairy J* 2006;**16**:1058–71.
69. Sharma N, Kapoor R, Gautam N, Kumari R. Purification and characterization of bacteriocin produced by *Bacillus subtilis* R75 isolated from fermented chunks of mung bean. *Food Technol Biotech* 2011;**49**:169–76.
70. Samelis J, Bleicher A, Delbes-Paus C, Kakouri A, Neuhaus K, Montel MC. FTIR-based polyphasic identification of lactic acid bacteria isolated from traditional Greek Graviera cheese. *Food Microbiol* 2010;**28**:76–83.
71. Siragusa GR, Cutter CN, Willett JL. Incorporation of bacteriocin in plastic retains activity and inhibits surface growth of bacteria on meat. *Food Microbiol* 1999;**16**:229–35.
72. Chen H, Hoover DG. Bacteriocins and their food applications. *Compr Rev Food Sci Food Saf* 2003;**2**:82–100.
73. Ercolini D, Storia A, Villani F, Mauriello G. Effect of a bacteriocin activated polythene film on *Listeria monocytogenes* as evaluated by viable staining and epifluorescence microscopy. *J Appl Microbiol* 2006;**100**:765–72.
74. Settanni L, Corsetti A. Application of bacteriocins in vegetable food biopreservation. *Int J Food Microbiol* 2008;**31**:123–38.
75. Ercolini D, Ferrocino I, La Storia A, Mauriello G, Gigli S, Masi P, et al. Development of spoilage microbiota in beef stored in nisin activated packaging. *Food Microbiol* 2010;**27**:137–43.
76. Simha BV, Sood SK, Kumariya R, Garsa AK. Simple and rapid purification of pediocin PA-1 from *Pediococcus pentosaceous* NCDC 273 suitable for industrial application. *Microbiol Res* 2012;**167** (9):544–9.
77. Danisco. *Antimicrobials.* <http://www.danisco.com/product-range/antimicrobials/>; 2013 [accessed 13.09.13].
78. Daview EA, Bevis HE, Delves-Broughton J. The use of the bacteriocin, nisin, as a preservative in ricotta type cheeses to control the food-borne pathogen *Listeria monocytogenes. Lett Appl Microbiol* 1997;**24**:343–6.
79. Snyder AB, Worobo RW. Chemical and genetic characterization of bacteriocins: antimicrobial peptides for food safety. *J Sci Food Agric* 2013;**94**:28–44.
80. Magiorakos AP, Srinivasan A, Carey RB, Carmeli Y, Falagas ME, Giske CG, et al. Multidrug-resistant, extensively drug-resistant and pandrug-resistant bacteria: an international expert proposal for interim standard definitions for acquired resistance. *Clin Microbiol Infect* 2012;**3**:268–81.
81. Falagas ME, Karageorgopoulos DE. Pandrug resistance (PDR), extensive drug resistance (XDR), and multidrug resistance (MDR) among Gram-negative bacilli: need for international harmonization in terminology. *Clin Infect Dis* 2008;**46**:1121–2.
82. CDC. Emergence of *Mycobacterium tuberculosis* with extensive resistance to second-line drugs— Worldwide, 2000–2004. *MMWR Morb Mortal Wkly Rep* 2006;**55**:301–5.

83. Kuo LC, Teng LJ, Yu CJ, Ho SW, Hsueh PR. Dissemination of a clone of unusual phenotype of pan drug-resistant *Acinetobacter baumannii* at a university hospital in Taiwan. *J Clin Microbiol* 2004;**42**:1759–63.
84. CDC. Notice to readers: revised definition of extensively drug-resistant tuberculosis. *MMWR Morb Mortal Wkly Rep* 2006;**55**:1176.
85. Freire-Morana L, Aronssona B, Manz C, Gyssensc IC, Sob AD, Monnete DL, et al. Critical shortage of new antibiotics in development against multidrug-resistant bacteria—time to react is now. *Drug Resist Updat* 2011;**14**:118–24.
86. WHO. *World Health Organization.* <http://www.who.int/mediacentre/news/releases/2010/drug_resistant_tb_20100318/es/>; 2015 [accessed 15.09.15].
87. Rath S, Padhy N. Monitoring in vitro antibacterial efficacy of *Terminalia alata* Heyne ex. Roth, against MDR enteropathogenic bacteria isolated from clinical samples. *J Acute Med* 2013;**3**:93–102.
88. Toussaint KA, Gallagher JC. Beta-lactam/beta-lactamase inhibitor combinations: from then to now. *Ann Pharmacother* 2015;**49**:86–8.
89. Lucasti C, Popescu I, Ramesh MK, Lipka J, Sable C. Comparative study of the efficacy and safety of ceftazidime/avibactam plus metronidazole versus meropenem in the treatment of complicated intra-abdominal infections in hospitalized adults: results of a randomized, double-blind, Phase II trial. *J Antimicrob Chemother* 2013;**68**:1183–92.
90. Viale P, Giannella M, Tedeschi S, Lewis R. Treatment of MDR-Gram negative infections in the 21st century: a never ending threat for clinicians. *Curr Opin Pharmacol* 2015;**24**:30–7.
91. Cotter PD, Ross RP, Hill C. Bacteriocins a viable alternative to antibiotics? *Nat Rev Microbiol* 2013;**11**:95–105.
92. Allen HK, Trachsel J, Looft T, Casey TA. Finding alternatives to antibiotics. *Ann NY Acad Sci* 2014;**1323**:91–100.
93. Amer EI, Mossallam SF, Mahrous H. Therapeutic enhancement of newly derived bacteriocins against *Giardia lamblia*. *Exp Parasitol* 2014;**146**:52–3.
94. Lancaster LE, Wintermeyer W, Rodnina MV. Colicins and their potential in cancer treatment. *Blood Cells Mol Dis* 2007;**38**:15–18.
95. Rea MC, Ross RP, Cotter PD, Hill C. Classification of bacteriocins from Gram-positive bacteria. In: Drider D, Rebuffat S, editors. *Prokaryotic antimicrobial peptides: from genes to applications*. New York, NY: Springer; 2011. p. 29–49.
96. Millet M, Cornut G, Dupont C, Shareck F, Archambault D, Lacroix M. Capacity of human nisin and pediocin-producing LAB to reduce vancomycin-resistant *Enterococci* (VRE) intestinal colonization. *Appl Environ Microbiol* 2008;**74**:1997–2003.
97. Ruiz FO, Pascual L, Giordano W, Barberis L. Bacteriocins and other bioactive substances of probiotic lactobacilli as biological weapons against *Neisseria gonorrhoeae*. *FEMS Pathog Dis* 2015;**73**(3):1–10.
98. Kaur G, Singh TP, Malik RK, Bhardwaj A, De S. Antibacterial efficacy of nisin, pediocin 34 and enterocin FH99 against *L. monocytogenes, E. faecium* and *E. faecalis* and bacteriocin cross resistance and antibiotic susceptibility of their bacteriocin resistant variants. *J Food Sci* 2014;**51**(2):233–44.
99. Gutiérrez-Chávez AJ, Martínez-Ortega EA, Valencia-Posadas M, León-Galván MF, de la Fuente-Salcido NM, Bideshi DK, et al. Potential use of *Bacillus thuringiensis* bacteriocins to control antibiotic-resistant bacteria associated with mastitis in dairy goats. *Folia Microbiol (Praha)* 2015;**60**(3):1–9.
100. Le Marrec C, Hyronimus B, Bressolier P, Verneuil B, Urdaci MC. Biochemical and genetic characterization of coagulin, a new antilisterial bacteriocin in the pediocin family of bacteriocins, produced by *Bacillus coagulans* I. *Appl Environ Microbiol* 2000;**66**:5213–20.

CHAPTER

MARINE NATURAL PRODUCTS IN FIGHTING MICROBIAL INFECTIONS

17

G.D. Mogoşanu[1], A.M. Grumezescu[2], L.E. Bejenaru[1] and C. Bejenaru[3]

[1]*Department of Pharmacognosy & Phytotherapy, Faculty of Pharmacy, University of Medicine and Pharmacy of Craiova, Craiova, Romania* [2]*Department of Science and Engineering of Oxide Materials and Nanomaterials, Faculty of Applied Chemistry and Materials Science, Politehnica University of Bucharest, Bucharest, Romania* [3]*Department of Vegetal & Animal Biology, Faculty of Pharmacy, University of Medicine and Pharmacy of Craiova, Craiova, Romania*

INTRODUCTION

Although it is a natural phenomenon, antibiotic resistance remains one of the most topical and pressing issues of the modern world because it limits the effectiveness of antibacterial agents to treat infections. Nowadays, approximately 90% of antibiotics approved for clinical use are natural substances or semisynthetic derivatives starting from a natural precursor. Therefore, the screening for novel antimicrobial compounds is turning to the exploration of natural resources, through combinatorial chemistry, data mining, microbial metagenomic analysis, improved microbial cultivation, and semisynthetic techniques.[1–3]

Marine resources biosynthesize a wide variety of active ingredients (bioactive secondary metabolites) with different pharmacological properties: cytostatic, anti-inflammatory, antibacterial, antifungal, antiviral, photoprotective, etc. Recent studies have revealed the importance of marine-derived bacteria, fungi, algae, sponges, and other invertebrates (eg, corals, dinoflagellates, mollusks, gastropods, and tubeworms) as producers of novel potent antibiotics useful in fighting microbial infections.[1,2] Due to their peculiar ecology, biosynthetic evolution, and structural complexity, marine natural compounds can penetrate cell membranes and interact with specific protein targets.[4]

Modern analytical methods that ensure the use of small amounts of natural products and easy sample preparation are very important for the isolation, purification, and physicochemical characterization of novel marine-derived antimicrobial compounds: for example, ultra-high-performance liquid chromatography (HPLC) coupled to high-resolution mass spectrometry, desorption electrospray ionization-mass spectrometry, and capillary probe nuclear magnetic resonance spectrometry.[5–7]

The most important marine-derived antimicrobial compounds include a wide range of chemical structures, depending on the taxonomic origin of species:

- *Marine bacteria*: Actinomycetes secondary metabolites (abyssomicin C, arenimycin, bahamaolides A and B, bonactin, chandrananimycins A–C, etc.), cyanobacterial alkaloids (hapalindoles), marine bacteria exopolysaccharides, glycolipids, bromopyrrole, chloropyrrole and bromophenyl derivatives, lipopeptides

Antibiotic Resistance. DOI: http://dx.doi.org/10.1016/B978-0-12-803642-6.00017-4

- *Marine fungi:* Cephalosporins, anthrones, anthraquinones, polyketides, alkaloids, peptides, sulfoalkylresorcinols, aflatoxins, diterpenoids, steroids, glicosphingolipids
- *Marine algae:* Monopolyunsaturated/polyunsaturated antimicrobial fatty acids, bromophenol derivatives, sesquiterpenoid quinones (peyssonoic acids A and B), diterpenoids (sargafuran), cyclic depsipeptides (massetolide A)
- *Marine sponges:* Fatty acids, brominated phenols, sesquiterpenoid quinones (puupehanol, puupehenone), diterpenoids (aplysin, aplysinol, agelasines A and B, kalihinol A), sesterterpenoids (hyrtiosal, heteronemin), macrolides (halichondramide, neopeltolide, spongistatin 1), sulfated sterols (eurysterols A and B, halistanol), meroterpenoids (fascioquinols A–F, alisiaquinones A–C), bromopyrrole alkaloids (ageliferin, bromoageliferin, benzosceptrin C, nagelamides, oroidin), steroidal alkaloids (plakinamines A and B), tetracyclic alkylpiperidine alkaloids (haliclonin A, halicyclamine A, haliclonacyclamine A and B), brominated spiro-cyclohexadienylisoxazolines (aerothionins), lipopeptides (trichoderins A, A1, and B, maribasins A and B), cyclodepsipeptides (mirabamides A–D)
- *Other marine invertebrates:* Corals (diterpenoids, 8-hydroxybriarane derivatives), dinoflagellates (goniodomin), mollusks (polypeptides, kahalalide A), gastropods (peptides, dolabellanin B2), tubeworms (cyclic depsipeptides, viscosin)

MARINE BACTERIA

Marine bacteria are natural sources for important antimicrobial active ingredients with potential therapeutic uses against human pathogens, such as actinomycetes secondary metabolites (abyssomicin C, arenimycin, bahamaolides A and B, bonactin, chandrananimycins A–C, chloro-dihydroquinones, frigocyclinone, glaciapyrroles A–C, gutingimycin, himalomycins A and B, lajollamycin, marinomycins A–D, trioxacarcins), cyanobacterial alkaloids (hapalindoles–fischambiguines A and B, ambiguine P, ambiguine G nitrite, ambiguine Q nitrite) and nonribosomal peptides, marine bacteria exopolysaccharides, glycolipids, bromopyrrole, chloropyrrole (pyrrolnitrin) and bromophenyl compounds, and lipopeptides (6-aminobutyric fengycin derivatives, maribasins A and B, tauramamide).

ACTINOMYCETES

Marine actinomycetes are spread widely across various marine ecosystems, as demonstrated by culture-dependent and culture-independent modern techniques.[8] In the last few decades, marine actinomycetes were recognized as predominantly deep-sea natural sources of active ingredients for therapeutic purposes, including novel antibacterial compounds, such as terpenoids, amino acid derivatives, polyenelike macrolides, and 20S proteasome inhibitors (salinosporamides A and B) isolated from *Salinispora* sp. (Micromonosporaceae).[3,9,10]

Modern molecular technologies (genetic engineering, genome mining) and improved cultivation procedures have revealed the great bio- and chemo-diversity of actinomycetes in terms of microbial secondary metabolites and their importance to drug discovery. Actually, the majority (c. 70%) of microbial secondary metabolites are biosynthesized by actinomycetes: for example, 50–55% of the known antibiotics derive mainly from *Streptomyces* sp.[11,12]

During the last several years, different novel antibacterial metabolites were isolated from marine actinomycetes: chandrananimycins A–C (*Actinomadura* sp.),[13] chloro-dihydroquinones (*Actinomyces* sp.),[14] helquinoline (*Janibacter limosus*),[15] marinomycins A–D (*Marinispora* sp.),[16] frigocyclinone (*Streptomyces griseus*),[17] lajollamycin (*Streptomyces nodosus*),[18] and also bonactin,[19] glaciapyrroles A–C,[20] gutingimycin,[21] himalomycins A and B,[22] and trioxacarcins[23] from different *Streptomyces* sp.

Salinipyrones A and B, novel polyketides isolated from *Salinispora pacifica*, which is a phylogenetically unique strain of the obligate marine actinomycete, exhibited lower antibacterial activity against antibiotic-resistant human pathogens.[24]

A novel polycyclic polyketide abyssomicin C and some of its derivatives, produced by a marine *Verrucosispora* strain (AB-18-032), highlighted potent antibacterial activity against Gram-positive bacteria, especially on clinical isolates of multiple-resistant and vancomycin-resistant *Staphylococcus aureus*.[25–27] Sporolides A and B, two macrolides with a large number (23 out of 24 carbon skeletons) of oxidized carbons, isolated from *Salinispora tropica* strains, showed lower in vitro antibacterial activity.[28] Marinomycins A and B, other polyenelike macrolides synthesized by marine actinomycetes (*Marinophilus* sp.), are antitumor antibiotics with antibacterial activity against vancomycin-resistant *Enterococcus faecium* (VREF) and methicillin-resistant *S. aureus* (MRSA), with minimum inhibitory concentrations (MICs) of 0.125–0.625 μg/mL.[29]

Three terpenoid chloro-dihydroquinones exhibiting antibacterial activity against MRSA and VREF were isolated from a new genus of actinomycete from deepwater sediment (La Jolla, CA).[1,14] Arenimycin, a novel benzo[α]naphthacene quinone derivative isolated from a *Salinispora arenicola* strain, an obligate marine actinomycete, was found to be active against MRSA and rifampin-resistant *S. aureus*, drug-resistant *Staphylococcus* sp. and various other Gram-positive human pathogens.[30] In addition, a farnesylated dibenzodiazepinone, diazepinomicin (ECO-4601), isolated from a *Micromonospora* strain, exhibited antibacterial, anti-inflammatory, and antitumor activity.[31]

In the last several years, other important antimicrobial compounds were isolated from marine actinomycetes, such as the following:

- Exopolysaccharides from coral-associated bacteria, actinomycetes growing on the mucus of the coral *Acropora digitifera* (Dana, 1846),[32–34] active against biofilm-forming human pathogens (*S. aureus*, *Streptococcus pyogenes*)[35]; benzopyran derivatives, 7-methylcoumarin, and two flavonoids (rhamnazin, cirsimaritin) from a marine *Streptomyces* sp.[36];
- Chroman derivatives, ammonificins A and B, from the marine hydrothermal vent bacterium *Thermovibrio ammonificans*[37];
- Kandenols A–E, eudesmane-type sesquiterpenoids from an endophytic *Streptomyces* sp. growing on mangrove stem *Kandelia candel* (L.) Druce, originate from Xiamen (Fujian, China), moderately active against *Bacillus subtilis* and *Mycobacterium vaccae*[38];
- Xiamycin A, dixiamycin A and B, oxiamycin and chloroxiamycin, antibacterial indolosesquiterpenes from a *Streptomyces* sp. (sediment, South China Sea)[39];
- Bahamaolides A and B, macrocyclic lactones from a *Streptomyces* sp. (sediment, North Cat Cay, Bahamas), strong inhibitors of *Candida albicans* isocitrate lyase[40];
- Marinopyrroles A and B, bromopyrrole alkaloids with uncommon bispyrrole structure, from an obligate marine *Streptomyces* strain, highly active against MRSA[41];

- Chitinase-active polysaccharides from a marine *Streptomyces* sp. DA11 associated with the South China Sea sponge *Craniella australiensis* (Carter, 1886), exhibiting antifungal effects against *Aspergillus niger* and *C. albicans*[42];
- Antimycins B1 and B2 from *Streptomyces lusitanus* growing on mangrove sediment, *Avicennia marina* (Forssk.) Vierh., collected in Fujian (China), with moderate activity against *S. aureus* and *Laribacter hongkongensis*.[43]

CYANOBACTERIA

Cyanobacteria, also known as *blue-green algae*, are marine-derived prokaryotic sources containing large, structurally diverse, active ingredients/secondary metabolites with antibacterial, antifungal (alkaloids, nonribosomal peptides), and cytotoxic/antiproliferative/antitumoral (apratoxins, aurilides, curacin A, dolastatin 10, hectochlorin, lyngbyabellins) properties against primary rat hepatocytes and HL-60 cells, and anti-inflammatory, antiparasitic (almiramides, dragonamide E, gallinamide A, viridamides A and B), antiviral, and sodium channel modulatory properties.[3,44–48] Considering that they are nitrogen-containing compounds, cyanobacterial alkaloids are biosynthesized by mixed biosynthetic pathways, mainly through large multimodular nonribosomal polypeptide (NRP) or mixed polyketide–NRP enzymatic systems.[46,47] Hapalindoles, such as fischambiguines A and B, ambiguine P, ambiguine Q nitrite, and ambiguine G nitrite, are isonitrile-containing indole alkaloids with polycyclic carbon skeletons biosynthesized by cyanobacteria *Fischerella ambigua* [(Kützing *ex* Bornet & Flahault) Gomont, 1895] through condensation of tryptophan derivatives and geranyl pyrophosphate. At a MIC value of 2 μM, fischambiguine B strongly inhibited biofilm formation and development of *Mycobacterium tuberculosis*.[49]

Antimicrobial effects against Gram-positive and Gram-negative bacteria exhibited nonribosomal peptides produced by Brazilian cyanobacteria[50] and by *Brevibacillus laterosporus* [(Laubach, 1916) Shida et al.] Lh-1 strain harvested from the sea sediment.[51] Pukeleimides A–F, produced by a toxic strain of *Lyngbya majuscula* (Harvey *ex* Gomont, 1892), exhibited antimicrobial activity against *Mycobacterium smegmatis* and *S. pyrogenes*.[52] In addition, antifungal properties against *Candida* sp. were highlighted for a novel compound with an unknown chemical structure isolated from a marine cyanobacteria *Oscillatoria laete-virians* BDU 20801.[53]

OTHER MARINE BACTERIA

Recently, marine bacteria exopolysaccharides have been described as promising antimicrobial natural compounds with a wide range of biotechnological applications.[54,55] Thus, exopolysaccharides from some marine bacteria exhibit antibiofilm activity against human pathogens (*Escherichia coli*, *Pseudomonas aeruginosa*, *Salmonella enterica*, *S. aureus*, *Staphylococcus epidermidis*, *S. pyogenes*): *Pseudoalteromonas* sp. 3J6,[56] antarctic *Pseudoalteromonas haloplanktis* TAC125,[57] *Vibrio* sp. QY101,[58] exopolysaccharides isolated from sponge-associated strains, such as *Bacillus licheniformis* strain associated with the Mediterranean marine organism *Spongia officinalis* L.[59]

Glycolipids isolated from some marine bacteria, such as *Brevibacterium casei* growing on the marine sponge *Dendrilla nigra* and tropical marine strain of *Serratia marcescens* from the hard coral, *Symphyllia* sp., disrupted the preformed biofilm under dynamic conditions. They also highlighted a broad spectrum of antimicrobial properties against mixed pathogens (*C. albicans* BH, *P. aeruginosa* PAO1, marine biofouling bacterium *Bacillus pumilus* TiO1).[60,61]

At a concentration of 0.06 mg/mL, using broth assay testing, a bromopyrrole compound isolated from *Pseudomonas bromoutilis* (shallow water near Puerto Rico) was found to be active against Gram-positive bacteria.[62] In addition, pyrrolnitrin, a chloropyrrole derivative [3-chloro-4-(2-nitro-3-chlorophenyl)-pyrrole] isolated from *Pseudomonas pyrrocinia* exhibited antifungal effects against *Trichophyton* sp. and especially against soil-borne fungal plant pathogens (*Fusarium sambucinum, Rhizoctonia solani*) and foliar fungal pathogens (*Fusarium culmorum* and *Fusarium graminearum*).[63–69] Used for the treatment of superficial dermatophytic infections, pyrrolnitrin (PYRO-ACE, Japan) is also active against mycobacterial pathogens.[70]

Antimicrobial bromophenyl compounds were isolated from different marine bacteria, such as 4,4′,6-tribromo-2,2′-biphenol from an extract of a marine *Pseudoalteromonas* sp. CMMED 290, highly active against MRSA[71]; 2,2′,3-tribromo-biphenyl-4,4′-dicarboxylic acid from *Pseudoalteromonas phenolica* O-BC30T, inactive against Gram-negative bacteria and fungi, but exhibiting significant activity against various MRSA clinical isolates (MICs 1–4 μg/mL), *B. subtilis*, and *Enterococcus serolicida*[72]; bromophenyl derivatives with incompletely elucidated structures from the marine bacterium *Pseudoalteromonas haloplanktis* INH strain.[73]

Anthraquinone derivatives biosynthesized by marine bacterium *Nocardia* sp. ALAA 2000 growing on the marine red alga *Laurencia spectabilis* (Postels & Ruprecht, 1840), harvested from the Ras-Ghârib (Egyptian Red Sea coast), exhibited antimicrobial properties against both Gram-positive and Gram-negative bacteria and fungi (MICs 0.1–10 μg/mL).[74]

From the marine-derived bacteria, antimicrobial peptides were also isolated and purified by bioactivity-guided fractionation:

- Maribasins A and B, two novel cyclic lipopeptides from the fermentation broth of *Bacillus marinus* B-9987 growing on *Suaeda salsa* (L.) Pallas, native on saline soils from Bohai coastline (China), with a broad-spectrum activity against phytopathogens (antifungal bioassay)[75];
- Two 6-aminobutyric fengycin lipopeptides from *Bacillus amyloliquefaciens* SH-B10 (deep-sea sediment), exhibiting antifungal effects against plant pathogens (paper-agar disk-diffusion testing) important for biological control and sustainable agriculture[76];
- Tauramamide and its ethyl ester derivative, a new lipopeptides from *Brevibacillus laterosporus* PNG276 (Papua New Guinea), which demonstrated potent and relatively selective inhibition of pathogenic *Enterococcus* sp.[77];
- In vitro antibacterial active thiopeptide compound with a rare aminoacetone moiety from the fermentation extracts of *Nocardiopsis* sp. TP-1161[78];
- Unnarmicines A and C, new antibacterial depsipeptides from *Photobacterium* MBIC06485, selectively inhibiting the growth of two strains of *Pseudovibrio*[79];
- Miuraenamides A and B, cyclic hybrid polyketide nonribosomal peptides, with macrocyclic structure and β-methoxyacrylate moiety, from a slightly halophilic myxobacteria native in Japan (*Paraliomyxa miuraensis*)[80];

- Ariakemicins A and B, unusual linear hybrid polyketide nonribosomal peptides, composed of threonine, two Ω-amino-(Ω-3)-methyl carboxylic acids with diene/triene units, and δ-isovanilloyl-butyric acid, from a marine gliding bacterium *Rapidithrix* sp., selectively inhibited the growth of Gram-positive bacteria[81];
- Peptidolipins B and E, lipopeptides from *Nocardia* sp. growing on ascidian *Trididemnum orbiculatum* (Van Name, 1902), collected in the Florida Keys (United States), exhibiting moderate activity against MRSA and methicillin-sensitive *S. aureus* (MSSA).[82]

Other antimicrobial compounds isolated from marine-derived bacteria included bile acid derivatives (steroids) from the sponge-associated bacterium *Psychrobacter* sp.,[83] various chemical structures, such as prodigiosin, a bright red tripyrrole pigment from *S. marcescens* (Bizio, 1819),[84–87] and also unknown structures from a variant of the ichthyotoxic *Pseudomonas piscicida* [(*ex* Bein, 1954) Gauthier et al., 1995].[88]

MARINE FUNGI

Marine fungi are also natural resources rich in various antimicrobial compounds, such as cephalosporins, anthrones, anthraquinones, polyketides, alkaloids, peptides, sulfoalkylresorcinols, aflatoxins, diterpenoids, steroids, and glicosphingolipids.

The fungus *Cephalosporium acremonium* (Corda, 1839) isolated from the sea near a sewage outfall of the coast of Sardinia produces some antibiotic substances, such as cephalosporin C (antibiotic N, different from cephalosporin N), active against Gram-negative bacteria[89–92] and cephalosporins P1, P2, P3, P4, and P5, exhibiting good effects in vitro against *Bacillus mesentericus*, *Mycobacterium phlei*, and *S. aureus*.[89,90,92]

Tetrahydrobostrycin and 1-deoxytetrahydrobostrycin, two novel hexahydroanthrones isolated from the marine-derived fungus *Aspergillus* sp. strain 05F16 (Manado Coral Reef, Indonesia), highlighted weak antibacterial activity against *E. coli* and *S. aureus*.[93]

Through bioactivity-guided fractionation, from the fungus *Aspergillus versicolor* [(Vuillemin) Tiraboschi, 1908] derived from a marine sponge *Petrosia* sp. were isolated five anthraquinones with antimicrobial activity against some Gram-positive clinically isolates (MIC values 0.78–6.25 μg/mL).[94] Also, during the last several years, two novel antimicrobial anthraquinones were isolated from marine fungi: monodictyquinone A (1,8-dihydroxy-2-methoxy-6-methylanthraquinone) from the sea urchin-derived fungus *Monodictys* sp.,[95] and isorhodoptilometrin-1-methyl ether from endophytic *A. versicolor* growing on green alga *Halimeda opuntia* [(L.) J.V. Lamouroux, 1816], harvested in Ras Mohamed National Park, South Sinai (Egypt), which exhibits moderate activity against *Bacillus cereus*, *B. subtilis*, and *S. aureus*.[96]

Different antimicrobial polyketides were isolated from marine-derived fungi, such as:

- Antifungal polyketides 7-*O*-methylkoninginin D and trichodermaketones A–D from *Trichoderma koningii* (Oudem, 1902), exhibiting synergistic antifungal activity against *C. albicans*, in combination with 0.05 μg/mL ketoconazole[97];
- Curvularin and α,β-dehydrocurvularin from the ethyl acetate extract of the fungus *Eupenicillium* sp. associated with the marine sponge *Axinella* sp.[98];

- Macrolides (+)-brefeldin A, (+)-brefeldin C and 7-oxobrefeldin A from *Penicillium* sp. PSU-F44, active against *Microsporum gypseum* [(E. Bodin) Guiart & Grigoraki, 1928] and MRSA[99];
- Nigrosporapyrones A–D and nigrospoxydons A–C, antimicrobial macrolides from *Nigrospora* sp. PSV-F18 and PSU-F5[100,101];
- Communols A–G, aromatic polyketides from *Penicillium commune* growing on gorgonian *Muricella abnormalis* (Nutting, 1912) (Danzhou, Hainan, China), moderately active against *Enterobacter aerogenes* and *E. coli*.[102]

Antimicrobial diketopiperazine alkaloids were isolated by bioassay-guided fractionation of extracts from marine-derived fungi: alternarosin A from the ethyl acetate extract of *Alternaria raphani* THW-18 halotolerant fungal strain (Hongdao Sea salt fields), having weak antibacterial activity against *B. subtilis*, *E. coli*, and *C. albicans*[103]; caboxamycin from the deep-sea strain *Streptomyces* sp. NTK937 (Canary Basin), showing inhibitory effects against Gram-positive bacteria[104]; and dimeric diketopiperazine brevianamide S from *A. versicolor* (Bohai Sea sediment, China), with a selective activity against *Mycobacterium bovis*.[105]

Marine-derived fungi can produce antimicrobial peptides, such as:

- Trichoderins A, A1, and B, anti-mycobacterial aminolipopeptides useful against active and dormant bacilli (*M. bovis*, *M. smegmatis*, and *M. tuberculosis*, with MIC values 0.02–2 μg/mL), isolated from a culture of marine sponge-derived fungus of *Trichoderma* sp.[106];
- Sclerotides A and B, two novel cyclic hexapeptides containing both anthranilic acid and dehydroamino acid units, and also sclerotiotides A, B, F, and I, from the marine-derived halotolerant *Aspergillus sclerotiorum* PT06-1, with antibacterial and antifungal action[107,108];
- Alternaramide, a cyclic depsipeptide from *Alternaria* sp. SF-5016, exhibiting weak antibacterial effect against *B. subtilis* and *S. aureus*.[109]

Other antimicrobial compounds biosynthesized by marine-derived fungi include:

- Sulfoalkylresorcinols (phenolic derivatives) from *Zygosporium* sp. KNC52, active against multidrug-resistant bacteria[110];
- Aflatoxin B2b isolated from endogenous *Aspergillus flavus* (Link, 1809) growing on *Hibiscus tiliaceus* L. mangrove root (Wenchang, Hainan, China), moderately active against *B. subtilis*, *E. aerogenes*, and *E. coli*[111];
- Dehydroxychlorofusarielin B, a polyoxygenated decalin (diterpenoid) derivative from *Aspergillus* sp., exhibiting antibacterial activity against MRSA and multidrug-resistant *S. aureus* (MDRSA)[112];
- Ring B aromatic steroids from endophytic fungus *Colletotrichum* sp., with antibacterial and antifungal properties against *Bacillus megaterium*, *E. coli*, and *Microbotryum violaceum* [(Pers.) G. Deml & Oberw., 1982], respectively[113];
- Cerebrosides (alternarosides A–C), glicosphingolipids isolated from the ethyl acetate extract of *Alternaria raphani* halotolerant fungal strain (THW-18) harvested from a marine sediment in the Hongdao Sea salt fields, exhibiting weak antibacterial activity against *B. subtilis*, *E. coli*, and *C. albicans* (MIC values 70–400 μM)[103];
- Asperamides A and B, sphingo-/glicosphingolipids based on 9-methyl-C^{20}-sphingosine, from the endophytic fungus *A. niger* EN-13 developed on the marine brown alga *Colpomenia sinuosa* [(Mertens *ex* Roth) Derbès & Solier, 1851], with antifungal properties against *C. albicans*.[114]

MARINE ALGAE

In the last several years, modern studies have highlighted the importance of marine algae for the isolation and characterization of novel antibacterial and antifungal compounds, with outstanding prospects for therapeutic use: monounsaturated/polyunsaturated antimicrobial fatty acids, bromophenol derivatives, sesquiterpenoid quinones (peyssonoic acids A and B), diterpenoids (sargafuran), and cyclic depsipeptides (massetolide A).

Biofilm formation/development of pathogenic fungi (*C. albicans*, *Candida dubliniensis*)[115,116] and bacteria (*S. aureus*, MRSA)[117] can be inhibited by monounsaturated/polyunsaturated antimicrobial fatty acids [stearidonic acid (18:4, *n*-3), eicosapentaenoic acid (20:5, *n*-3), docosapentaenoic acid (22:5, *n*-3), and docosahexaenoic acid (22:6, *n*-3)] and their brominated/acetylene derivatives extracted and purified from different marine algae. Also, monounsaturated (9*Z*)-hexadecenoic acid and polyunsaturated (6*Z*,9*Z*,12*Z*)-hexadecatrienoic acid and eicosapentaenoic acid, from the marine diatom *Phaeodactylum tricornutum* (Bohlin, 1897), are highly active against both Gram-positive and Gram-negative bacteria, including MDRSA.[118,119]

Through the shikimate pathway, different antimicrobial bromophenol derivatives are biosynthesized by green, brown, and red algae *Symphyocladia gracilis* [(Martens) Falkenberg, 1901], *Rhodomela larix* [(Turner) C. Agardh, 1822] and *Polysiphonia lanosa* [(L.) Tandy, 1931], respectively: 2,3-dibromobenzyl alcohol-4,5-disulfate dipotassium salt, 2,3-dibromo-4,5-dihydroxybenzaldehyde, 2,3-dibromo-4,5-dihydroxybenzyl alcohol, 3,5-dibromo-*p*-hydroxybenzyl alcohol and the 5-bromo-3,4-dihydroxybenzaldehyde.[120] In addition, several bromophenol compounds, together with 4,4′,5,5′-tetrahydroxydiphenylmethane, isolated by bioassay-guided technique from the crude extracts of the red alga *Odonthalia corymbifera* [(S.G. Gmelin) Greville, 1830], proved to be active against some human pathogenic fungi (*Aspergillus fumigatus*, *C. albicans*, *Trichophyton mentagrophytes*, and *Trichophyton rubrum*).[121]

Sesquiterpenoid quinones isolated from marine algae are also active as antimicrobials: peyssonoic acids A and B from red alga *Peyssonnelia* sp. inhibited biofilm formation/growth of some bacterial and fungal pathogens of marine algae, *Pseudoalteromonas bacteriolytica* and *Lindra thalassiae*, respectively[122]; 10-hydroxykahukuene B, a brominated metabolite from the red alga *Laurencia mariannensis* (Yamada, 1931)[123]; tiomanene and acetylmajapolenes A and B from Malaysian *Laurencia* sp.[124] Also, three novel antibacterial and antifungal laurene sesquiterpenoids were extracted from the red alga *Laurencia obtusa* [(Hudson) J.V. Lamouroux, 1813], collected from Red Sea sediments (Jeddah, KSA).[125]

Other important antibacterial and antifungal active ingredients isolated from marine algae include:

- Phlorotannins (phenolic derivatives) from the edible seaweed brown algae *Ecklonia cava* (Kjellman, 1885)[126];
- Sargafuran, a diterpenoid compound from the methanolic extract of the marine brown alga *Sargassum macrocarpum* (C. Agardh, 1820), collected from Japanese coastlines, bactericidal on *Propionibacterium acnes*[127];
- Massetolide A, a cyclic depsipeptide from *Pseudomonas* sp. cultures growing on a marine alga, active against *M. tuberculosis* (MICs 5–10 mg/mL) and *Mycobacterium avium-intracellulare* (MICs 2.5–5 mg/mL)[128];
- Sarganins A and B, two novel antibiotics from *Sargassum tenessimum*.[129]

MARINE SPONGES

Marine sponges are natural sources of remarkable chemical diversity with tremendous potential for therapeutic applications. Taking into account the biosynthesis pathways of secondary metabolites, various broad-spectrum antimicrobial compounds were isolated and purified from marine sponges:

- Fatty acids;
- Brominated phenols;
- Sesquiterpenoid quinones (puupehanol, puupehenone);
- Diterpenoids (aplysin, aplysinol, agelasines A and B, kalihinol A);
- Sesterterpenoids (hyrtiosal, heteronemin);
- Macrolides (halichondramide, neopeltolide, spongistatin 1);
- Sulfated sterols (eurysterols A and B, halistanol);
- Meroterpenoids (fascioquinols A–F, alisiaquinones A–C);
- Bromopyrrole alkaloids (ageliferin, bromoageliferin, benzosceptrin C, nagelamides A–N, Q, R, U–Z, oroidin);
- Steroidal alkaloids (plakinamines A and B);
- Tetracyclic alkylpiperidine alkaloids (haliclonin A, halicyclamine A, haliclonacyclamine A and B);
- Brominated spiro-cyclohexadienylisoxazolines (aerothionins);
- Lipopeptides (trichoderins A, A1, and B, maribasins A and B);
- Cyclodepsipeptides (mirabamides A–D).

Different antimicrobial fatty acids were also isolated from marine sponges:

- A new acetylenic fatty acid from the calcareous sponge *Paragrantia* cf. *waguensis*, active against *E. coli* and *S. aureus* (MICs 128 and 64 μg/mL, respectively)[130];
- Brominated unsaturated fatty acids from *Xestospongia* sp. (Papua New Guinea)[131];
- Motualevic acids A–F from *Siliquariaspongia* sp., inhibiting the growth of *S. aureus* and MRSA.[132]

Brominated phenols or dibromophenol derivatives isolated from the marine sponges of *Dysidea* genus collected from the coast of the Lakshadweep Islands in the Indian Ocean [2-(2′,4′-dibromophenoxy)-4,6-dibromophenol from *Dysidea granulosa*] and from the Federated States of Micronesia (polybrominated diphenyl ether) exhibited in vitro antibacterial activity against vancomycin-resistant and -sensitive enterococci, *Bacillus* sp., MRSA, and MSSA.[133–135] Also, polybrominated diphenyl ethers from the Indonesian sponge *Lamellodysidea herbacea* exert an inhibitory activity against *B. subtilis* and *Streptomyces* 85E (hyphae formation inhibition assay).[136] Other antimicrobial brominated phenols were isolated from the methanolic extract of the sponges *Dysidea herbacea*, Western Caroline Islands (brominated derivatives of 2-phenoxyphenol), *Verongia cauliformis* and *Verongia fistularis* (2,6-dibromo-4-acetamido-4-hydroxycyclohexadienone).[137,138]

Sesquiterpenoid quinones/(di)hydroquinones isolated from marine sponges exhibited antimicrobial properties mainly due to their common 1,4-benzoquinone moiety: puupehanol, puupehenone, and chloropuupehenone (*Hyrtios* sp.)[139] and nakijiquinones G–I from marine sponges of the *Spongillidae* family (Okinawa, Japan),[140] with antifungal properties against *Candida krusei* and *Cryptococcus neoformans* (MICs 1.25–2.5 μg/mL). In addition, puupehenone, 15-cyanopuupehenone, and 15α-cyanopuupehenol inhibited the growth of *M. tuberculosis* H37Rv (99%, 90%, and 96%, respectively).[141–143]

Marine sponges biosynthesized a wide range of antimicrobial diterpenoids, such as:

- Aplysin and aplysinol, a brominated bicyclic diterpenoids from the sea hare, *Aplysia kurodai* (Baba, 1937)[144,145];
- Antibacterial and antifungal tetracyclic furanoditerpenoids from the common bath sponge *Spongia officinalis*–spongia-13(16),14-dien-19-oic acid,[146–149] spongia-13(16)-14-dien-19-al, and spongia-13-(16)-14-diene[150,151];
- Agelasines A and B, with a purine or a 9-methyladenine unit, from an unidentified *Agelas* sp. (Pacific and Caribbean Islands), active against *B. subtilis*, *S. aureus*, *C. albicans*, and the marine bacterium B-392, by their Na^+, K^+-ATP-ase inhibitory effect[146–149];
- Tricyclic diterpenoids with isocyano, hydroxyl, tetrahydropyranyl, and chlorine groups from *Acanthella* sp., for example, kalihinol A, active in vitro against *B. subtilis*, *S. aureus* and *C. albicans*[152,153];
- Agelasidines B and C from the Okinawan sponge *Agelas nakamurai*.[152–154]

Antimicrobial sesterterpenoids are frequently isolated from the following marine sponge extracts:

- Sesterterpenoid sulfates (*Dysidea* sp.) and hyrtiosal (*Hyrtios erectus*) as potent inhibitors of *C. albicans* isocitrate lyase and active against *B. subtilis* and *Proteus vulgaris*[155–157];
- Heteronemin, a scalarin-type sesterterpenoid from the sponge *Heteronema erecta* (Keller, 1889), 99% inhibited *M. tuberculosis* H37Rv (MIC 6.25 mg/mL)[143,158];
- Phyllofolactore A and phyllofoloctone, 20,24-bishomoscalarane sesterterpenoids from the Pacific sponge *Phyllospongia foliascens*, with antifungal and anti-inflammatory properties[159–163];
- Antimycotic phyllofenone A, a 20,24-diethyl-25-norscalarane sesterterpenoid from *Phyllospongia foliascens*[164,165];
- Norsesterterpenoid peroxides from the Red Sea sponge *Prianos* sp., strongly inhibiting the growth of Gram-positive bacteria[166,167];
- Sesterterpenoid antibiotics from *Luffariella variabilis* (Poléjaeff, 1884).[168,169]

Antibacterial and antifungal macrolides (polyketides) were obtained from marine-derived sponges, such as: antifungal halichondramide from *Halichondria* sp.[170]; a novel 22-member macrolide from the Red Sea sponge *Theonella swinhoei*[171]; neopeltolide from a deepwater sponge (*Neopeltidae* family), inhibiting the growth of *C. albicans* (MIC 0.62 μg/mL)[172]; and spongistatin 1, a macrocyclic lactone polyether from *Hyrtios erecta* (Keller, 1889), active against different opportunistic yeasts and filamentous fungi.[173]

Isolated using bioassay-guided fractionation of some marine sponges extracts, novel sulfated sterols exhibited antifungal activity against amphotericin B-/fluconazole-resistant and wild-type strains of *C. albicans*: eurysterols A and B from *Euryspongia* sp. collected in Palau (MICs 15.6–62.5 μg/mL),[174] geodisterol-3-*O*-sulfite, and 29-demethyl-geodisterol-3-*O*-sulfite from

Topsentia sp.[175] Other compounds, such as halistanol from *Halichondria mooriei* and sterol derivatives from *Toxadocia zumi*, inhibited the growth of *B. subtilis* and *S. aureus* (50 and 100 μg/disk, respectively).[176–178]

Antimicrobial meroterpenoids were also isolated from marine sponges: fascioquinols A–F (C26) from a deepwater southern Australian marine sponge *Fasciospongia* sp., exhibiting selective antibacterial properties against *B. subtilis* and *S. aureus*,[179] alisiaquinones A–C and alisiaquinol from a New Caledonian deepwater sponge.[180]

In the last several years, modern studies have revealed that bromopyrrole alkaloids rank among the main antimicrobial components as the most common metabolites of marine sponges from *Agelas* genus (*Agelas conifera*, *Agelas oroides*, and related species): ageliferin,[181] bromoageliferin and its derivatives,[182] benzosceptrin C,[183] nagelamides A–N, Q, R, U–Z,[184–190] and oroidin.[191,192] Some of the abovementioned bromopyrrole alkaloids have peculiar structures: benzosceptrin C, a new dimeric bromopyrrole alkaloid with a benzocyclobutane ring[183]; nagelamide J, the first bromopyrrole alkaloid with a cyclopentane ring fused to an amino-imidazole ring[185]; nagelamide Q, a rare dimeric bromopyrrole alkaloid with a pyrrolidine ring; and nagelamide R, the first bromopyrrole alkaloid with an oxazoline ring.[187] Oroidin is the first marine-derived compound that inhibited enoyl-ACP (acyl carrier protein) reductase, a clinically relevant enzyme targeting the type II fatty acid pathway of several human pathogenic microorganisms (*E. coli*, *M. tuberculosis*).[191] Also, bromoageliferin and its derivatives inhibited *P. aeruginosa* biofilm.[182]

In addition, marine sponges biosynthesized a wide range of antimicrobial alkaloids, such as:

- Isoquinoline and isoindole derivatives from *Reniera* sp.[193];
- Manzamines, polycyclic β-carboline alkaloids from *Haliclona* sp. (Okinawa, Japan)[194];
- Plakinamines A and B, steroidal alkaloids from *Plakina* sp., inhibiting the growth of *S. aureus* and *C. albicans*[195];
- Two new discorhabdin-type pyrrolo-iminoquinone alkaloids from *Sceptrella* sp. (South Korea), exhibiting moderate to significant antibacterial activity and also inhibitory activity against sortase A, a key enzyme for cell wall protein anchoring and virulence of *S. aureus*[196];
- Bromotyrosine alkaloids with N-imidazolyl-quinolinone moiety (ceratinadins A–C) and their macrocyclic metabolites (pseudoceratins A and B) from Okinawan marine sponges *Pseudoceratina purpurea* (Carter, 1880) and related species, active against *C. albicans*[197,198];
- Fasciospongines A and B, 19-oxo-fasciospongine A and fasciospongine C, sulfated sesterterpenoid alkaloids from *Fasciospongia* sp., with potent inhibitory effect against *Streptomyces* 85E[199,200];
- Tetracyclic (macrocyclic) alkylpiperidine alkaloids from *Haliclona* sp., haliclonin A,[201] halicyclamine A,[202] haliclonacyclamine A and B, and 22-hydroxy-haliclonacyclamine B,[203] active against *M. tuberculosis*, *M. smegmatis*, and *M. bovis* (MICs 1–5 μg/mL);
- Aerothionins, brominated spiro-cyclohexadienylisoxazolines biosynthesized through condensation of 3,5-dibromotyrosine with a C^4–N^2 unit (possibly derived from ornithine),[204–206] from *Aplysia aerophoba* (Nardo, 1833) and *Verongia thiona*, inhibiting *M. tuberculosis* growth (70% 11-hydroxyaerothionin and 60% 11-oxo-12-*epi*-hydroxyaerothionin)[143,207–210];

- Topsentin and hamacanthin, bisindole alkaloids from the methanolic extract of *Spongosorites* sp.[211];
- Dysideanins A and B, two new indole alkaloids from the *Dysidea* sp. (South China)[212];
- 5-hydroxyindole-type alkaloids from the tropical sponge *Hyrtios* sp., inhibiting *C. albicans* isocitrate lyase[213];
- Hamacanthins, from a new deepwater sponge (*Hamacantha* sp.), inhibiting the growth of *C. albicans* and *C. neoformans*[214–217];
- Oceanapamine, the major metabolite of the Philippine marine sponge *Oceanapia* sp., active against *B. subtilis* and *E. coli* (25 μg/disk), *S. aureus* and *C. albicans* (50 μg/disk), and *P. aeruginosa* (100 μg/disk).[218]

Also, other nitrogen-containing antimicrobial compounds isolated from marine sponges include brominated tryptamines from *Smenospongia aurea* and *Smenospongia* (ident. *Polyfibrospongia*) *echina*[219] and tetramic acid glycosides (aurantosides D, E, and F) from *Siliquariaspongia japonica* (Hoshino, 1981), exhibiting antifungal activity against *A. fumigatus* and *C. albicans.*[220]

In recent years, novel antimicrobial and antibiofilm peptides directly involved in the defense mechanisms against multiple human pathogens (*B. subtilis, C. albicans, E. coli, Klebsiella* sp., *Mycobacterium* sp., *Proteus* sp., *P. aeruginosa, Shigella dysenteriae, S. aureus, Streptococcus mutans, Streptococcus pneumoniae, Streptococcus faecalis*) were isolated from marine sponges:

- Lipopeptides (trichoderins A, A1, and B, maribasins A and B), cyclic tripeptides (sclerotiotides A–K), cyclic hexapeptides (sclerotides A and B)[2,76,106];
- Cyclodepsipeptides (mirabamides A–D) from *Siliquariaspongia mirabilis*, inhibiting the growth of *B. subtilis* and *C. albicans* (1–5 μg/disk-diffusion assay)[221,222];
- Antifungal bicyclic peptides, callyaerins A–F and H, from the Indonesian marine sponge *Callyspongia aerizusa* (Desqueyroux-Faundez, 1984),[223] and theonellamides[224];
- Hymenamides A and B, antifungal bioactive peptides against *C. albicans* (MICs 33 and 66 μg/mL, respectively) and *C. neoformans* (MICs $<$133 and 33 μg/mL, respectively).[225]

OTHER MARINE INVERTEBRATES

Antibacterial and antimycotic compounds also have been isolated from different marine invertebrates, such as corals, dinoflagellates, mollusks, gastropods, tubeworms. Novel marine diterpenoids (8-hydroxybriarane derivatives) exhibiting antibiofilm, antibacterial (against *E. coli, Salmonella* Typhi, *Shigella flexneri, Vibrio cholerae*),[226] and antimycotic (against *A. niger, C. albicans, Penicillium notatum*)[227] effects were isolated from the methanolic extract of gorgonian coral *Junceella juncea* (Pallas, 1766). The microbiological techniques revealed biofilm-forming pathogens (sucrose-supplemented Congo red agar), in vitro biofilm reduction (microtitre plate assay based on quantitative spectrophotometric method), and antibacterial properties of the methanolic extract (crude methanol as reference for disk-diffusion assay). At a concentration of 12.5 mg/mL, pseudopteroxazole and *seco*-pseudopteroxazole, some novel benzoxazole diterpene alkaloids isolated from the West Indian gorgonian *Pseudopterogorgia elisabethae* (Bayer, 1961), inhibited the growth of *M. tuberculosis* H37Rv (97% and 66%, respectively), without significant cytotoxicity.[143,228] Also, C^{19}-hydroxysteroids (litosterol, nephalsterols B and C) isolated from a Red Sea soft corals *Nephthea* sp.

inhibited the growth of *M. tuberculosis* H37Rv (90%, 69%, and 96%, respectively) with MICs between 3.13 and 12.5 mg/mL.[143,229,230] Goniodomin, a new compound with particular structure (dihydrogeranyl side chain, lactone ring, five hydroxyl groups and four ether linkages), isolated from a marine dinoflagellate, *Goniodoma* sp., harvested in Puerto Rico, strongly inhibits *Cryptococcus*, *Trichophyton* and different fungi, with no antibacterial properties.[231] A polypeptide (kahalalide A) isolated from the sacoglossan mollusk *Elysia rufescens* (Pease, 1871), inhibited the growth of *M. tuberculosis* H37Rv (83%, at a concentration of 12 mg/mL).[143,232] In addition, hexadecylglycerol isolated from the nudibranch *Archidoris montereyensis* (Cooper, 1862) exhibited highly potent in vitro activity against *B. subtilis* and *S. aureus*.[233]

A novel peptide (dolabellanin B2) based on 33 amino acid residues, isolated from the body wall of the sea hare *Dolabella auricularia* (Lightfoot, 1786), a marine opisthobranch gastropod mollusk, was effective in various amounts (2.5–100 μg/mL) against some pathogenic microorganisms.[234] Also, a cyclic depsipeptide (viscosin) isolated from microbial cultures of *Pseudomonas* sp. growing on a marine tubeworm, in vitro inhibited *M. tuberculosis* and *M. avium-intracellulare* (MICs 10–20 and 5–10 mg/mL, respectively).[143,129]

CONCLUSIONS AND FUTURE PERSPECTIVES

During the last several decades, despite all the great achievements of antibiotherapy, infectious diseases remain the second-leading cause of death worldwide, mainly for children and the elderly. The increasingly resistance of microorganisms to common antibiotics has become an extremely serious and alarming matter. Therefore, the search for novel natural (renewable) resources in fighting microbial infections becomes more urgent. In the near future, marine-derived antimicrobial compounds, such as actinomycetes secondary metabolites, cyanobacterial alkaloids and non-ribosomal peptides, marine bacteria exopolysaccharides, marine fungi polyketides, alkaloids and peptides, marine algae fatty acids, bromophenol derivatives and sesquiterpenoid quinones, marine-derived sponges, brominated phenols, sesquiterpenoid quinones, diterpenoids, sesterterpenoids, macrolides, bromopyrrole alkaloids, and lipopeptides could become a reliable alternative for the prevention and control of microbial resistance.

ACKNOWLEDGMENTS

This paper is partially supported by the Sectoral Operational Programme Human Resources Development, financed from the European Social Fund and by the Romanian government under the contract number POSDRU/89/1.5/S/64153.

REFERENCES

1. Blunt JW, Copp BR, Hu WP, Munro MH, Northcote PT, Prinsep MR. Marine natural products. *Nat Prod Rep* 2007;**24**(1):31–86.
2. Blunt JW, Copp BR, Keyzers RA, Munro MH, Prinsep MR. Marine natural products. *Nat Prod Rep* 2014;**31**(2):160–258.

3. Harvey AL. Natural products in drug discovery. *Drug Discov Today* 2008;**13**(19–20):894–901.
4. Butler MS, Buss AD. Natural products—the future scaffolds for novel antibiotics? *Biochem Pharmacol* 2006;**71**(7):919–29.
5. Koehn FE, Carter GT. The evolving role of natural products in drug discovery. *Nat Rev Drug Discov* 2005;**4**(3):206–20.
6. Penesyan A, Kjelleberg S, Egan S. Development of novel drugs from marine surface associated microorganisms. *Mar Drugs* 2010;**8**(3):438–59.
7. Bugni TS, Richards B, Bhoite L, Cimbora D, Harper MK, Ireland CM. Marine natural product libraries for high-throughput screening and rapid drug discovery. *J Nat Prod* 2008;**71**(6):1095–8.
8. Lam KS. Discovery of novel metabolites from marine actinomycetes. *Curr Opin Microbiol* 2006;**9**(3):245–51.
9. Fenical W, Jensen PR. Developing a new resource for drug discovery: marine actinomycete bacteria. *Nat Chem Biol* 2006;**2**(12):666–73.
10. Bull AT, Stach JE. Marine actinobacteria: new opportunities for natural product search and discovery. *Trends Microbiol* 2007;**15**(11):491–9.
11. Subramani R, Aalbersberg W. Marine actinomycetes: an ongoing source of novel bioactive metabolites. *Microbiol Res* 2012;**167**(10):571–80.
12. Jensen PR, Moore BS, Fenical W. The marine actinomycete genus *Salinispora*: a model organism for secondary metabolite discovery. *Nat Prod Rep* 2015;**32**(5):738–51.
13. Maskey RP, Li FC, Qin S, Fiebig HH, Laatsch H. Chandrananimycins A∼C: production of novel anticancer antibiotics from a marine *Actinomadura* sp. isolate M048 by variation of medium composition and growth conditions. *J Antibiot (Tokyo)* 2003;**56**(7):622–9.
14. Soria-Mercado IE, Prieto-Davo A, Jensen PR, Fenical W. Antibiotic terpenoid chloro-dihydroquinones from a new marine actinomycete. *J Nat Prod* 2005;**68**(6):904–10.
15. Asolkar RN, Schröder D, Heckmann R, Lang S, Wagner-Döbler I, Laatsch H. Helquinoline, a new tetrahydroquinoline antibiotic from *Janibacter limosus* Hel 1+. *J Antibiot (Tokyo)* 2004;**57**(1):17–23.
16. Kwon HC, Kauffman CA, Jensen PR, Fenical W. Marinomycins A–D, antitumor-antibiotics of a new structure class from a marine actinomycete of the recently discovered genus *Marinispora*. *J Am Chem Soc* 2006;**128**(5):1622–32.
17. Bruntner C, Binder T, Pathom-aree W, Goodfellow M, Bull AT, Potterat O, et al. Frigocyclinone, a novel angucyclinone antibiotic produced by a *Streptomyces griseus* strain from Antarctica. *J Antibiot (Tokyo)* 2005;**58**(5):346–9.
18. Manam RR, Teisan S, White DJ, Nicholson B, Grodberg J, Neuteboom STC, et al. Lajollamycin, a nitro-tetraene spiro-β-lactone-γ-lactam antibiotic from the marine actinomycete *Streptomyces nodosus*. *J Nat Prod* 2005;**68**(2):240–3.
19. Schumacher RW, Talmage SC, Miller SA, Sarris KE, Davidson BS, Goldberg A. Isolation and structure determination of an antimicrobial ester from a marine-derived bacterium. *J Nat Prod* 2003;**66**(9):1291–3.
20. Macherla VR, Liu J, Bellows C, Teisan S, Lam KS, Potts BCM. Glaciapyrroles A, B, and C, pyrrolosesquiterpenes from a *Streptomyces* sp. isolated from an Alaskan marine sediment. *J Nat Prod* 2005;**68**(5):780–3.
21. Maskey RP, Sevvana M, Usón I, Helmke E, Laatsch H. Gutingimycin: a highly complex metabolite from a marine streptomycete. *Angew Chem Int Ed Engl* 2004;**43**(10):1281–3.
22. Maskey RP, Helmke E, Laatsch H. Himalomycin A and B: isolation and structure elucidation of new fridamycin type antibiotics from a marine *Streptomyces* isolate. *J Antibiot (Tokyo)* 2003;**56**(11):942–9.
23. Maskey RP, Helmke E, Kayser O, Fiebig HH, Maier A, Busche A, et al. Anti-cancer and antibacterial trioxacarcins with high anti-malaria activity from a marine streptomycete and their absolute stereochemistry. *J Antibiot (Tokyo)* 2004;**57**(12):771–9.

24. Oh DC, Gontang EA, Kauffman CA, Jensen PR, Fenical W. Salinipyrones and pacificanones, mixed-precursor polyketides from the marine actinomycete *Salinispora pacifica*. *J Nat Prod* 2008;**71**(4):570–5.
25. Riedlinger J, Reicke A, Zahner H, Krismer B, Bull AT, Maldonado LA, et al. Abyssomicins, inhibitors of the para-aminobenzoic acid pathway produced by the marine *Verrucosispora* strain AB-18-032. *J Antibiot (Tokyo)* 2004;**57**(4):271–9.
26. Bister B, Bischoff D, Strobele M, Riedlinger J, Reicke A, Wolter F, et al. Abyssomicin C – a polycyclic antibiotic from a marine *Verrucosispora* strain as an inhibitor of the *p*-aminobenzoic acid/tetrahydrofolate biosynthesis pathway. *Angew Chem Int Ed Engl* 2004;**43**(19):2574–6.
27. Rath JP, Kinast S, Maier ME. Synthesis of the full functionalized core structure of the antibiotic abyssomicin C. *Org Lett* 2005;**7**(14):3089–92.
28. Buchanan GO, Williams PG, Feling RH, Kauffman CA, Jensen PR, Fenical W. Sporolides A and B: structurally unprecedented halogenated macrolides from the marine actinomycete *Salinispora tropica*. *Org Lett* 2005;**7**(13):2731–4.
29. Jensen PR, Mincer TJ, Williams PG, Fenical W. Marine actinomycete diversity and natural product discovery. *Antonie Van Leeuwenhoek* 2005;**87**(1):43–8.
30. Asolkar RN, Kirkland TN, Jensen PR, Fenical W. Arenimycin, an antibiotic effective against rifampin- and methicillin-resistant *Staphylococcus aureus* from the marine actinomycete *Salinispora arenicola*. *J Antibiot (Tokyo)* 2010;**63**(1):37–9.
31. Charan RD, Schlingmann G, Janso J, Bernan V, Feng X, Carter GT. Diazepinomicin, a new antimicrobial alkaloid from marine *Micromonospora* sp. *J Nat Prod* 2004;**67**(8):1431–3.
32. Nithyanand P, Thenmozhi R, Rathna J, Pandian SK. Inhibition of *Streptococcus pyogenes* biofilm formation by coral-associated actinomycetes. *Curr Microbiol* 2010;**60**(6):454–60.
33. Bakkiyaraj D, Pandian SK. *In vitro* and *in vivo* antibiofilm activity of a coral associated actinomycete against drug resistant *Staphylococcus aureus* biofilms. *Biofouling* 2010;**26**(6):711–17.
34. Gowrishankar S, Mosioma ND, Pandian SK. Coral-associated bacteria as a promising antibiofilm agent against methicillin-resistant and -susceptible *Staphylococcus aureus* biofilms. *Evid Based Complement Alternat Med* 2012;**2012**:862374.
35. Thenmozhi R, Nithyanand P, Rathna J, Pandian SK. Antibiofilm activity of coral-associated bacteria against different clinical M serotypes of *Streptococcus pyogenes*. *FEMS Immunol Med Microbiol* 2009;**57**(3):284–94.
36. El-Gendy MM, Shaaban M, El-Bondkly AM, Shaaban KA. Bioactive benzopyrone derivatives from new recombinant fusant of marine *Streptomyces*. *Appl Biochem Biotechnol* 2008;**150**(1):85–96.
37. Andrianasolo EH, Haramaty L, Rosario-Passapera R, Bidle K, White E, Vetriani C, et al. Ammonificins A and B, hydroxyethylamine chroman derivatives from a cultured marine hydrothermal vent bacterium, *Thermovibrio ammonificans*. *J Nat Prod* 2009;**72**(6):1216–19.
38. Ding L, Maier A, Fiebig HH, Lin WH, Peschel G, Hertweck C. Kandenols A–E, eudesmenes from an endophytic *Streptomyces* sp. of the mangrove tree *Kandelia candel*. *J Nat Prod* 2012;**75**(12):2223–7.
39. Ding L, Münch J, Goerls H, Maier A, Fiebig HH, Lin WH, et al. Xiamycin, a pentacyclic indolosesquiterpene with selective anti-HIV activity from a bacterial mangrove endophyte. *Bioorg Med Chem Lett* 2010;**20**(22):6685–7.
40. Kim DG, Moon K, Kim SH, Park SH, Park S, Lee SK, et al. Bahamaolides A and B, antifungal polyene polyol macrolides from the marine actinomycete *Streptomyces* sp. *J Nat Prod* 2012;**75**(5):959–67.
41. Hughes CC, Prieto-Davo A, Jensen PR, Fenical W. The marinopyrroles, antibiotics of an unprecedented structure class from a marine *Streptomyces* sp. *Org Lett* 2008;**10**(4):629–31.
42. Han Y, Yang B, Zhang F, Miao X, Li Z. Characterization of antifungal chitinase from marine *Streptomyces* sp. DA11 associated with South China Sea sponge *Craniella australiensis*. *Mar Biotechnol (NY)* 2009;**11**(1):132–40.

43. Han Z, Xu Y, McConnell O, Liu L, Li Y, Qi S, et al. Two antimycin a analogues from marine-derived actinomycete *Streptomyces lusitanus*. *Mar Drugs* 2012;**10**(3):668–76.
44. Tan LT, Sitachitta N, Gerwick WH. The guineamides, novel cyclic depsipeptides from a Papua New Guinea collection of the marine cyanobacterium *Lyngbya majuscula*. *J Nat Prod* 2003;**66**(6):764–71.
45. Paul VJ, Thacker RW, Banks K, Golubic S. Benthic cyanobacterial bloom impacts the reefs of South Florida (Broward County, USA). *Coral Reefs* 2005;**24**(4):693–7.
46. Tan LT. Bioactive natural products from marine cyanobacteria for drug discovery. *Phytochemistry* 2007;**68**(7):954–79.
47. Nunnery JK, Mevers E, Gerwick WH. Biologically active secondary metabolites from marine cyanobacteria. *Curr Opin Biotechnol* 2010;**21**(6):787–93.
48. Uzair B, Tabassum S, Rasheed M, Rehman SF. Exploring marine cyanobacteria for lead compounds of pharmaceutical importance. *Sci World J* 2012;**2012**:179782.
49. Mo S, Krunic A, Santarsiero BD, Franzblau SG, Orjala J. Hapalindole-related alkaloids from the cultured cyanobacterium *Fischerella ambigua*. *Phytochemistry* 2010;**71**(17–18):2116–23.
50. Silva-Stenico ME, Silva CS, Lorenzi AS, Shishido TK, Etchegaray A, Lira SP, et al. Non-ribosomal peptides produced by Brazilian cyanobacterial isolates with antimicrobial activity. *Microbiol Res* 2011;**166**(3):161–75.
51. Ren ZZ, Zheng Y, Sun M, Liu JZ, Wang YJ. Purification and properties of an antimicrobial substance from marine *Brevibacillus laterosporus* Lh-1. *Wei Sheng Wu Xue Bao* 2007;**47**(6):997–1001.
52. Cardellina II JH, Moore RE. The structures of pukeleimides A, B, D, E, F, and G. *Tetrahedron Lett* 1979;**20**(22):2007–10.
53. Deth S.K. *Antimicrobial compounds from marine cyanobacteria with special reference to the bioactivity of a purified compound from Oscillatoria laete-virens BDU 20801*. PhD Thesis, Tiruchirappalli: Bharathidasan University; 1999.
54. Bernal P, Llamas MA. Promising biotechnological applications of antibiofilm exopolysaccharides. *Microb Biotechnol* 2012;**5**(6):670–3.
55. Rendueles O, Kaplan JB, Ghigo JM. Antibiofilm polysaccharides. *Environ Microbiol* 2013;**15**(2):334–6.
56. Dheilly A, Soum-Soutéra E, Klein GL, Bazire A, Compère C, Haras D, et al. Antibiofilm activity of the marine bacterium *Pseudoalteromonas* sp. strain 3J6. *Appl Environ Microbiol* 2010;**76**(11):3452–61.
57. Papa R, Parrilli E, Sannino F, Barbato G, Tutino ML, Artini M, et al. Anti-biofilm activity of the Antarctic marine bacterium *Pseudoalteromonas haloplanktis* TAC125. *Res Microbiol* 2013;**164**(5):450–6.
58. Jiang P, Li J, Han F, Duan G, Lu X, Gu Y, et al. Antibiofilm activity of an exopolysaccharide from marine bacterium *Vibrio* sp. QY101. *PLoS One* 2011;**6**(4):e18514.
59. Abu Sayem SM, Manzo E, Ciavatta L, Tramice A, Cordone A, Zanfardino A, et al. Anti-biofilm activity of an exopolysaccharide from a sponge-associated strain of *Bacillus licheniformis*. *Microb Cell Fact* 2011;**10**:74.
60. Kiran GS, Sabarathnam B, Selvin J. Biofilm disruption potential of a glycolipid biosurfactant from marine *Brevibacterium casei*. *FEMS Immunol Med Microbiol* 2010;**59**(3):432–8.
61. Dusane DH, Pawar VS, Nancharaiah YV, Venugopalan VP, Kumar AR, Zinjarde SS. Anti-biofilm potential of a glycolipid surfactant produced by a tropical marine strain of *Serratia marcescens*. *Biofouling* 2011;**27**(6):645–54.
62. Lovell FM. The structure of a bromine-rich marine antibiotic. *J Am Chem Soc* 1966;**88**(19):4510–11.
63. Arima K, Imanaka H, Kousaka M, Fukuta A, Tamura G. Pyrrolnitrin, a new antibiotic substance, produced by *Pseudomonas*. *Agric Biol Chem* 1964;**28**(8):575–6.
64. Aria K, Imanaka H, Kousaka M, Fukuda A, Tamura G. Studies on pyrrolnitrin, a new antibiotic. I. Isolation and properties of pyrrolnitrin. *J Antibiot (Tokyo)* 1965;**18**(5):201–4.

65. Imanaka H, Kousaka M, Tamura G, Arima K. Studies on pyrrolnitrin, a new antibiotic. 3. Structure of pyrrolnitrin. *J Antibiot (Tokyo)* 1965;**18**(5):207–10.
66. Nishida M, Matsubara T, Watanabe N. Pyrrolnitrin, a new antifungal antibiotic. Microbiological and toxicological observations. *J Antibiot (Tokyo)* 1965;**18**(5):211–19.
67. Burkhead KD, Schisler DA, Slininger PJ. Pyrrolnitrin production by biological control agent *Pseudomonas cepacia* B37w in culture and in colonized wounds of potatoes. *Appl Environ Microbiol* 1994;**60**(6):2031–9.
68. Di Santo R, Costi R, Artico M, Massa S, Lampis G, Deidda D, et al. Pyrrolnitrin and related pyrroles endowed with antibacterial activities against *Mycobacterium tuberculosis*. *Bioorg Med Chem Lett* 1998;**8** (20):2931–6.
69. van Pée KH, Ligon JM. Biosynthesis of pyrrolnitrin and other phenylpyrrole derivatives by bacteria. *Nat Prod Rep* 2000;**17**(2):157–64.
70. Lambowitz AM, Slayman CW. Effect of pyrrolnitrin on electron transport and oxidative phosphorylation in mitochondria isolated from *Neurospora crassa*. *J Bacteriol* 1972;**112**(2):1020–2.
71. Fehér D, Barlow R, McAtee J, Hemscheidt TK. Highly brominated antimicrobial metabolites from a marine *Pseudoalteromonas* sp. *J Nat Prod* 2010;**73**(11):1963–6.
72. Isnansetyo A, Kamei Y. Anti-methicillin-resistant *Staphylococcus aureus* (MRSA) activity of MC21-B, an antibacterial compound produced by the marine bacterium *Pseudoalteromonas phenolica* O-BC30T. *Int J Antimicrob Agents* 2009;**34**(2):131–5.
73. Hayashida-Soiza G, Uchida A, Mori N, Kuwahara Y, Ishida Y. Purification and characterization of antibacterial substances produced by a marine bacterium *Pseudoalteromonas haloplanktis* strain. *J Appl Microbiol* 2008;**105**(5):1672–7.
74. El-Gendy MM, Hawas UW, Jaspars M. Novel bioactive metabolites from a marine derived bacterium *Nocardia* sp. ALAA 2000. *J Antiobiot (Tokyo)* 2008;**61**(6):379–86.
75. Zhang DJ, Liu RF, Li YG, Tao LM, Tian L. Two new antifungal cyclic lipopeptides from *Bacillus marinus* B-9987. *Chem Pharm Bull (Tokyo)* 2010;**58**(12):1630–4.
76. Chen L, Wang N, Wang X, Hu J, Wang S. Characterization of two anti-fungal lipopeptides produced by *Bacillus amyloliquefaciens* SH-B10. *Bioresour Technol* 2010;**101**(22):8822–7.
77. Desjardine K, Pereira A, Wright H, Matainaho T, Kelly M, Andersen RJ. Tauramamide, a lipopeptide antibiotic produced in culture by *Brevibacillus laterosporus* isolated from a marine habitat: structure elucidation and synthesis. *J Nat Prod* 2007;**70**(12):1850–3.
78. Engelhardt K, Degnes KF, Kemmler M, Bredholt H, Fjaervik E, Klinkenberg G, et al. Production of a new thiopeptide antibiotic, TP-1161, by a marine *Nocardiopsis* species. *Appl Environ Microbiol* 2010;**76** (15):4969–76.
79. Oku N, Kawabata K, Adachi K, Katsuta A, Shizuri Y. Unnarmicins A and C, new antibacterial depsipeptides produced by marine bacterium *Photobacterium* sp. MBIC06485. *J Antibiot (Tokyo)* 2008;**61** (1):11–17.
80. Ojika M, Inukai Y, Kito Y, Hirata M, Iizuka T, Fudou R. Miuraenamides: antimicrobial cyclic depsipeptides isolated from a rare and slightly halophilic myxobacterium. *Chem Asian J* 2008;**3**(1):126–33.
81. Oku N, Adachi K, Matsuda S, Kasai H, Takatsuki A, Shizuri Y. Ariakemicins A and B, novel polyketide-peptide antibiotics from a marine gliding bacterium of the genus *Rapidithrix*. *Org Lett* 2008;**10** (12):2481–4.
82. Wyche TP, Hou Y, Vazquez-Rivera E, Braun D, Bugni TS. Peptidolipins B–F, antibacterial lipopeptides from an ascidian-derived *Nocardia* sp. *J Nat Prod* 2012;**75**(4):735–40.
83. Li H, Shinde PB, Lee HJ, Yoo ES, Lee CO, Hong J, et al. Bile acid derivatives from a sponge-associated bacterium *Psychrobacter* sp. *Arch Pharm Res* 2009;**32**(6):857–62.
84. Rapoport H, Holden KG. The synthesis of prodigiosin. *J Am Chem Soc* 1962;**84**(4):635–42.

85. Bennett JW, Bentley R. Seeing red: the story of prodigiosin. *Adv Appl Microbiol* 2000;**47**:1–32.
86. Fürstner A. Chemistry and biology of roseophilin and the prodigiosin alkaloids: a survey of the last 2500 years. *Angew Chem Int Ed Engl* 2003;**42**(31):3582–603.
87. Llagostera E, Soto-Cerrato V, Montaner B, Pérez-Tomás R. Prodigiosin induces apoptosis by acting on mitochondria in human lung cancer cells. *Ann NY Acad Sci* 2003;**1010**:178–81.
88. Buck JD, Meyers SP, Kamp KM. Marine bacteria with antiyeast activity. *Science* 1962;**138**(3547): 1339–40.
89. Crawford K, Heatley NG, Boyd PF, Hale CW, Kelly BK, Miller GA, et al. Antibiotic production by a species of *Cephalosporium*. *J Gen Microbiol* 1952;**6**(1–2):47–59.
90. Niss HF, Nash CH. Synthesis of cephalosporin C from sulfate by mutants of *Cephalosporium acremonium*. *Antimicrob Agents Chemother* 1973;**4**(4):474–8.
91. Zhang J, Demain AL. Regulation of ACV synthetase in penicillin- and cephalosporin-producing microorganisms. *Biotechnol Adv* 1991;**9**(4):623–41.
92. Cruz AJG, Pan T, Giordano RC, Araujo MLGC, Hokka CO. Cephalosporin C production by immobilized *Cephalosporium acremonium* cells in a repeated batch tower bioreactor. *Biotechnol Bioeng* 2004;**85**(1):96–102.
93. Xu J, Nakazawa T, Ukai K, Kobayashi H, Mangindaan RE, Wewengkang DS, et al. Tetrahydrobostrycin and 1-deoxytetrahydrobostrycin, two new hexahydroanthrone derivatives from a marine-derived fungus *Aspergillus* sp. *J Antibiot (Tokyo)* 2008;**61**(7):415–19.
94. Lee YM, Li H, Hong J, Cho HY, Bae KS, Kim MA, et al. Bioactive metabolites from the sponge-derived fungus *Aspergillus versicolor*. *Arch Pharm Res* 2010;**33**(2):231–5.
95. El-Beih AA, Kawabata T, Koimaru K, Ohta T, Tsukamoto S. Monodictyquinone a: a new antimicrobial anthraquinone from a sea urchin-derived fungus *Monodictys* sp. *Chem Pharm Bull (Tokyo)* 2007;**55**(7):1097–8.
96. Hawas UW, El-Beih AA, El-Halawany AM. Bioactive anthraquinones from endophytic fungus *Aspergillus versicolor* isolated from Red Sea algae. *Arch Pharm Res* 2012;**35**(10):1749–56.
97. Song F, Dai H, Tong Y, Ren B, Chen C, Sun N, et al. Trichodermaketones A–D and 7-O-methylkoninginin D from the marine fungus *Trichoderma koningii*. *J Nat Prod* 2010;**73**(5):806–10.
98. Xie LW, Ouyang YC, Zou K, Wang GH, Chen MJ, Sun HM, et al. Isolation and difference in anti-*Staphylococcus aureus* bioactivity of curvularin derivates from fungus *Eupenicillum* sp. *Appl Biochem Biotechnol* 2009;**159**(1):284–93.
99. Trisuwan K, Rukachaisirikul V, Sukpondma Y, Phongpaichit S, Preedanon S, Sakayaroj J. Lactone derivatives from the marine-derived fungus *Penicillium* sp. PSU-F44. *Chem Pharm Bull (Tokyo)* 2009;**57**(10):1100–2.
100. Trisuwan K, Rukachaisirikul V, Sukpondma Y, Preedanon S, Phongpaichit S, Rungjindamai N, et al. Epoxydons and a pyrone from the marine-derived fungus *Nigrospora* sp. PSU-F5. *J Nat Prod* 2008;**71**(8):1323–6.
101. Trisuwan K, Rukachaisirikul V, Sukpondma Y, Preedanon S, Phongpaichit S, Sakayaroj J. Pyrone derivatives from the marine-derived fungus *Nigrospora* sp. PSU-F18. *Phytochemistry* 2009;**70**(4): 554–7.
102. Wang J, Liu P, Wang Y, Wang H, Li J, Zhuang Y, et al. Antimicrobial aromatic polyketides from gorgonian-associated fungus, *Penicillium commune* 518. *Chin J Chem* 2012;**30**(6):1236–42.
103. Wang W, Wang Y, Tao H, Peng X, Liu P, Zhu W. Cerebrosides of the halotolerant fungus *Alternaria raphani* isolated from a sea salt field. *J Nat Prod* 2009;**72**(9):1695–8.
104. Hohmann C, Schneider K, Bruntner C, Irran E, Nicholson G, Bull AT, et al. Caboxamycin, a new antibiotic of the benzoxazole family produced by the deep-sea strain *Streptomyces* sp. NTK 937. *J Antibiot (Tokyo)* 2009;**62**(2):99–104.

105. Song F, Liu X, Guo H, Ren B, Chen C, Piggott AM, et al. Brevianamides with antitubercular potential from a marine-derived isolate of *Aspergillus versicolor*. *Org Lett* 2012;**14**(18):4770–3.
106. Pruksakorn P, Arai M, Kotoku N, Vilchèze C, Baughn AD, Moodley P, et al. Trichoderins, novel aminolipopeptides from a marine sponge-derived *Trichoderma* sp., are active against dormant mycobacteria. *Bioorg Med Chem Lett* 2010;**20**(12):3658–63.
107. Zheng J, Zhu H, Hong K, Wang Y, Liu P, Wang X, et al. Novel cyclic hexapeptides from marine-derived fungus, *Aspergillus sclerotiorum* PT06-1. *Org Lett* 2009;**11**(22):5262–5.
108. Zheng J, Xu Z, Wang Y, Hong K, Liu P, Zhu W. Cyclic tripeptides from the halotolerant fungus *Aspergillus sclerotiorum* PT06-1. *J Nat Prod* 2010;**73**(6):1133–7.
109. Kim MY, Sohn JH, Ahn JS, Oh H. Alternaramide, a cyclic depsipeptide from the marine-derived fungus *Alternaria* sp. SF-5016. *J Nat Prod* 2009;**72**(11):2065–8.
110. Kanoh K, Adachi K, Matsuda S, Shizuri Y, Yasumoto K, Kusumi T, et al. New sulfoalkylresorcinol from marine-derived fungus, *Zygosporium* sp. KNC52. *J Antibiot (Tokyo)* 2008;**61**(3):192–4.
111. Wang H, Lu Z, Qu HJ, Liu P, Miao C, Zhu T, et al. Antimicrobial aflatoxins from the marine-derived fungus *Aspergillus flavus* 092008. *Arch Pharm Res* 2012;**35**(8):1387–92.
112. Nguyen HP, Zhang D, Lee U, Kang JS, Choi HD, Son BW. Dehydroxychlorofusarielin B, an antibacterial polyoxygenated decalin derivative from the marine-derived fungus *Aspergillus* sp. *J Nat Prod* 2007; **70**(7):1188–90.
113. Zhang W, Draeger S, Schulz B, Krohn K. Ring B aromatic steroids from an endophytic fungus *Colletotrichum* sp. *Nat Prod Commun* 2009;**4**(11):1449–54.
114. Zhang Y, Wang S, Li XM, Cui CM, Feng C, Wang BC. New sphingolipids with a previously unreported 9-methyl-C^{20}-sphingosine moiety from a marine algous endophytic fungus *Aspergillus niger* EN-13. *Lipids* 2007;**42**(8):759–64.
115. Thibane VS, Kock JLF, Ells R, van Wyk PWJ, Pohl CH. Effect of marine polyunsaturated fatty acids on biofilm formation of *Candida albicans* and *Candida dubliniensis*. *Mar Drugs* 2010;**8**(10):2597–604.
116. Thibane VS, Ells R, Hugo A, Albertyn J, van Rensburg WJ, Van Wyk PW, et al. Polyunsaturated fatty acids cause apoptosis in *C. albicans* and *C. dubliniensis* biofilms. *Biochim Biophys Acta* 2012;**1820**(10):1463–8.
117. Dembitsky VM, Srebnik M. Natural halogenated fatty acids: their analogues and derivatives. *Prog Lipid Res* 2002;**41**(4):315–67.
118. Desbois AP, Lebl T, Yan L, Smith VJ. Isolation and structural characterization of two antibacterial free fatty acids from the marine diatom, *Phaeodactylum Tricornutum*. *Appl Microbiol Biotechnol* 2008;**81**(4):755–64.
119. Desbois AP, Mearns-Spragg A, Smith VJ. A fatty acid from the diatom *Phaeodactylum tricornutum* is antibacterial against diverse bacteria including multi-resistant *Staphylococcus aureus* (MRSA). *Mar Biotechnol (NY)* 2009;**11**(1):45–52.
120. Criagie JS, Gruening DE. Bromophenols from red algae. *Science* 1967;**157**(3792):1058–9.
121. Oh KB, Lee JH, Chung SC, Shin J, Shin HJ, Kim HK, et al. Antimicrobial activities of the bromophenols from the red alga *Odonthalia corymbifera* and some synthetic derivatives. *Bioorg Med Chem Lett* 2008;**18**(1):104–8.
122. Lane AL, Mular L, Drenkard EJ, Shearer TL, Engel S, Fredericq S, et al. Ecological leads for natural product discovery: novel sesquiterpene hydroquinones from the red macroalga *Peyssonnelia* sp. *Tetrahedron* 2010;**66**(2):455–61.
123. Ji NY, Li XM, Li K, Ding LP, Gloer JB, Wang BG. Diterpenes, sesquiterpenes, and a C15-acetogenin from the marine red alga *Laurencia mariannensis*. *J Nat Prod* 2007;**70**(12):1901–5.
124. Vairappan CS, Suzuki M, Ishii T, Okino T, Abe T, Masuda M. Antibacterial activity of halogenated sesquiterpenes from Malaysian *Laurencia* spp. *Phytochemistry* 2008;**69**(13):2490–4.

125. Alarif WM, Al-Lihaibi SS, Ayyad SE, Abdel-Rhman MH, Badria FA. Laurene-type sesquiterpenes from the Red Sea red alga *Laurencia obtusa* as potential antitumor-antimicrobial agents. *Eur J Med Chem* 2012;**55**:462–6.
126. Wijesekara I, Yoon NY, Kim SK. Phlorotannins from *Ecklonia cava* (*Phaeophyceae*): biological activities and potential health benefits. *Biofactors* 2010;**36**(6):408–14.
127. Kamei Y, Sueyoshi M, Hayashi K, Terada R, Nozaki H. The novel anti-*Propionibacterium acnes* compound, sargafuran, found in the marine brown alga *Sargassum macrocarpum*. *J Antibiot (Tokyo)* 2009; **62**(5):259–63.
128. Gerard J, Lloyd R, Barsby T, Haden P, Kelly MT, Andersen RJ. Massetolides A–H, antimycobacterial cyclic depsipeptides produced by two pseudomonads isolated from marine habitats. *J Nat Prod* 1997;**60**(3):223–9.
129. Conover JT, Sieburth JMcN. Effect of *Sargassum* distribution on its epibiota and antibacterial activity. *Bot Mar* 1964;**6**(1–2):147–57.
130. Tianero MD, Hanif N, de Voogd NJ, van Soest RW, Tanaka J. A new antimicrobial fatty acid from the calcareous sponge *Paragrantia* cf. *waguensis*. *Chem Biodivers* 2009;**6**(9):1374–7.
131. Taniguchi M, Uchio Y, Yasumoto K, Kusumi T, Ooi T. Brominated unsaturated fatty acids from marine sponge collected in Papua New Guinea. *Chem Pharm Bull (Tokyo)* 2008;**56**(3):378–82.
132. Keffer JL, Plaza A, Bewley CA. Motualevic acids A–F, antimicrobial acids from the sponge *Siliquariaspongia* sp. *Org Lett* 2009;**11**(5):1087–90.
133. Shridhar DM, Mahajan GB, Kamat VP, Naik CG, Parab RR, Thakur NR, et al. Antibacterial activity of 2-(2′,4′-dibromophenoxy)-4,6-dibromophenol from *Dysidea granulosa*. *Mar Drugs* 2009;**7**(3): 464–71.
134. Zhang H, Skildum A, Stromquist E, Rose-Hellekant T, Chang LC. Bioactive polybrominated diphenyl ethers from the marine sponge *Dysidea* sp. *J Nat Prod* 2008;**71**(2):262–4.
135. Thomas TR, Kavlekar DP, LokaBharathi PA. Marine drugs from sponge-microbe association—a review. *Mar Drugs* 2010;**8**(4):1417–68.
136. Hanif N, Tanaka J, Setiawan A, Trianto A, de Voogd NJ, Murni A, et al. Polybrominated diphenyl ethers from the Indonesian sponge *Lamellodysidea herbacea*. *J Nat Prod* 2007;**70**(3):432–5.
137. Sharma GM, Burkholder PR. Studies on the antimicrobial substances of sponges. II. Structure and synthesis of a bromine-containing antibacterial compound from a marine sponge. *Tetrahedron Lett* 1967;**8**(42):4147–50.
138. Sharma GM, Vig B, Burkholder PR. Studies on the antimicrobial substances of sponges. IV. Structure of a bromine-containing compound from a marine sponge. *J Org Chem* 1970;**35**(8):2823–6.
139. Xu WH, Ding Y, Jacob MR, Agarwal AK, Clark AM, Ferreira D, et al. Puupehanol, a sesquiterpene-dihydroquinone derivative from the marine sponge *Hyrtios* sp. *Bioorg Med Chem Lett* 2009;**19**(21):6140–3.
140. Takahashi Y, Kubota T, Ito J, Mikami Y, Fromont J, Kobayashi J. Nakijiquinones G–I, new sesquiterpenoid quinones from marine sponge. *Bioorg Med Chem* 2008;**16**(16):7561–4.
141. Hamann MT, Scheuer PJ, Kelly-Borges M. Biogenetically diverse, bioactive constituents of a sponge, order *Verongida*: bromotyramines and sesquiterpene-shikimate derived metabolites. *J Org Chem* 1993;**58**(24):6565–9.
142. Nasu SS, Yeung BKS, Hamann MT, Scheuer PJ, Kelly-Borges M, Goins KD. Puupehenone-related metabolites from two Hawaiian sponges, *Hyrtios* spp. *J Org Chem* 1995;**60**(22):7290–2.
143. El Sayed KA, Bartyzel P, Shen X, Perry TL, Zjawiony JK, Hamann MT. Marine natural products as antituberculosis agents. *Tetrahedron* 2000;**56**(7):949–53.
144. Yamamura S, Hirata Y. Structures of aplysin and aplysinol, naturally occurring bromo-compounds. *Tetrahedron* 1963;**19**(10):1485–96.

145. Matsuda H, Tomiie Y, Yamamura S, Hirata Y. The structure of aplysin-20. *Chem Commun (London)* 1967;**17**:898b–9b.
146. Nakamura H, Wu H, Ohizumi Y, Hirata Y. Agelasine-A, -B, -C and -D, novel bicyclic diterpenoids with a 9-methyladeninium unit possessing inhibitory effects on Na,K-ATPase from the Okinawa Sea sponge *Agelas* sp. *Tetrahedron Lett* 1984;**25**(28):2989–92.
147. Ishida K, Ishibashi M, Shigemori H, Sasaki T, Kobayashi J. Agelasine G, a new antileukemic alkaloid from the Okinawan marine sponge *Agelas* sp. *Chem Pharm Bull (Tokyo)* 1992;**40**(3):766–7.
148. Fu X, Schmitz FJ, Tanner RS, Kelly-Borges M. Agelasines H and I, 9-methyladenine-containing diterpenoids from an *Agelas* sponge. *J Nat Prod* 1998;**61**(4):548–50.
149. Mangalindan GC, Talaue MT, Cruz LJ, Franzblau SG, Adams LB, Richardson AD, et al. Agelasine F from a Philippine *Agelas* sp. sponge exhibits in vitro antituberculosis activity. *Planta Med* 2000;**66**(4):364–5.
150. Capelle N, Braekman JC, Daloze D, Tursch B. Chemical studies of marine invertebrates. XLIV. Three new spongian diterpenes from *Spongia officinalis*. *Bull Soc Chim Belg* 1980;**89**(5):399–404.
151. Gonzalez AG, Estrada DM, Martin JD, Martin VS, Perez C, Perez R. New antimicrobial diterpenes from the sponge *Spongia officinalis*. *Tetrahedron* 1984;**40**(20):4109–13.
152. Chang CWJ, Patra A, Roll DM, Scheuer PJ, Matsumoto GK, Clardy J. Kalihinol-A, a highly functionalized diisocyano diterpenoid antibiotic from a sponge. *J Am Chem Soc* 1984;**106**(16):4644–6.
153. Patra A, Chan CWJ, Scheuer PJ, Van Duyne GD, Matsumoto GK, Clardy J. An unprecedented triisocyano diterpenoid antibiotic from a sponge. *J Am Chem Soc* 1984;**106**(25):7981–3.
154. Nakamura H, Wu H, Kobayashi J, Kobayashi M, Ohizumi Y, Hirata Y. Physiologically active marine natural products from *Porifera*. VIII. Agelasidines. Novel hypotaurocyamine derivatives from the Okinawan Sea sponge *Agelas nakamurai* Hoshino. *J Org Chem* 1985;**50**(14):2494–7.
155. Lee D, Shin J, Yoon KM, Kim TI, Lee SH, Lee HS, et al. Inhibition of *Candida albicans* isocitrate lyase activity by sesterterpene sulfates from the tropical sponge *Dysidea* sp. *Bioorg Med Chem Lett* 2008;**18**(20):5377–80.
156. Du L, Shen L, Yu Z, Chen J, Guo Y, Tang Y, et al. Hyrtiosal, from the marine sponge *Hyrtios erectus*, inhibits HIV-1 integrase binding to viral DNA by a new inhibitor binding site. *ChemMedChem* 2008;**3**(1):173–80.
157. Ebada SS, Lin WH, Proksch P. Bioactive sesterterpenes and triterpenes from marine sponges: occurrence and pharmacological significance. *Mar Drugs* 2010;**8**(2):313–46.
158. Kazlauskas R, Murphy PT, Quinn RJ, Wells RJ. Heteronemin, a new scalarin type sesterterpene from the sponge *Heteronema erecta*. *Tetrahedron Lett* 1976;**17**(30):2631–4.
159. Kazlauskas R, Murphy PT, Wells RJ. Furodendin, a C_{22} degraded terpene from the sponge *Phyllospongia dendyi*. *Experientia* 1980;**36**(7):814–15.
160. Kazlauskas R, Murphy PT, Wells RJ, Daly JJ. Terpenoid constituents from two *Phyllospongia* spp. *Aust J Chem* 1980;**33**(8):1783–97.
161. Roy MC, Tanaka J, de Voogd N, Higa T. New scalarane class sesterterpenes from an Indonesian sponge, *Phyllospongia* sp. *J Nat Prod* 2002;**65**(12):1838–42.
162. Liu H, Namikoshi M, Meguro S, Nagai H, Kobayashi H, Yao X. Isolation and characterization of polybrominated diphenyl ethers as inhibitors of microtubule assembly from the marine sponge *Phyllospongia dendyi* collected at Palau. *J Nat Prod* 2004;**67**(3):472–4.
163. Ponomarenko LP, Kalinovsky AI, Stonik VA. New scalarane-based sesterterpenes from the sponge *Phyllospongia madagascarensis*. *J Nat Prod* 2004;**67**(9):1507–10.
164. Kikuchi H, Tsukitani Y, Shimizu I, Kobayashi M, Kitagawa I. Foliaspongin, an antiinflammatory bishomosesterterpene from the marine sponge *Phyllospongia foliascens* (Pallas). *Chem Pharm Bull (Japan)* 1981;**29**(5):1492–4.

165. Kikuchi H, Tsukitani Y, Shimizu I, Kobayashi M, Kitagawa I. Marine natural products. XI. An antiinflammatory scalarane-type bishomosesterterpene, foliaspongin, from the Okinawan marine sponge *Phyllospongia foliascens* (Pallas). *Chem Pharm Bull (Japan)* 1983;**31**(2):552–6.
166. Sokoloff S, Halevy S, Usieli V, Colorni A, Sarel S. Prianicin A and B, nor-sesterterpenoid peroxide antibiotics from Red Sea sponges. *Experientia* 1982;**38**(3):337–8.
167. Crews P, Bescansa P, Bakus GJ. A non-peroxide norsesterterpene from a marine sponge *Hyrtios erecta. Experientia* 1985;**41**(5):690–1.
168. de Silva ED, Scheuer PJ. Manoalide, an antibiotic sesterterpenoid from the marine sponge *Luffariella variabilis* (Polejaeff). *Tetrahedron Lett* 1980;**21**(17):1611–14.
169. de Silva ED, Scheuer PJ. Three new sesterterpenoid antibiotics from the marine sponge *Luffariella variabilis* (Polejaeff). *Tetrahedron Lett* 1981;**22**(33):3147–50.
170. Kernan MR, Faulkner DJ. Halichondramide, an antifungal macrolide from the sponge *Halichondria* sp. *Tetrahedron Lett* 1987;**28**(25):2809–12.
171. Carmely S, Kashman Y. Structure of swinholide-A, a new macrolide from the marine sponge *Theonella swinhoei. Tetrahedron Lett* 1985;**26**(4):511–14.
172. Wright AE, Botelho JC, Guzmán E, Harmody D, Linley P, McCarthy PJ, et al. Neopeltolide, a macrolide from a lithistid sponge of the family *Neopeltidae. J Nat Prod* 2007;**70**(3):412–16.
173. Pettit RK, McAllister SC, Pettit GR, Herald CL, Johnson JM, Cichacz ZA. A broad-spectrum antifungal from the marine sponge *Hyrtios erecta. Int J Antimicrob Agents* 1997;**9**(3):147–52.
174. Boonlarppradab C, Faulkner DJ. Eurysterols A and B, cytotoxic and antifungal steroidal sulfates from a marine sponge of the genus *Euryspongia. J Nat Prod* 2007;**70**(5):846–8.
175. Digirolamo JA, Li XC, Jacob MR, Clark AM, Ferreira D. Reversal of fluconazole resistance by sulfated sterols from the marine sponge *Topsentia* sp. *J Nat Prod* 2009;**72**(8):1524–8.
176. Fusetani N, Matsunaga S, Konosu S. Bioactive marine metabolites. II. Halistanol sulfate, an antimicrobial novel steroid sulfate from the marine sponge *Halichondria* cf. *moorei* Bergquist. *Tetrahedron Lett* 1981;**22**(21):1985–8.
177. Nakatsu T, Walker RP, Thompson JE, Faulkner DJ. Biologically-active sterol sulfates from the marine sponge *Toxadocia zumi. Experientia* 1983;**39**(7):759–61.
178. Slate DL, Lee RH, Rodriguez J, Crews P. The marine natural product, halistanol trisulfate, inhibits pp60v-src protein tyrosine kinase activity. *Biochem Biophys Res Commun* 1994;**203**(1):260–4.
179. Zhang H, Khalil ZG, Capon RJ. Fascioquinols A–F: bioactive meroterpenes from a deep-water southern Australian marine sponge, *Fasciospongia* sp. *Tetrahedron* 2011;**67**(14):2591–5.
180. Desoubzdanne D, Marcourt L, Raux R, Chevalley S, Dorin D, Doerig C, et al. Alisiaquinones and alisiaquinol, dual inhibitors of *Plasmodium falciparum* enzyme targets from a New Caledonian deep water sponge. *J Nat Prod* 2008;**71**(7):1189–92.
181. Stout EP, Wang YG, Romo D, Molinski TF. Pyrrole aminoimidazole alkaloid metabiosynthesis with marine sponges *Agelas conifera* and *Stylissa caribica. Angew Chem Int Ed Engl* 2012;**51**(20):4877–81.
182. Huigens RW, Richards JJ, Parise G, Ballard TE, Zeng W, Deora R, et al. Inhibition of *Pseudomonas aeruginosa* biofilm formation with bromoageliferin analogues. *J Am Chem Soc* 2007;**129**(22):6966–7.
183. Kubota T, Araki A, Yasuda T, Tsuda M, Fromont J, Aoyama K, et al. Benzosceptrin C, a new dimeric bromopyrrole alkaloid from sponge *Agelas* sp. *Tetrahedron Lett* 2009;**50**(52):7268–70.
184. Endo T, Tsuda M, Okada T, Mitsuhashi S, Shima H, Kikuchi K, et al. Nagelamides A–H, new dimeric bromopyrrole alkaloids from marine sponge *Agelas* species. *J Nat Prod* 2004;**67**(8):1262–7.
185. Araki A, Tsuda M, Kubota T, Mikami Y, Fromont J, Kobayashi J. Nagelamide J, a novel dimeric bromopyrrole alkaloid from a sponge *Agelas* species. *Org Lett* 2007;**9**(12):2369–71.
186. Araki A, Kubota T, Tsuda M, Mikami Y, Fromont J, Kobayashi J. Nagelamides K and L, dimeric bromopyrrole alkaloids from sponge *Agelas* species. *Org Lett* 2008;**10**(11):2099–102.

187. Araki A, Kubota T, Aoyama K, Mikami Y, Fromont J, Kobayashi J. Nagelamides Q and R, novel dimeric bromopyrrole alkaloids from sponges Agelas sp. *Org Lett* 2009;**11**(8):1785–8.
188. Kubota T, Araki A, Ito J, Mikami Y, Fromont J, Kobayashi J. Nagelamides M and N, new bromopyrrole alkaloids from sponge *Agelas* species. *Tetrahedron* 2008;**64**(48):10810–13.
189. Tanaka N, Kusama T, Takahashi-Nakaguchi A, Gonoi T, Fromont J, Kobayashi J. Nagelamides U–W, bromopyrrole alkaloids from a marine sponge *Agelas* sp. *Tetrahedron Lett* 2013;**54**(29):3794–6.
190. Tanaka N, Kusama T, Takahashi-Nakaguchi A, Gonoi T, Fromont J, Kobayashi J. Nagelamides X–Z, dimeric bromopyrrole alkaloids from a marine sponge *Agelas* sp. *Org Lett* 2013;**15**(13):3262–5.
191. Tasdemir D, Topaloglu B, Perozzo R, Brun R, O'Neill R, Carballeira NM, et al. Marine natural products from the Turkish sponge *Agelas oroides* that inhibit the enoyl reductases from *Plasmodium falciparum, Mycobacterium tuberculosis* and *Escherichia coli. Bioorg Med Chem* 2007;**15**(21):6834–45.
192. Ballard TE, Richards JJ, Aquino A, Reed CS, Melander C. Antibiofilm activity of a diverse oroidin library generated through reductive acylation. *J Org Chem* 2009;**74**(4):1755–8.
193. Nakagawa M, Edno M, Tanaka N, Gen-Pei L. Structures of xestospongin A, B, C and D, novel vasodilative compounds from marine sponge, *Xestospongia exigua. Tetrahedron Lett* 1984;**25**(30):3227–30.
194. Rao KV, Kasanah N, Wahyuono S, Tekwani BL, Schinazi RF, Hamann MT. Three new manzamine alkaloids from a common Indonesian sponge and their activity against infectious and tropical parasitic diseases. *J Nat Prod* 2004;**67**(8):1314–18.
195. Rosser RM, Faulkner DJ. Two steroidal alkaloids from a marine sponge, *Plakina* sp. *J Org Chem* 1984; **49**(26):5157–60.
196. Jeon JE, Na Z, Jung M, Lee HS, Sim CJ, Nahm K, et al. Discorhabdins from the Korean marine sponge *Sceptrella* sp. *J Nat Prod* 2010;**73**(2):258–62.
197. Jang JH, van Soest RW, Fusetani N, Matsunaga S. Pseudoceratins A and B, antifungal bicyclic bromotyrosine-derived metabolites from the marine sponge *Pseudoceratina purpurea. J Org Chem* 2007;**72**(4):1211–17.
198. Kon Y, Kubota T, Shibazaki A, Gonoi T, Kobayashi J. Ceratinadins A–C, new bromotyrosine alkaloids from an Okinawan marine sponge *Pseudoceratina* sp. *Bioorg Med Chem Lett* 2010;**20**(15):4569–72.
199. Yao G, Chang LC. Novel sulfated sesterterpene alkaloids from the marine sponge *Fasciospongia* sp. *Org Lett* 2007;**9**(16):3037–40.
200. Yao G, Kondratyuk TP, Tan GT, Pezzuto JM, Chang LC. Bioactive sulfated sesterterpene alkaloids and sesterterpene sulfates from the marine sponge *Fasciospongia* sp. *J Nat Prod* 2009;**72**(2):319–23.
201. Jang KH, Kang GW, Jeon JE, Lim C, Lee HS, Sim CJ, et al. Haliclonin A, a new macrocyclic diamide from the sponge *Haliclona* sp. *Org Lett* 2009;**11**(8):1713–16.
202. Arai M, Sobou M, Vilchéze C, Baughn A, Hashizume H, Pruksakorn P, et al. Halicyclamine A, a marine spongean alkaloid as a lead for anti-tuberculosis agent. *Bioorg Med Chem* 2008;**16**(14):6732–6.
203. Arai M, Ishida S, Setiawan A, Kobayashi M. Haliclonacyclamines, tetracyclic alkylpiperidine alkaloids, as anti-dormant mycobacterial substances from a marine sponge of *Haliclona* sp. *Chem Pharm Bull (Tokyo)* 2009;**57**(10):1136–8.
204. Fattorusso E, Minale L, Sodano G, Moody K, Thomson RH. Aerothionin, a tetrabromo-compound from *Aplysina aerophoba* and *Verongia thiona. J Chem Soc D* 1970;**12**:752–3.
205. Encarnación RD, Sandoval E, Malmstrøm J, Christophersen C. Calafianin, a bromotyrosine derivative from the marine sponge *Aplysina gerardogreeni. J Nat Prod* 2000;**63**(6):874–5.
206. Thoms C, Wolff M, Padmakumar K, Ebel R, Proksch P. Chemical defense of Mediterranean sponges *Aplysina cavernicola* and *Aplysina aerophoba. Z Naturforsch [C]* 2004;**59**(1–2):113–22.
207. Moody K, Thomson RH, Fattorusso E, Minale L, Sodano G. Aerothionin and homoaerothionin: two tetrabromo spirocyclohexadienylisoxazoles from *Verongia* sponges. *J Chem Soc Perkin Trans* 1972;**1**: 18–24.

208. Kernan MR, Cambie RC, Bergquist PR. Chemistry of sponges, VII. 11,19-Dideoxyfistularin 3 and 11-hydroxyaerothionin, bromotyrosine derivatives from *Pseudoceratina durissima*. *J Nat Prod* 1990;**53**(3):615–22.

209. Acosta AL, Rodriguez AD. 11-Oxoaerothionin: a cytotoxic antitumor bromotyrosine-derived alkaloid from the Caribbean marine sponge *Aplysina lacunosa*. *J Nat Prod* 1992;**55**(7):1007–12.

210. Gao H, Kelly M, Hamann MT. Bromotyrosine-derived metabolites from the sponge *Aiolochroia crassa*. *Tetrahedron* 1999;**55**(32):9717–26.

211. Bao B, Sun Q, Yao X, Hong J, Lee CO, Cho HY, et al. Bisindole alkaloids of the topsentin and hamacanthin classes from a marine sponge *Spongosorites* sp. *J Nat Prod* 2007;**70**(1):2–8.

212. Ren S, Ma W, Xu T, Lin X, Yin H, Yang B, et al. Two novel alkaloids from the South China Sea marine sponge *Dysidea* sp. *J Antibiot (Tokyo)* 2010;**63**(12):699–701.

213. Lee HS, Yoon KM, Han YR, Lee KJ, Chung SC, Kim TI, et al. 5-Hydroxyindole-type alkaloids, as *Candida albicans* isocitrate lyase inhibitors, from the tropical sponge *Hyrtios* sp. *Bioorg Med Chem Lett* 2009;**19**(4):1051–3.

214. Gunasekera SP, McCarthy PJ, Kelly-Borges M. Hamacanthins A and B, new antifungal bis indole alkaloids from the deep-water marine sponge, *Hamacantha* sp. *J Nat Prod* 1994;**57**(10):1437–41.

215. Jiang B, Yang CG, Wang J. Enantioselective synthesis for the (−)-antipode of the pyrazinone marine alkaloid, hamacanthin A. *J Org Chem* 2001;**66**(14):4865–9.

216. Jiang B, Yang CG, Wang J. Enantioselective synthesis of marine indole alkaloid hamacanthin B. *J Org Chem* 2002;**67**(4):1396–8.

217. Casapullo A, Bifulco G, Bruno I, Riccio R. New Bisindole alkaloids of the topsentin and hamacanthin classes from the Mediterranean marine sponge *Rhaphisia lacazei*. *J Nat Prod* 2000;**63**(4):447–51.

218. Boyd KG, Harper MK, Faulkner DJ. Oceanapamine, a sesquiterpene alkaloid from the Philippine sponge *Oceanapia* sp. *J Nat Prod* 1995;**58**(2):302–5.

219. Djura P, Stierle DB, Sullivan B, Faulkner DJ, Arnold EV, Clardy J. Some metabolites of the marine sponges *Smenospongia aurea* and *Smenospongia* (ident. *Polyfibrospongia*) *echina*. *J Org Chem* 1980;**45**(8):1435–41.

220. Sata NU, Matsunaga S, Fusetani N, van Soest RW. Aurantosides D, E, and F: new antifungal tetramic acid glycosides from the marine sponge *Siliquariaspongia japonica*. *J Nat Prod* 1999;**62**(7):969–71.

221. Plaza A, Gustchina E, Baker HL, Kelly M, Bewley CA. Mirabamides A–D, depsipeptides from the sponge *Siliquariaspongia mirabilis* that inhibit HIV-1 fusion. *J Nat Prod* 2007;**70**(11):1753–60.

222. Andavan GSB, Lemmens-Gruber R. Cyclodepsipeptides from marine sponges: natural agents for drug research. *Mar Drugs* 2010;**8**(3):810–34.

223. Ibrahim SR, Min CC, Teuscher F, Ebel R, Kakoschke C, Lin W, et al. Callyaerins A–F and H, new cytotoxic cyclic peptides from the Indonesian marine sponge *Callyspongia aerizusa*. *Bioorg Med Chem* 2010;**18**(14):4947–56.

224. Nishimura S, Arita Y, Honda M, Iwamoto K, Matsuyama A, Shirai A, et al. Marine antifungal theonellamides target 3β-hydroxysterol to activate Rho1 signaling. *Nat Chem Biol* 2010;**6**(7):519–26.

225. Kobayashi J, Tsuda M, Nakamura T, Mikami Y, Shigemori H. Hymenamides a and b, new proline-rich cyclic heptapeptides from the Okinawan marine sponge *Hymeniacidon* sp. *Tetrahedron* 1993;**49**(12):2391–402.

226. Kumar P, Selvi SS, Govindaraju M. In vitro anti-biofilm and anti-bacterial activity of *Junceella juncea* for its biomedical application. *Asian Pac J Trop Biomed* 2012;**2**(12):930–5.

227. Murthy YLN, Mallika D, Abdul R, Reddy GD. A new antifungal briarane diterpenoid from the gorgonian *Junceella juncea* Palas. *Bioorg Med Chem Lett* 2011;**21**(24):7522–5.

228. Rodríguez AD, Ramírez C, Rodríguez II, González E. Novel antimycobacterial benzoxazole alkaloids, from the west Indian Sea whip *Pseudopterogorgia elisabethae*. *Org Lett* 1999;**1**(3):527–30.

229. Iguchi K, Saitoh S, Yamada Y. Novel 19-oxygenated sterols from the Okinawan soft coral *Litophyton viridis. Chem Pharm Bull* 1989;**37**(9):2553–4.
230. Liu J, Zeng L, Wu D. A new sterol from the marine soft coral *Nephthea sinulata. Gaodeng Xuexiao Huaxue Xuebao (Chem J Chinese Univ)* 1992;**13**(3):355–7.
231. Sharma GM, Michael L, Burkholder PR. Goniodomin, a new antibiotic from a dinoflagellate. *J Antibiot (Japan)* 1968;**21**(11):659–64.
232. Bourel-Bonnet L, Rao KV, Hamann MT, Ganesan A. Solid-phase total synthesis of kahalalide A and related analogues. *J Med Chem* 2005;**48**(5):1330–5.
233. Gustafson K, Andersen RJ. Chemical studies of British Columbia nudibranchs. *Tetrahedron* 1985;**41**(6):1101–8.
234. Iijima R, Kisugi J, Yamazaki M. A novel antimicrobial peptide from the sea hare *Dolabella auricularia. Dev Comp Immunol* 2003;**27**(4):305–11.

CHAPTER

ALGAE AS A NOVEL SOURCE OF ANTIMICROBIAL COMPOUNDS: CURRENT AND FUTURE PERSPECTIVES

18

A. Parsaeimehr[1] and G.A. Lutzu[2]

[1]Laboratory of Biosystems Engineering, Key Laboratory of Agrobiology of Jiangsu Province, Institute of Biotechnology, Jiangsu Academy of Agricultural Sciences, Nanjing, Jiangsu, People's Republic of China [2]Key Laboratory of Biofuels, Qingdao Institute of Bioenergy and Bioprocess Technology, Chinese Academy of Sciences, Qingdao, Shandong, People's Republic of China

INTRODUCTION

Algae are documented as a polyphyletic, broad, and diverse group of unicellular-to-multicellular eukaryotic organisms.[1] The majorities of algal species are autotrophic and show a close metabolic pathway as well as a regulatory network to plant cells, even if many differentiated cells and tissue types (eg, xylem, phloem, phyllids, stomata, rhizoids, and roots) are not found in algal species.[2,3] Besides autotrophic algae, species such as *Cochlodinium polykrikoides*, *Chlorella protothecoides*, and *Dinophysis acuminata* support a mixotrophic metabolism; that is, they can gain energy both from photosynthesis and from the uptake of organic carbon.[4,5] Interestingly, other algal species such as *Prototheca* sp., *Crypthecodinium cohnii*, and *Schizochytrium* sp. are heterotrophs, and they can be characterized by limited or no photosynthetic activity.[6] Algae are currently classified into 10 major phyla (Heterokontophyta, Glaucophyta, Euglenophyta, Cryptophyta, Haptophyta, Rhodophyta, Dinophyta, Chlorophyta, prokaryotic Cyanophyta, and Prochlorophyta) via DNA sequence methods.[7]

Algae generate a range of valuable natural metabolites that may be used within the field of pharmacognosy and medicinal chemistry.[8] Comprehensive nutritional studies have indicated that some protein-rich algae like *Arthrospira platensis* and *Arthrospira maxima* are considered to be of high quality and comparable to vegetable proteins, and they also have the potential to be employed as dietary supplements in the food industry.[9] Antioxidants are known as molecules that inhibit or reduce the reactive oxygen species (ROS) in metabolism, and relevant research available to date shows that many algae species possess considerable antioxidant compounds that can be used by nutraceutical and pharmaceutical industries.[10] Currently, antioxidant compounds such as carotenoids (beta-carotene, fucoxanthin, lutein, antheraxanthin, and zeaxanthin), phenolics (stypodiol, isoepitaondiol, and taondiol), phycobilins (phycoerythrin), and sulfated polysaccharides (laminaran and sulfated galactans) are screened and isolated from different classes of algae[11,12].

Antibiotic Resistance. DOI: http://dx.doi.org/10.1016/B978-0-12-803642-6.00018-6

A recent study on three brown algae species, *Scytosiphon lomentaria*, *Papenfussiella kuromo*, and *Nemacystus decipiens* (Phaeophyceae), and *Porphyra* sp. (Rhodophyceae) demonstrated that phenolic compounds exist in a range about 2.2–9.4 mg catechin equivalent/g dry sample, and how these species contain significant properties of antioxidants.[13]

Vitamin B-12 (cobalamin) performs substantial functions in the regulation and synthesis of DNA, metabolism of cells, nervous system, and formation of blood, and as the antipernicious anemia factor in liver.[14] The Institute of Medicine (IOM) of the National Academies (United States) recommends that adults consume 2–3 μg of Vitamin B-12 per day.[15] Neither plants nor animals are able to produce Vitamin B-12 since they have no cobalamin-dependent enzymes, while only bacteria and archaea are known to possess the necessary enzyme for its production.[16] In particular, algae are able to incorporate this vitamin through a symbiotic relationship with bacteria, and algal phyla such as Chlorophyta, Glaucocystophyta, Rhodophyta, Cryptophyta, Dinophyta, Euglenophyta, Haptophyta, and Heterokontophyta are rich in Vitamin B-12 and can be a bioavailable source for mammals.[17] The worldwide incidence rates of cancer have been growing, and the improvement of cancer chemoprevention procedures using natural products has been documented as a field with significant potential to diminish the burden of cancer.[18] Considerably, algal bioactive compounds with the ability to induce apoptosis in cancer cells can be used as cancer chemopreventives. For example, fucoidan is a sulfated polysaccharide that is isolated from different brown algal species, such as *Fucus vesiculosus*, *Adenocystis utricularis*, or *Grateloupia filicina*, while *Spatoglossum schroederi* has been shown to exhibit a number of biological effects, including antitumor ones.[19,20] The antioxidant and anticancer activity of aqueous extracts of nine microalgal species, including *Nostoc muscorum* and *Oscillatoria* sp., have been investigated as well, and it was found that antioxidant and anticancer activities with the content of phycobilin pigment in both cyanobacterial species were improved by increasing the nitrate concentration in the algal BG-11 medium. However, the production of phycobilin pigments and allophycocyanin were reduced or completely stopped by reducing nitrate concentration in *Oscillatoria* sp.[21]

A range of algae types, including *Cladophora glomerata*, *Sargassum natans*, *Scinaia furcellata*, *Halopteris scoparia*, *Dictyota dichotoma*, and *Ulva lactuca*, were screened for antiprotozoal activity against *Leishmania donovani*, *Trypanosoma cruzi*, *Trypanosoma brucei rhodesiense*, and *Plasmodium falciparum*. Although all these tested algal species did not show antiprotozoal activity against *T. cruzi*, all of them displayed considerable trypanocidal activity against *T. brucei rhodesiense*, while *S. natans* showed the greatest effectiveness (IC_{50} 7.4 μg/mL). Furthermore, all these chosen algae except *H. scoparia* demonstrated antileishmanial activity; the strongest effect was observed by ethanolic extracts of *U. lactuca* and *Posidonia oceanica* (IC_{50} 5.9 and 8.0 μg/mL, respectively). In this study, FabI, a crucial enzyme of the fatty acid (FA) system of *P. falciparum*, was chosen as a target to assess the inhibitory influence of selected algae on tested microorganisms. The extracts of *C. glomerata* and *U. lactuca* efficiently inhibited the FabI enzyme (IC_{50} values of 1.0 and 4.0 μg/mL, respectively), while no cytotoxic of extracts were observed against mammalian L6 cells.[22]

A survey on 600 strains of Cyanophytes revealed how almost 10% of samples contained substances that caused significant reduction in cytopathic effect associated with viral infection. This study identified the order Chroococcales as the typical one producing antiviral agents. In another study on the lipids isolated and recovered from the Folch lower layer of Brazilian seaweed (*Sargassum vulgare*), the presence of sulfatides was revealed by the characteristic orcinol-positive band. The purified fraction exhibited considerable antiviral activity against herpes simplex virus type 1 and 2, and 1,2-di-*O*-palmitoyl-3-*O*-(6-sulfo-α-D-quinovopyranosyl)-glycerol compound was identified as a responsible antiviral substance.[23]

Cyanobacteria (blue-green algae) are recognized as a reliable source of natural products, and a range of bioactive compounds were isolated from them. For example, sulfonic acid–containing glycolipids are reported as a new class of HIV-1-inhibitory compounds against HIV-1 in cultured human lymphoblastoid CEM, MT-2, LDV-7, and C3-44 cell lines.[24] Leishmaniasis is a parasitic disease that is found in parts of the tropics, subtropics, and southern Europe. Studies of algal crude extracts from *Caulerpa racemosa*, *Osmundaria obtusiloba*, *Stypopodium zonale*, *Dictyota ciliolata*, and *Dictyota menstrualis* exhibited cytotoxicity of these extracts on macrophages showed CC50/24 h, with a concentration range of 18.2–240.0 μg/mL. In addition, the antileishmanial activity of the *O. obtusiloba* extract showed potential activity that was better than the effect obtained with pentamidine, a reference drug.[25]

ALGAE AS A UNIQUE SOURCE OF ANTIMICROBIALS

Humanity regularly faces problems connected with the increase in the occurrence of bacterial infections, and currently, synthetic drugs with considerable side effects are being applied to deal with these medical concerns. Meanwhile, overprescribing has made antibiotics less effective to the extent that microbes resistant to one or more antimicrobial agents are a growing problem in the world and are responsible of millions of casualties each year.[26,27] To address this issue, scientists have focused on renewable, naturally eco-friendly, and easily obtainable sources of bioactive compounds from natural products and, as previously mentioned, algae have attracted a great deal of attention.[28,29] Early reports related to 151 screened species of British marine algae for the production of antibiotics demonstrated that a number of them, including *Asparagopsis armata*, *Bonnemaisonia asparagoides*, *Bonnemaisonia hamifera*, *Chondrus crispus*, *Dilsea carnosa*, *Gloiosiphonia capillaris*, *Sphondylothamnion multifidum*, *Desmarestia aculeata*, *Dypsis ligulata*, *Laminaria digitata*, *Dictyopteris membranacea*, *D. dichotoma*, *Halidrys siliquosa*, and most members of the Rhodomelaceae family had considerable antibacterial activity. Moreover, these researchers claimed that even though the synthesis of antibiotics seems to be the characteristic of several families, it is not possible to set up a significant correlation between taxonomy and antibiotic production.[30]

In another study by Salvador et al.,[31] 82 different marine macroalgae from the Chlorophyceae, Phaeophyceae, and Rhodophyceae classes were screened for antimicrobial activity. Among the selected taxa, Phaeophyceae were reported to be as the most active, with 84% of the algae from this taxa showing antimicrobial activity.

These results were also comparable with those reported by Taskin et al.,[32] in which methanolic extracts of algae belonging to Rhodophyceae (*Corallina officinalis*), Phaeophyceae (*Cystoseira barbata*, *D. dichotoma*, *Halopteris filicina*, *Cladostephus spongiosus f. verticillatus*), and Chlorophyceae (*Ulva rigida*) from the North Aegean Sea (Turkey) showed antimicrobial activity against *Enterococcus faecalis*. The authors also claimed that extract from *C. barbata* has showed broader activity against a range of microorganisms such as *Escherichia coli*, *Enterobacter aerogenes*, *Staphylococcus aureus*, *Micrococcus luteus*, and *E. faecalis*.

More recently, Prakash et al.[33] evaluated the antimicrobial potential of some freshwater microalgae such as *Oscillatoria sancta* (Kuetz) Gomont, *Lyngbya birgei* Smith G.M., *Oedogonium echinospermum*, *Spirogyra decimina* (Muller) Kuetz, *Spirogyra grantiana* Transeau, *Spirogyra crassa*, *Spirogyra biformis* Jao, and *Spirogyra condensate* (Vaucher) Kuetz, collected from the Thamirabarani River (South India), against human bacterial pathogens including *E. coli* (ATCC 35218), *S. aureus* (ATCC 6538), *Salmonella* Typhi (MTCC 733), *Proteus vulgaris*, *Proteus mirabilis*, and *Streptococcus pyogenes*. The results indicated that the methanolic and ethanolic extracts of these microalgae showed antibacterial activity against most of the selected pathogens.

Different solvents such as ethanol, acetone, and methanol-toluene were used for the extraction of bioactive compounds from marine algal species belonging to different divisions (Chlorophyta, Phaeophyta, and Rhodophyta) and their respective extracts were screened for antibacterial and antifungal activity. Moreover, ethanol extracts revealed the most antimicrobial activity against selected microorganisms. Rhodophyta spp. (*Laurencia okamurai*, *Dasya scoparia*, *Grateloupia filicina*, and *Plocamium telfairiae*) showed a wide spectrum of antibacterial activity, and some of them were identified as having the strongest antimicrobial activity against all tested microorganisms. In particular, *Pseudomonas solancearum* and *Penicilium citrinum* were identified as the most sensitive microorganisms to the extracts of marine algae, and the extracts of the Rhodophyta spp. were identified as having the strongest antimicrobial activity against all the tested microorganisms, with great potential to be used for the development of new medicines.[34]

In another study, a range of solvents including hexane, chloroform, and ethanol were used to evaluate the antimicrobial activity of six algae (Rhodophyta and Chlorophyta) from the North Ceará coast (in northeastern Brazil). The results indicated that *Amansia multifida* (Rhodophyta) had considerable antimicrobial activity against Gram-negative strains such as *E. aerogenes*, *Klebsiella pneumoniae*, *Pseudomonas aeruginosa*, *Salmonella* Typhi, *Salmonella choleraesuis*, *Serratia marcescens*, *Vibrio cholerae*, and the Gram-positive bacteria *Bacillus subtilis* and *S. aureus*.[35] Similar results were found by Bhagavathy et al.,[36] who evaluated the existence of phytochemicals and their antimicrobial role of the green algae *Chlorococcum humicola*. Screening of extracts of *C. humicola* showed antimicrobial activity against virulent strains including *P. aeruginosa*, *E. coli*, *K. pneumoniae*, *V. cholerae*, *Salmonella* Typhimurium, *S. aureus*, *B. subtilis*, *Candida albicans*, *Aspergillus flavus*, and *Aspergillus niger*.

Analysis of various features of previous studies in the area of algae as a source of natural products reveals how many Rhodophyta spp. produce considerable bioactive compounds, such as halogenated terpenoids, acetogenins, and polyphenols with antimicrobial activity.[37] In particular, *S. aureus* is reported to be resistant to some extent to the antibiotic vancomycin. Bhateja et al.[38] evaluated aqueous and organic extracts of nine blue-green microalgae strains against in vitro generated vancomycin intermediate resistant *S. aureus* strains. They found out that aqueous extracts of all the blue-green microalgae cultures were inactive, while all the organic extracts of *Anabaena virabilis* and *Anabaena* sp. (hexane, chloroform, and methanolic) showed activity against these resistant strains with minimum inhibitory concentration (MIC) of 32–64 μg/mL.

The Canary Islands are recognized to be one of the most important areas of high biodiversity of the marine biota (both flora and fauna), and have attracted the interest of many biologists as a result. Researchers screened 44 species of seaweed from Gran Canaria (Canary Islands, Spain) for a better understanding of the antibacterial and antifungal compounds produced by seaweed. Results indicated that 28 species exhibited antibacterial activity, while *Asparagopsis taxiformis* and *Cymopolia barbata* showed the strongest antimicrobial activity against a wide-ranging spectrum of selected microbes. Indeed, each selected species with antibacterial activity showed antimicrobial activity against Gram-positive bacteria, while only *Asparagopsis taxiformis* and *Osmundea hybrid* showed antimicrobial activity against mycobacteria.[39]

A wide range of algae species, including *Ulva fasciata*, *Bryopsis plumosa*, *Chaetomorpha antennina*, *Acrosiphonia orientalis*, *Grateloupia filicina*, *Hypnea pannosa*, *Gracilaria corticata*, *Centroceras clavulatum*, *Portieria hornemannii*, *Cheilosporum spectabile*, *Sargassum wightii*, *Stocheospermum marginatum*, *Chnoospora bicanaliculata*, and *Padina tetrastromatica*, were screened for antimicrobial activity. Researchers stated that drying reduced the bioactive compounds in the chosen algal species, and they found that methanol:toluene (3:1) was the best solvent for extracting the antimicrobial compounds from fresh algae, and that the antimicrobial activity from chosen algae was due to a lipophilic compound that was stable over a broad range of temperatures (30–60°C). Furthermore, these selected algae were considerably active against Gram-negative bacteria than Gram-positive bacteria, and in particular, *Acrosiphonia orientalis* exhibited antibacterial activity against 70% of the tested microorganisms. *Stocheospermum marginatum* was the only seaweed that showed activity against *K. pneumoniae*, while extract from *Gracilaria corticata* was highly active against *P. mirabilis*.[40] Marine algae *Halimeda opuntia* (Order: Bryopsidales) and *Sarconema filiforme* (Order: Gigartinales) are described as potent algal species with considerable antimicrobial, antiplasmid, and cytotoxic activity. Antimicrobial bioassaying of these algae species revealed that the *Halimeda* extract has considerable antibacterial activity compared with commercial antibiotics against *S. aureus*, whereas the *Sarconema* extract can be considered as an antifungal with inhibitory influence on *C. albicans* growth.[41]

Interestingly, some reports point to the fact that the algal antimicrobial compounds change during algal growth and seasonal variations. For example, Hornsey and Hide[42] reported that seasonal variations in antibacterials occurred in four main patterns in tested algae:

1. The *Polysiphonia* type, in which antibiotic production occurs constantly throughout the year
2. The *Laminaria* type, which is characterized by maximum antibiotic production during the winter months and is also found in *Chondrus crispus*, *Laurencia pinnatifida*, and *U. lactuca*
3. The *Dictyota* type, showing a summer peak of activity, which also occurs in *Dilsea carnosa* and *Ascophyllum nodosum*
4. The *Codium* type, where there is a spring peak of activity, as exhibited by *Halidrys siliquosa.*

Nowadays, comprehensive methodologies (environmentally clean extraction techniques) and strategies (using advanced analytical instruments) are carried out to screen the novel natural products and to link the biological activity of extracts with a particular chemical composition.[43] For example, bioactive compounds (antioxidant and antimicrobials) in the algae *Himanthalia elongata* and the microalgae *Synechocystis* sp. were screened using gas chromatography–mass spectrometry (GC-MS) and high-performance liquid chromatography-diode array (HPLC-DAD) instruments and different FAs (palmitic, palmitoleic, and oleic acids) and volatile compounds (phytol, fucosterol, neophytadiene) with antimicrobial activity were identified against *E. coli*, *S. aureus*, *C. albicans*, and *A. niger.*[44]

ANTIMICROBIAL MOLECULES ISOLATED FROM ALGAL SPECIES

The algal phylum of Rhodophyta (also known as red algae) is considered as one of the oldest groups of algae; it includes more than 6500 identified species, characterized by the presence of phycoerythrin (a red protein-pigment complex), carrageenan (a linear, sulfated polysaccharide), and phlorotannins (a class of tannins).[45] Lim et al.[46] reported that the extracts of *Symphyocladia latiuscula,* a red alga, exhibit antimicrobial activity against a broad spectrum of microorganisms (15 strains of bacteria, 4 strains of fungi, and 17 strains of yeasts), while the strongest antimicrobial effect of red algae extract was observed against *Vibrio mimicus* (50 μg/mL) and *Vibrio vulnificus* (50 μg/mL). The antimicrobial compounds forming *Symphyocladia latiuscula* (2,3,6-tribromo-4,5-dihydroxybenzyl alcohol, 2,3,6-tribromo-4,5-dihydroxybenzyl methyl ether, and its isomer) were isolated and identified using solvent extractions via high-performance liquid chromatography (HPLC), nuclear magnetic resonance (NMR), mass spectrometry, and X-ray crystallographic analysis. Indeed, intraperitoneal (IP) injection in mice with 2,3,6-tribromo-4,5-dihydroxybenzyl alcohol (3 mg), and 2,3,6-tribromo-4,5-dihydroxybenzyl methyl ether (5 mg) had no acute toxicity response.

Oh et al.[47] reported that they successfully isolated a series of bromophenols from the red alga *Odonthalia corymbifera* in a substantial yield via reactions of bis (hydroxyphenyl) methanes with bromine. Among the isolated natural products 2,2′,3,3′-tetrabromo-4,4′,5,5′-tetrahydroxydiphenylmethane had a substantial antimicrobial activity against *C. albicans, Aspergillus fumigatus, Trichophyton rubrum,* and *Trichophyton mentagrophytes.* Conversely, the synthetic bromophenols 3,3′-dibromo-6,6′-dihydroxydiphenylmethane and 3,3′,5,5′-tetrabromo-6,6′-dihydroxydiphenylmethane exhibited effective antibacterial activity against *S. aureus, B. subtilis, Micrococcus luteus, P. vulgaris,* and *Salmonella* Typhimurium. In addition, 3,3′,5,5′-tetrabromo-2,2′,4,4′-tetrahydroxydiphenylmethane was isolated from *Odonthalia corymbifera* as a natural product compound.

Other isolated bioactive compounds, like novel laurene-type sesquiterpenes (12-hydroxy isolaurene, 8,11-dihydro-12-hydroxy isolaurene, and isolauraldehyde), were isolated from the alga *Laurencia obtuse* (order *Ceramiales*) and tested for antimicrobial activity, revealing a considerable amount against *B. subtilis* and *S. aureus* (MIC 35 and 27 μg/mL, respectively).[48]

A range of solvents (methanol, dichloromethane, hexane, and chloroform) were used to screen the antimicrobial activity of *Jania rubens* (Rhodophyta) extracts. Methanol and chloroform extracts (4 mg/disk) showed the strongest antimicrobial activity compared with hexane and dichloromethane extracts; moreover, GC-MS analysis of the volatile components identified 40 compounds, including *n*-docosane, *n*-eicosane, and *n*-tetratriacontane in *J. rubens.*[49]

Seaweeds such as *Laminaria saccharina, Laminaria digitata, H. elongata, Palmaria palmata, Chondrus crispus,* and *Enteromorpha spirulina* are recognized as edible algae, and evidence demonstrates that among them, *H. elongata* contains considerable amount of phenolics (151.3 mg GAE/g of seaweed extract), as well as of tannins (38.34 mg CE/g) and total flavonoids (42.5 mg QE/g). In addition, it has significant 2, 2-diphenyl-1-picrylhydrazyl (DPPH) scavenging activity with a 50% inhibition (EC50) level at 0.125 μg/mL of extract.[50]

In addition, the potential of *H. elongata* as a natural antimicrobial source for food preservation was assessed by Gupta et al.[51] Seaweed extract at a concentration of 6% inhibited the growth of *P. aeruginosa, E. faecalis, Listeria monocytogenes,* and *Salmonella abony,* while lower concentrations of seaweed extract extended the lag phase and decreased the exponential growth rate and final population densities of microorganisms in the culture. The microtiter technique was used by these researchers to detect the antimicrobial activity of seaweed extracts against food spoilage and food pathogenic bacteria (*Listeria monocytogenes, Salmonella abony, E. faecalis,* and *P. aeruginosa*). Among the tested algae species, methanolic extracts of *H. elongata* (Phaeophyceae, order Fucales) showed the most antimicrobial activity (100% inhibition), while antimicrobial activity of *Chondrus crispus* (Rhodophyta, order Gigartinales) extracts against *E. faecalis* was enhanced from 39.28% to 100% once ethanol and acetone were applied as solvents.

Cycloeudesmol ($C_{15}H_{26}O$; average mass 222.366 Da), laurinterol ($C_{15}H_{19}BrO$; average mass 295.215 Da), and debromolaurinterol ($C_{15}H_{20}O$; average mass 216.319 Da) are known as biologically active compounds with considerable antimicrobial activity. These compounds were isolated from marine algae and successfully tested at concentrations approaching that of streptomycin against *S. aureus*, *Salmonella choleraesuis*, *Mycobacterium smegmatis*, *C. albicans*, and *E. coli*.[52]

Indolocarbazoles represent a class of bioactive compounds with antibacterial, antifungal, and anticancer activity. *Nostoc sphaericum*, *Tolypothrix tjipanasensis*, and *Fischerella ambigua* strain 108b have been reported to contain indolocarbazoles and other bioactive molecules.[53] For example, bioassay-guided fraction of *Fischerella ambigua* led to the identification and isolation of a new antimicrobial compound, parsiguine; and tjipanazoles, a chlorine with antimicrobial activity, has been isolated from *Tolypothrix tjipanasensis*.[54]

Bromophenols are a group of phenolic compounds with antimicrobial activity that have been reported extensively in marine algae. A total of 49 species of marine algae from eastern Australia have been screened for bromophenols, and five types of these compounds, including 2- and 4-bromophenol, 2,4- and 2,6-dibromophenol, and 2,4,6-tribromophenol, were found in 62% of the samples, while 2,4,6-tribromophenol was found in all samples. The concentration of bromophenols ranged from 0.9 ng/g (green alga *Codium fragile*) to 2590 ng/g (red alga *Pterocladiella capillacea*).[55]

Elatol is a sesquiterpene alcohol that can be typically found in the *Laurencia* spp., and its involvements in antiherbivore and antimicroorganisms defense responses has been described in the literature.[56] It is known as an antimicrobial compound; in particular, sesquiterpene elatol as a diversity of elatol is found in Brazilian red seaweed *Laurencia dendroidea* (Hudson) J.V. Lamouroux. The authors reported that, sesquiterpene elatol had antileishmanial activity against *Leishmania amazonensis*, with an IC_{50} of 4.0 and 0.45 μM for promastigote and intracellular amastigote after 72 h, respectively.[57]

Karatungiol, a polyol (an alcohol containing multiple hydroxyl groups), is a novel antimicrobial compound that has been isolated from dinoflagellate *Amphidinium* sp. It exhibited antifungal activity against *A. niger* at 12 μg/disk and antiprotozoan activity against *Trichomonas foetus* at 1 μg/mL.[58] Colopsinol A–E are other bioactive molecules isolated from dinoflagellate *Amphidinium* sp., and they are considered as novel polyhydroxyl molecules (a new class of polyketide natural products possessing a gentiobioside moiety and a sulfate ester). Based on scientific evidence, Colopsinol A reduced the DNA activity of polymerases α and β, and colopsinol C, D, and E had cytotoxic effects, whereas colopsinol D showed less cytotoxicity (IC_{50}, 20 μg/mL).[59,60] H-pyrido[3,4-b]indole(norharman) and 4,4′-dihydroxybiphenyl are reported to be bioactive molecules with considerable anticyanobacterial activity (concentrations of 8–80 μg/mL) and moderate antibacterial (16–160 μg/mL) and antifungal (32–40 μg/mL) activity. These molecules have been isolated from *Nodularia harveyana* and *Nostoc insulare*, respectively, and their usage as allelopathic chemicals for the development of novel antifouling chemicals are suggested by researchers.[61] Terpenes are organic compounds derived from isoprene unites (C_5H_8). Currently, these compounds containing aryl groups have been attracting the attention of many researchers, since they showed a wide range of pharmacological activity with no destructive side effects.[62]

Cystoseira sp. (brown algae; order Fucales) and *Ulva* sp. (green algae; order Ulvales) are reported as natural sources of protein, carbohydrates, minerals, and vitamins, although more notably, these two species are a good source of terpenes with antimicrobial activity.[63] In particular, zosterdiol A, zosterdiol B, zosteronol, zosteronediol (tetraprenyltoluquinols), and prenyldiketones (arylterpenes) are antimicrobials isolated from the *Cystoseira* spp.[64] Also, labda-14-ene-8-ol, labda-14-ene-3α,8α-diol, labda-14-ene-8α-9α-diol, labda-14-ene-8α-hydroxy-3-one, ent-labda-13(16)14-diene-3-one, ent-labda-13(16),14-diene, and ent-labda-13(16),14-diene-3α-ol are diterpenes with potent antimicrobial activity, which are isolated from *U. fasciata* Delile, collected from Vizhinjam Harbor, on the southwestern coast of India.[65]

Bromoform and dibromoacetic acid are also reported as other bioactive molecules with antimicrobial activity against bacteria such as *Vibrio* sp., *E. coli*, *P. aeruginosa,* and *Staphylococcus* sp. These molecules can be found in *Asparagopsis armata*. Based on available evidence, bromoform and dibromoacetic acid are found in *Asparagopsis armata* in a range from 0.58% to 4.3% and 0.02% to 2.6% of algal dry weight, respectively.[66] (Table 18.1).

Lectins are introduced as a type of sugar-binding protein that is able to be bind to cell membranes in order to become the glycol portion of glycoconjugates on them. Lectins have the ability to connect the molecules, avoiding cell–cell interaction and getting the immune system involved.[88] They have been identified in algal species like *Eucheuma serra* and *Galaxaura marginata*, and isolated algal lectins have been demonstrated to be active against *V. vulnificus*.[89]

A comprehensive study by Kamei et al.[90] carried out on 342 species of collected marine algae from Japanese coastlines for antimicrobial activity, isolated diterpene sargafuran, a new antimicrobial molecule from the methanolic extract of the marine brown alga *Sargassum macrocarpum*. Based on these reports, argafuran completely inhibited the growth of *Propionibacterium acnes*. Moreover, viridamides A and B, two linear lipopeptides isolated from *Oscillatoria* nigro-viridis, proved to have antitrypanosomal activity ($IC_{50} = 1.1\ \mu M$) and antileishmanial activity ($IC_{50} = 1.5\ \mu M$).[91] Cyanovirin-N, borophycin, cryptophycin, lipopeptides, scyptolin, chondriol, cycloeudesmol, prepacifenol, laurinterol, and debromolaurinterol are other high-value metabolites derived from algal species that exhibit antimicrobial activity.[92,93]

ALGAL FAs AS ANTIMICROBIAL SUBSTANCES

FAs are known to be the most active antimicrobials in human skin lipid (10–15 $\mu g/m^2$ of free fatty acids exist). Among them, the most important are lauric acid (C12:0), myristic acid (C14:0), palmitic acid (C16:0), sapienic acid (C16:1*n*-10), and *cis*-8-24 octadecenoic acid (C18:1*n*-10).[94,95]

Algae are at the bottom of the food chain and tend to be the species that produce FA. Algal genera (such as Haptophyta, Cryptomonas, and Eustigmatophytes) typically contain a considerable amount of saturated fatty acids (SFAs) and polyunsaturated fatty acids (PUFAs). These essential FAs are important to human health and must be consumed regularly since they cannot be synthesized by humans.[96] Besides PUFAs, SFAs has been characterized as an antitumor compound, as well as being antibacterial agents in the red seaweed *Amphiroa zonata*.[97,98]

Palmitic acid is a SFA that exists in the FA profile of microalgal species. Evidence points to the fact that this SFA has antiviral activity; for example, Santoyo et al.[99] reported anti-herpes simplex

Table 18.1 A List of Bioactive Molecules Isolated from Algal Species

Bioactive Molecule	Algal Source	Mode of Action	Target	Reference
3-Formylindole	*Botryocladia leptopoda*	Antiviral and anticancer activity	*Vaccinia* virus	67,68
Polyhydroxylated	*Fucus vesiculosus*	Bactericidal	Gram-positive and Gram-negative bacteria	69,70
Fucophlorethol	*Ascophyllum nodosum*	Antitumor activity	HT-29 and HCT-116 human colon cancer	70,71
Fucoidan	*Symphyocladia latiuscula*	Bactericidal	*Staphylococcus aureus*, *Staphylococcus epidermidis*, and *Pseudomonas aeruginosa*	32,72
Sulfated galactans	*Botryocladia occidentalis*	Immunostimulatory activity, antiviral and anticoagulant activity	Herpes simplex virus, human metapneumovirus	73
Phycarine	*Laminaria digitata*	Antitumor activity	Stimulator of both hormonal and cellular immunity, and stimulator of phagocytosis by peripheral blood cells	74,75
Loliolide	*Undaria pinnatifida*	Antioxidant, bactericidal, and antitumor activity	*Bacillus subtilis*, *Escherichia coli*, *S. aureus*, and *Neisseria gonorrhoeae*	76,77
Bisebromoamide	*Lyngbya* sp.	Anticancer activity	Renal cell carcinoma	78
p-KG03 exopolysaccharide	*Gyrodinium impudicum*	Antifungal Antiprotozoan Antiviral	HSV-1 *Aspergillus niger* *Trichomonas foetus* *Encephalomyocarditis* virus	79,80
Allophycocyanin	*Spirulina platensis*	Antiviral	*Enterovirus* 71	81
Polysaccharide GA3P, D-galactan sulfate associated with L-(þ)-lactic acid	*Gymnodinium* sp.	Antitumor	Colon cancer cell lines	82
Calothrixin-A	*Calothrix* sp.	Antimalarial and anticancerous Antiviral	Human myeloid leukemia K562 cells	83–85
Sulfoquinovosyldiacylglycerol, KM043	*Gigartina tenella*	Antibacterial	Inhibitor of eukaryotic DNA polymerases and HIV reverse transcriptase type 1	86
Deschloroelatol	*Laurencia rigida*	Antifungal	*Mycotypha microspora*	87

type-1 (HSV-1) activity of palmitic acid from the *Dunaliella salina* and *Haematococcus pluvialis* spp. Recently, Lee et al.[100] demonstrated that palmitic acid purified from *Sargassum fusiforme* extract is able to bind to the CD4 cell receptor, blocking HIV-1 entry and infection.

Dichloromethane extraction of *Polysiphonia virgata* C. Agardh (Rhodomelaceae) exhibited antimicrobial activity against *M. smegmatis* and *Mycobacterium tuberculosis*. The results revealed that FAs, including oleic acid, linoleic acid, lauric acid, and myristic acid, are the major antimicrobial compounds found in *Polysiphonia virgata*. Oleic acid showed significant antibacterial activity against *M. smegmatis* with a MIC of 0.8 μg, while linoleic acid and lauric acid had MIC values of 1.56 and 3.125 μg, respectively. In addition, it has been reported that lauric acid, myristic acid, and linoleic acid had 100% inhibition at MIC values of 25 μg/mL, myristic acid and lauric acid showed 90% and 76% inhibition at 50 μg/mL, and linoleic acid showed moderate antimicrobial activity at 50 μg/mL against multidrug-resistant (MDR) *M. tuberculosis*.[101]

Dunialella salina is a halophile green microalga of which commercial cultivation for the production of β-carotene represents one of the successful and well-known cases of biotechnological approaches for the production of bioactive molecules. The antimicrobial activity of this microalga was screened against several microorganisms (*E. coli*, *S. aureus*, *C. albicans*, and *A. niger*). The extracts of *Dunaliella salina* were tested by GC-MS, and compounds that were earlier reported as antimicrobials, such as 15 volatile compounds including β-cyclocitral, α- and β-ionone, neophytadiene, and phytol and FA including palmitic, α-linolenic, and oleic acids, were identified.[102]

In another study, *Chlorococcum* strain HS-101 and *Dunaliella primolecta* were tested against methicillin-resistant *S. aureus* (MRSA). The methanol extracts of the samples showed antimicrobial activity, and the production of antibiotics improved by 1.8–2.3-fold in comparison to the control for *Chlorococcum* strain HS-101 and *Dunaliella primolecta*, respectively, in the improved BG-11 medium.[103]

Haematococcus pluvialis (of phylum Chlorophyta and family Haematococcaceae) is a freshwater microalga with high content of an antioxidant compound named astaxanthin. Rodríguez-Meizoso et al.[104] analyzed the extractions and compounds of *Haematococcus pluvialis* for antioxidant and antimicrobial activity using analytical techniques (HPLC-DAD, high-performance liquid chromatography–mass spectrometry (HPLC-MS), and GC-MS and in vitro assays for assessing antioxidant and antimicrobial activity). Based on the authors' reports, short-chain FAs were responsible for the antibacterial activity of *Haematococcus pluvialis*.

By the analysis of the *n*-hexane extract from *Oscillatoria redekei*, by silica gel and RP-18 column chromatography followed by HPLC, two mixtures of FAs were isolated. Further separation via HPLC resulted in the discovery of α-dimorphecolic acid, 9-hydroxy-10E, 12Z-octadecadienoic acid (9-HODE), coriolic acid, and 13-hydroxy-9Z, 11E-octadecadienoic acid (13-HODE). The identified FA inhibited the growth of the Gram-positive bacteria *B. subtilis* SBUG 14, *Micrococcus flavus* SBUG 16, *S. aureus* SBUG 11, and ATCC 25923; nevertheless, no antimicrobial activity was observed against MRSA strains.[105]

The hydroxyl group (−OH) of the carboxyl group of FA appears to be an important factor for the antimicrobial activity of these molecules, since methylated FA have reduced or no antimicrobial activity. Studies indicate that medium- and long-chain unsaturated fatty acids (UFAs) are reported to have more antimicrobial activity, and in general, they have superior antimicrobial activity compared to SFA with the same length of carbon chain.[106] Typically, there is a significant link between the number of double bonds in UFA carbon chains and FA antibacterial activity.[107] On the other hand, for the SFA, the most antimicrobial activity was reported from FAs with 10 or 12 carbons in the chain.[108] Evidence has revealed how FA inhibition acts as antimicrobials through a broad range of strategies, including disruption of electron transport chain, uncoupling of oxidative phosphorylation, cell lysis, inhibition of enzyme activity, impairment of nutrient uptake, and peroxidation and autoxidation.[109]

ALGAL ANTIMICROBIAL METABOLITES VERSUS MDR BACTERIA

MDR bacteria such as *S. aureus*, *K. pneumoniae*, *Acinetobacter baumannii*, *P. aeruginosa*, and *M. tuberculosis* are at the center of great attention in health-care institutions. The microalga *Phaeodactylum tricornutum* Bohlin (order *Naviculales*) has been reported to produce antimicrobial compounds. Via column chromatography, reverse-phase HPLC, and H-NMR spectroscopy, the PUFA eicosapentaenoic acid (EPA) was identified as an antimicrobial compound active against a broad spectrum of microorganisms, including MDR *S. aureus* bacterium.[110]

Postoperative infections can cause serious problems, including failure of the surgical procedure, other surgical complications, organ failure, and death. Based on scientific evidence, some species of algae possess bioactive molecules that can be used to fight against postoperative infectious drug-resistant bacteria. In this regard, different algal species (*Caulerpa cupressoides*, *Chaetomorpha linoides*, *Enteromorpha intestinalis*, *Gracilaria edulis*, *Colpomenia sinuosa*, *Hypnea musciformis*, *Laurencia cruciata*, *Padina boergesenii*, *Padina tetrastromatica*, *S. wightii*, and *U. lactuca*) were extracted by various solvents (acetone, *n*-butanol, propanol, water, and benzene), and were tested against postoperative infectious drug–resistant bacteria (*E. coli*, *K. pneumoniae*, *P. aeruginosa*, *S. pyogens*, and *S. aureus*). The results indicated that among all tested algal extracts, those obtained from *Caulerpa cupressoides* exhibited the most antimicrobial activity against *E. coli*, while propanol extracts of *Gracilaria edulis* had most inhibition on the growth of *K. pneumoniae*. In addition, acetone extracts of *Padina tetrastromatica* and *Laurencia cruciata* had the greatest inhibitory influence on *P. aeruginosa*, while butanol extracts from *Hypnea musciformis*, *Caulerpa cupressoides*, and *Chaetomorpha linoides* exhibited the most antimicrobial activity against *S. aureus*. *C. glomerata* is a reticulated filamentous green alga (Cladophorales) that methanol extracts and purified fractions showed antimicrobial activity against MDR *A. baumannii*. GC-MS analysis also revealed the existence of hydrocarbon compounds in purified fractions of *C. glomerata* with antibacterial activity.[111]

Monounsaturated fatty acid, (9*Z*)-hexadecenoic acid, and the relatively unusual PUFA (6*Z*,9*Z*,12*Z*)-hexadecatrienoic acid were isolated from *P. tricornutum*. These two novel FAs were reported to be active at μM concentrations against *Listonella anguillarum* and MRSA. Moreover, EPA with antibacterial activity on a broad spectrum of Gram-positive and Gram-negative bacteria was identified in the FA profile of *P. tricornutum*.[106]

MICROORGANISMS ASSOCIATED TO ALGAL SURFACE WITH ANTIMICROBIAL ACTIVITY

The competition of microorganisms and their interactions with each other for nutrition, growth, and space in the marine ecosystem is a brilliant opportunity for researchers to select microorganisms for the production of antimicrobials. Typically, microorganisms use three strategies for nutrition-access competition, including rapid growth rates, large population sizes, and production of antimicrobials.[112] These characteristics can be exploited for the selection of classes of microorganisms with considerable antimicrobial activity, and a successful case is represented by isolated *Streptomyces* spp. that are able to synthesize multiple antimicrobial compounds (polyketides and nonribosomally synthesized peptides).[113]

Burgess et al.[114] isolated over 400 strains of surface-associated bacteria from various algal species in Scottish coasts, and reported that antimicrobial activity was exhibited not only by 35% of the tested algae, but also by other marine epibionts and terrestrial human pathogens (*B. subtilis*, *E. coli*, and *P. aeruginosa*). Another study found that marine bacteria with the ability of synthesize antibacterials were isolated from marine alga *Fucus vesiculosus* and the isolated marine bacteria had antimicrobial activity against *E. coli*, *P. aeruginosa*, and *S. aureus*.[115]

Two epibiotic marine bacterial strains (EI-34-6 and II-111-5) were isolated from the surface of the marine alga *Palmaria palmate* and identified as *Bacillus* spp. These isolated microorganisms, grown as biofilm on the surface of nutrient glycerol ferric agar and marine Columbia glycerol agar, showed a range of antibacterials. The authors suggested that biofilm formation played a key role in the synthesis of antimicrobials by these isolated *Bacillus* spp.[116]

In another study, the bacteria associated with algae *Laminaria saccharina* isolated from the Kiel Fjord (Baltic Sea, Germany) were analyzed for antimicrobial activity. A total of 103 of 210 isolated strains showed antimicrobial activity against at least one tested microorganism. In general, 30% of these isolated bacteria showed inhibitory influence against *B. subtilis*, while 25% of them showed simultaneous antimicrobial activity against both *B. subtilis* and *Staphylococcus lentus*, and 11% showed antibacterial activity against *B. subtilis*, *S. lentus*, and *C. albicans*.[117]

Moreover, 6-oxo-de-*O*-methyllasiodiplodin, (*E*)-9-etheno-lasiodiplodin, lasiodiplodin, de-*O*-methyllasiodiplodin, and 5-hydroxy-de-*O*-methyllasiodiplodin were identified as bioactive molecules extracted from brown alga endophytic fungus (No. ZZF36) with antimicrobial activity.[118] In an experiment on the antimicrobial activity of microbial flora connected to seaweed (five species of green and brown marine algae), 224 bacterial strains were isolated and screened for antimicrobial activity. Based on the report, 38 out of 224 strains possessed antimicrobial molecules, while *Enteromorpha intestinalis* was the main source of strains with antimicrobials. The authors reported that each group of isolated bacteria with antimicrobial activity inhibits the growth of other groups.[119]

A total of 35 bacterial strains were isolated from the surface of *Egregia menziesii*, *Codium fragile*, *Sargassum muticum*, *Endarachne binghamiae*, *Centroceras clavulatum,* and *Laurencia pacifica*. Via 16S the rDNA sequencing technique, these strains were identified to belong to the phyla Firmicutes, Proteobacteria, and Actinobacteria and were studied for antimicrobial activity against *S. aureus*, *K. pneumoniae*, *P. mirabilis*, and *P. aeruginosa*, and for anticancer activity against HCT-116 colon cancer cells. Researchers reported that the strains Cc51 isolated from *Centroceras clavulatum*, Sm36 isolated from *S. muticum*, and Eb46 isolated from *Endarachne binghamiae* had anticancer activity (IC_{50} values of 6.492, 5.531, and 2.843 μg/mL, respectively), and antibacterial activity against *P. mirabilis*.[120]

Species of brown algae such as *Sargassum serratifolium*, *S. fusiforme*, *Sargassum filicinum*, *Padina arborescens*, *Undaria pinnatifida*, *Petalonia fascia*, *Colpomenia sinuosa*, *Scytosiphon lomentaria,* and *Ecklonia cava* are reported to be as a good source for the growth of epibiotic bacteria. Based on Kanagasabhapathy et al.,[121] 116 epibiotic bacteria were isolated from these algae, which 20% of them that exhibited antibacterial activity. The phylogenetic study via 16S rRNA sequencing revealed that all the bacteria with considerable antibacterial activity have significant correlation with the genus *Bacillus*.

CURRENT SITUATION AND PERSPECTIVES

Several high-value microalgae products are already well established in the marketplace, and there are clear opportunities for additional new products as well. Indeed, algae represent a promising and still unexplored source of bioactive compounds offering a wealth of chemicals with a wide range of applications. Their great potential exploitation, as algae or as extracts, in areas so diverse such as human nutrition (food, feed in aquaculture, biofertilizers in agriculture), environmental technology (biofilm, treatment of effluents coming from different human activities, capture of CO_2), bioenergy (biodiesel, bioethanol), and pharmaceutical products (anti-inflammatory, antiallergic, and analgesic agents), has recently been reviewed.[122–124] However, many other compounds produced by algae, such as carbohydrates, sterols, pigments, protein and enzymes, vitamins, and lipids, have already found or could have some health applications.[125]

In the fermentation industry, traditional crop carbohydrates as carbon sources can be easily replaced by algal carbohydrate,[126] while algal biomasses represent a potential feedstock for biodiesel production due to the lipids found inside algal cells.[127] In addition, algae can produce a wide range of biocompounds that can find applications in the health, food, and pharmaceutical industries, such as docosahexaenoic acid (DHA) and EPA, several proteins, and pigments. Another application of algae products is represented by the nutraceutical market. The term *nutraceutical* or *functional foods* was first developed in Japan in the early 1980s; since then, it has been adopted worldwide. A nutraceutical product can be broadly defined as a food whose beneficial effect on one or more target functions in the body has been satisfactorily demonstrated in a way that is relevant to either an improved state of health and wellbeing, a reduction of the risk of disease, or both.[128,129] The potential of algae products for functional foods was reviewed recently by Dominguez.[130] Therefore, for all these applications, algae could be regarded as a valuable option in producing biofuels and bio-based chemicals based on both their natural components and refined (or fermented) products.

The interest in algae-based chemicals and bio-products has increased over the past few years, as confirmed by the growing investments in research and development by a wide range of private companies from one side of the world to the other. Undoubtedly, a prevalent role in the algae revolution is played by the United States, where facilities and research projects represent a nationwide opportunity for a new and growing industry that can supply algae-derived renewable fuels, feeds, fertilizers, chemicals, and other products. An estimate by the Algae Biomass Organization based on a review of more than 470 algae business associates demonstrated that 67% of algae manufacturers intended to broaden their capability in 2013 and 95% of companies believe that algae-based fuel just might take on nonrenewable fuels as early as 2020.

For example, Solarvest Bioenergy has finished negotiations to purchase a 30,000-ft^2 center in Summerville, Prince Edward Island (Canada), where industrial nutraceutical supplements will be created from algae strains that are producing substantial amounts of ω-3 DHA/EPA. In addition, it was recently reported that Grovepoint Capital, an investment firm based in the United Kingdom, purchased a managing involvement in Algatechnologies, a rapidly growing, $50 million Israeli biotechnology company, to foster the production of astaxanthin, a cosmeceutical and nutritional supplement, from the microalga *Haematococcus pluvialis*. Solazyme Roquette, a well-known microalgal company based in the San Francisco area, has launched a new product, Almagine HL, as a natural, eco-friendly, whole food component abundant with nutritious fats and made up of protein, carbohydrates (polysaccharides, dietary fiber as well as simple sugars), emulsifiers, and various micronutrients, including lutein.[131] However, in a thorough analysis of the algae market, some important elements should be taken into consideration, such as the size of the potential market, possible competing nonalgae sources, and the time and cost of achieving approval for new products and their acceptance by customers. An obvious drawback to the viewpoint of developing such products as coproducts or by-products from algae, is represented by the very disparate sizes of the market for high-value products, and in particular by low-value commodity products such as animal feed or protein supplements. However, with the present greatly increased efforts to commercialize algae all over the world, new products are likely to be developed and marketed in the next decade.

CONCLUSION

This chapter demonstrated the potential of algae as a bio-producer of a wide range of high-value molecules. These valuable molecules have already proved their applications in different fields, including human health, food industry, biomedicine, and bioengineering. In fact, algae can be deployed to enhance food value, and they also can be considered as a source of pharmaceuticals and nutraceuticals. It should also be considered that the production of these compounds via algae can be coupled with the solution of environmental issues (eg, biodiesel production, wastewater treatment, and reduction of CO_2 emissions). In this way, it is very challenging for researchers to develop biological systems to explore the undoubted potentialities of algae. To reach this goal, it could be beneficial to screen and isolate new species and to introduce new technologies. Indeed, the cooperation between academic institutions and industrial sectors is undeniably a key factor.

REFERENCES

1. Delwiche CF, Andersen RA, Bhattacharya D, Mishler B, McCourt RM. Algal evolution and the early radiation of green plants. In: Cracraft J, Donoghue MJ, editors. *Assembling the tree of life*. New York, NY: Oxford University Press; 2004. p. 121–37.
2. Tolbert NE. Glycolate metabolism by higher plants and algae. In: Gibbs M, Latzko E, editors. *Photosynthesis II*. Berlin and Heidelberg: Springer; 1979. p. 338–52.
3. Popper ZA, Gurvan M, Hervé C, Domozych DS, Willats WGT, Tuohy MG, et al. Evolution and diversity of plant cell walls: from algae to flowering plants. *Annu Rev Plant Biol* 2011;**62**:567–90.
4. Jacobson DM, Andersen RA. The discovery of mixotrophy in photosynthetic species of *Dinophysis* (Dinophyceae): light and electron microscopical observations of food vacuoles in *Dinophysis acuminata*, *D. norvegica* and two heterotrophic dinophysoid dinoflagellates. *Phycologia* 1994;**33**(2):97–110.
5. Heredia-Arroyo T, Wei W, Hu B. Oil accumulation via heterotrophic/mixotrophic *Chlorella protothecoides*. *Appl Biochem Biotechnol* 2010;**162**(7):1978–95.
6. Ganuza E, Benítez-Santana T, Atalah E, Vega-Orellana O, Ganga R, Izquierdo MS. *Crypthecodinium cohnii* and *Schizochytrium* sp. as potential substitutes to fisheries-derived oils from seabream (*Sparus aurata*) microdiets. *Aquaculture* 2008;**277**(1):109–16.
7. Parsaeimehr A, Chen Y-F. Algal bioactive diversities against pathogenic microbes. *Microbiol Book Ser* 2013;**4**:796–802.

8. Pulz O, Gross W. Valuable products from biotechnology of microalgae. *Appl Biochem Biotechnol* 2004;**65**(6):635–48.
9. Becker EW. Microalgae as a source of protein. *Biotechnol Adv* 2007;**25**(2):207–10.
10. Sunda W, Kieber DJ, Kiene RP, Huntsman S. An antioxidant function for DMSP and DMS in marine algae. *Nature* 2002;**418**(6895):317–20.
11. Efterpi C, Bonos E, Giannenas I, Florou-Paneri P. Functional properties of carotenoids originating from algae. *J Sci Food Agric* 2013;**93**(1):5–11.
12. Jyotirmayee P, Das S, Kumar, Das B. Antibacterial activity of freshwater microalgae: a review. *Afr J Pharm Pharmacol* 2014;**8**(32):809–18.
13. Kuda T, Tsunekawa M, Goto H, Araki Y. Antioxidant properties of four edible algae harvested in the Noto Peninsula, Japan. *J Food Compos Anal* 2005;**18**(7):625–33.
14. Nielsen MJ, Rasmussen MR, Andersen CBF, Nexø E, Moestrup SK. Vitamin B12 transport from food to the body's cells-a sophisticated, multistep pathway. *Nat Rev Gastroenterol Hepatol* 2012;**9**(6):345–54.
15. Yates AA, Schlicker SA, Suitor CW. Dietary reference intakes: the new basis for recommendations for calcium and related nutrients, B vitamins, and choline. *J Am Diet Assoc* 1998;**98**(6):699–706.
16. Martens JH, Barg H, Warren M, Jahn D. Microbial production of vitamin B12. *Appl Microbiol Biotechnol* 2002;**58**(3):275–85.
17. Croft MT, Warren MJ, Smith AG. Algae need their vitamins. *Eukaryot Cell* 2006;**5**(8):1175–83.
18. Gullett NP, Ruhul Amin ARM, Bayraktar S, Pezzuto JM, Shin DM, Khuri FR, et al. Cancer prevention with natural compounds. *Semin Oncol* 2010;**37**(3):258–81.
19. Alekseyenko TV, Zhanayeva SY, Venediktova AA, Zvyagintseva TN, Kuznetsova TA, Besednova NN, et al. Antitumor and antimetastatic activity of fucoidan, a sulfated polysaccharide isolated from the Okhotsk Sea *Fucus evanescens* brown alga. *Bull Experim Biol Med* 2007;**143**(6):730–2.
20. Ngo D-H, Kim S-K. Sulfated polysaccharides as bioactive agents from marine algae. *Int J Biol Macromol* 2013;**62**:70–5.
21. Shanab SMM, Mostafa SSM, Shalaby EA, Mahmoud GI. Aqueous extracts of microalgae exhibit antioxidant and anticancer activities. *Asian Pac J Trop Biomed* 2012;**2**(8):608–15.
22. Orhan I, Sener B, Atıcı T, Brun R, Perozzo R, Tasdemir D. Turkish freshwater and marine macrophyte extracts show in vitro antiprotozoal activity and inhibit FabI, a key enzyme of *Plasmodium falciparum* fatty acid biosynthesis. *Phytomed* 2006;**13**(6):388–93.
23. Patterson GML, Baker KK, Baldwin CL, Bolis CM, Caplan FR, Larsen LK, et al. Antiviral activity of cultured blue-green algae (Cyanophya). *J Phycol* 1993;**29**(1):125–30.
24. Plouguerné E, de Souza ML, Sassaki GL, Figueiredo Cavalcanti J, Villela Romanos MT, da Gama BAP, et al. Antiviral sulfoquinovosyldiacylglycerols (SQDGs) from the Brazilian brown seaweed *Sargassum vulgare*. *Mar Drugs* 2013;**11**:4628–40.
25. Gustafson KR, Cardellina JH, Fuller RW, Weislow OS, Kiser RF, Snader KM, et al. AIDS-antiviral sulfolipids from cyanobacteria (blue-green algae). *J Nat Cancer Inst* 1989;**81**(16):1254–8.
26. Lira M-LF, Lopes R, Portes Gomes A, Barcellos G, Vercímo M, Osako K, et al. Anti-leishmanial activity of Brazilian green, brown, and red algae. *J Appl Phycol* 2015;**28**:1–8.
27. Harrison PF, Lederberg J. *Antimicrobial resistance: issues and options*. 1st ed. Washington, DC: National Academies Press; 1998.
28. Singh SB, Barrett JF. Empirical antibacterial drug discovery—foundation in natural products. *Biochem Pharmacol* 2006;**71**(7):1006–15.
29. Blunt JW, Copp BR, Munro MH, Northcote PT, Prinsep MR. Marine natural products. *Nat Prod Rep* 2010;**27**(2):165–237.
30. Hornsey IS, Hide D. The production of antimicrobial compounds by British marine algae I. Antibiotic-producing marine algae. *Brit Phycol J* 1974;**9**(4):353–61.

31. Salvador N, Gómez Garreta MA, Lavelli L, Ribera MA. Antimicrobial activity of Iberian macroalgae. *Sci Mar* 2007;**71**(1):101–13.
32. Taskin E, Ozturk M, Taskin E, Kurt O. Antibacterial activities of some marine algae from the Aegean Sea (Turkey). *Afr J Biotechnol* 2007;**6**(24):2746–51.
33. Prakash JW, Antonisamy JM, Jeeva S. Antimicrobial activity of certain fresh water microalgae from Thamirabarani River, Tamil Nadu, South India. *Asian Pac J Trop Biomed* 2011;**1**(2):S170–3.
34. Zheng Y, Chen Y-S, Hai-Sheng LU. Screening for antibacterial and antifungal activities in some marine algae from the Fujian coast of China with three different solvents. *Chin J Oceanol Limnol* 2001;**19**(4):327–31.
35. Lima-Filho JVM, Carvalho AFFU, Freitas SM, Melo VMM. Antibacterial activity of extracts of six macroalgae from the northeastern Brazilian coast. *Braz J Microbiol* 2002;**33**(4):311–14.
36. Bhagavathy S, Sumathy P, Bell IJS. Green algae *Chlorococcum humicola*—a new source of bioactive compounds with antimicrobial activity. *Asian Pac J Trop Biomed* 2011;**1**(1):S1–7.
37. Fenical W, Paul VJ. *Antimicrobial and cytotoxic terpenoids from tropical green algae of the family Udoteaceae. Eleventh international seaweed symposium.* Dordrecht: Springer Netherlands. 1984. p. 135–40.
38. Bhateja P, Mathur M, Pandya M, Fatma T, Rattan A. Activity of blue green microalgae extracts against in vitro generated *Staphylococcus aureus* with reduced susceptibility to vancomycin. *Fitoterapia* 2006;**77**(3):233–5.
39. Val A, Platas G, Basilio A, Cabello A, Gorrochategui J, Suay I, et al. Screening of antimicrobial activities in red, green and brown macroalgae from Gran Canaria (Canary Islands, Spain). *Int Microbiol* 2001;**4**(1):35–40.
40. Shanmughapriya S, Manilal A, Sujith S, Selvin J, Seghal Kiran G, Natarajaseenivasan K. Antimicrobial activity of seaweeds extracts against multiresistant pathogens. *Ann Microbiol* 2008;**58**(3):535–41.
41. Selim SA. Antimicrobial, antiplasmid and cytotoxicity potentials of marine algae *Halimeda opuntia* and *Sarconema filiforme* collected from Red Sea Coast. *World Acad Sci Eng Technol* 2012;**61**:1154–9.
42. Hornsey IS, Hide D. The production of antimicrobial compounds by British marine algae II. Seasonal variation in production of antibiotics. *Brit Phycol J* 1976;**11**(1):63–7.
43. Herrero M, Cifuentes A, Ibanez E. Sub-and supercritical fluid extraction of functional ingredients from different natural sources: plants, food-by-products, algae and microalgae: a review. *Food Chem* 2006;**98**(1):136–48.
44. Plaza M, Santoyo S, Jaime L, García-Blairsy Reina G, Herrero M, Señoráns FJ, et al. Screening for bioactive compounds from algae. *J Pharmac Biomed Anal* 2010;**51**(2):450–5.
45. Bourgougno N, Stiger-Pouvreau V. Chemiodiversity and bioactivity within red and brown macroalgae along the French coasts, metropole and overseas departments and territories. In: Kim S-L, editor. *Handbook of marine macroalgae: biotechnology and applied phycology*. Chichester: John Wiley & Sons, Ltd; 2012. p. 58–105.
46. Lim C-W, Lee J-S, Cho Y-J. Structures and some properties of the antimicrobial compounds in the red alga, *Symphyocladia latiuscula. Kor J Fish Aquatic Sci* 2000;**33**(4):280–7.
47. Oh K-B, Lee JH, Chung S-C, Shin J, Shin HJ, Kim H-K, et al. Antimicrobial activities of the bromophenols from the red alga *Odonthalia corymbifera* and some synthetic derivatives. *Bioorg Med Chem Lett* 2008;**18**(1):104–8.
48. Alarif WM, Al-Lihaibi SS, Ayyad S-EN, Abdel-Rhman MH, Badria FA. Laurene-type sesquiterpenes from the Red Sea red alga *Laurencia obtusa* as potential antitumor–antimicrobial agents. *Eur J Med Chem* 2012;**55**:462–6.
49. Karabay-Yavasoglu NU, Sukatar A, Ozdemir G, Horzum Z. Antimicrobial activity of volatile components and various extracts of the red alga *Jania rubens*. *Phytother Res* 2007;**21**(2):153–6.
50. Cox S, Abu-Ghannam N, Gupta S. An assessment of the antioxidant and antimicrobial activity of six species of edible Irish seaweeds. *Int Food Res J* 2010;**17**:205–20.
51. Gupta S, Cox S, Rajauria G, Kumar Jaiswal A, Abu-Ghannam N. Growth inhibition of common food spoilage and pathogenic microorganisms in the presence of brown seaweed extracts. *Food Bioproc Technol* 2012;**5**(5):1907–16.

52. Sims JJ, Donnell MS, Leary JV, Lacy GH. Antimicrobial agents from marine algae. *Antimicrob Agents Chemother* 1975;**7**(3):320–1.
53. Sánchez C, Méndez C, Salas JA. Indolocarbazole natural products: occurrence, biosynthesis, and biological activity. *Nat Product Rep* 2006;**23**(6):1007–45.
54. Borowitzka MA. Microalgae as sources of pharmaceuticals and other biologically active compounds. *J Appl Phycol* 1995;**7**(1):3–15.
55. Whitfield FB, Helidoniotis F, Shaw KJ, Svoronos D. Distribution of bromophenols in species of marine algae from eastern Australia. *J Agric Food Chem* 1999;**47**(6):2367–73.
56. Veiga-Santos P, Pelizzaro-Rocha KJ, Santos AO, Ueda-Nakamura T, Silva SO, Sudatti DB, et al. In vitro anti-trypanosomal activity of elatol isolated from red seaweed *Laurencia dendroidea*. *Parasitology* 2010;**137**(11):1661–70.
57. dos Santos AO, Veiga-Santos P, Ueda-Nakamura T, Bueno Sudatti D, Bianco EM, Crespo Pereira R, et al. Effect of elatol, isolated from red seaweed *Laurencia dendroidea*, on *Leishmania amazonensis*. *Mar Drugs* 2010;**8**(11):2733–43.
58. Washida K, Koyama T, Yamada K, Kita M, Uemura D. Karatungiols A and B, two novel antimicrobial polyol compounds, from the symbiotic marine dinoflagellate *Amphidinium* sp. *Tetrahedron Lett* 2006;**47**(15):2521–5.
59. Kobayashi J, Kubota T, Takahashi M, Ishibashi M, Tsuda M, Naoki H. Colopsinol A, a novel polyhydroxyl metabolite from marine dinoflagellate *Amphidinium* sp. *J Org Chem* 1999;**64**(5):1478–82.
60. Kubota T, Tsuda M, Takahashi M, Ishibashi M, Naoki H, Kobayashi J. Colopsinols B and C, new long chain polyhydroxy compounds from cultured marine dinoflagellate *Amphidinium* sp. *J Chem Soc Perkin Trans 1* 1999;**23**:3483–7.
61. Volk R-B, Furkert FH. Antialgal, antibacterial and antifungal activity of two metabolites produced and excreted by cyanobacteria during growth. *Microbiol Res* 2006;**161**(2):180–6.
62. Gonzalez AG, Darias V, Estevez E. Chemotherapeutic activity of polyhalogenated terpenes Spanish algae. *Planta Med* 1982;**44**:44–6.
63. Plaza M, Cifuentes A, Ibáñez E. In the search of new functional food ingredients from algae. *Trends Food Sci Technol* 2008;**19**(1):31–9.
64. Kukovinets OS, Zainullin RA, Kislitsyn MI. Natural arylterpenes and their biological activity. *Chem Nat Comp* 2006;**42**(1):1–15.
65. Chakraborty K, Lipton AP, Raj RP, Vijayan KK. Antibacterial labdane diterpenoids of *Ulva fasciata* Delile from southwestern coast of the Indian Peninsula. *Food Chem* 2010;**119**(4):1399–408.
66. Paul NA, de Nys R, Steinberg PD. Chemical defence against bacteria in the red alga *Asparagopsis armata*: linking structure with function. *Mar Ecol Prog Ser* 2006;**306**:87–101.
67. Bano S, Bano N, Ahmad VU, Shameel M, Amjad S. Marine natural products: 3-formylindole from the red algae *Botryocladia leptopoda*. *J Nat Prod* 1986;**49**(3). 549-549
68. Andreani A, Bonazzi D, Cavrini V, Gatti R, Giovanninetti G, Franchi L, et al. Research on substances with antiviral activity. VII. Activity and lipophilic of property N-alkyl-2-chloro-3-formylindole thiosemicarbazone. *Il Farmaco Edizione Sci* 1977;**32**(10):703–12.
69. Sandsdalen E, Haug T, Stensvåg K, Styrvold OB. The antibacterial effect of a polyhydroxylated fucophlorethol from the marine brown alga *Fucus vesiculosus*. *World J Microbiol Biotechnol* 2003;**19**(8):777–82.
70. Holdt SL, Kraan S. Bioactive compounds in seaweed: functional food applications and legislation. *J Appl Phycol* 2011;**23**(3):543–97.
71. Li Y, Qian Z-J, Kim M-M, Kim S-K. Cytotoxic activities of phlorethol and fucophlorethol derivatives isolated from Laminariaceae *Ecklonia cava*. *J Food Biochem* 2011;**35**(2):357–69.
72. Tajbakhsh S, Pouyan M, Zandi K, Bahramian P, Sartavi K, Fouladvand M, et al. In vitro study of antibacterial activity of the alga *Sargassum oligocystum* from the Persian Gulf. *Eur Rev Med Pharmacol Sci* 2011;**15**(3):293–8.

73. Jiao G, Yu G, Zhang J, Ewart HS. Chemical structures and bioactivities of sulfated polysaccharides from marine algae. *Mar Drugs* 2011;**9**(2):196–223.
74. Vetvicka V, Yvin J-C. Effects of marine β − 1,3 glucan on immune reactions. *Int Immunopharmacol* 2004;**4**(6):721–30.
75. Vetvick V, Dvorak B, Vetvickova J, Richter J, Krizan J, Sima P, et al. Orally administered marine (1→ 3)-β-D-glucan phycarine stimulates both humoral and cellular immunity. *Int J Biol Macromol* 2007;**40**(4):291–8.
76. Kimura J, Maki N. New loliolide derivatives from the brown alga *Undaria pinnatifida*. *J Nat Prod* 2002;**65**(1):57–8.
77. Grabarczyk M, Wińska K, Mączka W, Potaniec B, Anioł M. Loliolide—the most ubiquitous lactone. *Folia Biol Oecol* 2015;11 ISSN (Online) 1730-2366, http://dx.doi.org/10.1515/fobio-2015-0001
78. Toshiaki T, Sasaki H, Fukazawa H, Suenaga K. Bisebromoamide, a potent cytotoxic peptide from the marine cyanobacterium *Lyngbya* sp.: isolation, stereostructure, and biological activity. *Org Lett* 2009;**11** (21):5062–5.
79. Yim JH, Kim SJ, Ahn SH, Lee CK, Rhie KT, Lee HK. Antiviral effects of sulfated exopolysaccharide from the marine microalga *Gyrodinium impudicum* strain KG03. *Mar Biotechnol* 2004;**6**(1):17–25.
80. Amaro HM, Guedes AC, Malcata FX. Antimicrobial activities of microalgae: an invited review. *Sci Against Microb Pathog: Commun Curr Res Technol Adv* 2011;**3**:1272–84.
81. Shih S-R, Tsai K-N, Li Y-S, Chueh C-C, Chan E-C. Inhibition of enterovirus 71-induced apoptosis by allophycocyanin isolated from a blue-green alga *Spirulina platensis*. *J Med Virol* 2003;**70**(1):119–25.
82. Umemura K, Yanase K, Suzuki M, Okutani K, Yamori T, Andoh T. Inhibition of DNA topoisomerases I and II, and growth inhibition of human cancer cell lines by a marine microalgal polysaccharide. *Biochem Pharmacol* 2003;**66**(3):481–7.
83. Doan NT, Rickards RW, Rothschild JM, Smith GD. Allelopathic actions of the alkaloid 12-epi-hapalindole E isonitrile and calothrixin A from cyanobacteria of the genera *Fischerella* and *Calothrix*. *J Appl Phycol* 2000;**12**(3–5):409–16.
84. McPhail KL, Correa J, Linington RG, González J, Ortega-Barría E, Capson TL, et al. Antimalarial linear lipopeptides from a Panamanian strain of the marine cyanobacterium *Lyngbya majuscula*. *J Nat Prod* 2007;**70**(6):984–8.
85. Mandal S, Rath J. *Anticancer drug development from Cyanobacteria. Extremophilic cyanobacteria for novel drug development*. Cham: Springer International Publishing; 2015. p. 63–78.
86. Ohta K, Mizushima Y, Hirata N, Takemura M, Sugawara F, Matsukage A, et al. Sulfoquinovosyldiacylglycerol, KM043, a new potent inhibitor of eukaryotic DNA polymerases and HIV-reverse transcriptase type 1 from a marine red alga, *Gigartina tenella*. *Chem Pharm Bull* 1998;**46** (4):684–6.
87. Nagle DG, Wedge DE. *Antifungal properties of cyanobacteria and algae: ecological and agricultural implications. Chemical ecology of plants: allelopathy in aquatic and terrestrial ecosystems*. Basel: Birkhäuser; 2002. p. 7–32.
88. Van Damme EJM, Peumans WJ, Pusztai A, Bardocz S. *Handbook of plant lectins: properties and biomedical applications*. Chichester and New York, NY: John Wiley & Sons; 1998.
89. Liao W-R, Lin J-Y, Shieh W-Y, Jeng W-L, Huang R. Antibiotic activity of lectins from marine algae against marine vibrios. *J Ind Microbiol Biotechnol* 2003;**30**(7):433–9.
90. Kamei Y, Sueyoshi M, Hayashi K-I, Terada R, Nozaki H. The novel anti-*Propionibacterium acnes* compound, Sargafuran, found in the marine brown alga *Sargassum macrocarpum*. *J Antibiot* 2009;**62**(5):259–63.
91. Simmons LT, Engene N, Ureña LD, Romero LI, Ortega-Barría E, Gerwick L, et al. Viridamides A and B, lipodepsipeptides with antiprotozoal activity from the marine cyanobacterium *Oscillatoria nigro-viridis*. *J Nat Prod* 2008;**71**(9):1544–50.

92. Singh S, Kate BN, Banerjee UC. Bioactive compounds from cyanobacteria and microalgae: an overview. *Crit Rev Biotechnol* 2005;**25**(3):73–95.
93. Faid M. Antimicrobials from marine algae. In: Rai M, Chikindas M, editors. *Natural antimicrobials in food safety and quality*. Wallingford: CABI International; 2011. p. 95–103.
94. Simopoulos AP. Essential fatty acids in health and chronic disease. *Am J Clin Nutr* 1999;**70**(3):560s–9s.
95. Stabili L, Acquaviva MI, Biandolino F, Cavallo RA, De Pascali, Fanizzi FP, et al. The lipidic extract of the seaweed *Gracilariopsis longissima* (Rhodophyta, Gracilariales): a potential resource for biotechnological purposes? *New Biotechnol* 2012;**29**(3):443–50.
96. Ryckebosch E, Bruneel C, Termote-Verhalle R, Goiris K, Muylaert K, Foubert I. Nutritional evaluation of microalgae oils rich in omega-3 long chain polyunsaturated fatty acids as an alternative for fish oil. *Food Chem* 2014;**160**:393–400.
97. Kabara JJ, Swieczkowski DM, Conley AJ, Truant JP. Fatty acids and derivatives as antimicrobial agents. *Antimicrob Agents Chemother* 1972;**2**(1):23–8.
98. Harada H, Yamashita U, Kurihara H, Fukushi E, Kawabata J, Kamei Y. Antitumor activity of palmitic acid found as a selective cytotoxic substance in a marine red alga. *Anticancer Res* 2001;**22**(5):2587–90.
99. Santoyo S, Jaime L, Plaza M, Herrero M, Rodriguez-Meizoso I, Ibañez E, et al. Antiviral compounds obtained from microalgae commonly used as carotenoid sources. *J Appl Phycol* 2012;**24**(4):731–41.
100. Lee DY-W Lin X, Paskaleva EE, Liu Y, Puttamadappa SS, Thornber C, et al. Palmitic acid is a novel CD4 fusion inhibitor that blocks HIV entry and infection. *AIDS Res Hum Retroviruses* 2009;**25**(12):1231–41.
101. Saravanakumar DEM, Folb PI, Campbell BW, Smith P. Antimycobacterial activity of the Red Alga *Polysiphonia virgata*. *Pharm Biol* 2008;**46**(4):254–60.
102. Herrero M, Ibáñez E, Cifuentes A, Reglero G, Santoyo S. *Dunaliella salina* microalga pressurized liquid extracts as potential antimicrobials. *J Food Prot* 2006;**69**(10):2471–7.
103. Ohta S, Shiomi Y, Kawashima A, Aozasa O, Nakao T, Nagate T, et al. Antibiotic effect of linolenic acid from *Chlorococcum* strain HS-101 and *Dunaliella primolecta* on methicillin-resistant *Staphylococcus aureus*. *J Appl Phycol* 1995;**7**(2):121–7.
104. Rodríguez-Meizoso I, Jaime L, Santoyo S, Señoráns FJ, Cifuentes A, Ibáñez E. Subcritical water extraction and characterization of bioactive compounds from *Haematococcus pluvialis* microalga. *J Pharm Biomed Anal* 2010;**51**(2):456–63.
105. Mundt S, Kreitlow S, Jansen R. Fatty acids with antibacterial activity from the cyanobacterium *Oscillatoria redekei* HUB 051. *J Appl Phycol* 2003;**15**(2–3):263–7.
106. Desbois AP, Lebl T, Yan L, Smith VJ. Isolation and structural characterisation of two antibacterial free fatty acids from the marine diatom *Phaeodactylum tricornutum*. *Appl Microbiol Biotechnol* 2008;**81**(4):755–64.
107. Feldlaufer MD, Knox DA, Lusby WR, Shimanuki H. Antimicrobial activity of fatty acids against *Bacillus* larvae, the causative agent of American foulbrood disease. *Apidologie* 1993;**24**(2):95–9.
108. Wille JJ, Kydonieus A. Palmitoleic acid isomer (C16:1Δ6) in human skin sebum is effective against Gram-positive bacteria. *Skin Pharmacol Physiol* 2003;**16**(3):176–87.
109. Desbois AP, Smith VJ. Antibacterial free fatty acids: activities, mechanisms of action and biotechnological potential. *Appl Microbiol Biotechnol* 2010;**85**(6):1629–42.
110. Desbois AP, Mearns-Spragg A, Smith VJ. A fatty acid from the diatom *Phaeodactylum tricornutum* is antibacterial against diverse bacteria including multi-resistant *Staphylococcus aureus* (MRSA). *Mar Biotechnol* 2009;**11**(1):45–52.
111. Ravikumar S, Anburajan L, Ramanathan G, Kaliaperumal N. Screening of seaweed extracts against antibiotic resistant post operative infectious pathogens. *Seaweed Res Utilisation* 2002;**24**(1):95–9.

112. Hibbing ME, Fuqua C, Parsek MP, Peterson SB. Bacterial competition: surviving and thriving in the microbial jungle. *Nature Rev Microbiol* 2010;**8**(1):15–25.
113. Challis GL, Hopwood DA. Synergy and contingency as driving forces for the evolution of multiple secondary metabolite production by *Streptomyces* species. *Proc Natl Acad Sci* 2003;**100**(2):14555–61.
114. Burgess JG, Jordan EM, Bregu M, Mearns-Spragg A, Boyd KG. Microbial antagonism: a neglected avenue of natural products research. *J Biotechnol* 1999;**70**(1):27–32.
115. Mearns-Spragg A, Bregu M, Boyd KG, Burgess JG. Cross-species induction and enhancement of antimicrobial activity produced by epibiotic bacteria from marine algae and invertebrates, after exposure to terrestrial bacteria. *Lett Appl Microbiol* 1998;**27**(3):142–6.
116. Yan L, Boyd KG, Burgess JG. Surface attachment induced production of antimicrobial compounds by marine epiphytic bacteria using modified roller bottle cultivation. *Mar Biotechnol* 2002;**4**(4):356–66.
117. Wiese J, Thiel V, Nagel K, Staufenberger T, Imhoff JF. Diversity of antibiotic-active bacteria associated with the brown alga *Laminaria saccharina* from the Baltic Sea. *Mar Biotechnol* 2009;**11**(2):287–300.
118. Yang R-Y, Li C-Y, Lin Y-C, Peng G-T, She Z-G, Zhou S-N. Lactones from a brown alga endophytic fungus (No. ZZF36) from the South China Sea and their antimicrobial activities. *Bioorg Med Chem Lett* 2006;**16**:4205–8.
119. Lemos ML, Toranzo AE, Barja JL. Antibiotic activity of epiphytic bacteria isolated from intertidal seaweeds. *Microb Ecol* 1985;**11**(2):149–63.
120. Villarreal-Gómez LJ, Soria-Mercado IE, Guerra-Rivas G, Ayala-Sánchez NE. Antibacterial and anticancer activity of seaweeds and bacteria associated with their surface. *Rev Biol Mar Oceanogr* 2010;**45**(2):267–75.
121. Kanagasabhapathy M, Sasaki H, Haldar S, Yamasaki S, Nagata S. Antibacterial activities of marine epibiotic bacteria isolated from brown algae of Japan. *Ann Microbiol* 2006;**56**(2):167–73.
122. Concas A, Lutzu GA, Locci AM, Cao G. *Nannochloris eucaryotum* growth: kinetics analysis and use of 100% (v/v) CO_2. *Adv Environ Res Int J* 2013;**2**:19–33.
123. Concas A, Lutzu GA, Pisu M, Cao G. Experimental analysis and novel modelling of semi-batch photobioreactors operated with *Chlorella vulgaris* and fed with 100% (v/v) CO_2. *Chem Eng J* 2012;**213**:203–13.
124. Ji B, Zhang W, Zhang N, Wang J, Lutzu GA, Liu T. Biofilm cultivation of the oleaginous microalgae specie *Pseudochlorococcum* sp. *Bioproc Biosyst Eng* 2014;**37**:1369–75.
125. Parsaeimehr A, Zhilan S, Xiao D, Chen Y. Simultaneous improvement in production of microalgal biodiesel and high-value alpha-linolenic acid by a single regulator acetylcholine. *Biotechnol Biofuels* 2015;**8**:11. Available from: http://dx.doi.org/10.1186/s13068-015-0196-0
126. Mata TM, Martins AA, Caetano NS. Microalgae for biodiesel production and other application: a review. *Renew Sustain Energy Rev* 2010;**14**:217–32.
127. Wijffels RH, Barbosa MJ, Eppink MHM. Microalgae for the production of bulk chemicals and biofuels. *Biofuels Bioprod Biorefin* 2010;**4**:287–95.
128. Borowitzka MA. High-value products from microalgae—their development and commercialization. *J Appl Phycol* 2013;**25**:743–56.
129. Bahadar A, Khan MB. Progress in energy from microalgae: a review. *Rene Sustain Energy Rev* 2013;**27**:128–48.
130. Dominguez H. *Functional ingredients from algae for foods and nutraceuticals*. Oxford: Woodhead Publishing; 2013, p. 768.
131. Map of algae projects in the USA—Algamoil. <http://www.algamoil.com/map-of-algae-projects/>; 2015. [accessed 21.06.15].

Index

Note: Page numbers followed by "*f*" and "*t*" refer to figures and tables, respectively.

A

B

D

E

T

FSC
www.fsc.org
MIX
Papier | Fördert
gute Waldnutzung
FSC® C083411

Kolibri 360 GmbH
Ferdinand-Jühlke-Straße 7
99095 Erfurt, Deutschland
produktsicherheit@kolibri360.de